Control and Nonlinearity

Mathematical
Surveys
and
Monographs

Volume 136

Control and Nonlinearity

Jean-Michel Coron

American Mathematical Society

EDITORIAL COMMITTEE

Jerry L. Bona Peter S. Landweber
Michael G. Eastwood Michael P. Loss

J. T. Stafford, Chair

2000 *Mathematics Subject Classification*. Primary 93B05, 93B52, 93C10, 93C15, 93C20, 93D15; Secondary 35K50, 35L50, 35L60, 35Q30, 35Q53, 35Q55, 76B75.

For additional information and updates on this book, visit
www.ams.org/bookpages/surv-136

Library of Congress Cataloging-in-Publication Data

Coron, Jean-Michel, 1956–
  Control and nonlinearity / Jean-Michel Coron.
    p. cm. — (Mathematical surveys and monographs, ISSN 0076-5376 ; v. 136)
  Includes bibliographical references and index.
  ISBN-13: 978-0-8218-3668-2 (alk. paper)
  ISBN-10: 0-8218-3668-4 (alk. paper)
  1. Control theory. 2. Nonlinear control theory. I. Title.

QA402.3.C676 2007
515′.642–dc22
                                                            2006048031

Copying and reprinting. Individual readers of this publication, and nonprofit libraries acting for them, are permitted to make fair use of the material, such as to copy a chapter for use in teaching or research. Permission is granted to quote brief passages from this publication in reviews, provided the customary acknowledgment of the source is given.

Republication, systematic copying, or multiple reproduction of any material in this publication is permitted only under license from the American Mathematical Society. Requests for such permission should be addressed to the Acquisitions Department, American Mathematical Society, 201 Charles Street, Providence, Rhode Island 02904-2294, USA. Requests can also be made by e-mail to reprint-permission@ams.org.

© 2007 by the American Mathematical Society. All rights reserved.
Printed in the United States of America.

∞ The paper used in this book is acid-free and falls within the guidelines
established to ensure permanence and durability.
Visit the AMS home page at http://www.ams.org/

10 9 8 7 6 5 4 3 2 1   12 11 10 09 08 07

Contents

Preface	ix
Part 1. Controllability of linear control systems	1
Chapter 1. Finite-dimensional linear control systems	3
1.1. Definition of controllability	3
1.2. An integral criterion for controllability	4
1.3. Kalman's type conditions for controllability	9
1.4. The Hilbert Uniqueness Method	19
Chapter 2. Linear partial differential equations	23
2.1. Transport equation	24
2.2. Korteweg-de Vries equation	38
2.3. Abstract linear control systems	51
2.4. Wave equation	67
2.5. Heat equation	76
2.6. A one-dimensional Schrödinger equation	95
2.7. Singular optimal control: A linear 1-D parabolic-hyperbolic example	99
2.8. Bibliographical complements	118
Part 2. Controllability of nonlinear control systems	121
Chapter 3. Controllability of nonlinear systems in finite dimension	125
3.1. The linear test	126
3.2. Iterated Lie brackets and the Lie algebra rank condition	129
3.3. Controllability of driftless control affine systems	134
3.4. Bad and good iterated Lie brackets	141
3.5. Global results	150
3.6. Bibliographical complements	156
Chapter 4. Linearized control systems and fixed-point methods	159
4.1. The Linear test: The regular case	159
4.2. The linear test: The case of loss of derivatives	165
4.3. Global controllability for perturbations of linear controllable systems	177
Chapter 5. Iterated Lie brackets	181
Chapter 6. Return method	187
6.1. Description of the method	187
6.2. Controllability of the Euler and Navier-Stokes equations	192

6.3. Local controllability of a 1-D tank containing a fluid modeled by the Saint-Venant equations — 203

Chapter 7. Quasi-static deformations — 223
7.1. Description of the method — 223
7.2. Application to a semilinear heat equation — 225

Chapter 8. Power series expansion — 235
8.1. Description of the method — 235
8.2. Application to a Korteweg-de Vries equation — 237

Chapter 9. Previous methods applied to a Schrödinger equation — 247
9.1. Controllability and uncontrollability results — 247
9.2. Sketch of the proof of the controllability in large time — 252
9.3. Proof of the nonlocal controllability in small time — 263

Part 3. Stabilization — 271

Chapter 10. Linear control systems in finite dimension and applications to nonlinear control systems — 275
10.1. Pole-shifting theorem — 275
10.2. Direct applications to the stabilization of finite-dimensional control systems — 279
10.3. Gramian and stabilization — 282

Chapter 11. Stabilization of nonlinear control systems in finite dimension — 287
11.1. Obstructions to stationary feedback stabilization — 288
11.2. Time-varying feedback laws — 295
11.3. Output feedback stabilization — 305
11.4. Discontinuous feedback laws — 311

Chapter 12. Feedback design tools — 313
12.1. Control Lyapunov function — 313
12.2. Damping feedback laws — 314
12.3. Homogeneity — 328
12.4. Averaging — 332
12.5. Backstepping — 334
12.6. Forwarding — 337
12.7. Transverse functions — 340

Chapter 13. Applications to some partial differential equations — 347
13.1. Gramian and rapid exponential stabilization — 347
13.2. Stabilization of a rotating body-beam without damping — 351
13.3. Null asymptotic stabilizability of the 2-D Euler control system — 356
13.4. A strict Lyapunov function for boundary control of hyperbolic systems of conservation laws — 361

Appendix A. Elementary results on semigroups of linear operators — 373

Appendix B. Degree theory — 379

Bibliography — 397

List of symbols 421

Index 423

Preface

A control system is a dynamical system on which one can act by using suitable *controls*. There are a lot of problems that appear when studying a control system. But the most common ones are the *controllability* problem and the *stabilization* problem.

The controllability problem is, roughly speaking, the following one. Let us give two states. Is it possible to move the control system from the first one to the second one? We study this problem in Part 1 and in Part 2. Part 1 studies the controllability of *linear* control systems, where the situation is rather well understood, even if there are still quite challenging open problems in the case of linear partial differential control systems. In Part 2 we are concerned with the controllability of *nonlinear* control systems. We start with the case of finite-dimensional control systems, a case where quite powerful geometrical tools are known. The case of nonlinear partial differential equations is much more complicated to handle. We present various methods to treat this case as well as many applications of these methods. We emphasize control systems for which the nonlinearity plays a crucial role, in particular, for which it is the nonlinearity that gives the controllability or prevents achieving some specific interesting motions.

The stabilization problem is the following one. We have an equilibrium which is unstable without the use of the control. Let us give a concrete example. One has a stick that is placed vertically on one of his fingers. In principle, if the stick is exactly vertical with a speed exactly equal to 0, it should remain vertical. But, due to various small errors (the stick is not exactly vertical, for example), in practice, the stick falls down. In order to avoid this, one moves the finger in a suitable way, depending on the position and speed of the stick; we use a *"feedback law"* (or *"closed-loop control"*) which stabilizes the equilibrium. The problem of the stabilization is the existence and construction of such stabilizing feedback laws for a given control system. We study it in Part 3, both for finite-dimensional control systems and for systems modeled by partial differential equations. Again we emphasize the case where the nonlinear terms play a crucial role.

Let us now be more precise on the contents of the different parts of this book.

Part 1: Controllability of linear control systems

This first part is devoted to the controllability of *linear* control systems. It has two chapters: The first one deals with finite-dimensional control systems, the second one deals with infinite-dimensional control systems modeled by partial differential equations.

Let us detail the contents of these two chapters.

Chapter 1. This chapter focuses on the *controllability* of *linear finite-dimensional control systems*. We first give an integral necessary and sufficient condition for a linear time-varying finite-dimensional control system to be controllable. For a special quadratic cost, it leads to the optimal control. We give some examples of applications.

These examples show that the use of this necessary and sufficient condition can lead to computations which are somewhat complicated even for very simple control systems. In particular, it requires integrating linear differential equations. We present the famous Kalman rank condition for the controllability of linear *time-invariant* finite-dimensional control systems. This new condition, which is also necessary and sufficient for controllability, is purely algebraic: it does not require integrations of linear differential equations. We turn then to the case of linear *time-varying* finite-dimensional control systems. For these systems we give a sufficient condition for controllability, which turns out to be also necessary for analytic control systems. This condition only requires computing derivatives; again no integrations are needed.

We describe, in the framework of linear time-varying finite-dimensional control systems, the Hilbert Uniqueness Method (HUM), due to Jacques-Louis Lions. This method is quite useful in infinite dimension for finding numerically optimal controls for linear control systems.

Chapter 2. The subject of this chapter is the controllability of some classical linear partial differential equations. For the reader who is familiar with this subject, a large part of this chapter can be omitted; most of the methods detailed here are very well known. One can find much more advanced material in some references given throughout this chapter. We study a transport equation, a Korteweg-de Vries equation, a heat equation, a wave equation and a Schrödinger equation.

We prove the well-posedness of the Cauchy problem associated to these equations. The controllability of these equations is studied by means of various methods: explicit methods, extension method, moments theory, flatness, Hilbert Uniqueness Method, duality between controllability and observability. This duality shows that the controllability can be reduced to an observability inequality. We show how to prove this inequality by means of the multiplier method or Carleman inequalities. We also present a classical abstract setting which allows us to treat the well-posedness and the controllability of many partial differential equations in the same framework.

Part 2: Controllability of nonlinear control systems

This second part deals with the controllability of *nonlinear* control systems.

We start with the case of nonlinear *finite-dimensional* control systems. We recall the linear test and explain some geometrical methods relying on iterated Lie brackets when this test fails.

Next we consider nonlinear partial differential equations. For these *infinite-dimensional* control systems, we begin with the case where the linearized control system is controllable. Then we get local controllability results and even global controllability results if the nonlinearity is not too big. The case where the linearized control system is not controllable is more difficult to handle. In particular, the tool of iterated Lie brackets, which is quite useful for treating this case in finite dimension, turns out to be useless for many interesting infinite-dimensional control systems. We present three methods to treat some of these systems, namely the return method, quasi-static deformations and power series expansions. On various examples, we show how these three methods can be used.

Let us give more details on the contents of the seven chapters of Part 2.

Chapter 3. In this chapter we study the local controllability of *finite-dimensional nonlinear* control systems around a given equilibrium. One does not know any interesting necessary and sufficient condition for small-time local controllability, even for analytic control systems. However, one knows powerful necessary conditions and powerful sufficient conditions.

We recall the classical "linear test": If the linearized control system at the equilibrium is controllable, then the nonlinear control system is locally controllable at this equilibrium.

When the linearized control system is not controllable, the situation is much more complicated. We recall the Lie algebra condition, a necessary condition for local controllability of (analytic) control systems. It relies on iterated Lie brackets. We explain why iterated Lie brackets are natural for the problem of controllability.

We study in detail the case of the driftless control systems. For these systems, the above Lie algebra rank condition turns out to be sufficient, even for global controllability.

Among the iterated Lie brackets, we describe some of them which are "good" and give the small-time local controllability, and some of them which are "bad" and lead to obstructions to small-time local controllability.

Chapter 4. In this chapter, we first consider the problem of the controllability around an equilibrium of a *nonlinear partial differential equation* such that the linearized control system around the equilibrium is controllable. In finite dimension, we have already seen that, in such a situation, the nonlinear control system is locally controllable around the equilibrium. Of course in infinite dimension one expects that a similar result holds. We prove that this is indeed the case for various equations: A nonlinear Korteweg-de Vries equation, a nonlinear hyperbolic equation and a nonlinear Schrödinger equation. For the first equation, one uses a natural fixed-point method. For the two other equations, the situation is more involved due to a problem of loss of derivatives. For the hyperbolic equation, one uses, to take care of this problem, an ad-hoc fixed-point method, which is specific to hyperbolic systems. For the case of the Schrödinger equation, this problem is overcome by the use of a Nash-Moser method.

Sometimes these methods, which lead to *local* controllability results, can be adapted to give a *global* controllability result if the nonlinearity is not too big at infinity. We present an example for a nonlinear one-dimensional wave equation.

Chapter 5. We present an application of the use of iterated Lie brackets for a nonlinear partial differential equation (a nonlinear Schrödinger equation). We also explain why iterated Lie brackets are less powerful in infinite dimension than in finite dimension.

Chapter 6. This chapter deals with the *return method*. The idea of the return method goes as follows: If one can find a trajectory of the nonlinear control system such that:
- it starts and ends at the equilibrium,
- the linearized control system around this trajectory is controllable,

then, in general, the implicit function theorem implies that one can go from every state close to the equilibrium to every other state close to the equilibrium. In Chapter 6, we sketch some results in flow control which have been obtained by this method, namely:
- global controllability results for the Euler equations of incompressible fluids,
- global controllability results for the Navier-Stokes equations of incompressible fluids,
- local controllability of a 1-D tank containing a fluid modeled by the shallow water equations.

Chapter 7. This chapter develops the *quasi-static deformation method*, which allows one to prove in some cases that one can move from a given equilibrium to another given equilibrium if these two equilibria are connected in the set of equilibria. The idea is just to move very slowly the control (quasi-static deformation) so that at each time the state is close to the curve of equilibria connecting the two given equilibria. If some of these equilibria are unstable, one also uses suitable feedback laws in order to stabilize them; without these feedback laws the quasi-static deformation method would not work. We present an application to a semilinear heat equation.

Chapter 8. This chapter is devoted to the *power series expansion method*: One makes some power series expansion in order to decide whether the nonlinearity allows us to move in every (oriented) direction which is not controllable for the linearized control system around the equilibrium. We present an application to a nonlinear Korteweg-de Vries equation.

Chapter 9. The previous three methods (return, quasi-static deformations, power series expansion) can be used together. We present in this chapter an example for a nonlinear Schrödinger control equation.

Part 3: Stabilization

The two previous parts were devoted to the controllability problem, which asks if one can move from a first given state to a second given state. The control that one gets is an open-loop control: it depends on time and on the two given states, but it *does not* depend on the state during the evolution of the control system. In many practical situations one prefers closed-loop controls, i.e., controls which do not depend on the initial state but depend, at time t, on the state x at this time. One requires that these closed-loop controls (asymptotically) stabilize the point one wants to reach. Usually such closed-loop controls (or feedback laws) have the advantage of being be more robust to disturbances (recall the experiment of the stick on the finger). The main issue discussed in this part is the problem of deciding whether a controllable system can be (asymptotically) stabilized.

This part is divided into four chapters: Chapter 10, Chapter 11, Chapter 12 and Chapter 13, which we now briefly describe.

Chapter 10. This chapter is mainly concerned with the stabilization of *finite-dimensional linear* control systems. We first start by recalling the classical pole-shifting theorem. A consequence of this theorem is that every controllable linear system can be stabilized by means of linear feedback laws. This implies that, if the linearized control system at an equilibrium of a nonlinear control system is controllable, then this equilibrium can be stabilized by smooth feedback laws.

Chapter 11. This chapter discusses the stabilization of *finite-dimensional nonlinear* control systems, mainly in the case where the nonlinearity plays a key role. In particular, it deals with the case where the linearized control system around the equilibrium that one wants to stabilize is no longer controllable. Then there are obstructions to stabilizability by smooth feedback laws even for controllable systems. We recall some of these obstructions. There are two ways to enlarge the class of feedback laws in order to recover stabilizability properties. The first one is the use of discontinuous feedback laws. The second one is the use of time-varying feedback laws. We give only comments and references on the first method, but we give details on the second one. We also show the interest of time-varying feedback laws for output stabilization: In this case the feedback laws depend only on the output, which is only part of the state.

Chapter 12. In this chapter, we present important tools for constructing explicit stabilizing feedback laws, namely:

1. control Lyapunov function,
2. damping,
3. homogeneity,
4. averaging,
5. backstepping,
6. forwarding,
7. transverse functions.

These methods are illustrated on various control systems, in particular, the stabilization of the attitude of a rigid spacecraft.

Chapter 13. In this chapter, we give examples of how some tools introduced for the stabilization of *finite-dimensional* control systems can be used to stabilize some partial differential equations. We treat the following four examples:

1. rapid exponential stabilization by means of Gramians for linear time-reversible partial differential equations,
2. stabilization of a rotating body-beam without damping,
3. stabilization of the Euler equations of incompressible fluids,
4. stabilization of hyperbolic systems.

Appendices
This book has two appendices. In the first one (Appendix A), we recall some classical results on semigroups generated by linear operators and classical applications to evolution equations. We omit the proofs but we give precise references where they can be found. In the second appendix (Appendix B), we construct the degree of a map and prove the properties of the degree we use in this book. As an application of the degree, we also prove the Brouwer and Schauder fixed-point theorems which are also used in this book.

Acknowledgments. I thank the Rutgers University Mathematics Department, especially Eduardo Sontag and Héctor Sussmann, for inviting me to give the 2003 Dean Jacqueline B. Lewis Memorial Lectures. This book arose from these lectures. I am grateful to the Institute Universitaire de France for providing ideal working conditions for writing this book.

It is a pleasure to thank Azgal Abichou, Claude Bardos, Henry Hermes, Vilmos Komornik, Marius Tucsnak, and José Urquiza, for useful discussions. I am also especially indebted to Karine Beauchard, Eduardo Cerpa, Yacine Chitour, Emmanuelle Crépeau, Olivier Glass, Sergio Guerrero, Thierry Horsin, Rhouma Mlayeh, Christophe Prieur, Lionel Rosier, Emmanuel Trélat, and Claire Voisin who recommended many modifications and corrections. I also thank Claire Voisin for important suggestions and for her constant encouragement.

It is also a pleasure to thank my former colleagues at the Centre Automatique et Systèmes, Brigitte d'Andréa-Novel, François Chaplais, Michel Fliess, Yves Lenoir, Jean Lévine, Philippe Martin, Nicolas Petit, Laurent Praly and Pierre Rouchon, for convincing me to work in control theory.

Jean-Michel Coron
October, 2006

Part 1

Controllability of linear control systems

This first part is devoted to the controllability of *linear* control systems. It has two chapters: The first one deals with finite-dimensional control systems, the second one deals with infinite-dimensional control systems modeled by partial differential equations.

Let us detail the contents of these two chapters.

Chapter 1. This chapter focuses on the controllability of linear *finite-dimensional* control systems. We first give an integral necessary and sufficient condition (Theorem 1.11 on page 6) for a linear time-varying finite-dimensional control system to be controllable. For a special quadratic cost, it leads to the optimal control (Proposition 1.13 on page 8). We give examples of applications.

These examples show that the use of this necessary and sufficient condition can lead to computations which are somewhat complicated even for very simple control systems. In particular, it requires integrating linear differential equations. In Section 1.3 we first give the famous Kalman rank condition (Theorem 1.16 on page 9) for the controllability of linear *time-invariant* finite-dimensional control systems. This new condition, which is also necessary and sufficient for controllability, is purely algebraic; it does not require integrations of linear differential equations. We turn then to the case of linear *time-varying* finite-dimensional control systems. For these systems we give a sufficient condition for controllability (Theorem 1.18 on page 11), which turns out to be also necessary for analytic control systems. This condition only requires computing derivatives. Again no integrations are needed.

In Section 1.4, we describe, in the framework of linear time-varying finite-dimensional control systems, the Hilbert Uniqueness Method, due to Jacques-Louis Lions. This method is quite useful in infinite dimension to find numerically optimal controls for linear control systems.

Chapter 2. The subject of this chapter is the controllability of some classical *partial differential equations*. For the reader who is familiar with this subject, a large part of this chapter can be omitted; most of the methods detailed here are very well known. The linear partial differential equations which are treated are the following:

1. a transport equation (Section 2.1),
2. a Korteweg-de Vries equation (Section 2.2),
3. a one-dimensional wave equation (Section 2.4),
4. a heat equation (Section 2.5),
5. a one-dimensional Schrödinger equation (Section 2.6),
6. a family of one-dimensional heat equations depending on a small parameter (Section 2.7).

For these equations, after proving the well-posedness of the Cauchy problem, we study their controllability by means of various methods (explicit method, extension method, duality between controllability and observability, observability inequalities, multiplier method, Carleman inequalities, moment theory, Laplace transform and harmonic analysis ...). We also present in Section 2.3 a classical abstract setting which allows us to treat the well-posedness and the controllability of many partial differential equations in the same framework.

CHAPTER 1

Finite-dimensional linear control systems

This chapter focuses on the controllability of linear finite-dimensional control systems. It is organized as follows.

- In Section 1.2 we give an integral necessary and sufficient condition (Theorem 1.11 on page 6) for a linear time-varying finite-dimensional control system to be controllable. For a special quadratic cost, it leads to the optimal control (Proposition 1.13 on page 8). We give examples of applications.
- These examples show that the use of this necessary and sufficient condition can lead to computations which are somewhat complicated even for very simple control systems. In particular, it requires integrating linear differential equations. In Section 1.3 we first give the famous Kalman rank condition (Theorem 1.16 on page 9) for the controllability of linear *time-invariant* finite-dimensional control systems. This new condition, which is also necessary and sufficient for controllability, is purely algebraic; it does not require integrations of linear differential equations. We turn then to the case of linear *time-varying* finite-dimensional control systems. For these systems we give a sufficient condition for controllability (Theorem 1.18 on page 11), which turns out to be also necessary for analytic control systems. This condition only requires computing derivatives. Again no integrations are needed. We give two proofs of Theorem 1.18. The first one is the classical proof. The second one is new.
- In Section 1.4, we describe, in the framework of linear time-varying finite-dimensional control systems, the Hilbert Uniqueness Method (HUM), due to Jacques-Louis Lions. This method, which in finite dimension is strongly related to the integral necessary and sufficient condition of controllability given in Section 1.2, is quite useful in infinite dimension to find numerically optimal controls for linear control systems.

1.1. Definition of controllability

Let us start with some notations. For $k \in \mathbb{N} \setminus \{0\}$, $\mathbb{R}^k$ denotes the set of k-dimensional real column vector. For $k \in \mathbb{N} \setminus \{0\}$ and $l \in \mathbb{N} \setminus \{0\}$, we denote by $\mathcal{L}(\mathbb{R}^k; \mathbb{R}^l)$ the set of linear maps from $\mathbb{R}^k$ into $\mathbb{R}^l$. We often identify, in the usual way, $\mathcal{L}(\mathbb{R}^k; \mathbb{R}^l)$ with the set, denoted $\mathcal{M}_{k,l}(\mathbb{R})$, of $k \times l$ matrices with real coefficients. We denote by $\mathcal{M}_{k,l}(\mathbb{C})$ the set of $k \times l$ matrices with complex coefficients. Throughout this chapter, T_0, T_1 denote two real numbers such that $T_0 < T_1$, $A : (T_0, T_1) \to \mathcal{L}(\mathbb{R}^n; \mathbb{R}^n)$ denotes an element of $L^\infty((T_0, T_1); \mathcal{L}(\mathbb{R}^n; \mathbb{R}^n))$ and $B : (T_0, T_1) \to \mathcal{L}(\mathbb{R}^m; \mathbb{R}^n)$ denotes an element of $L^\infty((T_0, T_1); \mathcal{L}(\mathbb{R}^m; \mathbb{R}^n))$. We consider the time-varying linear control system

(1.1) $$\dot{x} = A(t)x + B(t)u,\ t \in [T_0, T_1],$$

where, at time $t \in [T_0, T_1]$, the state is $x(t) \in \mathbb{R}^n$ and the control is $u(t) \in \mathbb{R}^m$, and $\dot{x} := dx/dt$.

We first define the solution of the Cauchy problem

$$\dot{x} = A(t)x + B(t)u(t), \ x(T_0) = x^0, \tag{1.2}$$

for given u in $L^1((T_0, T_1); \mathbb{R}^m)$ and given x^0 in $\mathbb{R}^n$.

DEFINITION 1.1. Let $b \in L^1((T_0, T_1); \mathbb{R}^n)$. A map $x : [T_0, T_1] \to \mathbb{R}^n$ is a *solution of*

$$\dot{x} = A(t)x + b(t), \ t \in (T_0, T_1), \tag{1.3}$$

if $x \in C^0([T_0, T_1]; \mathbb{R}^n)$ and satisfies

$$x(t_2) = x(t_1) + \int_{t_1}^{t_2} (A(t)x(t) + b(t))dt, \ \forall (t_1, t_2) \in [T_0, T_1]^2.$$

In particular, for $x^0 \in \mathbb{R}^n$, a solution to the Cauchy problem

$$\dot{x} = A(t)x + b(t), \ t \in (T_0, T_1), \ x(T_0) = x^0, \tag{1.4}$$

is a function $x \in C^0([T_0, T_1]; \mathbb{R}^n)$ such that

$$x(\tau) = x^0 + \int_{T_0}^{\tau} (A(t)x(t) + b(t))dt, \ \forall \tau \in [T_0, T_1].$$

It is well known that, for every $b \in L^1((T_0, T_1); \mathbb{R}^n)$ and for every $x^0 \in \mathbb{R}^n$, the Cauchy problem (1.4) has a unique solution.

Let us now define the controllability of system (1.1). (This concept goes back to Rudolph Kalman [**263, 264**].)

DEFINITION 1.2. The linear time-varying control system (1.1) is *controllable* if, for every $(x^0, x^1) \in \mathbb{R}^n \times \mathbb{R}^n$, there exists $u \in L^\infty((T_0, T_1); \mathbb{R}^m)$ such that the solution $x \in C^0([T_0, T_1]; \mathbb{R}^n)$ of the Cauchy problem (1.2) satisfies $x(T_1) = x^1$.

REMARK 1.3. One could replace in this definition $u \in L^\infty((T_0, T_1); \mathbb{R}^m)$ by $u \in L^2((T_0, T_1); \mathbb{R}^m)$ or by $u \in L^1((T_0, T_1); \mathbb{R}^m)$. It follows from the proof of Theorem 1.11 on page 6 that these changes of spaces do not lead to different controllable systems. We have chosen to consider $u \in L^\infty((T_0, T_1); \mathbb{R}^m)$ since this is the natural space for general nonlinear control system (see, in particular, Definition 3.2 on page 125).

1.2. An integral criterion for controllability

We are now going to give a necessary and sufficient condition for the controllability of system (1.1) in terms of the resolvent of the time-varying linear system $\dot{x} = A(t)x$. This condition is due to Rudolph Kalman, Yu-Chi Ho and Kumpati Narendra [**265**, Theorem 5]. Let us first recall the definition of the resolvent of the time-varying linear system $\dot{x} = A(t)x$.

DEFINITION 1.4. The *resolvent* R of the time-varying linear system $\dot{x} = A(t)x$ is the map

$$R: \ [T_0, T_1]^2 \ \to \ \mathcal{L}(\mathbb{R}^n; \mathbb{R}^n)$$
$$(t_1, t_2) \ \mapsto \ R(t_1, t_2)$$

such that, for every $t_2 \in [T_0, T_1]$, the map $R(\cdot, t_2) : [T_0, T_1] \to \mathcal{L}(\mathbb{R}^n; \mathbb{R}^n)$, $t_1 \mapsto R(t_1, t_2)$, is the solution of the Cauchy problem

(1.5) $$\dot{M} = A(t)M, \ M(t_2) = \mathrm{Id}_n,$$

where Id_n denotes the identity map of $\mathbb{R}^n$.

One has the following classical properties of the resolvent.

PROPOSITION 1.5. *The resolvent R is such that*

(1.6) $$R \in C^0([T_0, T_1]^2; \mathcal{L}(\mathbb{R}^n; \mathbb{R}^n)),$$

(1.7) $$R(t_1, t_1) = \mathrm{Id}_n, \ \forall t_1 \in [T_0, T_1],$$

(1.8) $$R(t_1, t_2)R(t_2, t_3) = R(t_1, t_3), \ \forall (t_1, t_2, t_3) \in [T_0, T_1]^3.$$

In particular,

(1.9) $$R(t_1, t_2)R(t_2, t_1) = \mathrm{Id}_n, \ \forall (t_1, t_2) \in [T_0, T_1]^2.$$

Moreover, if $A \in C^0([T_0, T_1]; \mathcal{L}(\mathbb{R}^n; \mathbb{R}^n))$, then $R \in C^1([T_0, T_1]^2; \mathcal{L}(\mathbb{R}^n; \mathbb{R}^n))$ and one has

(1.10) $$\frac{\partial R}{\partial t_1}(t, \tau) = A(t)R(t, \tau), \ \forall (t, \tau) \in [T_0, T_1]^2,$$

(1.11) $$\frac{\partial R}{\partial t_2}(t, \tau) = -R(t, \tau)A(\tau), \ \forall (t, \tau) \in [T_0, T_1]^2.$$

REMARK 1.6. Equality (1.10) follows directly from the definition of the resolvent. Equality (1.11) can be obtained from (1.10) by differentiating (1.9) with respect to t_2.

EXERCISE 1.7. *Let us define A by*

$$A(t) = \begin{pmatrix} t & -1 \\ 1 & t \end{pmatrix}.$$

Compute the resolvent of $\dot{x} = A(t)x$.

Answer. *Let x_1 and x_2 be the two components of x in the canonical basis of $\mathbb{R}^2$. Let $z = x_1 + ix_2 \in \mathbb{C}$. One gets*

$$\dot{z} = (t + i)z,$$

which leads to

(1.12) $$z(t_1) = z(t_2) \exp\left(\frac{t_1^2}{2} - \frac{t_2^2}{2} + it_1 - it_2\right).$$

From (1.12), one gets

$$R(t_1, t_2) = \begin{pmatrix} \cos(t_1 - t_2) \exp\left(\frac{t_1^2}{2} - \frac{t_2^2}{2}\right) & -\sin(t_1 - t_2) \exp\left(\frac{t_1^2}{2} - \frac{t_2^2}{2}\right) \\ \sin(t_1 - t_2) \exp\left(\frac{t_1^2}{2} - \frac{t_2^2}{2}\right) & \cos(t_1 - t_2) \exp\left(\frac{t_1^2}{2} - \frac{t_2^2}{2}\right) \end{pmatrix}.$$

EXERCISE 1.8. *Let A be in $C^1([T_0, T_1]; \mathcal{L}(\mathbb{R}^n, \mathbb{R}^n))$.*

1. *For $k \in \mathbb{N}$, compute the derivative of the map*

$$\begin{array}{rcl} [T_0, T_1] & \to & \mathcal{L}(\mathbb{R}^n, \mathbb{R}^n) \\ t & \mapsto & A(t)^k. \end{array}$$

2. Let us assume that
$$A(t)A(\tau) = A(\tau)A(t), \ \forall (t,\tau) \in [T_0, T_1]^2.$$

Show that

(1.13) $$R(t_1, t_2) = \exp\left(\int_{t_2}^{t_1} A(t)dt\right), \ \forall (t_1, t_2) \in [T_0, T_1]^2.$$

3. Give an example of A such that (1.13) does not hold.
Answer. Take $n = 2$ and
$$A(t) := \begin{pmatrix} 0 & t \\ 0 & 1 \end{pmatrix}.$$

One gets, for every $(t_1, t_2) \in [T_0, T_1]^2$,
$$R(t_1, t_2) = \begin{pmatrix} 1 & (t_1 - 1)e^{t_1 - t_2} - t_2 + 1 \\ 0 & e^{t_1 - t_2} \end{pmatrix},$$
$$\exp\left(\int_{t_2}^{t_1} A(t)dt\right) = \begin{pmatrix} 1 & \frac{t_1 + t_2}{2}(e^{t_1 - t_2} - 1) \\ 0 & e^{t_1 - t_2} \end{pmatrix}.$$

Of course, the main property of the resolvent is the fact that it gives the solution of the Cauchy problem (1.4). Indeed, one has the following classical proposition.

PROPOSITION 1.9. *The solution of the Cauchy problem (1.4) satisfies*

(1.14) $$x(t_1) = R(t_1, t_0)x(t_0) + \int_{t_0}^{t_1} R(t_1, \tau)b(\tau)d\tau, \ \forall (t_0, t_1) \in [T_0, T_1]^2.$$

In particular,

(1.15) $$x(t) = R(t, T_0)x^0 + \int_{T_0}^{t} R(t, \tau)b(\tau)d\tau, \ \forall t \in [T_0, T_1].$$

Equality (1.14) is known as Duhamel's principle.

Let us now define the controllability Gramian of the control system $\dot{x} = A(t)x + B(t)u$.

DEFINITION 1.10. *The* controllability Gramian *of the control system*
$$\dot{x} = A(t)x + B(t)u$$
is the symmetric $n \times n$-matrix

(1.16) $$\mathfrak{C} := \int_{T_0}^{T_1} R(T_1, \tau)B(\tau)B(\tau)^{\text{tr}} R(T_1, \tau)^{\text{tr}} d\tau.$$

In (1.16) and throughout the whole book, for a matrix M or a linear map M from $\mathbb{R}^k$ into $\mathbb{R}^l$, M^{tr} denotes the transpose of M.

Our first condition for the controllability of $\dot{x} = A(t)x + B(t)u$ is given in the following theorem [**265**, Theorem 5] due to Rudolph Kalman, Yu-Chi Ho and Kumpati Narendra.

THEOREM 1.11. *The linear time varying control system $\dot{x} = A(t)x + B(t)u$ is controllable if and only if its controllability Gramian is invertible.*

REMARK 1.12. Note that, for every $x \in \mathbb{R}^n$,
$$x^{\text{tr}}\mathfrak{C}x = \int_{T_0}^{T_1} |B(\tau)^{\text{tr}}R(T_1,\tau)^{\text{tr}}x|^2 d\tau.$$

Hence the controllability Gramian $\mathfrak{C}$ is a *nonnegative* symmetric matrix. In particular, $\mathfrak{C}$ is invertible if and only if there exists $c > 0$ such that
$$(1.17) \qquad x^{\text{tr}}\mathfrak{C}x \geqslant c|x|^2, \ \forall x \in \mathbb{R}^n.$$

Proof of Theorem 1.11. We first assume that $\mathfrak{C}$ is invertible and prove that $\dot{x} = A(t)x + B(t)u$ is controllable. Let x^0 and x^1 be in $\mathbb{R}^n$. Let $\bar{u} \in L^\infty((T_0, T_1); \mathbb{R}^m)$ be defined by
$$(1.18) \qquad \bar{u}(\tau) := B(\tau)^{\text{tr}}R(T_1,\tau)^{\text{tr}}\mathfrak{C}^{-1}(x^1 - R(T_1,T_0)x^0), \ \tau \in (T_0, T_1).$$

(In (1.18) and in the following, the notation "$\tau \in (T_0, T_1)$" stands for "for almost every $\tau \in (T_0, T_1)$" or in the distribution sense in $\mathcal{D}'(T_0, T_1)$, depending on the context.) Let $\bar{x} \in C^0([T_0, T_1]; \mathbb{R}^n)$ be the solution of the Cauchy problem
$$(1.19) \qquad \dot{\bar{x}} = A(t)\bar{x} + B(t)\bar{u}(t), \ \bar{x}(T_0) = x^0.$$

Then, by Proposition 1.9,
$$\begin{aligned}\bar{x}(T_1) &= R(T_1, T_0)x^0 \\ &\quad + \int_{T_0}^{T_1} R(T_1, \tau)B(\tau)B(\tau)^{\text{tr}}R(T_1,\tau)^{\text{tr}}\mathfrak{C}^{-1}(x^1 - R(T_1,T_0)x^0)d\tau \\ &= R(T_1,T_0)x^0 + x^1 - R(T_1,T_0)x^0 \\ &= x^1.\end{aligned}$$

Hence $\dot{x} = A(t)x + B(t)u$ is controllable.

Let us now assume that $\mathfrak{C}$ is not invertible. Then there exists $y \in \mathbb{R}^n \setminus \{0\}$ such that $\mathfrak{C}y = 0$. In particular, $y^{\text{tr}}\mathfrak{C}y = 0$, that is,
$$(1.20) \qquad \int_{T_0}^{T_1} y^{\text{tr}}R(T_1,\tau)B(\tau)B(\tau)^{\text{tr}}R(T_1,\tau)^{\text{tr}}y \, d\tau = 0.$$

But the left hand side of (1.20) is equal to
$$\int_{T_0}^{T_1} |B(\tau)^{\text{tr}}R(T_1,\tau)^{\text{tr}}y|^2 d\tau.$$

Hence (1.20) implies that
$$(1.21) \qquad y^{\text{tr}}R(T_1,\tau)B(\tau) = 0, \ \tau \in (T_0, T_1).$$

Now let $u \in L^1((T_0, T_1); \mathbb{R}^m)$ and $x \in C^0([T_0, T_1]; \mathbb{R}^n)$ be the solution of the Cauchy problem
$$\dot{x} = A(t)x + B(t)u(t), \ x(T_0) = 0.$$

Then, by Proposition 1.9 on the previous page,
$$x(T_1) = \int_{T_0}^{T_1} R(T_1, \tau)B(\tau)u(\tau)d\tau.$$

In particular, by (1.21),
$$(1.22) \qquad y^{\text{tr}}x(T_1) = 0.$$

Since $y \in \mathbb{R}^n \setminus \{0\}$, there exists $x^1 \in \mathbb{R}^n$ such that $y^{\text{tr}}x^1 \neq 0$ (for example $x^1 := y$). It follows from (1.22) that, whatever u is, $x(T_1) \neq x^1$. This concludes the proof of Theorem 1.11. ∎

Let us point out that the control $\bar{u}$ defined by (1.18) has the quite interesting property given in the following proposition, also due to Rudolph Kalman, Yu-Chi Ho and Kumpati Narendra [**265**, Theorem 8].

PROPOSITION 1.13. *Let $(x^0, x^1) \in \mathbb{R}^n \times \mathbb{R}^n$ and let $u \in L^2((T_0, T_1); \mathbb{R}^m)$ be such that the solution of the Cauchy problem*

(1.23) $$\dot{x} = A(t)x + B(t)u, \quad x(T_0) = x^0,$$

satisfies
$$x(T_1) = x^1.$$

Then
$$\int_{T_0}^{T_1} |\bar{u}(t)|^2 dt \leqslant \int_{T_0}^{T_1} |u(t)|^2 dt$$

with equality (if and) only if
$$u(t) = \bar{u}(t) \text{ for almost every } t \in (T_0, T_1).$$

Proof of Proposition 1.13. Let $v := u - \bar{u}$. Then, $\bar{x}$ and x being the solutions of the Cauchy problems (1.19) and (1.23), respectively, one has

$$\begin{aligned}\int_{T_0}^{T_1} R(T_1, t)B(t)v(t)dt &= \int_{T_0}^{T_1} R(T_1, t)B(t)u(t)dt - \int_{T_0}^{T_1} R(T_1, t)B(t)\bar{u}(t)dt \\ &= (x(T_1) - R(T_1, T_0)x(T_0)) \\ &\quad - (\bar{x}(T_1) - R(T_1, T_0)\bar{x}(T_0)).\end{aligned}$$

Hence

(1.24) $$\int_{T_0}^{T_1} R(T_1, t)B(t)v(t)dt = (x^1 - R(T_1, T_0)x^0) - (x^1 - R(T_1, T_0)x^0) = 0.$$

One has

(1.25) $$\int_{T_0}^{T_1} |u(\tau)|^2 d\tau = \int_{T_0}^{T_1} |\bar{u}(\tau)|^2 d\tau + \int_{T_0}^{T_1} |v(\tau)|^2 d\tau + 2\int_{T_0}^{T_1} \bar{u}^{\mathrm{tr}}(\tau)v(\tau)d\tau.$$

From (1.18) (note also that $\mathfrak{C}^{\mathrm{tr}} = \mathfrak{C}$),

$$\int_{T_0}^{T_1} \bar{u}^{\mathrm{tr}}(\tau)v(\tau)d\tau = (x^1 - R(T_1, T_0)x^0)^{\mathrm{tr}} \mathfrak{C}^{-1} \int_{T_0}^{T_1} R(T_1, \tau)B(\tau)v(\tau)d\tau,$$

which, together with (1.24), gives

(1.26) $$\int_{T_0}^{T_1} \bar{u}^{\mathrm{tr}}(\tau)v(\tau)d\tau = 0.$$

Proposition 1.13 then follows from (1.25) and (1.26). ∎

EXERCISE 1.14. *Let us consider the control system*

(1.27) $$\dot{x}_1 = u, \quad \dot{x}_2 = x_1 + tu,$$

where the control is $u \in \mathbb{R}$ and the state is $x := (x_1, x_2)^{tr} \in \mathbb{R}^2$. Let $T > 0$. We take $T_0 := 0$ and $T_1 := T$. Compute $\mathfrak{C}$. Check that the control system (1.27) is not controllable.

Answer. *One finds*
$$\mathfrak{C} = \begin{pmatrix} T & T^2 \\ T^2 & T^3 \end{pmatrix},$$

which is a matrix of rank 1.

EXERCISE 1.15. *Let us consider the control system*

(1.28) $$\dot{x}_1 = x_2, \ \dot{x}_2 = u,$$

where the control is $u \in \mathbb{R}$ and the state is $x := (x_1, x_2)^{tr} \in \mathbb{R}^2$. Let $T > 0$. Compute the $\bar{u} \in L^2(0, T)$ which minimizes $\int_0^T |u(\tau)|^2 d\tau$ under the constraint that the solution x of (1.28) such that $x(0) = (-1, 0)^{tr}$ satisfies $x(T) = (0, 0)^{tr}$.

Answer. Let $T_0 = 0$, $T_1 = T$, $x^0 = (-1, 0)^{tr}$ and $x^1 = (0, 0)^{tr}$. One gets

$$R(T_1, \tau) = \begin{pmatrix} 1 & T - \tau \\ 0 & 1 \end{pmatrix}, \ \forall \tau \in [T_0, T_1],$$

$$\mathfrak{C} = \begin{pmatrix} \frac{T^3}{3} & \frac{T^2}{2} \\ \frac{T^2}{2} & T \end{pmatrix}, \ \mathfrak{C}^{-1} = \frac{12}{T^3} \begin{pmatrix} 1 & -\frac{T}{2} \\ -\frac{T}{2} & \frac{T^2}{3} \end{pmatrix},$$

which lead to

$$\bar{u}(t) = -\frac{12}{T^3}\left(t - \frac{T}{2}\right), \ x_2(t) = -\frac{12}{T^3}\left(\frac{t^2}{2} - \frac{T}{2}t\right), \ x_1(t) = -1 - \frac{12}{T^3}\left(\frac{t^3}{6} - \frac{T}{4}t^2\right).$$

1.3. Kalman's type conditions for controllability

The necessary and sufficient condition for controllability given in Theorem 1.11 on page 6 requires computing the matrix $\mathfrak{C}$, which might be quite difficult (and even impossible) in many cases, even for simple linear control systems. In this section we give a new criterion for controllability which is much simpler to check.

For simplicity, we start with the case of a time invariant system, that is, the case where $A(t)$ and $B(t)$ do not depend on time. Note that a priori the controllability of $\dot{x} = Ax + Bu$ could depend on T_0 and T_1 (more precisely, it could depend on $T_1 - T_0$). (This is indeed the case for some important linear partial differential control systems; see for example Section 2.1.2 and Section 2.4.2 below.) So we shall speak about the controllability of the time invariant linear control system $\dot{x} = Ax + Bu$ on $[T_0, T_1]$.

The famous Kalman rank condition for controllability is given in the following theorem.

THEOREM 1.16. *The time invariant linear control system $\dot{x} = Ax + Bu$ is controllable on $[T_0, T_1]$ if and only if*

(1.29) $$\mathrm{Span}\,\{A^i Bu;\ u \in \mathbb{R}^m, i \in \{0, \ldots, n-1\}\} = \mathbb{R}^n.$$

In particular, whatever $T_0 < T_1$ and $\tilde{T}_0 < \tilde{T}_1$ are, the time invariant linear control system $\dot{x} = Ax + Bu$ is controllable on $[T_0, T_1]$ if and only if it is controllable on $[\tilde{T}_0, \tilde{T}_1]$.

REMARK 1.17. Theorem 1.16 is Theorem 10 of [**265**], a joint paper by Rudolph Kalman, Yu-Chi Ho and Kumpati Narendra. The authors of [**265**] say on page 201 of their paper that Theorem 1.16 is the "simplest and best known" criterion for controllability. As they mention in [**265**, Section 11], Theorem 1.16 has previously appeared in the paper [**296**] by Joseph LaSalle. By [**296**, Theorem 6 and page 15] one has that the rank condition (1.29) implies that the time invariant linear control system $\dot{x} = Ax + Bu$ is controllable on $[T_0, T_1]$; moreover, it follows from [**296**, page 15] that, if the rank condition (1.29) is not satisfied, then there exists $\eta \in \mathbb{R}^n$ such that, for every $u \in L^\infty((T_0, T_1); \mathbb{R}^m)$,

$$(\dot{x} = Ax + Bu(t),\ x(T_0) = 0) \Rightarrow (\eta^{\mathrm{tr}} x(T_1) = 0)$$

and therefore the control system $\dot{x} = Ax + Bu$ is not controllable on $[T_0, T_1]$. The paper [**265**] also refers, in its Section 11, to the paper [**389**] by Lev Pontryagin, where a condition related to condition (1.29) appears, but without any connection stated to the controllability of the control system $\dot{x} = Ax + Bu$.

Proof of Theorem 1.16. Since A does not depend on time, one has
$$R(t_1, t_2) = e^{(t_1 - t_2)A}, \, \forall (t_1, t_2) \in [T_0, T_1]^2.$$

Hence

(1.30) $$\mathfrak{C} = \int_{T_0}^{T_1} e^{(T_1 - \tau)A} B B^{\mathrm{tr}} e^{(T_1 - \tau)A^{\mathrm{tr}}} d\tau.$$

Let us first assume that the time invariant linear control system $\dot{x} = Ax + Bu$ is not controllable on $[T_0, T_1]$. Then, by Theorem 1.11 on page 6, the linear map $\mathfrak{C}$ is not invertible. Hence there exists $y \in \mathbb{R}^n \setminus \{0\}$ such that

(1.31) $$\mathfrak{C}y = 0,$$

which implies that

(1.32) $$y^{\mathrm{tr}} \mathfrak{C} y = 0.$$

From (1.30) and (1.32), one gets
$$\int_{T_0}^{T_1} |B^{\mathrm{tr}} e^{(T_1 - \tau)A^{\mathrm{tr}}} y|^2 d\tau = 0,$$

from which we get

(1.33) $$k(\tau) = 0, \, \forall \tau \in [T_0, T_1],$$

with

(1.34) $$k(\tau) := y^{\mathrm{tr}} e^{(T_1 - \tau)A} B, \, \forall \tau \in [T_0, T_1].$$

Differentiating i-times (1.34) with respect to τ, one easily gets

(1.35) $$k^{(i)}(T_1) = (-1)^i y^{\mathrm{tr}} A^i B,$$

which, together with (1.33) gives

(1.36) $$y^{\mathrm{tr}} A^i B = 0, \, \forall i \in \mathbb{N}.$$

In particular,

(1.37) $$y^{\mathrm{tr}} A^i B = 0, \, \forall i \in \{0, \ldots, n - 1\}.$$

But, since $y \neq 0$, (1.37) implies that (1.29) does not hold.

In order to prove the converse it suffices to check that, for every $y \in \mathbb{R}^n$,

(1.38) $$(1.37) \Rightarrow (1.36),$$

(1.39) $$(k^{(i)}(T_1) = 0, \, \forall i \in \mathbb{N}) \Rightarrow (k = 0 \text{ on } [T_0, T_1]),$$

(1.40) $$(1.32) \Rightarrow (1.31).$$

Implication (1.38) follows from the Cayley-Hamilton theorem. Indeed, let P_A be the characteristic polynomial of the matrix A
$$P_A(z) := \det(z \mathrm{Id}_n - A) = z^n - \alpha_n z^{n-1} - \alpha_{n-1} z^{n-2} \ldots - \alpha_2 z - \alpha_1.$$

Then the Cayley-Hamilton theorem states that $P_A(A) = 0$, that is,

(1.41) $$A^n = \alpha_n A^{n-1} + \alpha_{n-1} A^{n-2} \ldots + \alpha_2 A + \alpha_1 \mathrm{Id}_n.$$

In particular,
$$(1.37) \Rightarrow y^{\mathrm{tr}} A^n B = 0.$$
Then a straightforward induction argument using (1.41) gives (1.38). Implication (1.39) follows from the fact that the map $k : [T_0, T_1] \to \mathcal{M}_{1,m}(\mathbb{R})$ is analytic. Finally, in order to get (1.40), it suffices to use the Cauchy-Schwarz inequality for the semi-definite positive bilinear form $q(a,b) = a^{\mathrm{tr}} \mathfrak{C} b$, that is,
$$|y^{\mathrm{tr}} \mathfrak{C} z| \leqslant (y^{\mathrm{tr}} \mathfrak{C} y)^{1/2} (z^{\mathrm{tr}} \mathfrak{C} z)^{1/2}.$$
This concludes the proof of Theorem 1.16. ∎

Let us now turn to the case of time-varying linear control systems. We assume that A and B are of class C^∞ on $[T_0, T_1]$. Let us define, by induction on i a sequence of maps $B_i \in C^\infty([T_0, T_1]; \mathcal{L}(\mathbb{R}^m; \mathbb{R}^n))$ in the following way:

(1.42) $\quad B_0(t) := B(t),\ B_i(t) := \dot{B}_{i-1}(t) - A(t) B_{i-1}(t),\ \forall t \in [T_0, T_1].$

Then one has the following theorem (see, in particular, the papers [**86**] by A. Chang and [**447**] by Leonard Silverman and Henry Meadows).

THEOREM 1.18. *Assume that, for some $\bar{t} \in [T_0, T_1]$,*

(1.43) $\quad \mathrm{Span}\ \{B_i(\bar{t}) u;\ u \in \mathbb{R}^m,\ i \in \mathbb{N}\} = \mathbb{R}^n.$

Then the linear control system $\dot{x} = A(t) x + B(t) u$ is controllable (on $[T_0, T_1]$).

Before giving two different proofs of Theorem 1.18, let us make two simple remarks. Our first remark is that the Cayley-Hamilton theorem (see (1.41)) can no longer be used: there are control systems $\dot{x} = A(t) x + B(t) u$ and $\bar{t} \in [T_0, T_1]$ such that
(1.44)
$$\mathrm{Span}\ \{B_i(\bar{t}) u;\ u \in \mathbb{R}^m,\ i \in \mathbb{N}\} \neq \mathrm{Span}\ \{B_i(\bar{t}) u;\ u \in \mathbb{R}^m,\ i \in \{0, \ldots, n-1\}\}.$$
For example, let us take $T_0 = 0$, $T_1 = 1$, $n = m = 1$, $A(t) = 0$, $B(t) = t$. Then
$$B_0(t) = t,\ B_1(t) = 1,\ B_i(t) = 0,\ \forall i \in \mathbb{N} \setminus \{0, 1\}.$$
Therefore, if $\bar{t} = 0$, the left hand side of (1.44) is $\mathbb{R}$ and the right hand side of (1.44) is $\{0\}$.

However, one has the following proposition, which, as far as we know, is new and that we shall prove later on (see pages 15–19; see also [**448**] by Leonard Silverman and Henry Meadows for prior related results).

PROPOSITION 1.19. *Let $\bar{t} \in [T_0, T_1]$ be such that (1.43) holds. Then there exists $\varepsilon > 0$ such that, for every $t \in ([T_0, T_1] \cap (\bar{t} - \varepsilon, \bar{t} + \varepsilon)) \setminus \{\bar{t}\}$,*

(1.45) $\quad \mathrm{Span}\ \{B_i(t) u;\ u \in \mathbb{R}^m,\ i \in \{0, \ldots, n-1\}\} = \mathbb{R}^n.$

Our second remark is that the sufficient condition for controllability given in Theorem 1.18 is not a necessary condition (unless $n = 1$, or A and B are assumed to be analytic; see Exercise 1.23 on page 19). Indeed, let us take $n = 2$, $m = 1$, $A = 0$. Let $f \in C^\infty([T_0, T_1])$ and $g \in C^\infty([T_0, T_1])$ be such that

(1.46) $\quad f = 0$ on $[(T_0 + T_1)/2, T_1],\ g = 0$ on $[T_0, (T_0 + T_1)/2],$

(1.47) $\quad f(T_0) \neq 0,\ g(T_1) \neq 0.$

Let B be defined by

(1.48) $$B(t) := \begin{pmatrix} f(t) \\ g(t) \end{pmatrix}, \forall t \in [T_0, T_1].$$

Then
$$\mathfrak{C} = \begin{pmatrix} \int_{T_0}^{T_1} f(t)^2 dt & 0 \\ 0 & \int_{T_0}^{T_1} g(t)^2 dt \end{pmatrix}.$$

Hence, by (1.47), $\mathfrak{C}$ is invertible. Therefore, by Theorem 1.11 on page 6, the linear control system $\dot{x} = A(t)x + B(t)u$ is controllable (on $[T_0, T_1]$). Moreover, one has

$$B_i(t) = \begin{pmatrix} f^{(i)}(t) \\ g^{(i)}(t) \end{pmatrix}, \forall t \in [T_0, T_1], \forall i \in \mathbb{N}.$$

Hence, by (1.46),

Span $\{B_i(t)u; u \in \mathbb{R}, i \in \mathbb{N}\} \subset \{(a, 0)^{\text{tr}}; a \in \mathbb{R}\}, \forall t \in [T_0, (T_0+T_1)/2]$,

Span $\{B_i(t)u; u \in \mathbb{R}, i \in \mathbb{N}\} \subset \{(0, a)^{\text{tr}}; a \in \mathbb{R}\}, \forall t \in [(T_0+T_1)/2, T_1]$.

Therefore, for every $\bar{t} \in [T_0, T_1]$, (1.43) does not hold.

EXERCISE 1.20. *Prove that there exist* $f \in C^\infty([T_0, T_1])$ *and* $g \in C^\infty([T_0, T_1])$ *such that*

(1.49) $$\text{Support } f \cap \text{Support } g = \{T_0\}.$$

Let us fix such f *and* g. *Let* $B \in C^\infty([T_0, T_1]; \mathbb{R}^2)$ *be defined by (1.48). We take* $n = 2$, $m = 1$, $A = 0$. *Prove that:*

1. *For every* $T \in (T_0, T_1)$, *the control system* $\dot{x} = B(t)u$ *is controllable on* $[T_0, T]$.
2. *For every* $\bar{t} \in [T_0, T_1]$, *(1.43) does not hold.*

Let us now give a first proof of Theorem 1.18. We assume that the linear control system $\dot{x} = A(t)x + B(t)u$ is not controllable. Then, by Theorem 1.11 on page 6, $\mathfrak{C}$ is not invertible. Therefore, there exists $y \in \mathbb{R}^n \setminus \{0\}$ such that $\mathfrak{C}y = 0$. Hence

$$0 = y^{\text{tr}} \mathfrak{C} y = \int_{T_0}^{T_1} |B(\tau)^{\text{tr}} R(T_1, \tau)^{\text{tr}} y|^2 d\tau = 0,$$

which implies, using (1.8), that

(1.50) $$K(\tau) := z^{\text{tr}} R(\bar{t}, \tau) B(\tau) = 0, \forall \tau \in [T_0, T_1],$$

with
$$z := R(T_1, \bar{t})^{\text{tr}} y.$$

Note that, by (1.9), $R(T_1, \bar{t})^{\text{tr}}$ is invertible (its inverse is $R(\bar{t}, T_1)^{\text{tr}}$). Hence z, as y, is not 0. Using (1.11), (1.42) and an induction argument on i, one gets

(1.51) $$K^{(i)}(\tau) = z^{\text{tr}} R(\bar{t}, \tau) B_i(\tau), \forall \tau \in [T_0, T_1], \forall i \in \mathbb{N}.$$

By (1.7), (1.50) and (1.51),

$$z^{\text{tr}} B_i(\bar{t}) = 0, \forall i \in \mathbb{N}.$$

As $z \neq 0$, this shows that (1.43) does not hold. This concludes our first proof of Theorem 1.18. ∎

Our second proof, which is new, is more delicate but may have applications for some controllability issues for infinite-dimensional control systems. Let $p \in \mathbb{N}$ be such that
$$\text{Span } \{B_i(\bar{t})u; \; u \in \mathbb{R}^m, \; i \in \{0, \ldots, p\}\} = \mathbb{R}^n.$$
By continuity there exists $\varepsilon > 0$ such that, with $[t_0, t_1] := [T_0, T_1] \cap [\bar{t} - \varepsilon, \bar{t} + \varepsilon]$,

(1.52) $\quad \text{Span } \{B_i(t)u; \; u \in \mathbb{R}^m, \; i \in \{0, \ldots, p\}\} = \mathbb{R}^n, \; \forall t \in [t_0, t_1].$

Let us first point out that

(1.53) $\quad \displaystyle\sum_{i=0}^{p} B_i(t)B_i(t)^{\mathrm{tr}}$ is invertible for every $t \in [t_0, t_1]$.

Indeed, if this is not the case, there exist $t^* \in [t_0, t_1]$ and

(1.54) $\quad a \in \mathbb{R}^n \setminus \{0\}$

such that
$$\sum_{i=0}^{p} B_i(t^*)B_i(t^*)^{\mathrm{tr}} a = 0.$$
In particular,
$$\sum_{i=0}^{p} a^{\mathrm{tr}} B_i(t^*)B_i(t^*)^{\mathrm{tr}} a = 0,$$
which implies that

(1.55) $\quad B_i(t^*)^{\mathrm{tr}} a = 0, \; \forall i \in \{0, \ldots, p\}.$

From (1.55), we get that
$$\left(\sum_{i=0}^{p} B_i(t^*) y_i\right)^{\mathrm{tr}} a = 0, \; \forall (y_0, \ldots, y_p) \in (\mathbb{R}^m)^{p+1},$$
which is in contradiction with (1.52) and (1.54). Hence (1.53) holds, which allows us to define, for $j \in \{0, \ldots, p\}$, $Q_j \in C^\infty([t_0, t_1]; \mathcal{L}(\mathbb{R}^n; \mathbb{R}^m))$ by
$$Q_j(t) := B_j(t)^{\mathrm{tr}} \left(\sum_{i=0}^{p} B_i(t)B_i(t)^{\mathrm{tr}}\right)^{-1}, \; \forall t \in [t_0, t_1].$$
We then have

(1.56) $\quad \displaystyle\sum_{i=0}^{p} B_i(t) Q_i(t) = \mathrm{Id}_n, \; \forall t \in [t_0, t_1].$

Let $(x^0, x^1) \in \mathbb{R}^n \times \mathbb{R}^n$. Let $\gamma^0 \in C^\infty([T_0, T_1]; \mathbb{R}^n)$ be the solution of the Cauchy problem

(1.57) $\quad \dot{\gamma}^0 = A(t)\gamma^0, \; \gamma^0(T_0) = x^0.$

Similarly, let $\gamma^1 \in C^\infty([T_0, T_1]; \mathbb{R}^n)$ be the solution of the Cauchy problem

(1.58) $\quad \dot{\gamma}^1 = A(t)\gamma^1, \; \gamma^1(T_1) = x^1.$

Let $d \in C^\infty([T_0, T_1])$ be such that

(1.59) $\quad d = 1$ on a neighborhood of $[T_0, t_0]$ in $[T_0, T_1]$,

(1.60) $\quad d = 0$ on a neighborhood of $[t_1, T_1]$ in $[T_0, T_1]$.

Let $\Gamma \in C^\infty([T_0, T_1]; \mathbb{R}^n)$ be defined by

(1.61) $$\Gamma(t) := d(t)\gamma^0(t) + (1 - d(t))\gamma^1(t), \, \forall t \in [T_0, T_1].$$

From (1.57), (1.58), (1.59), (1.60) and (1.61), one has

(1.62) $$\Gamma(T_0) = x^0, \, \Gamma(T_1) = x^1.$$

Let $q \in C^\infty([T_0, T_1]; \mathbb{R}^n)$ be defined by

(1.63) $$q(t) := -\dot{\Gamma}(t) + A(t)\Gamma(t), \, \forall t \in [T_0, T_1].$$

It readily follows from (1.57), (1.58), (1.59) and (1.60) that

(1.64) $$q = 0 \text{ on a neighborhood of } [T_0, t_0] \cup [t_1, T_1] \text{ in } [T_0, T_1].$$

Let us now define a sequence $(u_i)_{i \in \{0, \ldots, p-1\}}$ of elements of $C^\infty([t_0, t_1]; \mathbb{R}^m)$ by requiring (decreasing induction on i) that

(1.65) $$u_{p-1}(t) := Q_p(t)q(t), \, \forall t \in [t_0, t_1],$$
(1.66) $$u_{i-1}(t) := -\dot{u}_i(t) + Q_i(t)q(t), \, \forall i \in \{1, \ldots, p-1\}, \, \forall t \in [t_0, t_1].$$

Finally, we define $u : [T_0, T_1] \to \mathbb{R}^m$, $r : [T_0, T_1] \to \mathbb{R}^n$ and $x : [T_0, T_1] \to \mathbb{R}^n$ by

(1.67) $$u := 0 \text{ on } [T_0, t_0] \cup [t_1, T_1] \text{ and } u(t) := \dot{u}_0(t) - Q_0(t)q(t), \, \forall t \in (t_0, t_1),$$

(1.68) $$r := 0 \text{ on } [T_0, t_0] \cup [t_1, T_1] \text{ and } r(t) := \sum_{i=0}^{p-1} B_i(t)u_i(t), \, \forall t \in (t_0, t_1),$$

(1.69) $$x(t) := \Gamma(t) + r(t), \, \forall t \in [T_0, T_1].$$

It readily follows from (1.64), (1.65), (1.66), (1.67) and (1.68) that u, r and x are of class C^∞. From (1.62), (1.68) and (1.69), one has

(1.70) $$x(T_0) = x^0, \, x(T_1) = x^1.$$

Let $\theta \in C^\infty([T_0, T_1]; \mathbb{R}^n)$ be defined by

(1.71) $$\theta(t) := \dot{x}(t) - (A(t)x(t) + B(t)u(t)), \, \forall t \in [T_0, T_1].$$

From (1.63), (1.64), (1.67), (1.68), (1.69) and (1.71), one has

(1.72) $$\theta = 0 \text{ on } [T_0, t_0] \cup [t_1, T_1].$$

Let us check that

(1.73) $$\theta = 0 \text{ on } (t_0, t_1).$$

From (1.63), (1.68), (1.69) and (1.71), one has on (t_0, t_1),

$$\begin{aligned}\theta &= \dot{\Gamma} + \dot{r} - A(\Gamma + r) - Bu \\ &= -q + \left(\sum_{i=0}^{p-1} \dot{B}_i u_i\right) + \left(\sum_{i=0}^{p-1} B_i \dot{u}_i\right) - \left(\sum_{i=0}^{p-1} AB_i u_i\right) - Bu,\end{aligned}$$

which, together with (1.42), (1.65), (1.66) and (1.67), leads to

(1.74)
$$\begin{aligned}
0 &= -q + \left(\sum_{i=0}^{p-1}(B_{i+1} + AB_i)u_i\right) + \left(\sum_{i=1}^{p-1} B_i(-u_{i-1} + Q_i q)\right) \\
&\quad + B\dot{u}_0 - \left(\sum_{i=0}^{p-1} AB_i u_i\right) - Bu \\
&= -q + B_p u_{p-1} + \left(\sum_{i=1}^{p-1} B_i Q_i q\right) + B(u + Q_0 q) - Bu \\
&= -q + \sum_{i=0}^{p} B_i Q_i q.
\end{aligned}$$

From (1.56) and (1.74), one has (1.73). Finally, our second proof of Theorem 1.18 on page 11 follows from (1.70), (1.71), (1.72) and (1.73). ∎

REMARK 1.21. The above proof is related to the following property: A generic under-determined linear differential operator L has a right inverse M (i.e., an operator M satisfying $(L \circ M)q = q$, for every q) which is also a linear differential operator. This general result is due to Mikhael Gromov; see [**206**, (B), pages 150–151] for ordinary differential equations and [**206**, Theorem, page 156] for general differential equations. Here we consider the following differential operator:

$$L(x, u) := \dot{x} - A(t)x - B(t)u, \forall x \in C^\infty([t_0, t_1]; \mathbb{R}^n), \forall u \in C^\infty([t_0, t_1]; \mathbb{R}^m).$$

Even though the linear differential operator L is given and not generic, Property (1.52) implies the existence of such a right inverse M, which is in fact defined above: It suffices to take

$$M(q) := (x, u),$$

with

$$x(t) := \sum_{i=0}^{p-1} B_i(t) u_i(t), \forall t \in [t_0, t_1],$$

$$u(t) := \dot{u}_0(t) - Q_0(t)q(t), \forall t \in [t_0, t_1],$$

where $(u_i)_{i \in \{0,\ldots,p-1\}}$ is the sequence of functions in $C^\infty([t_0, t_1]; \mathbb{R}^m)$ defined by requiring (1.65) and (1.66).

Proof of Proposition 1.19 on page 11. Our proof is divided into three steps. In Step 1, we explain why we may assume that $A = 0$. In Step 2, we take care of the case of a scalar control ($m = 1$). Finally in Step 3, we reduce the multi-input case ($m > 1$) to the case of a scalar control.

Step 1. Let $R \in C^\infty([T_0, T_1] \times [T_0, T_1]; \mathcal{L}(\mathbb{R}^n; \mathbb{R}^n))$ be the resolvent of the time-varying linear system $\dot{x} = A(t)x$ (See Definition 1.4 on page 4). Let

$$\tilde{B} \in C^\infty([T_0, T_1]; \mathcal{L}(\mathbb{R}^m; \mathbb{R}^n))$$

be defined by

$$\tilde{B}(t) = R(\bar{t}, t) B(t), \forall t \in [T_0, T_1].$$

Let us define, by induction on $i \in \mathbb{N}$, $\tilde{B}_i \in C^\infty([T_0, T_1]; \mathcal{L}(\mathbb{R}^m; \mathbb{R}^n))$ by

$$\tilde{B}_0 = \tilde{B} \text{ and } \tilde{B}_i = \dot{\tilde{B}}_{i-1}, \forall i \in \mathbb{N} \setminus \{0\}.$$

In other words

(1.75) $$\tilde{B}_i = \tilde{B}^{(i)}.$$

Using (1.11), one readily gets, by induction on $i \in \mathbb{N}$,

$$\tilde{B}_i(t) = R(\bar{t}, t) B_i(t), \forall t \in [T_0, T_1], \forall i \in \mathbb{N}.$$

In particular, since $R(\bar{t}, \bar{t}) = \mathrm{Id}_n$ (see (1.7)) and $R(\bar{t}, t)$ is invertible for every $t \in [T_0, T_1]$ (see (1.9)),

$$\mathrm{Span}\ \{B_i(\bar{t})u;\ u \in \mathbb{R}^m,\ i \in \mathbb{N}\} = \mathrm{Span}\ \{\tilde{B}_i(\bar{t})u;\ u \in \mathbb{R}^m,\ i \in \mathbb{N}\},$$

$$\dim \mathrm{Span}\ \{B_i(t)u;\ u \in \mathbb{R}^m,\ i \in \{0, \ldots, n-1\}\}$$
$$= \dim \mathrm{Span}\ \{\tilde{B}_i(t)u;\ u \in \mathbb{R}^m,\ i \in \{0, \ldots, n-1\}\}, \forall t \in [T_0, T_1].$$

Hence, replacing B by $\tilde{B}$ and using (1.75), it suffices to consider the case where

(1.76) $$A = 0.$$

In the following two steps we assume that (1.76) holds. In particular,

(1.77) $$B_i = B^{(i)}, \forall i \in \mathbb{N}.$$

Step 2. In this step, we treat the case of a scalar control; we assume that $m = 1$. Let us assume, for the moment, that the following lemma holds.

LEMMA 1.22. *Let $B \in C^\infty([T_0, T_1]; \mathbb{R}^n)$ and $\bar{t} \in [T_0, T_1]$ be such that*

(1.78) $$\mathrm{Span}\ \{B^{(i)}(\bar{t});\ i \in \mathbb{N}\} = \mathbb{R}^n.$$

Then there exist n integers $p_i \in \mathbb{N}$, $i \in \{1, \ldots, n\}$, n functions $a_i \in C^\infty([T_0, T_1])$, $i \in \{1, \ldots, n\}$, n vectors $f_i \in \mathbb{R}^n$, $i \in \{1, \ldots, n\}$, such that

(1.79) $$p_i < p_{i+1}, \forall i \in \{1, \ldots, n-1\},$$

(1.80) $$a_i(\bar{t}) \neq 0, \forall i \in \{1, \ldots, n\},$$

(1.81) $$B(t) = \sum_{i=1}^n a_i(t)(t - \bar{t})^{p_i} f_i, \forall t \in [T_0, T_1],$$

(1.82) $$\mathrm{Span}\ \{f_i;\ i \in \{1, \ldots, n\}\} = \mathbb{R}^n.$$

From (1.77) and (1.81), one gets that, as $t \to \bar{t}$,

(1.83) $$\det(B_0(t), B_1(t), \ldots, B_{n-1}(t)) = K(t - \bar{t})^{-(n(n-1)/2) + \sum_{i=1}^n p_i}$$
$$+ O\left((t - \bar{t})^{1 - (n(n-1)/2) + \sum_{i=1}^n p_i}\right),$$

for the constant K defined by

(1.84) $$K := K(p_1, \ldots, p_n, a_1(\bar{t}), \ldots, a_n(\bar{t}), f_1, \ldots, f_m).$$

Let us compute K. Let $\bar{B} \in C^\infty(\mathbb{R}; \mathbb{R}^n)$ be defined by

(1.85) $$\bar{B}(t) = \sum_{i=1}^n a_i(\bar{t}) t^{p_i} f_i, \forall t \in \mathbb{R}.$$

One has

(1.86) $$\det(\bar{B}^{(0)}(t), \bar{B}^{(1)}(t), \ldots, \bar{B}^{(n-1)}(t)) = K t^{-(n(n-1)/2) + \sum_{i=1}^n p_i}, \forall t \in \mathbb{R}.$$

But, as one easily sees, for fixed t, fixed real numbers $a_i(\bar{t})$ and fixed integers p_i satisfying (1.79), the map
$$(f_1,\ldots,f_n) \in \mathbb{R}^n \times \ldots \times \mathbb{R}^n \mapsto \det\left(\bar{B}^{(0)}(t), \bar{B}^{(1)}(t), \ldots, \bar{B}^{(n-1)}(t)\right) \in \mathbb{R}$$
is multilinear and vanishes if the vectors $f_1,\ldots,f_n$ are dependent. Therefore K can be written in the following way:

(1.87) $$K := F(p_1,\ldots p_n)\left(\prod_{i=1}^{n} a_i(\bar{t})\right) \det(f_1,\ldots,f_n).$$

Taking for $(f_1,\ldots,f_n)$ the canonical basis of $\mathbb{R}^n$ and $a_i(\bar{t}) = 1$ for every $i \in \{1,\ldots,n\}$, one gets

(1.88) $$\det\left(\bar{B}^{(0)}(t), \bar{B}^{(1)}(t), \ldots, \bar{B}^{(n-1)}(t)\right) = t^{-(n(n-1)/2)+\sum_{i=1}^{n} p_i} \det M,$$

with
$$M := \begin{pmatrix} 1 & p_1 & p_1(p_1-1) & \cdots & p_1(p_1-1)(p_1-2)\ldots(p_1-n+1) \\ 1 & p_2 & p_2(p_2-1) & \cdots & p_2(p_2-1)(p_2-2)\ldots(p_2-n+1) \\ \vdots & \vdots & \vdots & \vdots & \vdots \\ 1 & p_n & p_n(p_n-1) & \cdots & p_n(p_n-1)(p_n-2)\ldots(p_n-n+1) \end{pmatrix}.$$

The determinant of M can be computed thanks to the Vandermonde determinant. One has

(1.89) $$\det M = \det \begin{pmatrix} 1 & p_1 & p_1^2 & \cdots & p_1^{n-1} \\ 1 & p_2 & p_2^2 & \cdots & p_2^{n-1} \\ \vdots & \vdots & \vdots & \vdots & \vdots \\ 1 & p_n & p_n^2 & \cdots & p_n^{n-1} \end{pmatrix} = \prod_{1 \leqslant i < j \leqslant n} (p_j - p_i).$$

Hence, from (1.87), (1.88) and (1.89), we have

(1.90) $$K := \left(\prod_{1 \leqslant i < j \leqslant n} (p_j - p_i)\right) \left(\prod_{i=1}^{n} a_i(\bar{t})\right) \det(f_1,\ldots,f_n).$$

From (1.79), (1.80), (1.82), and (1.90), it follows that

(1.91) $$K \neq 0.$$

From (1.83) and (1.91), one gets the existence of $\varepsilon > 0$ such that, for every $t \in ([T_0, T_1] \cap (\bar{t}-\varepsilon, \bar{t}+\varepsilon)) \setminus \{\bar{t}\}$, (1.45) holds.

Let us now prove Lemma 1.22 by induction on n. This lemma clearly holds if $n = 1$. Let us assume that it holds for every integer less than or equal to $(n-1) \in \mathbb{N} \setminus \{0\}$. We want to prove that it holds for n. Let $p_1 \in \mathbb{N}$ be such that

(1.92) $$B^{(i)}(\bar{t}) = 0, \forall i \in \mathbb{N} \cap [0, p_1-1],$$

(1.93) $$B^{(p_1)}(\bar{t}) \neq 0.$$

(Property (1.78) implies the existence of such a p_1.) Let

(1.94) $$f_1 := B^{(p_1)}(\bar{t}).$$

By (1.93) and (1.94),
$$f_1 \neq 0.$$

Let E be the orthogonal complement of f_1 in $\mathbb{R}^n$:
$$E := f_1^\perp \simeq \mathbb{R}^{n-1}. \tag{1.95}$$
Let $\Pi_E : \mathbb{R}^n \to E$ be the orthogonal projection on E. Let $C \in C^\infty([T_0, T_1]; E)$ be defined by
$$C(t) := \Pi_E B(t), \forall t \in [T_0, T_1]. \tag{1.96}$$
From (1.78) and (1.96), one gets that
$$\text{Span } \{C^{(i)}(\bar{t})u; u \in \mathbb{R}^m, i \in \mathbb{N}\} = E. \tag{1.97}$$
Hence, by the induction assumption, there are $(n-1)$ integers $p_i \in \mathbb{N}$, $i \in \{2, \ldots, n\}$, $(n-1)$ functions $a_i \in C^\infty([T_0, T_1])$, $i \in \{2, \ldots, n\}$, $(n-1)$ vectors $f_i \in E$, $i \in \{2, \ldots, n\}$, such that
$$p_i < p_{i+1}, \forall i \in \{2, \ldots, n-1\}, \tag{1.98}$$
$$a_i(\bar{t}) \neq 0, \forall i \in \{2, \ldots, n\}, \tag{1.99}$$
$$C(t) = \sum_{i=2}^{n} a_i(t)(t - \bar{t})^{p_i} f_i, \forall t \in [T_0, T_1], \tag{1.100}$$
$$\text{Span } \{f_i; i \in \{2, \ldots, n\}\} = E. \tag{1.101}$$
Let $g \in C^\infty([T_0, T_1])$ be such that (see (1.95), (1.96) and (1.100))
$$B(t) = g(t)f_1 + \sum_{i=2}^{n} a_i(t)(t - \bar{t})^{p_i} f_i, \forall t \in [T_0, T_1]. \tag{1.102}$$
Using (1.92), (1.93), (1.94), (1.95), (1.99), (1.101) and (1.102), one gets
$$p_1 < p_i, \forall i \in \{2, \ldots, n\},$$
$\exists a_1 \in C^\infty([T_0, T_1])$ such that $a_1(\bar{t}) \neq 0$ and $g(t) = (t - \bar{t})^{p_1} a_1(t), \forall t \in [T_0, T_1]$.

This concludes the proof of Lemma 1.22 and the proof of Proposition 1.19 on page 11 for $m = 1$.

Step 3. Here we still assume that (1.76) holds. We explain how to reduce the case $m > 1$ to the case $m = 1$. Let, for $i \in \{1, \ldots, m\}$, $b_i \in C^\infty([T_0, T_1]; \mathbb{R}^n)$ be such that
$$B(t) = (b_1(t), \ldots, b_m(t)), \forall t \in [T_0, T_1]. \tag{1.103}$$
We define, for $i \in \{1, \ldots, m\}$, a linear subspace E_i of $\mathbb{R}^n$ by
$$E_i := \text{Span } \left\{b_k^{(j)}(\bar{t}); k \in \{1, \ldots, i\}, j \in \mathbb{N}\right\}. \tag{1.104}$$
From (1.43), (1.77), (1.103) and (1.104), we have
$$E_m = \mathbb{R}^n. \tag{1.105}$$
Let $q \in \mathbb{N} \setminus \{0\}$ be such that, for every $i \in \{1, \ldots, m\}$,
$$E_i = \text{Span } \left\{b_k^{(j)}(\bar{t}); k \in \{1, \ldots, i\}, j \in \{0, \ldots, q-1\}\right\}. \tag{1.106}$$
Let $b \in C^\infty([T_0, T_1]; \mathbb{R}^n)$ be defined by
$$b(t) := \sum_{i=1}^{m}(t - \bar{t})^{(i-1)q} b_i(t), \forall t \in [T_0, T_1]. \tag{1.107}$$

From (1.106) and (1.107), one readily gets, by induction on $i \in \{1, \ldots, m\}$,
$$E_i = \text{Span}\left\{b^{(j)}(\bar{t}); j \in \{0, \ldots iq - 1\}\right\}.$$
In particular, taking $i = m$ and using (1.105), we get

(1.108) $$\text{Span}\left\{b^{(j)}(\bar{t}); j \in \mathbb{N}\right\} = \mathbb{R}^n.$$

Hence, by Proposition 1.19 applied to the case $m = 1$ (i.e., Step 2), there exists $\varepsilon > 0$ such that, for every $t \in ([T_0, T_1] \cap (\bar{t} - \varepsilon, \bar{t} + \varepsilon)) \setminus \{\bar{t}\}$,

(1.109) $$\text{Span}\left\{b^{(j)}(t); i \in \{0, \ldots, n-1\}\right\} = \mathbb{R}^n.$$

Since, by (1.107),
$$\text{Span}\left\{b^{(j)}(t); j \in \{0, \ldots, n-1\}\right\}$$
$$\subset \text{Span}\left\{b_i^{(j)}(t); i \in \{1, \ldots, m\}, j \in \{0, \ldots, n-1\}\right\},$$
this concludes the proof of Proposition 1.19 on page 11. ∎

Let us end this section with an exercise.

EXERCISE 1.23. *Let us assume that the two maps A and B are analytic and that the linear control system $\dot{x} = A(t)x + B(t)u$ is controllable (on $[T_0, T_1]$). Prove that:*

1. *For every $t \in [T_0, T_1]$,*
$$Span \ \{B_i(t)u; u \in \mathbb{R}^m, i \in \mathbb{N}\} = \mathbb{R}^n.$$

2. *The set*
$$\{t \in [T_0, T_1]; Span \ \{B_i(t)u; u \in \mathbb{R}^m, i \in \{0, \ldots, n-1\}\} \neq \mathbb{R}^n\}$$
is finite.

1.4. The Hilbert Uniqueness Method

In this section, our goal is to describe, in the framework of finite-dimensional linear control systems, a method, called the Hilbert Uniqueness Method (HUM), introduced by Jacques-Louis Lions in [**325, 326**] to solve controllability problems for linear partial differential equations. This method is closely related to duality between controllability and observability. This duality is classical in finite dimension. For infinite-dimensional control systems, this duality has been proved by Szymon Dolecki and David Russell in [**146**]. (See the paper [**293**] by John Lagnese and the paper [**48**] by Alain Bensoussan for a detailed description of the HUM together with its connection to prior works.) The HUM is also closely related to Theorem 1.11 on page 6 and Proposition 1.13 on page 8. In fact it provides a method to compute the control $\bar{u}$ defined by (1.18) for quite general control systems in infinite dimension.

We consider again the time-varying linear control system

(1.110) $$\dot{x} = A(t)x + B(t)u, \ t \in [T_0, T_1],$$

where, at time $t \in [T_0, T_1]$, the state is $x(t) \in \mathbb{R}^n$ and the control is $u(t) \in \mathbb{R}^m$. Let us also recall that $A \in L^\infty((T_0, T_1); \mathcal{L}(\mathbb{R}^n; \mathbb{R}^n))$ and $B \in L^\infty((T_0, T_1); \mathcal{L}(\mathbb{R}^m; \mathbb{R}^n))$. We are interested in the set of states which can be reached from 0 during the time interval $[T_0, T_1]$. More precisely, let $\mathcal{R}$ be the set of $x^1 \in \mathbb{R}^n$ such that there exists $u \in L^2((T_0, T_1); \mathbb{R}^m)$ such that the solution of the Cauchy problem

(1.111) $$\dot{x} = A(t)x + B(t)u(t),\ x(T_0) = 0,$$

satisfies $x(T_1) = x^1$. For ϕ^1 given in $\mathbb{R}^n$, we consider the solution $\phi : [T_0, T_1] \to \mathbb{R}^n$ of the following backward Cauchy linear problem:

(1.112) $$\dot{\phi} = -A(t)^{\mathrm{tr}}\phi,\ \phi(T_1) = \phi^1,\ t \in [T_0, T_1].$$

This linear system is called the adjoint system of the control system (1.110). This terminology is justified by the following proposition.

PROPOSITION 1.24. *Let $u \in L^2((T_0, T_1); \mathbb{R}^m)$. Let $x : [T_0, T_1] \to \mathbb{R}^n$ be the solution of the Cauchy problem (1.111). Let $\phi^1 \in \mathbb{R}^n$ and let $\phi : [T_0, T_1] \to \mathbb{R}^n$ be the solution of the Cauchy problem (1.112). Then*

(1.113) $$x(T_1) \cdot \phi^1 = \int_{T_0}^{T_1} u(t) \cdot B(t)^{\mathrm{tr}}\phi(t)\,dt.$$

In (1.113) and throughout the whole book, for a and b in $\mathbb{R}^l$, $l \in \mathbb{N} \setminus \{0\}$, $a \cdot b$ denotes the usual scalar product of a and b. In other words $a \cdot b := a^{\mathrm{tr}}b$.

Proof of Proposition 1.24. We have

$$\begin{aligned} x(T_1) \cdot \phi^1 &= \int_{T_0}^{T_1} \frac{d}{dt}(x(t) \cdot \phi(t))\,dt \\ &= \int_{T_0}^{T_1} ((A(t)x(t) + B(t)u(t)) \cdot \phi(t) - x(t) \cdot A(t)^{\mathrm{tr}}\phi(t))\,dt \\ &= \int_{T_0}^{T_1} u(t) \cdot B(t)^{\mathrm{tr}}\phi(t)\,dt. \end{aligned}$$

This concludes the proof of Proposition 1.24. ∎

Let us now denote by Λ the following map

$$\begin{aligned} \mathbb{R}^n &\to \mathbb{R}^n \\ \phi^1 &\mapsto x(T_1) \end{aligned}$$

where $x : [T_0, T_1] \to \mathbb{R}^n$ is the solution of the Cauchy problem

(1.114) $$\dot{x} = A(t)x + B(t)\bar{u}(t),\ x(T_0) = 0,$$

with

(1.115) $$\bar{u}(t) := B(t)^{\mathrm{tr}}\phi(t).$$

Here $\phi : [T_0, T_1] \to \mathbb{R}^n$ is the solution of the adjoint problem

(1.116) $$\dot{\phi} = -A(t)^{\mathrm{tr}}\phi,\ \phi(T_1) = \phi^1,\ t \in [T_0, T_1].$$

Then the following theorem holds.

THEOREM 1.25. *One has*

(1.117) $$\mathcal{R} = \Lambda(\mathbb{R}^n).$$

Moreover, if $x^1 = \Lambda(\phi^1)$ and if $u^* \in L^2((T_0, T_1); \mathbb{R}^m)$ is a control which steers the control system (1.110) from 0 to x^1 during the time interval $[T_0, T_1]$, then

$$\text{(1.118)} \quad \int_{T_0}^{T_1} |\bar{u}(t)|^2 dt \leqslant \int_{T_0}^{T_1} |u^*(t)|^2 dt,$$

(where $\bar{u}$ is defined by (1.115)-(1.116)), with equality if and only if $u^* = \bar{u}$.

Of course Theorem 1.25 on the previous page follows from the proofs of Theorem 1.11 on page 6 and of Proposition 1.13 on page 8; but we provide a new (slightly different) proof, which is more suitable to treat the case of linear control systems in infinite dimension.

By the definition of Λ,

$$\text{(1.119)} \quad \Lambda(\mathbb{R}^n) \subset \mathcal{R}.$$

Let x^1 be in $\mathcal{R}$. Let $u^* \in L^2((T_0, T_1); \mathbb{R}^m)$ be such that the solution x^* of the Cauchy problem

$$\dot{x}^* = A(t)x^* + B(t)u^*(t), \ x^*(T_0) = 0,$$

satisfies

$$\text{(1.120)} \quad x^*(T_1) = x^1.$$

Let $\mathcal{U} \subset L^2((T_0, T_1); \mathbb{R}^m)$ be the set of maps of the form $t \in [T_0, T_1] \mapsto B(t)^{\text{tr}}\phi(t)$ where $\phi : [T_0, T_1] \to \mathbb{R}^n$ satisfies (1.112) for some $\phi^1 \in \mathbb{R}^n$. This set $\mathcal{U}$ is a vector subspace of $L^2((T_0, T_1); \mathbb{R}^m)$. By its definition, $\mathcal{U}$ is of finite dimension (its dimension is less than or equal to n). Hence $\mathcal{U}$ is a *closed* vector subspace of $L^2((T_0, T_1); \mathbb{R}^m)$. Let $\tilde{u}$ be the orthogonal projection of u^* on $\mathcal{U}$. One has

$$\text{(1.121)} \quad \int_{T_0}^{T_1} u^*(t) \cdot u(t) = \int_{T_0}^{T_1} \tilde{u}(t) \cdot u(t) dt, \ \forall u \in \mathcal{U}.$$

Let $\tilde{x} : [T_0, T_1] \to \mathbb{R}^n$ be the solution of the Cauchy problem

$$\text{(1.122)} \quad \dot{\tilde{x}} = A(t)\tilde{x} + B(t)\tilde{u}(t), \ \tilde{x}(T_0) = 0.$$

From Proposition 1.24, (1.120), (1.121) and (1.122),

$$x^1 \cdot \phi^1 = \tilde{x}(T_1) \cdot \phi^1, \ \forall \phi^1 \in \mathbb{R}^n,$$

which implies that

$$\text{(1.123)} \quad x^1 = \tilde{x}(T_1).$$

Since $\tilde{u} \in \mathcal{U}$, there exists $\tilde{\phi}^1$ such that the solution $\tilde{\phi}$ of the Cauchy problem

$$\dot{\tilde{\phi}} = -A(t)^{\text{tr}}\tilde{\phi}, \ \tilde{\phi}(T_1) = \tilde{\phi}^1, \ t \in [T_0, T_1],$$

satisfies

$$\tilde{u}(t) = B(t)^{\text{tr}}\tilde{\phi}(t), \ t \in [T_0, T_1].$$

By the definition of Λ and (1.122),

$$\Lambda(\tilde{\phi}^1) = \tilde{x}(T_1),$$

which, together with (1.123), implies that $x^1 = \Lambda(\tilde{\phi}^1)$ and concludes the proof of (1.117).

Finally, let $x^1 = \Lambda(\phi^1)$, let $\bar{u}$ be defined by (1.115)-(1.116) and let

$$u^* \in L^2((T_0, T_1); \mathbb{R}^m)$$

be a control which steers the control system (1.110) from 0 to x^1 during the time interval $[T_0, T_1]$. Note that, by the definition of $\mathcal{U}$, $\bar{u} \in \mathcal{U}$. Moreover, using Proposition 1.24 on page 20 once more, we get that
$$\int_{T_0}^{T_1} u^*(t) \cdot u(t) dt = \int_{T_0}^{T_1} \bar{u}(t) \cdot u(t) dt, \, \forall u \in \mathcal{U}.$$
Hence $\bar{u}$ is the orthogonal projection of u^* on $\mathcal{U}$ and we have
$$\int_{T_0}^{T_1} |u^*(t)|^2 dt = \int_{T_0}^{T_1} |\bar{u}(t)|^2 dt + \int_{T_0}^{T_1} |u^*(t) - \bar{u}|^2 dt.$$
This concludes the proof of Theorem 1.25. ∎

CHAPTER 2

Linear partial differential equations

The subject of this chapter is the controllability of some classical partial differential equations. For the reader who is familiar with this subject, a large part of this chapter can be omitted; most of the methods detailed here are very well known. One can find much more advanced material in some references given throughout this chapter. The organization of this chapter is as follows.

- Section 2.1 concerns a transport equation. We first prove the well-posedness of the Cauchy problem (Theorem 2.4 on page 27). Then we study the controllability by different methods, namely:
    - An explicit method: for this simple transport equation, one can give explicitly a control steering the control system from every given state to every other given state (if the time is large enough, a necessary condition for this equation).
    - The extension method. This method turns out to be useful for many hyperbolic (even nonlinear) equations.
    - The duality between controllability and observability. The idea is the following one. Controllability for a linear control system is equivalent to the surjectivity of a certain linear map $\mathcal{F}$ from a Hilbert space H_1 to another Hilbert space H_2. The surjectivity of $\mathcal{F}$ is equivalent to the existence of $c > 0$ such that

(2.1) $$\|\mathcal{F}^*(x_2)\|_{H_1} \geqslant c\|x_2\|_{H_2}, \, \forall x_2 \in H_2,$$

    where $\mathcal{F}^* : H_2 \to H_1$ is the adjoint of $\mathcal{F}$. So, one first computes $\mathcal{F}^*$ and then proves (2.1). Inequality (2.1) is called the observability inequality. This method is nowadays the most popular one to prove the controllability of a linear control partial differential equation. The difficult part of this approach is to prove the observability inequality (2.1). There are many methods to prove such an inequality. Here we use the multiplier method. We also present, in this chapter, other methods for other equations.
- Section 2.2 is devoted to a linear Korteweg-de Vries (KdV) control equation. We first prove the well-posedness of the Cauchy problem (Theorem 2.23 on page 39). The proof relies on the classical semigroup approach. Then we prove a controllability result, namely Theorem 2.25 on page 42. The proof is based on the duality between controllability and observability. The observability inequality (2.1) ((2.156) for our KdV equation) uses a smoothing effect, a multiplier method and a compactness argument.
- In Section 2.3, we present a classical general framework which includes as special cases the study of the previous equations and their controllability as well as of many other equations.

- Section 2.4 is devoted to a time-varying linear one-dimensional wave equation. We first prove the well-posedness of the Cauchy problem (Theorem 2.53 on page 68). The proof also relies on the classical semigroup approach. Then we prove a controllability result, namely Theorem 2.55 on page 72. The proof is based again on the duality between controllability and observability. We prove the observability inequality (Proposition 2.60 on page 74) by means of a multiplier method (in a special case only).
- Section 2.5 concerns a linear heat equation. Again, we first take care of the well-posedness of the Cauchy problem (Theorem 2.63 on page 77) by means of the abstract approach. Then we prove the controllability of this equation (Theorem 2.66 on page 79). The proof is based on the duality between controllability and observability, but some new phenomena appear due to the irreversibility of the heat equation. In particular, the observability inequality now takes a new form, namely (2.398). This new inequality is proved by establishing a global Carleman inequality. We also give in this section a method, based on the flatness approach, to solve a motion planning problem for a one-dimensional heat equation. Finally we prove that one cannot control a heat equation in dimension larger than one by means of a finite number of controls. To prove this result, we use a Laplace transform together with a classical restriction theorem on the zeroes of an entire function of exponential type.
- Section 2.6 is devoted to the study of a one-dimensional linear Schrödinger equation. For this equation, the controllability result (Theorem 2.87 on page 96) is obtained by the moments theory method, a method which is quite useful for linear control systems with a finite number of controls.
- In Section 2.7, we consider a singular optimal control. We have a family of linear control one-dimensional heat equations depending on a parameter $\varepsilon > 0$. As $\varepsilon \to 0$, the heat equations degenerate into a transport equation. The heat equation is controllable for every time, but the transport equation is controllable only for large time. Depending on the time of controllability, we study the behavior of the optimal controls as $\varepsilon \to 0$. The lower bounds (Theorem 2.95 on page 104) are obtained by means of a Laplace transform together with a classical representation of entire functions of exponential type in $\mathbb{C}_+$. The upper bounds (Theorem 2.96) are obtained by means of an observability inequality proved with the help of global Carleman inequalities.
- Finally in Section 2.8 we give some bibliographical complements on the subject of the controllability of infinite-dimensional linear control systems.

2.1. Transport equation

Let $T > 0$ and $L > 0$. We consider the linear control system

(2.2) $$y_t + y_x = 0,\ t \in (0, T),\ x \in (0, L),$$
(2.3) $$y(t, 0) = u(t),$$

where, at time t, the control is $u(t) \in \mathbb{R}$ and the state is $y(t, \cdot) : (0, L) \to \mathbb{R}$. Our goal is to study the controllability of the control system (2.2)-(2.3); but let us first study the Cauchy problem associated to (2.2)-(2.3).

2.1.1. Well-posedness of the Cauchy problem.
Let us first recall the usual definition of solutions of the Cauchy problem

(2.4) $$y_t + y_x = 0, \ t \in (0,T), \ x \in (0,L),$$

(2.5) $$y(t,0) = u(t), \ t \in (0,T),$$

(2.6) $$y(0,x) = y^0(x), \ x \in (0,L).$$

where $T > 0$, $y^0 \in L^2(0,L)$ and $u \in L^2(0,T)$ are given. In order to motivate this definition, let us first assume that there exists a function y of class C^1 on $[0,T] \times [0,L]$ satisfying (2.4)-(2.5)-(2.6) in the usual sense. Let $\tau \in [0,T]$. Let $\phi \in C^1([0,\tau] \times [0,L])$. We multiply (2.4) by ϕ and integrate the obtained identity on $[0,\tau] \times [0,L]$. Using (2.5), (2.6) and integrations by parts, one gets

$$-\int_0^\tau \int_0^L (\phi_t + \phi_x) y \, dx \, dt + \int_0^\tau y(t,L)\phi(t,L) \, dt - \int_0^\tau u(t)\phi(t,0) \, dt$$
$$+ \int_0^L y(\tau,x)\phi(\tau,x) \, dx - \int_0^L y^0(x)\phi(0,x) \, dx = 0.$$

This equality leads to the following definition.

DEFINITION 2.1. *Let $T > 0$, $y^0 \in L^2(0,L)$ and $u \in L^2(0,T)$ be given. A solution of the Cauchy problem (2.4)-(2.5)-(2.6) is a function $y \in C^0([0,T]; L^2(0,L))$ such that, for every $\tau \in [0,T]$ and for every $\phi \in C^1([0,\tau] \times [0,L])$ such that*

(2.7) $$\phi(t,L) = 0, \ \forall t \in [0,\tau],$$

one has

(2.8) $$-\int_0^\tau \int_0^L (\phi_t + \phi_x) y \, dx \, dt - \int_0^\tau u(t)\phi(t,0) \, dt$$
$$+ \int_0^L y(\tau,x)\phi(\tau,x) \, dx - \int_0^L y^0(x)\phi(0,x) \, dx = 0.$$

This definition is also justified by the following proposition.

PROPOSITION 2.2. *Let $T > 0$, $y^0 \in L^2(0,L)$ and $u \in L^2(0,T)$ be given. Let us assume that y is a solution of the Cauchy problem (2.4)-(2.5)-(2.6) which is of class C^1 in $[0,T] \times [0,L]$. Then*

(2.9) $$y^0 \in C^1([0,L]),$$

(2.10) $$u \in C^1([0,T]),$$

(2.11) $$y(0,x) = y^0(x), \ \forall x \in [0,L],$$

(2.12) $$y(t,0) = u(t), \ \forall t \in [0,T],$$

(2.13) $$y_t(t,x) + y_x(t,x) = 0, \ \forall (t,x) \in [0,T] \times [0,L].$$

Proof of Proposition 2.2. Let $\phi \in C^1([0,T] \times [0,L])$ vanish on $(\{0,T\} \times [0,L]) \cap ([0,T] \times \{0,L\})$. From Definition 2.1, we get, taking $\tau := T$,

$$\int_0^T \int_0^L (\phi_t + \phi_x) y \, dx \, dt = 0,$$

which, using integrations by parts, gives

$$\int_0^T \int_0^L (y_t + y_x)\phi \, dx \, dt = 0. \tag{2.14}$$

Since the set $\phi \in C^1([0,T] \times [0,L])$ vanishing on $(\{0,T\} \times [0,L]) \cap ([0,T] \times \{0,L\})$ is dense in $L^1((0,T) \times (0,L))$, (2.14) implies that

$$\int_0^T \int_0^L (y_t + y_x)\phi \, dx \, dt = 0, \quad \forall \phi \in L^1((0,T) \times (0,L)). \tag{2.15}$$

Taking $\phi \in L^1((0,T) \times (0,L))$ defined by

$$\phi(t,x) := 1 \text{ if } y_t(t,x) + y_x(t,x) \geqslant 0, \tag{2.16}$$

$$\phi(t,x) := -1 \text{ if } y_t(t,x) + y_x(t,x) < 0, \tag{2.17}$$

one gets

$$\int_0^T \int_0^L |y_t + y_x| \, dx \, dt = 0, \tag{2.18}$$

which gives (2.13). Now let $\phi \in C^1([0,T] \times [0,L])$ be such that

$$\phi(t,L) = 0, \forall t \in [0,T]. \tag{2.19}$$

From (2.8) (with $\tau = T$), (2.13), (2.19) and integrations by parts, we get

$$\int_0^T (y(t,0) - u(t))\phi(t,0) \, dt + \int_0^T (y(0,x) - y^0(x))\phi(0,x) \, dx = 0. \tag{2.20}$$

Let $\mathcal{B} : C^1([0,T] \times [0,L]) \to L^1(0,T) \times L^1(0,L)$ be defined by

$$\mathcal{B}(\phi) := (\phi(\cdot,0), \phi(0,\cdot))$$

One easily checks that

(2.21) $\mathcal{B}(\{\phi \in C^1([0,T] \times [0,L]); (2.19) \text{ holds}\})$ is dense in $L^1(0,T) \times L^1(0,L)$.

Proceeding as for the proof of (2.18), we deduce from (2.20) and (2.21)

$$\int_0^T |y(t,0) - u(t)| \, dt + \int_0^T |y(0,x) - y^0(x)| \, dx = 0.$$

This concludes the proof of Proposition 2.2. ∎

EXERCISE 2.3. Let $T > 0$, $y^0 \in L^2(0,L)$ and $u \in L^2(0,T)$ be given. Let $y \in C^0([0,T]; L^2(0,L))$. Prove that y is a solution of the Cauchy problem (2.4)-(2.5)-(2.6) if and only if, for every $\phi \in C^1([0,T] \times [0,L])$ such that

$$\phi(t,L) = 0, \forall t \in [0,T],$$
$$\phi(T,x) = 0, \forall x \in [0,L],$$

one has

$$\int_0^T \int_0^L (\phi_t + \phi_x) y \, dx \, dt + \int_0^T u(t)\phi(t,0) \, dt + \int_0^L y^0(x)\phi(0,x) \, dx = 0.$$

With Definition 2.1, one has the following theorem.

THEOREM 2.4. Let $T > 0$, $y^0 \in L^2(0, L)$ and $u \in L^2(0, T)$ be given. Then the Cauchy problem (2.4)-(2.5)-(2.6) has a unique solution. This solution satisfies

(2.22) $$\|y(\tau, \cdot)\|_{L^2(0,L)} \leqslant \|y^0\|_{L^2(0,L)} + \|u\|_{L^2(0,T)}, \ \forall \tau \in [0, T].$$

Proof of Theorem 2.4. Let us first prove the uniqueness of the solution to the Cauchy problem (2.4)-(2.5)-(2.6). Let us assume that y_1 and y_2 are two solutions of this problem. Let $y := y_2 - y_1$. Then $y \in C^0([0, T]; L^2(0, L))$ and, if $\tau \in [0, T]$ and $\phi \in C^1([0, \tau] \times [0, L])$ satisfy (2.7), one has

(2.23) $$-\int_0^\tau \int_0^L (\phi_t + \phi_x) y\, dx\, dt + \int_0^L y(\tau, x)\phi(\tau, x)\, dx = 0.$$

Let $\tau \in [0, T]$. Let $(f_n)_{n \in \mathbb{N}}$ be a sequence of functions in $C^1(\mathbb{R})$ such that

(2.24) $$f_n = 0 \text{ on } [L, +\infty), \ \forall n \in \mathbb{N},$$

(2.25) $$f_n|_{(0,L)} \to y(\tau, \cdot) \text{ in } L^2(0, L) \text{ as } n \to +\infty.$$

For $n \in \mathbb{N}$, let $\phi_n \in C^1([0, \tau] \times [0, L])$ be defined by

(2.26) $$\phi_n(t, x) = f_n(\tau + x - t), \ \forall (t, x) \in [0, \tau] \times [0, L].$$

By (2.24) and (2.26), (2.7) is satisfied for $\phi := \phi_n$. Moreover,

$$\phi_{nt} + \phi_{nx} = 0.$$

Hence, from (2.23) with $\phi := \phi_n$ and from (2.26), we get

(2.27) $$\int_0^L y(\tau, x) f_n(x)\, dx = \int_0^L y(\tau, x)\phi_n(\tau, x)\, dx = 0.$$

Letting $n \to \infty$ in (2.27), we get, using (2.25),

$$\int_0^L |y(\tau, x)|^2\, dx = 0.$$

Hence, for every $\tau \in [0, T]$, $y(\tau, \cdot) = 0$.

Let us now give two proofs of the existence of a solution. The first one relies on the fact that one is able to give an explicit solution! Let us define $y : [0, T] \times [0, L] \to \mathbb{R}$ by

(2.28) $$y(t, x) := y^0(x - t), \ \forall (t, x) \in [0, T] \times (0, L) \text{ such that } t \leqslant x,$$

(2.29) $$y(t, x) := u(t - x), \ \forall (t, x) \in [0, T] \times (0, L) \text{ such that } t > x.$$

Then one easily checks that this y is a solution of the Cauchy problem (2.4)-(2.5)-(2.6). Moreover, this y satisfies (2.22).

Let us now give a second proof of the existence of a solution. This second proof is longer than the first one, but can be used for much more general situations (see, for example, Sections 2.2.1, 2.4.1, 2.5.1).

Let us first treat the case where

(2.30) $$u \in C^2([0, T]) \text{ and } u(0) = 0,$$

(2.31) $$y^0 \in H^1(0, L) \text{ and } y^0(0) = 0.$$

Let $A : \mathcal{D}(A) \subset L^2(0, L) \to L^2(0, L)$ be the linear operator defined by

(2.32) $$\mathcal{D}(A) := \{f \in H^1(0, L); f(0) = 0\},$$

(2.33) $$Af := -f_x, \ \forall f \in \mathcal{D}(A).$$

Note that

(2.34) $$\mathcal{D}(A) \text{ is dense in } L^2(0,L),$$

(2.35) $$A \text{ is closed.}$$

Let us recall that (2.35) means that $\{(f, Af); f \in \mathcal{D}(A)\}$ is a closed subspace of $L^2(0,L) \times L^2(0,L)$; see Definition A.1 on page 373. Moreover, A has the property that, for every $f \in \mathcal{D}(A)$,

(2.36) $$(Af, f)_{L^2(0,L)} = -\int_0^L f f_x dx = -\frac{f(L)^2}{2} \leqslant 0.$$

One easily checks that the adjoint A^* of A is defined (see Definition A.3 on page 373) by

$$\mathcal{D}(A^*) := \{f \in H^1(0,L); f(L) = 0\},$$
$$A^* f := f_x, \forall f \in \mathcal{D}(A^*).$$

In particular, for every $f \in \mathcal{D}(A^*)$,

(2.37) $$(A^* f, f)_{L^2(0,L)} = \int_0^L f f_x dx = -\frac{f(0)^2}{2} \leqslant 0.$$

Then, by a classical result on inhomogeneous initial value problems (Theorem A.7 on page 375), (2.30), (2.31), (2.34), (2.35), (2.36) and (2.37), there exists

$$z \in C^1([0,T]; L^2(0,L)) \cap C^0([0,T]; H^1(0,L))$$

such that

(2.38) $$z(t, 0) = 0, \forall t \in [0,T],$$

(2.39) $$\frac{dz}{dt} = Az - \dot{u},$$

(2.40) $$z(0, \cdot) = y^0.$$

Let $y \in C^1([0,T]; L^2(0,L)) \cap C^0([0,T]; H^1(0,L))$ be defined by

(2.41) $$y(t, x) = z(t, x) + u(t), \forall (t, x) \in [0,T] \times [0,L].$$

Let $\tau \in [0,T]$. Let $\phi \in C^1([0,\tau]; L^2(0,L)) \cap C^0([0,\tau]; H^1(0,L))$. Using (2.30), (2.38), (2.39), (2.40) and (2.41), straightforward integrations by parts show that

(2.42) $$-\int_0^\tau \int_0^L (\phi_t + \phi_x) y \, dx \, dt + \int_0^\tau y(t, L) \phi(t, L) dt - \int_0^\tau u(t) \phi(t, 0) dt$$
$$+ \int_0^L y(\tau, x) \phi(\tau, x) dx - \int_0^L y^0(x) \phi(0, x) dx = 0.$$

In particular, if we take $\phi := y|_{[0,\tau] \times [0,L]}$, then

(2.43) $$\int_0^\tau |y(t, L)|^2 dt - \int_0^\tau |u(t)|^2 dt + \int_0^L |y(\tau, x)|^2 dx - \int_0^L |y^0(x)|^2 dx = 0,$$

which implies that

(2.44) $$\|y(\tau, \cdot)\|_{L^2(0,L)} \leqslant \|u\|_{L^2(0,T)} + \|y^0\|_{L^2(0,L)}, \forall \tau \in [0,T],$$

that is (2.22).

Now let $y^0 \in L^2(0,L)$ and let $u \in L^2(0,T)$. Let $(y_n^0)_{n\in\mathbb{N}}$ be a sequence of functions in $\mathcal{D}(A)$ such that

(2.45) $\qquad\qquad\qquad y_n^0 \to y^0$ in $L^2(0,L)$ as $n \to +\infty$.

Let $(u_n)_{n\in\mathbb{N}}$ be a sequence of functions in $C^2([0,T])$ such that $u_n(0) = 0$ and

(2.46) $\qquad\qquad\qquad u_n \to u$ in $L^2(0,T)$ as $n \to +\infty$.

For $n \in \mathbb{N}$, let $z_n \in C^1([0,T]; L^2(0,L)) \cap C^0([0,T]; H^1(0,L))$ be such that

(2.47) $\qquad\qquad\qquad z_n(t,0) = 0, \forall t \in [0,T],$

(2.48) $\qquad\qquad\qquad \dfrac{dz_n}{dt} = Az_n - \dot{u}_n,$

(2.49) $\qquad\qquad\qquad z_n(0,\cdot) = y_n^0,$

and let $y_n \in C^1([0,T]; L^2(0,L)) \cap C^0([0,T]; H^1(0,L))$ be defined by

(2.50) $\qquad\qquad y_n(t,x) = z_n(t,x) + u_n(t), \forall (t,x) \in [0,T] \times [0,L].$

Let $\tau \in [0,T]$. Let $\phi \in C^1([0,\tau] \times [0,L])$ be such that $\phi(\cdot, L) = 0$. From (2.42) (applied with $y^0 := y_n^0$, $u := u_n$ and $y := y_n$), one gets

(2.51) $\quad -\displaystyle\int_0^\tau \int_0^L (\phi_t + \phi_x) y_n \, dx \, dt - \int_0^\tau u_n(t) \phi(t,0) \, dt$
$$+ \int_0^L y_n(\tau, x) \phi(\tau, x) \, dx - \int_0^L y_n^0(x) \phi(0, x) \, dx = 0, \forall n \in \mathbb{N}.$$

From (2.44) (applied with $y^0 := y_n^0$, $u := u_n$ and $y := y_n$),

(2.52) $\qquad\qquad \|y_n\|_{C^0([0,T]; L^2(0,L))} \leqslant \|u_n\|_{L^2(0,T)} + \|y_n^0\|_{L^2(0,L)}.$

Let $(n,m) \in \mathbb{N}^2$. From (2.44) (applied with $y^0 := y_n^0 - y_m^0$, $u := u_n - u_m$ and $y := y_n - y_m$),

(2.53) $\qquad \|y_n - y_m\|_{C^0([0,T]; L^2(0,L))} \leqslant \|u_n - u_m\|_{L^2(0,T)} + \|y_n^0 - y_m^0\|_{L^2(0,L)}.$

From (2.45), (2.46) and (2.53), $(y_n)_{n\in\mathbb{N}}$ is a Cauchy sequence in $C^0([0,T]; L^2(0,L))$. Hence there exists $y \in C^0([0,T]; L^2(0,L))$ such that

(2.54) $\qquad\qquad\qquad y_n \to y$ in $C^0([0,T]; L^2(0,L))$ as $n \to +\infty$.

From (2.51) and (2.54), one gets (2.8). From (2.52) and (2.54), one gets (2.22). This concludes our second proof of the existence of a solution to the Cauchy problem (2.4)-(2.5)-(2.6) satisfying (2.22). ∎

2.1.2. Controllability. Let us now turn to the controllability of the control system (2.2)-(2.3). We start with a natural definition of controllability.

DEFINITION 2.5. *Let $T > 0$. The control system (2.2)-(2.3) is controllable in time T if, for every $y^0 \in L^2(0,L)$ and every $y^1 \in L^2(0,L)$, there exists $u \in L^2(0,T)$ such that the solution y of the Cauchy problem (2.4)-(2.5)-(2.6) satisfies $y(T,\cdot) = y^1$.*

With this definition we have the following theorem.

THEOREM 2.6. *The control system (2.2)-(2.3) is controllable in time T if and only if $T \geqslant L$.*

REMARK 2.7. Let us point out that Theorem 2.6 on the previous page shows that there is a condition on the time T in order to have controllability. Such a phenomenon never appears for linear control systems in finite dimension. See Theorem 1.16 on page 9. However, note that such a phenomenon can appear for nonlinear control systems in finite dimension (see for instance Example 6.4 on page 190).

We shall provide three different proofs of Theorem 2.6:

1. A proof based on the explicit solution y of the Cauchy problem (2.4)-(2.5)-(2.6).
2. A proof based on the extension method.
3. A proof based on the duality between the controllability of a linear control system and the observability of its adjoint.

2.1.2.1. *Explicit solutions.* Let us start the proof based on the explicit solution of the Cauchy problem (2.4)-(2.5)-(2.6). We first take $T \in (0, L)$ and check that the control system (2.2)-(2.3) is not controllable in time T. Let us define y^0 and y^1 by

$$y^0(x) = 1 \text{ and } y^1(x) = 0, \ \forall x \in [0, L].$$

Let $u \in L^2(0, T)$. Then, by (2.28), the solution y of the Cauchy problem (2.4)-(2.5)-(2.6) satisfies

$$y(T, x) = 1, \ x \in (T, L).$$

In particular, $y(T, \cdot) \neq y^1$. This shows that the control system (2.2)-(2.3) is not controllable in time T.

We now assume that $T \geqslant L$ and show that the control system (2.2)-(2.3) is controllable in time T. Let $y^0 \in L^2(0, L)$ and $y^1 \in L^2(0, L)$. Let us define $u \in L^2(0, T)$ by

$$u(t) = y^1(T - t), \ t \in (T - L, T),$$
$$u(t) = 0, \ t \in (0, T - L).$$

Then, by (2.29), the solution y of the Cauchy problem (2.4)-(2.5)-(2.6) satisfies

$$y(T, x) = u(T - x) = y^1(x), \ x \in (0, L).$$

This shows that the control system (2.2)-(2.3) is controllable in time T. ∎

2.1.2.2. *Extension method.* This method has been introduced in [**425**] by David Russell. See also [**426**, Proof of Theorem 5.3, pages 688–690] by David Russell and [**332**] by Walter Littman. We explain it on our control system (2.2)-(2.3) to prove its controllability in time T if $T \geqslant L$. Let us first introduce a new definition.

DEFINITION 2.8. Let $T > 0$. The control system (2.2)-(2.3) is *null controllable in time T* if, for every $y^0 \in L^2(0, L)$, there exists $u \in L^2(0, T)$ such that the solution y of the Cauchy problem (2.4)-(2.5)-(2.6) satisfies $y(T, \cdot) = 0$.

One has the following lemma.

LEMMA 2.9. *The control system (2.2)-(2.3) is controllable in time T if and only if it is null controllable in time T.*

Proof of Lemma 2.9. The "only if" part is obvious. For the "if" part, let us assume that the control system (2.2)-(2.3) is null controllable in time T. Let $y^0 \in L^2(0, L)$ and let $y^1 \in L^2(0, L)$. Let us assume, for the moment, that there exist $\bar{y}^0 \in L^2(0, L)$ and $\bar{u} \in L^2(0, T)$ such that the solution $\bar{y} \in C^0([0, T]; L^2(0, L))$ of the Cauchy problem

(2.55) $$\bar{y}_t + \bar{y}_x = 0, \ t \in (0, T), \ x \in (0, L),$$
(2.56) $$\bar{y}(t, 0) = \bar{u}(t), \ t \in (0, T),$$
(2.57) $$\bar{y}(0, x) = \bar{y}^0(x), \ x \in (0, L),$$

satisfies

(2.58) $$\bar{y}(T, x) := y^1(x), \ x \in (0, L).$$

Since the control system (2.2)-(2.3) is null controllable in time T, there exists $\tilde{u} \in L^2(0, T)$ such that the solution $\tilde{y} \in C^0([0, T]; L^2(0, L))$ of the Cauchy problem

(2.59) $$\tilde{y}_t + \tilde{y}_x = 0, \ t \in (0, T), \ x \in (0, L),$$
(2.60) $$\tilde{y}(t, 0) = \tilde{u}(t), \ t \in (0, T),$$
(2.61) $$\tilde{y}(0, x) = y^0(x) - \bar{y}^0(x), \ x \in (0, L),$$

satisfies

(2.62) $$\tilde{y}(T, x) := 0, \ x \in (0, L).$$

Let us define $u \in L^2(0, T)$ by

(2.63) $$u := \bar{u} + \tilde{u}.$$

From (2.55) to (2.57) and (2.59) to (2.61), the solution $y \in C^0([0, T]; L^2(0, L))$ of the Cauchy problem

$$y_t + y_x = 0, \ t \in (0, T), \ x \in (0, L),$$
$$y(t, 0) = u(t), \ t \in (0, T),$$
$$y(0, x) = y^0(x), \ x \in (0, L),$$

is $y = \bar{y} + \tilde{y}$. By (2.58) and by (2.62), $y(T, \cdot) = y^1$. Hence, as desired, the control u steers the control system (2.2)-(2.3) from the state y^0 to the state y^1 during the time interval $[0, T]$. It remains to prove the existence of $\bar{y}^0 \in L^2(0, L)$ and $\bar{u} \in L^2(0, T)$. Let $z \in C^0([0, T]; L^2(0, L))$ be the solution of the Cauchy problem

(2.64) $$z_t + z_x = 0, \ t \in (0, T), \ x \in (0, L),$$
(2.65) $$z(t, 0) = 0, \ t \in (0, T),$$
(2.66) $$z(0, x) = y^1(L - x), \ x \in (0, L).$$

Note that from (2.64) we get $z \in H^1((0, L); H^{-1}(0, T))$. In particular, $z(\cdot, L)$ is well defined and

(2.67) $$z(\cdot, L) \in H^{-1}(0, T).$$

In fact $z(\cdot, L)$ has more regularity than the one given by (2.67): one has

(2.68) $$z(\cdot, L) \in L^2(0, T).$$

Property (2.68) can be seen by the two following methods.

- For the first one, one just uses the explicit expression of z; see (2.28) and (2.29).
- For the second one, we start with the case where $y^1 \in H^1(0,L)$ satisfies $y^1(L) = 0$. Then $z \in C^1([0,T]; L^2(0,L)) \cap C^0([0,T]; H^1(0,L))$. We multiply (2.64) by z and integrate on $[0,T] \times [0,L]$. Using (2.65) and (2.66), we get

$$\int_0^L z(T,x)^2 dx - \int_0^L y^1(x)^2 dx + \int_0^T z(t,L)^2 dt = 0.$$

In particular,

(2.69) $$\|z(\cdot, L)\|_{L^2(0,T)} \leqslant \|y^1\|_{L^2(0,L)}.$$

By density, (2.69) also holds if y^1 is only in $L^2(0,L)$, which completes the second proof of (2.68).

We define $\bar{y}^0 \in L^2(0,L)$ and $\bar{u} \in L^2(0,T)$ by

$$\bar{y}^0(x) = z(T, L-x),\ x \in (0,L),$$
$$\bar{u}(t) = z(T-t, L),\ t \in (0,T).$$

Then one easily checks that the solution $\bar{y}$ of the Cauchy problem (2.55)-(2.56)-(2.57) is

(2.70) $$\bar{y}(t,x) = z(T-t, L-x),\ t \in (0,T),\ x \in (0,L).$$

From (2.66) and (2.70), we get (2.58). This concludes the proof of Lemma 2.9. ∎

REMARK 2.10. The fact that $z(\cdot, L) \in L^2(0,T)$ is sometimes called a hidden regularity property; it does not follow directly from the regularity required on z, i.e., $z \in C^0([0,T]; L^2(0,L))$. Such a priori unexpected extra regularity properties appear, as it is now well known, for hyperbolic equations (see in particular (2.307) on page 68, [**285**] by Heinz-Otto Kreiss, [**431, 432**] by Reiko Sakamoto, [**88**, Théorème 4.4, page 378] by Jacques Chazarain and Alain Piriou, [**297, 298**] by Irena Lasiecka and Roberto Triggiani, [**324**, Théorème 4.1, page 195] by Jacques-Louis Lions, [**473**] by Daniel Tataru and [**277**, Chapter 2] by Vilmos Komornik. It also appears for our Korteweg-de Vries control system studied below; see (2.140) due to Lionel Rosier [**407**, Proposition 3.2, page 43].

Let us now introduce the definition of a solution to the Cauchy problem

(2.71) $$y_t + y_x = 0,\ (t,x) \in (0,+\infty) \times \mathbb{R},\ y(0,x) = y^0(x),$$

where y^0 is given in $L^2(\mathbb{R})$. With the same motivation as for Definition 2.1 on page 25, one proposes the following definition.

DEFINITION 2.11. Let $y^0 \in L^2(\mathbb{R})$. A *solution of the Cauchy problem (2.71)* is a function $y \in C^0([0,+\infty); L^2(\mathbb{R}))$ such that, for every $\tau \in [0, +\infty)$, for every $R > 0$ and for every $\phi \in C^1([0,\tau] \times \mathbb{R})$ such that

$$\phi(t,x) = 0,\ \forall t \in [0,\tau],\ \forall x \in \mathbb{R}\ \text{such that}\ |x| \geqslant R,$$

one has

$$-\int_0^\tau \int_{-R}^R (\phi_t + \phi_x) y\, dx\, dt + \int_{-R}^R y(\tau, x) \phi(\tau, x) dx - \int_{-R}^R y^0(x) \phi(0, x) dx = 0.$$

Then, adapting the proof of Theorem 2.4 on page 27, one has the following proposition.

PROPOSITION 2.12. *For every $y^0 \in L^2(\mathbb{R})$, the Cauchy problem (2.71) has a unique solution. This solution y satisfies*
$$\|y(\tau, \cdot)\|_{L^2(\mathbb{R})} = \|y^0\|_{L^2(\mathbb{R})}, \; \forall \tau \in [0, +\infty).$$

In fact, as in the case of the Cauchy problem (2.4)-(2.5)-(2.6), one can give y explicitly:
$$(2.72) \qquad y(t, x) = y^0(x - t), \; t \in (0, +\infty), \; x \in \mathbb{R}.$$

Then the extension method goes as follows. Let $y^0 \in L^2(0, L)$. Let $R \geqslant 0$. Let $\bar{y}^0 \in L^2(\mathbb{R})$ be such that
$$(2.73) \qquad \bar{y}^0(x) = y^0(x), \; x \in (0, L),$$
$$(2.74) \qquad \bar{y}^0(x) = 0, \; x \in (-\infty, -R).$$

Let $\bar{y} \in C^0([0, +\infty); L^2(\mathbb{R}))$ be the solution of the Cauchy problem
$$\bar{y}_t + \bar{y}_x = 0, \; t \in (0, +\infty), \; x \in \mathbb{R},$$
$$\bar{y}(0, x) = \bar{y}^0(x), \; x \in \mathbb{R}.$$

Using (2.74) and the explicit expression (2.72) of the solution of the Cauchy problem (2.71), one sees that
$$(2.75) \qquad \bar{y}(t, x) = 0 \text{ if } x < t - R.$$

Adapting the proofs of (2.68), one gets that
$$\bar{y}(\cdot, 0) \in L^2(0, +\infty).$$

Let $T \geqslant L$. Then one easily checks that the solution $y \in C^0([0, T]; L^2(0, L))$ of the Cauchy problem
$$y_t + y_x = 0, \; t \in (0, T), \; x \in (0, L),$$
$$y(t, 0) = \bar{y}(t, 0), \; t \in (0, T),$$
$$y(0, x) = y^0(x), \; x \in (0, L),$$
is given by
$$(2.76) \qquad y(t, x) = \bar{y}(t, x), \; t \in (0, T), \; x \in (0, L).$$

From (2.75), one gets $y(T, \cdot) = 0$ if $R \leqslant T - L$. Hence the control $t \in (0, T) \mapsto \bar{y}(t, 0)$ steers the control system (2.2)-(2.3) from the state y^0 to 0 during the time interval $[0, T]$.

Of course, for our simple control system (2.2)-(2.3), the extension method seems to be neither very interesting nor very different from the explicit method detailed in Section 2.1.2.1 (taking $R = 0$ leads to the same control as in Section 2.1.2.1). However, the extension method has some quite interesting advantages compared to the explicit method for more complicated hyperbolic equations where the explicit method cannot be easily applied; see, in particular, the paper [332] by Walter Littman.

Let us also point out that the extension method is equally useful for our simple control system (2.2)-(2.3) if one is interested in more regular solutions. Indeed, let $m \in \mathbb{N}$, let us assume that $y^0 \in H^m(0, L)$ and that we want to steer the control system (2.2)-(2.3) from y^0 to 0 in time $T > L$ in such a way that the state always remains in $H^m(0, L)$. Then it suffices to take $R := T - L$ and to impose on $\bar{y}^0$ to be in $H^m(\mathbb{R})$. Note that if $m \geqslant 1$, one cannot take $T = L$, as shown in the following exercise.

EXERCISE 2.13. *Let $m \in \mathbb{N} \setminus \{0\}$. Let $T = L$. Let $y^0 \in H^m(0, L)$. Prove that there exists $u \in L^2(0, T)$ such that the solution $y \in C^0([0, T]; L^2(0, L))$ of the Cauchy problem (2.4)-(2.5)-(2.6) satisfies*

$$y(T, x) = 0, \, x \in (0, L),$$
$$y(t, \cdot) \in H^m(0, L), \, \forall t \in [0, T],$$

if and only if

$$(y^0)^{(j)}(0) = 0, \, \forall j \in \{0, \ldots, m-1\}.$$

REMARK 2.14. Let us emphasize that there are strong links between the regularity of the states and the regularity of the control: For $r \geqslant 0$, if the states are in $H^r(0, L)$, one cannot have controllability with control $u \in H^s(0, T)$ for

(2.77) $$s > r.$$

Moreover, (2.77) is optimal: For every $T > L$, for every $y^0 \in H^r(0, L)$ and for every $y^1 \in H^r(0, L)$, there exists $u \in H^r(0, T)$ such that the solution y of the Cauchy problem (2.4)-(2.5)-(2.6) satisfies $y(T, \cdot) = y^1$ and $y \in C^0([0, T]; H^r(0, L))$. Of course this problem of the links between the regularity of the states and the regularity of the control often appears for partial differential equations; see, in particular, Theorem 9.4 on page 248. The optimal links often are still open problems; see, for example, the Open Problem 9.6 on page 251.

2.1.2.3. *Duality between controllability and observability.* Let $T > 0$. Let us define a linear map $\mathcal{F}_T : L^2(0, T) \to L^2(0, L)$ in the following way. Let $u \in L^2(0, T)$. Let $y \in C^0([0, T]; L^2(0, L))$ be the solution of the Cauchy problem (2.4)-(2.5)-(2.6) with $y^0 := 0$. Then

$$\mathcal{F}_T(u) := y(T, \cdot).$$

One has the following lemma.

LEMMA 2.15. *The control system (2.2)-(2.3) is controllable in time T if and only if $\mathcal{F}_T$ is onto.*

Proof of Lemma 2.15. The "only if" part is obvious. Let us assume that $\mathcal{F}_T$ is onto and prove that the control system (2.2)-(2.3) is controllable in time T. Let $y^0 \in L^2(0, L)$ and $y^1 \in L^2(0, L)$. Let $\tilde{y}$ be the solution of the Cauchy problem (2.4)-(2.5)-(2.6) with $u := 0$. Since $\mathcal{F}_T$ is onto, there exists $u \in L^2(0, T)$ such that $\mathcal{F}_T(u) = y^1 - \tilde{y}(T, \cdot)$. Then the solution y of the Cauchy problem (2.4)-(2.5)-(2.6) satisfies $y(T, \cdot) = \tilde{y}(T, \cdot) + y^1 - \tilde{y}(T, \cdot) = y^1$, which concludes the proof of Lemma 2.15. ∎

In order to decide whether $\mathcal{F}_T$ is onto or not, we use the following classical result of functional analysis (see e.g. [**419**, Theorem 4.15, page 97], or [**71**, Théorème II.19, pages 29–30] for the more general case of unbounded operators).

PROPOSITION 2.16. *Let H_1 and H_2 be two Hilbert spaces. Let $\mathcal{F}$ be a linear continuous map from H_1 into H_2. Then $\mathcal{F}$ is onto if and only if there exists $c > 0$ such that*

(2.78) $$\|\mathcal{F}^*(x_2)\|_{H_1} \geqslant c\|x_2\|_{H_2}, \forall x_2 \in H_2.$$

Moreover, if (2.78) holds for some $c > 0$, there exists a linear continuous map $\mathcal{G}$ from H_2 into H_1 such that

$$\mathcal{F} \circ \mathcal{G}(x_2) = x_2, \forall x_2 \in H_2,$$

$$\|\mathcal{G}(x_2)\|_{H_1} \leqslant \frac{1}{c}\|x_2\|_{H_2}, \forall x_2 \in H_2.$$

In control theory, inequality (2.78) is called an "observability inequality". In order to apply this proposition, we make explicit $\mathcal{F}_T^*$ in the following lemma.

LEMMA 2.17. *Let $z^T \in H^1(0, L)$ be such that*

(2.79) $$z^T(L) = 0.$$

Let $z \in C^1([0,T]; L^2(0,L)) \cap C^0([0,T]; H^1(0,L))$ be the (unique) solution of

(2.80) $$z_t + z_x = 0,$$
(2.81) $$z(t, L) = 0, \forall t \in [0, T],$$
(2.82) $$z(T, \cdot) = z^T.$$

Then

(2.83) $$\mathcal{F}_T^*(z^T) = z(\cdot, 0).$$

Proof of Lemma 2.17. Let us first point out that the proof of the existence and uniqueness of $z \in C^1([0,T]; L^2(0,L)) \cap C^0([0,T]; H^1(0,L))$ satisfying (2.80) to (2.82) is the same as the proof of the existence and uniqueness of the solution of (2.38) to (2.40). In fact, if $\tilde{z} \in C^1([0,T]; L^2(0,L)) \cap C^0([0,T]; H^1(0,L))$ is the solution of

$$\tilde{z}(t, 0) = 0, \forall t \in [0, T],$$
$$\frac{d\tilde{z}}{dt} = A\tilde{z},$$
$$\tilde{z}(0, x) = z^T(L - x), \forall x \in [0, L],$$

then

$$z(t, x) = \tilde{z}(T - t, L - x), \forall (t, x) \in [0, T] \times [0, L].$$

Let $u \in C^2([0, T])$ be such that $u(0) = 0$. Let

$$y \in C^1([0,T]; L^2(0,L)) \cap C^0([0,T]; H^1(0,L))$$

be such that

(2.84) $$y_t + y_x = 0,$$
(2.85) $$y(t, 0) = u(t), \forall t \in [0, T],$$
(2.86) $$y(0, \cdot) = 0.$$

(Again the existence of such a y is proved above; see page 28.) Then, from (2.80), (2.81), (2.82), (2.84), (2.85) and (2.86), we get, using integrations by parts,

$$\int_0^L z^T \mathcal{F}_T(u)dx = \int_0^L z^T y(T,x)dx$$
$$= \int_0^T \int_0^L (zy)_t \, dx \, dt$$
$$= -\int_0^T \int_0^L (z_x y + z y_x) \, dx \, dt$$
$$= \int_0^T z(t,0) u(t) \, dt,$$

which, since the set of $u \in C^2([0,T])$ such that $u(0) = 0$ is dense in $L^2(0,T)$, concludes the proof of Lemma 2.17. ∎

From Lemma 2.17, it follows that inequality (2.78) is equivalent to

(2.87) $$\int_0^T z(t,0)^2 dt \geqslant c^2 \int_0^L z^T(x)^2 dx,$$

for every $z^T \in H^1(0,L)$ such that (2.79) holds, z being the (unique) solution of (2.80)-(2.81)-(2.82) in $C^1([0,T]; L^2(0,L)) \cap C^0([0,T]; H^1(0,L))$.

Let us now present two methods to prove (2.87). The first one relies on the explicit solution of (2.80)-(2.81)-(2.82), the second one on the so-called multiplier method.

Proof of (2.87) by means of explicit solutions. We assume that $T \geqslant L$. One notices that the solution of (2.80)-(2.81)-(2.82) is given (see also (2.28) and (2.29)) by

$$z(t,x) = z^T(x + T - t) \text{ if } 0 < x < L + t - T,$$
$$z(t,x) = 0, if L + t - T < x < L.$$

In particular, since $T \geqslant L$,

$$\int_0^T z(t,0)^2 dt = \int_0^L z^T(x)^2 dx,$$

showing that (2.87) holds with $c = 1$. ∎

Proof of (2.87) by means of the multiplier method. We now assume that $T > L$ and prove that the observability inequality (2.87) indeed holds with

(2.88) $$c := \sqrt{\frac{T-L}{T}}.$$

Let $z^T \in H^1(0,L)$ be such that (2.79) holds. Let

$$z \in C^1([0,T]; L^2(0,L)) \cap C^0([0,T]; H^1(0,L))$$

be the (unique) solution of (2.80) to (2.82). Let us multiply (2.80) by z and integrate the obtained equality on $[0, L]$. Using (2.81), one gets

$$\frac{\mathrm{d}}{\mathrm{d}t}\left(\int_0^L |z(t,x)|^2 dx\right) = |z(t,0)|^2. \tag{2.89}$$

Let us now multiply (2.80) by xz and integrate the obtained equality on $[0, L]$. Using (2.81), one gets

$$\frac{\mathrm{d}}{\mathrm{d}t}\left(\int_0^L x|z(t,x)|^2 dx\right) = \int_0^L |z(t,x)|^2 dx. \tag{2.90}$$

For $t \in [0, T]$, let $e(t) := \int_0^L |z(t,x)|^2 dx$. From (2.90), we have

$$\int_0^T e(t)dt = \int_0^L x|z(T,x)|^2 dx - \int_0^L x|z(0,x)|^2 dx$$

$$\leqslant L\int_0^L |z(T,x)|^2 dx = Le(T). \tag{2.91}$$

From (2.89), we get

$$e(t) = e(T) - \int_t^T |z(\tau,0)|^2 d\tau \geqslant e(T) - \int_0^T |z(\tau,0)|^2 d\tau. \tag{2.92}$$

From (2.82), (2.91) and (2.92), we get

$$(T-L)\|z^T\|_{L^2(0,L)}^2 \leqslant T\int_0^T |z(\tau,0)|^2 d\tau, \tag{2.93}$$

which, together with Lemma 2.17 on page 35 and the density in $L^2(0, L)$ of the functions $z^T \in H^1(0, L)$ such that (2.79) holds, proves the observability inequality (2.87) with c given by (2.88). ∎

REMARK 2.18. Since

$$\sqrt{\frac{T-L}{T}} < 1, \forall T \geqslant L > 0,$$

the multiplier method gives a weaker observability inequality (2.87) than the method based on explicit solutions, which gives the optimal inequality. (Note also that the multiplier method, in contrast with the method based explicit solutions, does not allow us to prove the controllability in the limiting case $T = L$.) However, the multiplier method is quite flexible and gives interesting controllability results for many partial differential linear control systems; see, for example, Section 2.4.2 and the references in Remark 2.19.

REMARK 2.19. The multiplier method goes back to the paper [362] by Cathleen Morawetz. In the framework of control systems, it has been introduced in [234] by Lop Fat Ho, and in [326, 325] by Jacques-Louis Lions. For more details and results on this method, see the book [277] by Vilmos Komornik and the references therein.

REMARK 2.20. One can find much more general results on the controllability of one-dimensional hyperbolic linear systems in [423, 424, 426] by David Russell.

2.2. Korteweg-de Vries equation

This section is borrowed from the paper [**407**] by Lionel Rosier. Let $L > 0$ and $T > 0$. We consider the linear control system

$$y_t + y_x + y_{xxx} = 0, \, t \in (0,T), \, x \in (0,L), \tag{2.94}$$
$$y(t,0) = y(t,L) = 0, \, y_x(t,L) = u(t), \, t \in (0,T), \tag{2.95}$$

where, at time t, the control is $u(t) \in \mathbb{R}$ and the state is $y(t,\cdot) : (0,L) \mapsto \mathbb{R}$. Our goal is to study the controllability of the control system (2.94)-(2.95). Again we start by studying the well-posedness of the Cauchy problem associated to (2.94)-(2.95).

2.2.1. Well-posedness of the Cauchy problem.
Let us first give a natural definition of solutions of the Cauchy problem

$$y_t + y_x + y_{xxx} = 0, \, t \in (0,T), \, x \in (0,L), \tag{2.96}$$
$$y(t,0) = y(t,L) = 0, y_x(t,L) = u(t), \, t \in (0,T), \tag{2.97}$$
$$y(0,x) = y^0(x), \, x \in (0,L), \tag{2.98}$$

where $T > 0$, $y^0 \in L^2(0,L)$ and $u \in L^2(0,T)$ are given. In order to motivate this definition, let us first assume that there exists a function y of class C^3 on $[0,T] \times [0,L]$ satisfying (2.96) to (2.98) in the usual sense. Let $\tau \in [0,T]$ and let $\phi \in C^3([0,\tau] \times [0,L])$ be such that

$$\phi(t,0) = \phi(t,L) = 0, \, \forall t \in [0,\tau]. \tag{2.99}$$

We multiply (2.96) by ϕ and integrate the obtained equality on $[0,\tau] \times [0,L]$. Using (2.97), (2.98), (2.99) and integrations by parts, one gets

$$-\int_0^\tau \int_0^L (\phi_t + \phi_x + \phi_{xxx}) y \, dx \, dt - \int_0^\tau u(t)\phi_x(t,L) dt + \int_0^\tau y_x(t,0)\phi_x(t,0) dt$$
$$+ \int_0^L y(\tau,x)\phi(\tau,x) dx - \int_0^L y^0(x)\phi(0,x) dx = 0.$$

This equality leads to the following definition.

DEFINITION 2.21. *Let $T > 0$, $y^0 \in L^2(0,L)$ and $u \in L^2(0,T)$ be given. A solution of the Cauchy problem (2.96)-(2.97)-(2.98) is a function $y \in C^0([0,T]; L^2(0,L))$ such that, for every $\tau \in [0,T]$ and for every $\phi \in C^3([0,\tau] \times [0,L])$ such that,*

$$\phi(t,0) = \phi(t,L) = \phi_x(t,0) = 0, \, \forall t \in [0,\tau], \tag{2.100}$$

one has

$$-\int_0^\tau \int_0^L (\phi_t + \phi_x + \phi_{xxx}) y \, dx \, dt - \int_0^\tau u(t)\phi_x(t,L) dt \tag{2.101}$$
$$+ \int_0^L y(\tau,x)\phi(\tau,x) dx - \int_0^L y^0(x)\phi(0,x) dx = 0.$$

Proceeding as in the proof of Proposition 2.2 on page 25, one gets the following proposition, which also justifies Definition 2.21.

2.2. KORTEWEG-DE VRIES EQUATION

PROPOSITION 2.22. *Let $T > 0$, $y^0 \in L^2(0, L)$ and $u \in L^2(0, T)$ be given. Let us assume that y is a solution of the Cauchy problem (2.96)-(2.97)-(2.98) which is of class C^3 in $[0, T] \times [0, L]$. Then*

$$y^0 \in C^3([0, L]),$$
$$u \in C^2([0, T]),$$
$$y(0, x) = y^0(x), \forall x \in [0, L],$$
$$y(t, 0) = y(t, L) = 0, \; y_x(t, L) = u(t), \forall t \in [0, T],$$
$$y_t(t, x) + y_x(t, x) + y_{xxx}(t, x) = 0, \forall (t, x) \in [0, T] \times [0, L].$$

With Definition 2.21, one has the following theorem.

THEOREM 2.23. *Let $T > 0$, $y^0 \in L^2(0, L)$ and $u \in L^2(0, T)$ be given. Then the Cauchy problem (2.96)-(2.97)-(2.98) has a unique solution. This solution satisfies*

(2.102) $$\|y(\tau, \cdot)\|_{L^2(0,L)} \leqslant \|y^0\|_{L^2(0,L)} + \|u\|_{L^2(0,T)}, \forall \tau \in [0, T].$$

Proof of Theorem 2.23. We start with the existence statement. We proceed as in the second proof of the existence statement in Theorem 2.4 on page 27. Let $T > 0$. Let us first treat the case where

(2.103) $$u \in C^2([0, T]),$$
(2.104) $$u(0) = 0,$$
(2.105) $$y^0 \in H^3(0, L), \; y^0(0) = y^0(L) = y^0_x(L) = 0.$$

Let $A : \mathcal{D}(A) \subset L^2(0, L) \to L^2(0, L)$ be the linear operator defined by

(2.106) $$\mathcal{D}(A) := \{f \in H^3(0, L); \; f(0) = f(L) = f_x(L) = 0\},$$
(2.107) $$Af := -f_x - f_{xxx}, \forall f \in \mathcal{D}(A).$$

Note that

(2.108) $$\mathcal{D}(A) \text{ is dense in } L^2(0, L),$$
(2.109) $$A \text{ is closed.}$$

Moreover,

(2.110) $$(Af, f)_{L^2(0,L)} = -\int_0^L (f_x + f_{xxx}) f dx = -\frac{f_x(0)^2}{2} \leqslant 0.$$

One easily checks that the adjoint A^* of A is defined by

$$\mathcal{D}(A^*) := \{f \in H^3(0, L); \; f(0) = f_x(0) = f(L) = 0\},$$
$$A^* f := f_x + f_{xxx}, \forall f \in \mathcal{D}(A^*).$$

In particular,

(2.111) $$(A^* f, f)_{L^2(0,L)} = \int_0^L (f_x + f_{xxx}) f dx = -\frac{f_x(L)^2}{2} \leqslant 0.$$

Then, by a classical result on inhomogeneous initial value problems (Theorem A.7 on page 375), (2.103), (2.105), (2.108), (2.109), (2.110) and (2.111), there exists

$$z \in C^1([0, T]; L^2(0, L)) \cap C^0([0, T]; H^3(0, L))$$

such that

(2.112) $$z(t,0) = z(t,L) = z_x(t,L) = 0, \ \forall t \in [0,T],$$

(2.113) $$\frac{dz}{dt} = Az + \dot{u}(t)\frac{x(L-x)}{L} + u(t)\frac{L-2x}{L},$$

(2.114) $$z(0,\cdot) = y^0.$$

Let $y \in C^1([0,T]; L^2(0,L)) \cap C^0([0,T]; H^3(0,L))$ be defined by

(2.115) $$y(t,x) = z(t,x) - u(t)\frac{x(L-x)}{L}, \ \forall (t,x) \in [0,T] \times [0,L].$$

Note that, from (2.112) and (2.115),

(2.116) $$y(t,0) = y(t,L), \ \forall t \in [0,T].$$

Let $\tau \in [0,T]$. Let $\phi \in C^1([0,\tau]; L^2(0,L)) \cap C^0([0,\tau]; H^3(0,L))$ be such that

(2.117) $$\phi(t,0) = \phi(t,L) = 0, \ \forall t \in [0,T].$$

Using (2.104), (2.112), (2.113), (2.114) and (2.115), straightforward integrations by parts show that

(2.118)
$$-\int_0^\tau \int_0^L (\phi_t + \phi_x + \phi_{xxx})y\,dx\,dt - \int_0^\tau u(t)\phi_x(t,L)\,dt + \int_0^\tau y_x(t,0)\phi_x(t,0)\,dt \\ + \int_0^L y(\tau,x)\phi(\tau,x)\,dx - \int_0^L y^0(x)\phi(0,x)\,dx = 0.$$

In particular, if we take $\phi := y|_{[0,\tau]\times[0,L]}$ (see also (2.116)), then

(2.119) $$\int_0^\tau |y_x(t,0)|^2\,dt - \int_0^\tau |u(t)|^2\,dt + \int_0^L |y(\tau,x)|^2\,dx - \int_0^L |y^0(x)|^2\,dx = 0,$$

which implies that

(2.120) $$\|y\|_{C^0([0,T];L^2(0,L))} \leqslant \|u\|_{L^2(0,T)} + \|y^0\|_{L^2(0,L)}.$$

Now let $y^0 \in L^2(0,L)$ and let $u \in L^2(0,T)$. Let $(y_n^0)_{n\in\mathbb{N}}$ be a sequence of functions in $\mathcal{D}(A)$ such that

(2.121) $$y_n^0 \to y^0 \text{ in } L^2(0,L) \text{ as } n \to +\infty.$$

Let $(u_n)_{n\in\mathbb{N}}$ be a sequence of functions in $C^2([0,T])$ such that $u_n(0) = 0$ and

(2.122) $$u_n \to u \text{ in } L^2(0,T) \text{ as } n \to +\infty.$$

For $n \in \mathbb{N}$, let $z_n \in C^1([0,T]; L^2(0,L)) \cap C^0([0,T]; H^3(0,L))$ be such that

(2.123) $$z_n(t,0) = z_n(t,L) = z_{nx}(t,L) = 0, \ \forall t \in [0,T],$$

(2.124) $$\frac{dz_n}{dt} = Az_n + \dot{u}_n(t)\frac{x(L-x)}{L} + u_n(t)\frac{L-2x}{L},$$

(2.125) $$z_n(0,\cdot) = y_n^0,$$

and let $y_n \in C^1([0,T]; L^2(0,L)) \cap C^0([0,T]; H^3(0,L))$ be defined by

(2.126) $$y_n(t,x) = z_n(t,x) - u_n(t)\frac{x(L-x)}{L}, \ \forall (t,x) \in [0,T] \times [0,L].$$

Let $\tau \in [0,T]$. Let $\phi \in C^3([0,\tau] \times [0,L])$ be such that $\phi(\cdot, 0) = \phi_x(\cdot, 0) = \phi(\cdot, L) = 0$. From (2.118) (applied to the case where $y^0 := y_n^0$, $u := u_n$ and $y := y_n$), one gets

$$(2.127) \quad -\int_0^\tau \int_0^L (\phi_t + \phi_x + \phi_{xxx}) y_n \, dx \, dt - \int_0^\tau u_n(t) \phi_x(t, L) \, dt$$
$$+ \int_0^L y_n(\tau, x) \phi(\tau, x) \, dx - \int_0^L y_n^0(x) \phi(0, x) \, dx = 0.$$

From (2.120) (applied to the case where $y^0 := y_n^0$, $u := u_n$ and $y := y_n$),

$$(2.128) \quad \|y_n\|_{C^0([0,T];L^2(0,L))} \leqslant \|u_n\|_{L^2(0,T)} + \|y_n^0\|_{L^2(0,L)}.$$

Let $(n,m) \in \mathbb{N}^2$. From (2.120) (applied to the case where $y^0 := y_n^0 - y_m^0$, $u := u_n - u_m$ and $y := y_n - y_m$),

$$(2.129) \quad \|y_n - y_m\|_{C^0([0,T];L^2(0,L))} \leqslant \|u_n - u_m\|_{L^2(0,T)} + \|y_n^0 - y_m^0\|_{L^2(0,L)}.$$

From (2.121), (2.122) and (2.129), one gets that the sequence $(y_n)_{n \in \mathbb{N}}$ is a Cauchy sequence in $C^0([0,T]; L^2(0,L))$. Hence there exists $y \in C^0([0,T]; L^2(0,L))$ such that

$$(2.130) \quad y_n \to y \text{ in } C^0([0,T]; L^2(0,L)) \text{ as } n \to +\infty.$$

From (2.127) and (2.130), one gets (2.101). From (2.128) and (2.130), one gets (2.102). This concludes our proof of the existence of a solution to the Cauchy problem (2.96)-(2.97)-(2.98) satisfying (2.102).

Let us now turn to the proof of uniqueness. Let us assume that, for some $T > 0$, $y^0 \in L^2(0,L)$ and $u \in L^2(0,T)$, the Cauchy problem (2.96)-(2.97)-(2.98) has two solutions y_1 and y_2. Let $y := y_2 - y_1 \in C^0([0,T]; L^2(0,L))$. Let $\tau \in [0,T]$ and $\phi \in C^3([0,\tau] \times [0,L])$ be such that (2.100) holds. Then, by Definition 2.21 on page 38,

$$(2.131) \quad -\int_0^\tau \int_0^L (\phi_t + \phi_x + \phi_{xxx}) y \, dx \, dt + \int_0^L y(\tau, x) \phi(\tau, x) \, dx = 0.$$

By an easy density argument, (2.131) in fact holds for every

$$\phi \in C^1([0,\tau]; L^2(0,L)) \cap C^0([0,\tau]; H^3(0,L))$$

such that (2.100) holds.

Let $\tau \in [0,L]$. Let $(f_n)_{n \in \mathbb{N}}$ be a sequence of functions in $H^3(0,L)$ such that

$$(2.132) \quad f_n(0) = f_n(L) = f_{nx}(0) = 0,$$
$$(2.133) \quad f_n \to y(\tau, \cdot) \text{ in } L^2(0,L) \text{ as } n \to +\infty.$$

For $n \in \mathbb{N}$, let

$$(2.134) \quad \psi_n \in C^1([0,\tau]; L^2(0,L)) \cap C^0([0,\tau]; \mathcal{D}(A))$$

be the solution of

$$(2.135) \quad \frac{d\psi_n}{dt} = A\psi_n,$$
$$(2.136) \quad \psi_n(0, x) = f_n(L - x).$$

Let $\phi_n \in C^1([0,\tau]; L^2(0,L)) \cap C^0([0,\tau]; H^3(0,L))$ be defined by

$$(2.137) \quad \phi_n(t, x) := \psi_n(\tau - t, L - x), \, \forall (t,x) \in [0,\tau] \times [0,L].$$

By (2.134) and (2.137), (2.100) is satisfied for $\phi := \phi_n$. Moreover,
$$\phi_{nt} + \phi_{nx} + \phi_{nxxx} = 0.$$
Hence, from (2.131) with $\phi := \phi_n$ and from (2.137), we get

(2.138) $$\int_0^L y(\tau, x) f_n(x) dx = \int_0^L y(\tau, x) \phi_n(\tau, x) dx = 0.$$

Letting $n \to \infty$ in (2.138), we get, using (2.133),
$$\int_0^L |y(\tau, x)|^2 dx = 0.$$

Hence, for every $\tau \in [0, T]$, $y(\tau, \cdot) = 0$. This concludes the proof of Theorem 2.23. ∎

2.2.2. Controllability. Let us now turn to the controllability of the control system (2.94)-(2.95). We start with a natural definition of controllability.

DEFINITION 2.24. Let $T > 0$. The control system (2.94)-(2.95) is *controllable in time T* if, for every $y^0 \in L^2(0, L)$ and every $y^1 \in L^2(0, L)$, there exists $u \in L^2(0, T)$ such that the solution y of the Cauchy problem (2.96)-(2.97)-(2.98) satisfies $y(T, \cdot) = y^1$.

With this definition we have the following theorem, due to Lionel Rosier [**407**, Theorem 1.2, page 35].

THEOREM 2.25. *Let*

(2.139) $$\mathcal{N} := \left\{ 2\pi \sqrt{\frac{j^2 + l^2 + jl}{3}}; \, j, l \in \mathbb{N} \setminus \{0\} \right\}.$$

Let $T > 0$. The control system (2.94)-(2.95) is controllable in time T if and only if $L \notin \mathcal{N}$.

Proof of Theorem 2.25. We follow [**407**] by Lionel Rosier. For $y^0 \in L^2(0, L)$, let $y \in C^0([0, +\infty); L^2(0, L))$ be the solution of the Cauchy problem
$$y_t + y_x + y_{xxx} = 0,$$
$$y(\cdot, 0) = y(\cdot, L) = y_x(\cdot, L) = 0,$$
$$y(0, \cdot) = y^0.$$
(This means that, for every $T > 0$, y restricted to $[0, T] \times (0, L)$ is a solution of this Cauchy problem on $[0, T] \times (0, L)$ in the sense of Definition 2.21 on page 38.) We denote by $S(t) y^0$ the function $y(t, \cdot)$. In other words, $S(t)$, $t \in [0, +\infty)$, is the semigroup of continuous linear operators associated to the linear operator A; see page 374.

One of the main ingredients of the proof is the regularizing effect given in (2.140) and (2.141) of the following proposition due to Lionel Rosier.

PROPOSITION 2.26 ([**407**, Proposition 3.2, page 43]). *For $y^0 \in L^2(0, L)$, let $T > 0$, and $y(t, \cdot) := S(t) y^0$ for $t \in [0, T]$. Then $y_x(\cdot, 0)$ makes sense in $L^2(0, T)$,*

$y \in L^2((0,T); H^1(0,L))$ and

(2.140) $$\|y_x(\cdot,0)\|_{L^2(0,T)} \leqslant \|y^0\|_{L^2(0,L)},$$

(2.141) $$\|y\|_{L^2((0,T);H^1(0,L))} \leqslant \left(\frac{4T+L}{3}\right)^{1/2} \|y^0\|_{L^2(0,L)},$$

(2.142) $$\|y^0\|^2_{L^2(0,L)} \leqslant \frac{1}{T}\|y\|^2_{L^2((0,T)\times(0,L))} + \|y_x(\cdot,0)\|^2_{L^2(0,T)}.$$

The meaning of (2.140) is the following. By (2.96) (which holds in the distributions sense) and the fact that $y \in C^0([0,T]; L^2(0,L))$ we have that

$$y \in H^3((0,L); H^{-1}(0,T))$$

and therefore $y_x(\cdot,0)$ makes sense and is in $H^{-1}(0,T)$. Inequality (2.140) tells us that $y_x(\cdot,0)$ is in fact in $L^2(0,T)$ (and that the $L^2(0,T)$-norm is less than or equal to the right hand side of (2.140)). This is a hidden regularity property; see Remark 2.10 on page 32. Let us remark that (2.141) is the Kato smoothing effect [**266**, Theorem 6.2]).

Proof of Proposition 2.26. By density of $\mathcal{D}(A)$ in $L^2(0,L)$, we just need to prove the inequalities (2.140), (2.141) and (2.142) if $y^0 \in \mathcal{D}(A)$. So, let us take $y^0 \in \mathcal{D}(A)$. Then

(2.143) $$y \in C^1([0,T]; L^2(0,L)) \cap C^0([0,T]; \mathcal{D}(A)),$$

(2.144) $$y_t + y_x + y_{xxx} = 0,$$

(2.145) $$y(\cdot,0) = y(\cdot,L) = y_x(\cdot,L) = 0,$$

(2.146) $$y(0,\cdot) = y^0.$$

We multiply (2.144) by y and integrate on $[0,T] \times [0,L]$. Using (2.144), (2.145), (2.146) and simple integrations by parts, one gets

$$\int_0^T |y_x(t,0)|^2 dt = \int_0^L |y^0(x)|^2 dx - \int_0^L |y(T,x)|^2 dx \leqslant \int_0^L |y^0(x)|^2 dx,$$

which proves (2.140).

Let us now multiply (2.144) by xy and integrate on $[0,T] \times [0,L]$. Using (2.144), (2.145), (2.146) and simple integrations by parts, one gets

(2.147) $$-\int_0^T \int_0^L |y(t,x)|^2 dx dt + \int_0^L x|y(T,x)|^2 dx$$
$$- \int_0^L x|y^0(x)|^2 dx + 3\int_0^T \int_0^L |y_x(t,x)|^2 dx dt = 0.$$

Using (2.102), we get

$$\|y\|_{L^2((0,T)\times(0,L))} \leqslant T^{1/2} \|y\|_{C^0([0,T];L^2(0,L))} \leqslant T^{1/2} \|y^0\|_{L^2(0,L)},$$

which, together with (2.147), gives (2.141).

Finally, we multiply (2.144) by $(T-t)y$ and integrate on $[0,T] \times [0,L]$. Using (2.144), (2.145), (2.146) and simple integrations by parts, one gets

$$\int_0^T \int_0^L |y(t,x)|^2 dx dt - \int_0^L T|y^0(x)|^2 dx + \int_0^T (T-t)|y_x(t,0)|^2 dt = 0,$$

which proves (2.142). This concludes the proof of Proposition 2.26. ∎

In order to prove Theorem 2.25, we use the duality between controllability and observability (as in Section 2.1.2.3 for the transport equation). Let $T > 0$. Let us define a linear map $\mathcal{F}_T : L^2(0,T) \mapsto L^2(0,L)$ in the following way. Let $u \in L^2(0,T)$. Let $y \in C^0([0,T]; L^2(0,L))$ be the solution of the Cauchy problem (2.96)-(2.97)-(2.98) with $y^0 := 0$. Then
$$\mathcal{F}_T(u) := y(T,\cdot).$$
One has the following lemma, the proof of which is similar to the proof of Lemma 2.15 on page 34.

LEMMA 2.27. *The control system (2.94)-(2.95) is controllable in time T if and only if $\mathcal{F}_T$ is onto.*

Again, in order to decide whether $\mathcal{F}_T$ is onto, we use Proposition 2.16 on page 35. In order to apply this proposition, we make explicit $\mathcal{F}_T^*$ in the following lemma.

LEMMA 2.28. *Let $z^T \in H^3(0,L)$ be such that*
$$z^T(0) = z_x^T(0) = z^T(L) = 0. \tag{2.148}$$
Let $z \in C^1([0,T]; L^2(0,L)) \cap C^0([0,T]; H^3(0,L))$ be the (unique) solution of
$$z_t + z_x + z_{xxx} = 0, \tag{2.149}$$
$$z(t,0) = z_x(t,0) = z(t,L) = 0, \forall t \in [0,T], \tag{2.150}$$
$$z(T,\cdot) = z^T. \tag{2.151}$$
Then
$$\mathcal{F}_T^*(z^T) = z_x(\cdot, L). \tag{2.152}$$

Proof of Lemma 2.28. Let us first point out that the proof of the existence and uniqueness of $z \in C^1([0,T]; L^2(0,L)) \cap C^0([0,T]; H^3(0,L))$ satisfying (2.149) to (2.151) is similar to the proof of the existence and uniqueness of the solution of (2.112)-(2.113)-(2.114). In fact, if $\tilde{z} \in C^1([0,T]; L^2(0,L)) \cap C^0([0,T]; H^3(0,L))$ is the solution of
$$\tilde{z}(t,0) = \tilde{z}(t,L) = \tilde{z}_x(t,L) = 0, \forall t \in [0,T],$$
$$\frac{d\tilde{z}}{dt} = A\tilde{z},$$
$$\tilde{z}(0,x) = z^T(L-x), \forall x \in [0,L],$$
then
$$z(t,x) = \tilde{z}(T-t, L-x), \forall (t,x) \in [0,T] \times [0,L].$$
Now let $u \in C^2([0,T])$ be such that $u(0) = 0$. Let
$$y \in C^1([0,T]; L^2(0,L)) \cap C^0([0,T]; H^3(0,L))$$
be such that
$$y_t + y_x + y_{xxx} = 0, \tag{2.153}$$
$$y(t,0) = y(t,L) = 0, y_x(t,L) = u(t), \forall t \in [0,T], \tag{2.154}$$
$$y(0,\cdot) = 0. \tag{2.155}$$

(Again the existence of such a y is proved above.) Then, from (2.149), (2.150), (2.151), (2.153), (2.154) and (2.155), we get, using integrations by parts,

$$\int_0^L z^T \mathcal{F}_T(u)dx = \int_0^L z^T y(T,x)dx$$
$$= \int_0^T \int_0^L (zy)_t dx dt$$
$$= -\int_0^T \int_0^L (z_x + z_{xxx})y + z(y_x + y_{xxx})dx dt$$
$$= \int_0^T z_x(t,L)u(t)dt,$$

which, since the set of $u \in C^2([0,T])$ such that $u(0) = 0$ is dense in $L^2(0,T)$, concludes the proof of Lemma 2.28. ∎

Let us now assume that $L \notin \mathcal{N}$ and prove that the observability inequality (2.78) holds in this case. Replacing x by $L - x$ and t by $T - t$ and using Lemma 2.28, one sees that observability inequality (2.78) is equivalent to the inequality

(2.156) $$c\|y^0\|_{L^2(0,L)} \leqslant \|y_x(\cdot,0)\|_{L^2(0,T)}, \forall y^0 \in \mathcal{D}(A),$$

with $y(t,\cdot) := S(t)y^0$. We argue by contradiction and therefore assume that (2.156) does not hold whatever $c > 0$ is. Then there exists a sequence $(y_n^0)_{n \in \mathbb{N}}$ of functions in $\mathcal{D}(A)$ such that

(2.157) $$\|y_n^0\|_{L^2(0,L)} = 1, \forall n \in \mathbb{N},$$

(2.158) $$y_{nx}(\cdot,0) \to 0 \text{ in } L^2(0,T) \text{ as } n \to +\infty,$$

with $y_n(t,\cdot) := S(t)y_n^0$. By (2.141) and (2.157),

(2.159) the sequence $(y_n)_{n \in \mathbb{N}}$ is bounded in $L^2((0,T);H^1(0,L))$.

Moreover, since $y_{nt} = -y_{nx} - y_{nxxx}$, (2.159) implies that

(2.160) the sequence $(y_{nt})_{n \in \mathbb{N}}$ is bounded in $L^2((0,T);H^{-2}(0,L))$.

Let us recall the following compactness result [21] due to Jean-Pierre Aubin (see also the paper [449] by Jacques Simon for more general results).

THEOREM 2.29. *Let X, B, Y be three Banach spaces such that*

$$X \subset B \subset Y \text{ with compact embedding } X \hookrightarrow B.$$

Let $T > 0$. Let K be a bounded subset of $L^2((0,T);X)$. We assume that there exists $C > 0$ such that

$$\left\|\frac{\partial f}{\partial t}\right\|_{L^2((0,T);Y)} \leqslant C, \forall f \in K.$$

Then K is relatively compact in $L^2((0,T);B)$.

We apply this theorem with $X := H^1(0,L)$, $B := L^2(0,L)$, $Y := H^{-2}(0,L)$ and $K := \{y_n; n \in \mathbb{N}\}$. By (2.159) and (2.160), the assumptions of Theorem 2.29 hold. From this theorem, one gets that the set $\{y_n; n \in \mathbb{N}\}$ is relatively compact in $L^2((0,T); L^2(0,L))$. Hence, without loss of generality, we can assume that, for some $y \in L^2((0,T); L^2(0,L))$,

(2.161) $\qquad y_n \to y$ in $L^2((0,T); L^2(0,L))$ as $n \to +\infty$.

From (2.142), (2.158) and (2.161) applied to $y_n - y_m$, we get that $(y_n^0)_{n \in \mathbb{N}}$ is a Cauchy sequence in $L^2(0,L)$. Hence there exists $y^0 \in L^2(0,L)$ such that

(2.162) $\qquad y_n^0 \to y^0$ in $L^2(0,L)$ as $n \to \infty$.

Clearly, from (2.161) and (2.162),

(2.163) $$y(t, \cdot) = S(t) y^0.$$

By (2.157) and (2.162),

(2.164) $$\|y^0\|_{L^2(0,L)} = 1.$$

By the continuity of the map
$$z^0 \in L^2(0,L) \mapsto (t \mapsto (S(t)z^0)_x(0)) \in L^2(0,T)$$
(see (2.140)), (2.158), (2.162) and (2.163),

(2.165) $$y_x(\cdot, 0) = 0 \text{ in } L^2(0,T).$$

Finally (2.163), (2.164) and (2.165) are in contradiction with the following lemma.

LEMMA 2.30. *Let $T > 0$ and let $y^0 \in L^2(0,L)$ such that (2.165) holds for y defined by (2.163). If $L \notin \mathcal{N}$, then $y^0 = 0$.*

Proof of Lemma 2.30. One uses a method due to Claude Bardos, Gilles Lebeau and Jeffrey Rauch [34, page 1063]. Let $T > 0$. For $T' > 0$, let $N_{T'}$ be the set of $y^0 \in L^2(0,L)$ such that, if $y \in C^0([0, +\infty); L^2(0,L))$ is the solution of the Cauchy problem (2.96)-(2.97)-(2.98) with $u := 0$, then
$$y_x(\cdot, 0) = 0 \text{ in } L^2(0,T').$$
Clearly

(2.166) $\qquad (0 < T' < T'') \Rightarrow (N_{T''} \subset N_{T'}).$

Of course, due to (2.140), for every $T' > 0$, $N_{T'}$ is a closed linear subspace of $L^2(0,L)$. Let us prove that

(2.167) $\qquad$ for every $T' > 0$, the vector space $N_{T'}$ is of finite dimension.

Let $T' > 0$ and let $(y_n^0)_{n \in \mathbb{N}}$ be a sequence of elements in $N_{T'}$ such that $\|y_n^0\|_{L^2(0,L)} = 1$. The same proof as for (2.162) gives us that one can extract from $(y_n^0)_{n \in \mathbb{N}}$ a subsequence converging in $L^2(0,L)$. Hence the unit ball of $N_{T'}$ is compact. This proves (2.167).

From (2.166) and (2.167) and an argument of cardinality, there exist $T' > 0$ and $\eta > 0$ such that $T' + \eta < T$ and

(2.168) $\qquad N_{T'+t} = N_{T'}, \forall t \in [0, \eta].$

Let

(2.169) $\quad M := \{f : t \in [0, \eta/2] \mapsto S(t) y^0; \, y^0 \in N_{T'}\} \subset C^0([0, \eta/2]; L^2(0,L)).$

Note that, by (2.167), M is a vector space of finite dimension, which implies that

(2.170) $\qquad M$ is a closed subspace of $L^2((0, \eta/2); H^{-2}(0,L))$.

Let $y^0 \in N_{T'}$. Since $S(\tau)S(t)y^0 = S(\tau+t)y^0$, we get from (2.168),

(2.171) $$S(t)y^0 \in N_{T'}, \forall t \in [0, \eta].$$

Let $y \in C^0([0, \eta]; L^2(0, L))$ be defined by

$$y(t, \cdot) = S(t)y^0.$$

Since $y \in H^1((0, \eta); H^{-2}(0, L))$, there exists $z \in L^2((0, \eta/2); H^{-2}(0, L))$ such that

(2.172) $$\lim_{\varepsilon \to 0^+} \frac{y(\varepsilon + \cdot, \cdot) - y(\cdot, \cdot)}{\varepsilon} = z \text{ in } L^2((0, \eta/2); H^{-2}(0, L)).$$

By (2.169) and (2.171),

(2.173) $$\left(\tau \in [0, \eta/2] \mapsto \frac{y(\varepsilon + \tau, \cdot) - y(\tau, \cdot)}{\varepsilon}\right) \in M, \forall \varepsilon \in [0, \eta/2].$$

By (2.170), (2.172) and (2.173),

$$z \in M,$$

which implies that $z \in C^0([0, \eta/2]; L^2(0, L))$ and $z(0) \in N_{T'}$. Hence

$$y \in C^1([0, \eta/2]; L^2(0, L)).$$

Therefore, using (2.163), $y \in C^0([0, \eta/2], H^3(0, L))$ and y^0 is such that

(2.174) $$y^0 \in H^3(0, L),$$
(2.175) $$y^0(0) = y^0_x(0) = y^0(L) = y^0_x(L) = 0,$$
(2.176) $$Ay^0 = z(0) \in N_{T'}.$$

Hence one can define a unique $\mathbb{C}$-linear map $\mathcal{A} : \mathbb{C}N_{T'} \to \mathbb{C}N_{T'}$ by requiring

$$\mathcal{A}(\zeta \varphi) = \zeta A\varphi, \forall \zeta \in \mathbb{C}, \forall \varphi \in N_{T'}.$$

Then (see also (2.167)), if $N_{T'} \neq \{0\}$, this linear map has an eigenvector. Hence Lemma 2.30 is a consequence of (2.166) and of the following lemma.

LEMMA 2.31 ([**407**, Lemma 3.5, page 45]). *There exist* $\lambda \in \mathbb{C}$ *and*

$$\varphi \in H^3((0, L); \mathbb{C}) \setminus \{0\}$$

such that

(2.177) $$-\varphi_x - \varphi_{xxx} = \lambda \varphi,$$
(2.178) $$\varphi(0) = \varphi_x(0) = \varphi(L) = \varphi_x(L) = 0,$$

if and only if

$$L \in \mathcal{N}.$$

One can prove Lemma 2.31 by looking at the Fourier transform of φ extended to $\mathbb{R}$ by 0 outside $[0, L]$; see [**407**, pages 45–46] for a detailed proof. This concludes the proof of Lemma 2.30 and the proof of the "if" part of Theorem 2.25 on page 42. ∎

Let us now assume that

(2.179) $$L \in \mathcal{N}$$

and prove that the control system (2.94)-(2.95) is not controllable in time $T > 0$, whatever $T > 0$ is. By (2.179) and by Lemma 2.31, there exist $\lambda \in \mathbb{C}$ and $\varphi \in H^3((0, L); \mathbb{C})$ such that (2.177) and (2.178) hold and

(2.180) $$\varphi \neq 0.$$

Let $(y, u) \in C^0([0, T]; L^2(0, L)) \times L^2(0, T)$ be a trajectory of the control system (2.94)-(2.95). Let us first assume that $y(0, \cdot) \in \mathcal{D}(A)$ and $u \in C^2([0, L])$ satisfies $u(0) = 0$. Then $y \in C^1([0, T]; L^2(0, L)) \cap C^0([0, T]; D(A))$. We multiply (2.94) by φ and integrate on $[0, L]$. Using (2.95), (2.177) and (2.178) and integrations by parts, one gets

(2.181) $$\frac{d}{dt} \int_0^L y\varphi dx = -\lambda \int_0^L y\varphi dx.$$

An easy density argument shows that (2.181) also holds if $y(0, \cdot)$ is only in $L^2(0, L)$ and u is any function in $L^2(0, T)$. From (2.181), one gets

(2.182) $$\int_0^L y(T, x)\varphi(x) dx = e^{-\lambda T} \int_0^L y(0, x)\varphi(x) dx,$$

which, together with (2.180), shows that (2.94)-(2.95) is not controllable in time $T > 0$. This concludes the proof of Theorem 2.25 on page 42. ∎

REMARK 2.32. Let us point out that, if $\lambda \in \mathbb{C}$ and $\varphi \in H^3((0, L); \mathbb{C})$ are such that (2.177), (2.178) and (2.180) hold, then

(2.183) $$\lambda \in i\mathbb{R}.$$

Indeed, straightforward integrations by parts together with (2.178) give

(2.184) $$\int_0^L \overline{\varphi}\varphi_x dx = -\int_0^L \varphi\overline{\varphi}_x dx = i\Im \int_0^L \overline{\varphi}\varphi_x dx,$$

(2.185) $$\int_0^L \overline{\varphi}\varphi_{xxx} dx = -\int_0^L \varphi\overline{\varphi}_{xxx} dx = i\Im \int_0^L \overline{\varphi}\varphi_{xxx} dx,$$

where, for $z \in \mathbb{C}$, $\Im z$ denotes the imaginary part of z and $\overline{z}$ the complex conjugate of z. We multiply (2.177) by $\overline{\varphi}$ and integrate on $[0, L]$. Using (2.184) and (2.185), we get

(2.186) $$\lambda \int_0^L \varphi\overline{\varphi} dx = -i\Im \int_0^L \overline{\varphi}(\varphi_x + \varphi_{xxx}) dx.$$

From (2.180) and (2.186), we get (2.183). ∎

REMARK 2.33. Let us fix $L \in (0, +\infty) \setminus \mathcal{N}$ and $T > 0$. Let $y^0 \in L^2(0, L)$ and let $y^1 \in L^2(0, L)$. Let $\mathcal{U}$ be the set of $u \in L^2(0, T)$ such that the solution $y \in C^0([0, T]; L^2(0, L))$ of the Cauchy problem

$$y_t + y_x + y_{xxx} = 0, \ (t, x) \in (0, T) \times (0, L),$$
$$y(t, 0) = y(t, L) = 0, \ y_x(t, L) = u(t), \ t \in (0, T),$$
$$y(0, x) = y^0(x), \ x \in (0, L),$$

satisfies
$$y(T, x) = y^1(x), \, x \in (0, L).$$
Clearly $\mathcal{U}$ is a closed affine subspace of $L^2(0, T)$. By Theorem 2.25 on page 42, $\mathcal{U}$ is not empty. Hence there exists a unique $\bar{u} = \bar{u}_{y^0, y^1} \in \mathcal{U}$ such that
$$\bar{u} \in \mathcal{U} \text{ and } \|\bar{u}\|_{L^2(0,T)} = \inf\{\|u\|_{L^2(0,T)}; \, u \in \mathcal{U}\}.$$
By the proof of Theorem 2.25 on page 42, there exists $C > 0$ (depending on $T > 0$ and $L > 0$ but not on $y^0 \in L^2(0, L)$ and $y^1 \in L^2(0, L)$) such that
$$\|\bar{u}_{y^0, y^1}\|_{L^2(0,L)} \leqslant C(\|y^0\|_{L^2(0,L)} + \|y^1\|_{L^2(0,L)}). \tag{2.187}$$
Straightforward arguments show that the map $(y^0, y^1) \in L^2(0, L)^2 \mapsto \bar{u}_{y^0, y^1} \in L^2(0, T)$ is linear and therefore, by (2.187), continuous.

EXERCISE 2.34. *Let $L > 0$ and $T > 0$. We consider the following linear control system*
$$y_t + y_x + \int_0^x y(t, s) ds = 0, \, t \in (0, T), \, x \in (0, L), \tag{2.188}$$
$$y(t, 0) = u(t), \, t \in (0, T), \tag{2.189}$$
where, at time $t \in [0, T]$, the state is $y(t, \cdot) \in L^2(0, L)$ and the control is $u(t) \in \mathbb{R}$.
1. *Check that, if $y \in C^1([0, T] \times [0, L])$ satisfies (2.188)-(2.189), then, for every $\tau \in [0, T]$ and for every $\phi \in C^1([0, \tau] \times [0, L])$ such that*
$$\phi(t, L) = 0, \, \forall t \in [0, \tau], \tag{2.190}$$
one has
$$-\int_0^\tau \int_0^L (\phi_t + \phi_x - \int_x^L \phi(t, s) ds) y \, dx \, dt - \int_0^\tau u(t) \phi(t, 0) dt \tag{2.191}$$
$$+ \int_0^L y(\tau, x) \phi(\tau, x) dx - \int_0^L y^0(x) \phi(0, x) dx = 0,$$
with $y^0(x) := y(0, x)$, $\forall x \in [0, L]$.
2. *Let $A : \mathcal{D}(A) \subset L^2(0, L) \to L^2(0, L)$ be the linear (unbounded) operator on $L^2(0, L)$ defined by*
$$\mathcal{D}(A) := \{y \in H^1(0, L); \, y(0) = 0\}, \tag{2.192}$$
$$Ay = -y_x - \int_0^x y(s) ds. \tag{2.193}$$
   2.a. *Prove that A is a closed operator (Definition A.1 on page 373) and a dissipative operator (Definition A.2 on page 373).*
   2.b. *Compute the adjoint A^* of A. Check that A^* is also a dissipative operator.*
3. **Well-posedness of the Cauchy problem.** *Let $u \in L^2(0, T)$ and $y^0 \in L^2(0, L)$. Until the end of this exercise one says that $y : (0, T) \times (0, L) \to \mathbb{R}$ is a solution of the Cauchy problem*
$$y_t + y_x + \int_0^x y(t, s) ds = 0, \, t \in (0, T), \, x \in (0, L), \tag{2.194}$$
$$y(t, 0) = u(t), \, t \in (0, T), \tag{2.195}$$
$$y(0, x) = y^0(x), \, x \in (0, L), \tag{2.196}$$

if $y \in C^0([0,T]; L^2(0,L))$ and if (2.191) holds, for every $\tau \in [0,T]$ and for every $\phi \in C^1([0,\tau] \times [0,L])$ satisfying (2.190).

3.a. Prove that the Cauchy problem (2.194), (2.195) and (2.196) has at least one solution.

3.b. Prove that the Cauchy problem (2.194), (2.195) and (2.196) has at most one solution.

4. **Expression of the adjoint.** Let us define a linear map $\mathcal{F}_T : L^2(0,T) \to L^2(0,L)$ in the following way. Let $u \in L^2(0,T)$. Let $y \in C^0([0,T]; L^2(0,L))$ be the solution of the Cauchy problem (2.194), (2.195) and (2.196) with $y^0 := 0$. Then

$$\mathcal{F}_T(u) := y(T, \cdot).$$

Compute $\mathcal{F}_T^*$.

5. **Observability inequality.** In this question, we assume that $u = 0$.

5.a. Prove that, for every $y^0 \in L^2(0,L)$, the solution y of the Cauchy problem (2.194), (2.195) and (2.196) satisfies

$$y(\cdot, L) \in L^2(0,T).$$

5.b. We now assume that $T > L$. Prove that there exists $C_1 > 0$ such that, for every $y^0 \in L^2(0,L)$, the solution y of the Cauchy problem (2.194), (2.195) and (2.196) satisfies

(2.197) $$\|y^0\|_{L^2(0,L)}^2 \leqslant C_1 \left(\|y(\cdot, L)\|_{L^2(0,T)}^2 + \int_0^T \left(\int_0^L y(t,x) dx \right)^2 dt \right).$$

(**Hint.** Proceed as in the proof of (2.93): multiply (2.194) successively by y and by $(L-x)y$ and integrate on $[0,L]$.)

5.c. Prove that, if $\lambda \in \mathbb{C}$ and $z \in H^1((0,L); \mathbb{C})$ are such that

$$z_x + \int_0^x z(s) ds = \lambda z \text{ in } L^2(0,L),$$

$$z(0) = 0,$$

then $z = 0$.

5.d. We again assume that $T > L$. Prove that there exists $C_2 > 0$ such that, for every $y^0 \in L^2(0,L)$, the solution y of the Cauchy problem (2.194), (2.195) and (2.196) satisfies

$$\|y^0\|_{L^2(0,L)} \leqslant C_2 \|y(\cdot, L)\|_{L^2(0,T)}.$$

6. **Controllability.** We again assume that $T > L$. Prove the controllability of the control system (2.188)-(2.189). More precisely, prove that there exists $C_3 > 0$ such that, for every $y^0 \in L^2(0,L)$ and for every $y^1 \in L^2(0,L)$, there exists $u \in L^2(0,T)$ such that

$$\|u\|_{L^2(0,T)} \leqslant C_3(\|y^0\|_{L^2(0,L)} + \|y^1\|_{L^2(0,L)})$$

and such that the solution y of the Cauchy problem (2.194), (2.195) and (2.196) satisfies $y(T, \cdot) = y^1$ in $L^2(0,L)$.

2.3. Abstract linear control systems

Looking at Section 2.1 and Section 2.2, one sees that some proofs for the transport equation and for the Korteweg-de Vries equation are quite similar. This is, for example, the case for the well-posedness of the associated Cauchy problems (compare Section 2.1.1 with Section 2.2.1). This is also the case for the fact that a suitable observability ((2.78) for the transport equation and (2.156) for the Korteweg-de Vries equation) is equivalent to the controllability. The goal of this section is to present a general framework which includes as special cases the study of these equations and their controllability as well as many other equations. This general framework is now very classical. For pioneer papers, let us mention, in particular, [**434**] by Dietmar Salamon, [**499, 500**] by George Weiss, and [**138**] by Ruth Curtain and George Weiss. For more recent references and more advanced results, let us mention, for example,

- [**49**, Chapter 3] by Alain Bensoussan, Giuseppe Da Prato, Michel Delfour and Sanjoy Mitter,
- [**93**] by Yacine Chitour and Emmanuel Trélat,
- [**280**, Section 2.2] by Vilmos Komornik and Paola Loreti,
- [**302**, Chapter 7] by Irena Lasiecka and Roberto Triggiani,
- [**412**] by Lionel Rosier,
- [**462**, Chapter 1 to Chapter 5] by Olof Staffans,
- [**486**, Section 2.4.1] by Marius Tucsnak.

REMARK 2.35. Note, however, that, as it is often the case for partial differential equations, this general framework is far from being the "end of the story". For example it gives the observability inequality that one needs to prove but does not provide any method to prove it.

For two normed linear spaces H_1 and H_2, we denote by $\mathcal{L}(H_1; H_2)$ the set of continuous linear maps from H_1 into H_2 and denote by $\|\cdot\|_{\mathcal{L}(H_1;H_2)}$ the usual norm in this space.

Let H and U be two Hilbert spaces. Just to simplify the notations, these Hilbert spaces are assumed to be real Hilbert spaces. The space H is the state space and the space U is the control space. We denote by $(\cdot,\cdot)_H$ the scalar product in H, by $(\cdot,\cdot)_U$ the scalar product in U, by $\|\cdot\|_H$ the norm in H and by $\|\cdot\|_U$ the norm in U.

Let $S(t)$, $t \in [0, +\infty)$, be a strongly continuous semigroup of continuous linear operators on H (see Definition A.5 and Definition A.6 on page 374). Let A be the infinitesimal generator of the semigroup $S(t)$, $t \in [0, +\infty)$, (see Definition A.9 on page 375). As usual, we denote by $S(t)^*$ the adjoint of $S(t)$. Then $S(t)^*$, $t \in [0, +\infty)$, is a strongly continuous semigroup of continuous linear operators and the infinitesimal generator of this semigroup is the adjoint A^* of A (see Theorem A.11 on page 375). The domain $D(A^*)$ is equipped with the usual graph norm $\|\cdot\|_{D(A^*)}$ of the unbounded operator A^*:

$$\|z\|_{D(A^*)} := \|z\|_H + \|A^*z\|_H, \ \forall z \in D(A^*).$$

This norm is associated to the scalar product in $D(A^*)$ defined by

$$(z_1, z_2)_{D(A^*)} := (z_1, z_2)_H + (A^*z_1, A^*z_2)_H, \ \forall (z_1, z_2) \in D(A^*)^2.$$

With this scalar product, $D(A^*)$ is a Hilbert space. Let $D(A^*)'$ be the dual of $D(A^*)$ with the pivot space H. In particular,
$$D(A^*) \subset H \subset D(A^*)'.$$
Let
(2.198) $$B \in \mathcal{L}(U, D(A^*)').$$
In other words, B is a linear map from U into the set of linear functions from $D(A^*)$ into $\mathbb{R}$ such that, for some $C > 0$,
$$|(Bu)z| \leqslant C\|u\|_U \|z\|_{D(A^*)}, \; \forall u \in U, \forall z \in D(A^*).$$
We also assume the following regularity property (also called admissibility condition):

(2.199) $\quad \forall T > 0, \exists C_T > 0$ such that $\displaystyle\int_0^T \|B^* S(t)^* z\|_U^2 dt \leqslant C_T \|z\|_H^2, \; \forall z \in D(A^*).$

In (2.199) and in the following, $B^* \in \mathcal{L}(D(A^*); U)$ is the adjoint of B. It follows from (2.199) that the operators
$$\big(z \in D(A^*)\big) \mapsto \big((t \mapsto B^* S(t)^* z) \in C^0([0,T]; U)\big),$$
$$\big(z \in D(A^*)\big) \mapsto \big((t \mapsto B^* S(T-t)^* z) \in C^0([0,T]; U)\big)$$
can be extended in a unique way as continuous linear maps from H into $L^2((0,T); U)$. We use the same symbols to denote these extensions.

Note that, using the fact that $S(t)^*$, $t \in [0, +\infty)$, is a strongly continuous semigroup of continuous linear operators on H, it is not hard to check that (2.199) is equivalent to
$$\exists T > 0, \exists C_T > 0 \text{ such that } \int_0^T \|B^* S(t)^* z\|_U^2 dt \leqslant C_T \|z\|_H^2, \; \forall z \in D(A^*).$$

The control system we consider here is

(2.200) $$\dot{y} = Ay + Bu, \; t \in (0, T),$$

where, at time t, the control is $u(t) \in U$ and the state is $y(t) \in H$.

The remaining part of this section is organized as follows:
- In Section 2.3.1, we study the well-posedness of the Cauchy problem associated to (2.200).
- In Section 2.3.2, we study the controllability of the control system (2.200). We give the observability inequality which is equivalent to the controllability of (2.200).
- Finally, in Section 2.3.3, we revisit the transport equation and the Korteweg-de Vries equation, studied above, in the framework of this abstract setting.

2.3.1. Well-posedness of the Cauchy problem. Let $T > 0$, $y^0 \in H$ and $u \in L^2((0,T); U)$. We are interested in the Cauchy problem

(2.201) $$\dot{y} = Ay + Bu(t), \; t \in (0, T),$$
(2.202) $$y(0) = y^0.$$

We first give the definition of a solution to (2.201)-(2.202). We mimic what we have done in Section 2.1.1 for a linear transport equation and in Section 2.2.1 for a linear Korteweg-de Vries equation.

Let $\tau \in [0,T]$ and $\varphi : [0,\tau] \to H$. We take the scalar product in H of (2.201) with φ and integrate on $[0,\tau]$. At least formally, we get, using an integration by parts together with (2.202),

$$(y(\tau), \varphi(\tau))_H - (y^0, \varphi(0))_H - \int_0^\tau (y(t), \dot{\varphi}(t) + A^*\varphi(t))_H dt = \int_0^\tau (u(t), B^*\varphi(t))_U dt.$$

Taking $\varphi(t) = S(\tau-t)^* z^\tau$, for every given $z^\tau \in H$, we have formally $\dot{\varphi}(t) + A^*\phi(t) = 0$, which leads to the following definition.

DEFINITION 2.36. *Let $T > 0$, $y^0 \in H$ and $u \in L^2((0,T);U)$. A solution of the Cauchy problem (2.201)-(2.202) is a function $y \in C^0([0,T];H)$ such that*

(2.203)
$$(y(\tau), z^\tau)_H - (y^0, S(\tau)^* z^\tau)_H = \int_0^\tau (u(t), B^* S(\tau - t)^* z^\tau)_U dt, \forall \tau \in [0,T], \forall z^\tau \in H.$$

Note that, by the regularity property (2.199), the right hand side of (2.203) is well defined (see page 52).

With this definition one has the following theorem.

THEOREM 2.37. *Let $T > 0$. Then, for every $y^0 \in H$ and for every $u \in L^2((0,T);U)$, the Cauchy problem (2.201)-(2.202) has a unique solution y. Moreover, there exists $C = C(T) > 0$, independent of $y^0 \in H$ and $u \in L^2((0,T);U)$, such that*

(2.204) $$\|y(\tau)\|_H \leqslant C(\|y^0\|_H + \|u\|_{L^2((0,T);U)}), \forall \tau \in [0,T].$$

Proof of Theorem 2.37. Let $T > 0$. Also let $y^0 \in H$ and $u \in L^2((0,T);U)$. Then (see in particular the regularity property (2.199)), for every $\tau > 0$, the linear form

$$\begin{array}{rcl} H & \to & \mathbb{R} \\ z^\tau & \mapsto & (y^0, S(\tau)^* z^\tau)_H + \int_0^\tau (u, B^* S(\tau - t)^* z^\tau)_U dt, \end{array}$$

is continuous. Therefore, by Riesz's theorem, there exists one and only one $y^\tau \in H$ such that

(2.205) $$(y^\tau, z^\tau)_H = (y^0, S(\tau)^* z^\tau)_H + \int_0^\tau (u, B^* S(\tau - t)^* z^\tau)_U dt, \forall z^\tau \in H.$$

This shows the uniqueness of the solution to the Cauchy problem (2.201)-(2.202).

Concerning the existence of a solution satisfying (2.204), let $y : [0,T] \to H$ be defined by

(2.206) $$y(\tau) = y^\tau, \forall \tau \in [0,T].$$

From (2.205) and (2.206), we get (2.203). By Theorem A.8 on page 375 and Theorem A.11 on page 375, there exists $C' > 0$ such that

(2.207) $$\|S(t)^*\|_{\mathcal{L}(H;H)} \leqslant C', \forall t \in [0,T].$$

From (2.199), (2.205), (2.206) and (2.207), we get that

$$\|y(\tau)\|_H \leqslant C'\|y^0\|_H + C_T^{1/2}\|u\|_{L^2((0,T);U)}, \forall \tau \in [0,T],$$

which gives (2.204) with $C := \text{Max}\,\{C', C_T^{1/2}\}$.

It just remains to check that $y \in C^0([0,T]; H)$. Let $\tau \in [0,T]$ and $(\tau_n)_{n \in \mathbb{N}}$ be a sequence of real numbers in $[0,T]$ such that

(2.208) $$\tau_n \to \tau \text{ as } n \to +\infty.$$

Let $z^\tau \in H$ and let $(z^{\tau_n})_{n \in \mathbb{N}}$ be a sequence of elements in H such that

(2.209) $$z^{\tau_n} \rightharpoonup z^\tau \text{ weakly in } H \text{ as } n \to +\infty.$$

From (2.209), and since $S(t)$, $t \in [0, +\infty)$, is a *strongly continuous* (see Definition A.6 on page 374) semigroup of continuous linear operators on H,

(2.210)
$$\lim_{n \to +\infty} (y^0, S(\tau_n)^* z^{\tau_n})_H = \lim_{n \to +\infty} (S(\tau_n) y^0, z^{\tau_n})_H = (S(\tau) y^0, z^\tau)_H = (y^0, S(\tau)^* z^\tau)_H.$$

We extend u to an element in $L^2((-T,T); H)$ by requiring

$$u(t) := 0, \, t \in (-T, 0).$$

Note that, letting $s = \tau_n - t$,

(2.211) $$\int_0^{\tau_n} (u(t), B^* S(\tau_n - t)^* z^{\tau_n})_U \, dt = \int_0^T (\chi_n(s) u(\tau_n - s), B^* S(s)^* z^{\tau_n})_U \, ds,$$

(2.212) $$\int_0^\tau (u(t), B^* S(\tau - t)^* z^\tau)_U \, dt = \int_0^T (\chi(s) u(\tau - s), B^* S(s)^* z^\tau)_U \, ds,$$

where $\chi_n : [0,T] \to \mathbb{R}$ and $\chi : [0,T] \to \mathbb{R}$ are defined by

(2.213) $\quad\quad\quad\quad \chi_n = 1$ on $[0, \tau_n]$ and $\chi_n = 0$ outside $[0, \tau_n]$,

(2.214) $\quad\quad\quad\quad \chi = 1$ on $[0, \tau]$ and $\chi = 0$ outside $[0, \tau]$.

From (2.208), (2.213) and (2.214), one gets

(2.215) $$\chi_n(\cdot) u(\tau_n - \cdot) \to \chi(\cdot) u(\tau - \cdot) \text{ in } L^2((0,T); U) \text{ as } n \to \infty.$$

From (2.199) and (2.209), one gets

(2.216) $$B^* S(\cdot)^* z^{\tau_n} \rightharpoonup B^* S(\cdot)^* z^\tau \text{ weakly in } L^2((0,T); U) \text{ as } n \to \infty.$$

(Let us recall that a continuous linear map between two Hilbert spaces is weakly continuous; see, for example, [**71**, Théorème III.9, page 39].) From (2.211), (2.212), (2.215) and (2.216), we have

(2.217) $$\lim_{n \to +\infty} \int_0^{\tau_n} (u, B^* S(\tau_n - t)^* z^{\tau_n})_U \, dt = \int_0^\tau (u, B^* S(\tau - t)^* z^\tau)_U \, dt.$$

From (2.205), (2.210) and (2.217), one gets that

$$(y(\tau_n), z^{\tau_n})_H \to (y(\tau), z^\tau)_H \text{ as } n \to +\infty,$$

which implies that

$$y(\tau_n) \to y(\tau) \text{ in } H \text{ as } n \to +\infty.$$

This concludes the proof of Theorem 2.37. ∎

2.3.2. Controllability and observability inequality.

In this section we are interested in the controllability of the control system (2.200). In contrast to the case of linear finite-dimensional control systems, many types of controllability are possible and interesting. We define here three types of controllability.

DEFINITION 2.38. Let $T > 0$. The control system (2.200) is *exactly controllable in time T* if, for every $y^0 \in H$ and for every $y^1 \in H$, there exists $u \in L^2((0,T);U)$ such that the solution y of the Cauchy problem

(2.218) $$\dot{y} = Ay + Bu(t),\; y(0) = y^0,$$

satisfies $y(T) = y^1$.

DEFINITION 2.39. Let $T > 0$. The control system (2.200) is *null controllable in time T* if, for every $y^0 \in H$ and for every $\tilde{y}^0 \in H$, there exists $u \in L^2((0,T);U)$ such that the solution of the Cauchy problem (2.218) satisfies $y(T) = S(T)\tilde{y}^0$.

Let us point out that, by linearity, we get an equivalent definition of "null controllable in time T" if, in Definition 2.39, one assumes that $\tilde{y}^0 = 0$. This explains the usual terminology "null controllability".

DEFINITION 2.40. Let $T > 0$. The control system (2.200) is *approximately controllable in time T* if, for every $y^0 \in H$, for every $y^1 \in H$, and for every $\varepsilon > 0$, there exists $u \in L^2((0,T);U)$ such that the solution y of the Cauchy problem (2.218) satisfies $\|y(T) - y^1\|_H \leqslant \varepsilon$.

Clearly

(exact controllability) $\Rightarrow$ (null controllability and approximate controllability).

The converse is false in general (see, for example, the control system (2.374)-(2.375) below). However, the converse holds if S is a strongly continuous group of linear operators (see Definition A.12 on page 376 and Definition A.13 on page 376). More precisely, one has the following theorem.

THEOREM 2.41. *Assume that $S(t)$, $t \in \mathbb{R}$, is a strongly continuous group of linear operators. Let $T > 0$. Assume that the control system (2.200) is null controllable in time T. Then the control system (2.200) is exactly controllable in time T.*

Proof of Theorem 2.41. Let $y^0 \in H$ and $y^1 \in H$. From the null controllability assumption applied to the initial data $y^0 - S(-T)y^1$, there exists $u \in L^2((0,T);U)$ such that the solution $\tilde{y}$ of the Cauchy problem

$$\dot{\tilde{y}} = A\tilde{y} + Bu(t),\; \tilde{y}(0) = y^0 - S(-T)y^1,$$

satisfies

(2.219) $$\tilde{y}(T) = 0.$$

One easily sees that the solution y of the Cauchy problem

$$\dot{y} = Ay + Bu(t),\; y(0) = y^0,$$

is given by

(2.220) $$y(t) = \tilde{y}(t) + S(t-T)y^1,\; \forall t \in [0,T].$$

In particular, from (2.219) and (2.220),

$$y(T) = y^1.$$

This concludes the proof of Theorem 2.41. ∎

Let us now introduce some "optimal control maps". Let us first deal with the case where the control system (2.200) is exactly controllable in time T. Then, for every y^1, the set $U^T(y^1)$ of $u \in L^2((0,T);U)$ such that
$$(\dot{y} = Ay + Bu(t), y(0) = 0) \Rightarrow (y(T) = y^1)$$
is nonempty. Clearly the set $U^T(y^1)$ is a closed affine subspace of $L^2((0,T);U)$. Let us denote by $\mathcal{U}^T(y^1)$ the projection of 0 on this closed affine subspace, i.e., the element of $U^T(y^1)$ of the smallest $L^2((0,T);U)$-norm. Then it is not hard to see that the map
$$\mathcal{U}^T : \begin{array}{rcl} H & \to & L^2((0,T);U) \\ y^1 & \mapsto & \mathcal{U}^T(y^1) \end{array}$$
is a linear map. Moreover, using the closed graph theorem (see, for example, [**419**, Theorem 2.15, page 50]) one readily checks that this linear map is continuous.

Let us now deal with the case where the control system (2.200) is null controllable in time T. Then, for every y^0, the set $U_T(y^0)$ of $u \in L^2((0,T);U)$ such that
$$(\dot{y} = Ay + Bu(t), y(0) = y^0) \Rightarrow (y(T) = 0)$$
is nonempty. Clearly the set $U_T(y^0)$ is a closed affine subspace of $L^2((0,T);U)$. Let us denote by $\mathcal{U}_T(y^0)$ the projection of 0 on this closed affine subspace, i.e., the element of $U_T(y^0)$ of the smallest $L^2((0,T);U)$-norm. Then, again, it is not hard to see that the map
$$\mathcal{U}_T : \begin{array}{rcl} H & \to & L^2((0,T);U) \\ y^0 & \mapsto & \mathcal{U}_T(y^0) \end{array}$$
is a continuous linear map.

The main results of this section are the following ones.

THEOREM 2.42. *Let $T > 0$. The control system (2.200) is exactly controllable in time T if and only if there exists $c > 0$ such that*

$$(2.221) \qquad \int_0^T \|B^*S(t)^*z\|_U^2 \, dt \geqslant c\|z\|_H^2, \ \forall z \in D(A^*).$$

Moreover, if such a $c > 0$ exists and if c^T is the maximum of the set of $c > 0$ such that (2.221) holds, one has

$$(2.222) \qquad \|\mathcal{U}^T\|_{\mathcal{L}(H;L^2((0,T);U))} = \frac{1}{\sqrt{c^T}}.$$

THEOREM 2.43. *The control system (2.200) is approximately controllable in time T if and only if, for every $z \in H$,*
$$(B^*S(\cdot)^*z = 0 \text{ in } L^2((0,T);U)) \Rightarrow (z = 0).$$

THEOREM 2.44. *Let $T > 0$. The control system (2.200) is null controllable in time T if and only if there exists $c > 0$ such that*

$$(2.223) \qquad \int_0^T \|B^*S(t)^*z\|_U^2 \, dt \geqslant c\|S(T)^*z\|_H^2, \ \forall z \in D(A^*).$$

Moreover, if such a $c > 0$ exists and if c_T is the maximum of the set of $c > 0$ such that (2.223) holds, then

$$(2.224) \qquad \|\mathcal{U}_T\|_{\mathcal{L}(H;L^2((0,T);U))} = \frac{1}{\sqrt{c_T}}.$$

THEOREM 2.45. *Assume that, for every $T > 0$, the control system (2.200) is null controllable in time T. Then, for every $T > 0$, the control system (2.200) is approximately controllable in time T.*

Inequalities (2.221) and (2.223) are usually called observability inequalities for the abstract linear control system $\dot{y} = Ay + Bu$.

Proof of Theorem 2.42 and Theorem 2.43. Let $T > 0$. Let us define a linear map $\mathcal{F}_T : L^2((0,T);U) \to H$ in the following way. Let $u \in L^2((0,T);U)$. Let $y \in C^0([0,T];H)$ be the solution of the Cauchy problem (2.218) with $y^0 := 0$. Then

$$\mathcal{F}_T(u) := y(T, \cdot).$$

One has the following lemma, the proof of which is similar to the proof of Lemma 2.15 on page 34.

LEMMA 2.46. *The control system (2.200) is exactly controllable in time T if and only if $\mathcal{F}_T$ is onto. The control system (2.200) is approximately controllable in time T if and only if $\mathcal{F}_T(L^2((0,T);U))$ is dense in H.*

It is a classical result of functional analysis that $\mathcal{F}_T(L^2((0,T);U))$ is dense in H if and only if $\mathcal{F}_T^*$ is one-to-one (see, e.g., [**419**, Corollaries (b), page 94] or [**71**, Corollaire II.17 (b), page 28]). Let us recall on the other hand that, by Proposition 2.16 on page 35, $\mathcal{F}_T$ is onto if and only if there exists $c > 0$ such that

$$\|\mathcal{F}_T^*(z^T)\|_{L^2((0,T);U)} \geqslant c\|z^T\|_H, \ \forall z^T \in H.$$

Hence, the first part of Theorem 2.42 as well as Theorem 2.43 are a consequence of the following lemma.

LEMMA 2.47. *For every $z^T \in H$,*

$$(\mathcal{F}_T^*(z^T))(t) = B^*S(T-t)^*z^T.$$

Proof of Lemma 2.47. Let $z^T \in D(A^*)$. Let us also recall that $D(A^*)$ is dense in H; see Theorem A.11 on page 375. Let $u \in L^2((0,T);U)$. Let $y \in C^0([0,T];H)$ be the solution of the Cauchy problem

$$\dot{y} = Ay + Bu(t), \ y(0) = 0.$$

By the definition of $\mathcal{F}_T$ and Definition 2.36 on page 53,

$$(u, \mathcal{F}_T^*(z^T))_{L^2((0,T);U)} = (\mathcal{F}_T u, z^T)_H = (y(T), z^T)_H = \int_0^T (u, B^*S(T-t)^*z^T)_U dt.$$

This concludes the proof of Lemma 2.47 and also the proof of the first part of Theorem 2.42 and Theorem 2.43. ∎

Let us now turn to the second part of Theorem 2.42, i.e., equality (2.222). By the definition of c^T and Lemma 2.47

$$(2.225) \qquad \|\mathcal{F}_T^*(z^T)\|_{L^2((0,T);U)}^2 \geqslant c^T\|z^T\|_H^2, \ \forall z^T \in H.$$

By Proposition 2.16 on page 35, (2.225) implies the existence of a continuous linear map $\mathcal{U} : H \to L^2((0,T);U)$ such that

$$\mathcal{F}_T \mathcal{U} y^1 = y^1, \ \forall y^1 \in H, \tag{2.226}$$

$$\|\mathcal{U}\|_{\mathcal{L}(H;L^2((0,T);U))} \leqslant \frac{1}{\sqrt{c^T}}. \tag{2.227}$$

From (2.226), (2.227) and the definition of $\mathcal{U}^T$, one gets

$$\|\mathcal{U}^T\|_{\mathcal{L}(H;L^2((0,T);U))} \leqslant \frac{1}{\sqrt{c^T}}. \tag{2.228}$$

Finally, from the definition of $\mathcal{U}^T$, one has

$$\mathcal{F}_T \mathcal{U}^T y^1 = y^1, \ \forall y^1 \in H,$$

which implies that

$$(\mathcal{F}_T^* z^T, \mathcal{U}^T z^T)_{L^2((0,T);U)} = \|z^T\|_H^2, \ \forall z^T \in H. \tag{2.229}$$

From (2.229), one has

$$\begin{aligned}\|z^T\|_H^2 &\leqslant \|\mathcal{F}_T^* z^T\|_{L^2((0,T);U)} \|\mathcal{U}^T z^T\|_{L^2((0,T);U)} \\ &\leqslant \|\mathcal{F}_T^* z^T\|_{L^2((0,T);U)} \|\mathcal{U}^T\|_{\mathcal{L}(H;L^2((0,T);U))} \|z^T\|_H, \ \forall z^T \in H,\end{aligned}$$

which, together with the definition of c^T and Lemma 2.47, gives that

$$c^T \geqslant \frac{1}{\|\mathcal{U}^T\|_{\mathcal{L}(H;L^2((0,T);U))}^2}.$$

This concludes the proof of (2.222). ∎

Proof of Theorem 2.44. The key tool for the proof of Theorem 2.44 is the following lemma due to Ronald Douglas [148] and Szymon Dolecki and David Russell [146, pages 194–195] and which will be proved later on.

LEMMA 2.48. *Let H_1, H_2 and H_3 be three Hilbert spaces. Let C_2 be a continuous linear map from H_2 into H_1 and let C_3 be a densely defined closed linear operator from $\mathcal{D}(C_3) \subset H_3$ into H_1. Then the two following properties are equivalent:*

(i) *There exists $M \geqslant 0$ such that*

$$\|C_2^* h_1\|_{H_2} \leqslant M \|C_3^* h_1\|_{H_3}, \ \forall h_1 \in \mathcal{D}(C_3^*). \tag{2.230}$$

(ii) *One has the following inclusion:*

$$C_2(H_2) \subset C_3(\mathcal{D}(C_3)). \tag{2.231}$$

Moreover, if $M \geqslant 0$ is such that (2.230) holds, there exists a continuous linear map C_1 from H_2 into H_3 such that

$$C_1(H_2) \subset \mathcal{D}(C_3), \ C_2 = C_3 C_1, \tag{2.232}$$

$$\|C_1\|_{\mathcal{L}(H_2;H_3)} \leqslant M. \tag{2.233}$$

REMARK 2.49. In fact, it is assumed in [148] that the linear operator C_3 is continuous from H_3 int H_2. However, the proof given in [148] can be easily adapted to treat the case where C_3 is only a a densely defined closed linear operator. (See Definition A.1 on page 373 for the definition of a closed linear operator.)

Let us apply Lemma 2.48 with the following spaces and maps.

- The spaces H_1, H_2 and H_3 are given by
$$H_1 := H, \ H_2 := H, \ H_3 := L^2((0,T); U).$$

- The linear map C_2 maps $y^0 \in H$ to $y(T) \in H$, where $y \in C^0([0,T]; H)$ is the solution of the Cauchy problem
$$\dot{y} = Ay, \ y(0) = y^0.$$

In other words,

(2.234) $$C_2 = S(T).$$

- The linear map C_3 maps $u \in L^2((0,T); U)$ to $y(T) \in H$, where $y \in C^0([0,T]; H)$ is the solution of the Cauchy problem
$$\dot{y} = Ay + Bu(t), \ y(0) = 0.$$

In other words,

(2.235) $$C_3 = \mathcal{F}_T.$$

Note that C_3, in this case, is a continuous linear operator from $L^2((0,T); U)$ into H. (For an application with unbounded C_3, see Remark 2.98 on page 110.)

With these choices, the null controllability of the linear control system $\dot{y} = Ay + Bu$ is equivalent to the inclusion (2.231). Indeed, let us first assume that (2.231) holds and prove the null controllability. Let y^0 and $\tilde{y}^0$ be both in H. Let $u \in L^2((0,T); U)$ be such that $C_3 u = C_2(\tilde{y}^0 - y^0)$, i.e., $\mathcal{F}_T u = S(T)(\tilde{y}^0 - y^0)$. Let $y \in C^0([0,T]; H)$ be the solution of the Cauchy problem
$$\dot{y} = Ay + Bu(t), \ y(0) = y^0.$$

Then $y = y_1 + y_2$, where $y_1 \in C^0([0,T]; H)$ is the solution of the Cauchy problem
$$\dot{y}_1 = Ay_1, \ y_1(0) = y^0,$$

and where $y_2 \in C^0([0,T]; H)$ is the solution of the Cauchy problem
$$\dot{y}_2 = Ay_2 + Bu, \ y_2(0) = 0.$$

In particular,
$$y(T) = y_1(T) + y_2(T) = C_2 y^0 + C_3 u = S(T) y^0 + S(T)(\tilde{y}^0 - y^0) = S(T)\tilde{y}^0,$$
which proves the null controllability. The proof of the converse is similar.

Let us now interpret (2.230). Using Lemma 2.47, (2.234) and (2.235), one gets that (2.230) is equivalent to (2.223) with $c = 1/M^2$.

Finally, concerning equality (2.224), let us define, with the notations of Lemma 2.48,
$$\mathcal{U} := -C_1 \in \mathcal{L}(H, L^2((0,T); U)).$$

One has

(2.236) $$((\dot{y} = Ay + B\mathcal{U}y^0, \ y(0) = y^0) \Rightarrow (y(T) = 0)), \ \forall y^0 \in H,$$

(2.237) $$\|\mathcal{U}\|_{\mathcal{L}(H; L^2((0,T); U))} \leqslant \frac{1}{\sqrt{c_T}}.$$

From (2.236), (2.237) and the definition of $\mathcal{U}_T$, one has

(2.238) $$\|\mathcal{U}_T\|_{\mathcal{L}(H; L^2((0,T); U))} \leqslant \frac{1}{\sqrt{c_T}}.$$

It just remains to prove that

(2.239) $$\|\mathcal{U}_T\|_{\mathcal{L}(H;L^2((0,T);U))} \geqslant \frac{1}{\sqrt{c_T}}.$$

From the definition of $\mathcal{U}_T$, one has

(2.240) $$\mathcal{F}_T \mathcal{U}_T = -S(T),$$

From (2.240), one has, for every $z \in H$,

$$\begin{aligned} \|S(T)^*z\|_H^2 &= -(\mathcal{U}_T^* \mathcal{F}_T^* z, S(T)^* z)_H \\ &\leqslant \|\mathcal{U}_T\|_{\mathcal{L}(H;L^2((0,T);U))} \|\mathcal{F}_T^* z\|_{L^2((0,T);U)} \|S(T)^*z\|_H, \end{aligned}$$

which implies that

(2.241) $$\|\mathcal{F}_T^* z\|_{L^2((0,T);U)} \geqslant \frac{1}{\|\mathcal{U}_T\|_{\mathcal{L}(H;L^2((0,T);U))}} \|S(T)^*z\|_H, \forall z \in H.$$

Inequality (2.239) follows from (2.241), the definition of c_T and Lemma 2.47. This concludes the proof of Theorem 2.44 assuming Lemma 2.48. ∎

Proof of Lemma 2.48. Let us first prove that (ii) implies (i). So we assume that (2.231) holds. Let $h_2 \in H_2$. By (2.231), there exists one and only one $h_3 \in \mathcal{D}(C_3)$, orthogonal to the kernel of C_3, such that $C_2 h_2 = C_3 h_3$. Let us denote by $C_1 : H_2 \to H_3$ the map defined by $C_1(h_2) := h_3$. It is not hard to check that C_1 is linear. Moreover, by the construction of C_1, (2.232) holds. Note that, for every $h_1 \in \mathcal{D}(C_3^*) \subset H_1$ and for every $x \in H_2$,

(2.242) $$\begin{aligned} (C_2^* h_1, x)_{H_2} &= (h_1, C_2 x)_{H_1} \\ &= (h_1, C_3 C_1 x)_{H_1} \\ &= (C_3^* h_1, C_1 x)_{H_3}. \end{aligned}$$

Hence, if C_1 is continuous, (2.232) implies (2.230) with $M := \|C_1\|_{\mathcal{L}(H_2;H_3)}$. So, let us check that C_1 is continuous. By the closed graph theorem (see, for example, [**419**, Theorem 2.15, page 50]), it suffices to check that the graph of C_1 is closed. Let $(h_2^n)_{n\in\mathbb{N}}$, $(h_3^n)_{n\in\mathbb{N}}$, $h_2 \in H_2$ and $h_3 \in H_3$ be such that

(2.243) $$h_2^n \in H_2, h_3^n \in H_3, \forall n \in \mathbb{N},$$

(2.244) $$h_3^n = C_1 h_2^n, \forall n \in \mathbb{N},$$

(2.245) $$h_2^n \to h_2 \text{ in } H_2 \text{ and } h_3^n \to h_3 \text{ in } H_3 \text{ as } n \to +\infty.$$

From (2.244), one has

(2.246) $$C_2 h_2^n = C_3 h_3^n, h_3^n \in \text{Ker}(C_3)^\perp, \forall n \in \mathbb{N}.$$

Letting $n \to +\infty$ in (2.246) and using (2.245), one gets (let us recall that C_3 is a closed operator)

$$h_3 \in \mathcal{D}(C_3), C_2 h_2 = C_3 h_3 \text{ and } h_3 \in \text{Ker}(C_3)^\perp,$$

which tells us that $h_3 = C_1 h_2$. This concludes the proof of the continuity of C_1 and also the proof of (ii)⇒(i).

Let us now prove the converse. Hence we assume that (2.230) holds for some $M \geqslant 0$. Let us define a map $K : C_3^*(\mathcal{D}(C_3^*)) \subset H_3 \to H_2$ by requiring that

$$K(C_3^* h_1) = C_2^* h_1, \forall h_1 \in \mathcal{D}(C_3^*) \subset H_1.$$

The map K is well defined since, by (2.230),

$$(C_3^* h_1 = C_3^* \tilde{h}_1) \Rightarrow (C_3^*(h_1 - \tilde{h}_1) = 0) \Rightarrow (C_2^*(h_1 - \tilde{h}_1) = 0) \Rightarrow (C_2^* h_1 = C_2^* \tilde{h}_1).$$

for every $(h_1, \tilde{h}_1) \in \mathcal{D}(C_3^*) \times \mathcal{D}(C_3^*)$. Moreover, K is linear. The vector space $C_3^*(\mathcal{D}(C_3^*))$, which is a vector subspace of H_3, is of course equipped with the norm of H_3. Then K is also continuous since, by (2.230),

$$\|K(C_3^* h_1)\|_{H_2} = \|C_2^* h_1\|_{H_2} \leqslant M \|C_3^* h_1\|_{H_3}, \forall h_1 \in \mathcal{D}(C_3^*) \subset H_1,$$

which implies that $K \in \mathcal{L}(C_3^*(\mathcal{D}(C_3^*)); H_2)$ and that

(2.247) $$\|K\|_{\mathcal{L}(C_3^*(\mathcal{D}(C_3^*)); H_2)} \leqslant M.$$

By (2.247), K can be uniquely extended as a continuous linear map from $\overline{C_3^*(\mathcal{D}(C_3^*))}$ into H_2. Then we extend K to a linear map from H_3 into H_2 by requiring that K vanishes on $\overline{C_3^*(\mathcal{D}(C_3^*))}^\perp$. Clearly $K \in \mathcal{L}(H_3; H_2)$ and

(2.248) $$\|K\|_{\mathcal{L}(H_3; H_2)} \leqslant M,$$

(2.249) $$KC_3^* = C_2^*.$$

From (2.249), one easily gets

(2.250) $$K^*(H_2) \subset \mathcal{D}(C_3) \text{ and } C_3 K^* = C_2,$$

which readily implies (2.231). Note that, if we let $C_1 := K^*$, (2.232) follows from (2.250), and (2.233) follows from (2.248). This concludes the proof of Lemma 2.48. ∎

Proof of Theorem 2.45 on page 57. Let $y^0 \in H$, $y^1 \in H$, $T > 0$ and $\varepsilon > 0$. Since the semigroup $S(t)$, $t \in [0, +\infty)$, is strongly continuous, there exists $\eta \in (0, T)$ such that

(2.251) $$\|S(\eta) y^1 - y^1\|_H \leqslant \varepsilon,$$

(see Definition A.6 on page 374). Since the control system (2.200) is null controllable in time η, there exists $\bar{u} \in L^2((0, \eta); U)$ such that the solution $\bar{y} \in C^0([0, \eta]; H)$ of the Cauchy problem

$$\dot{\bar{y}} = A\bar{y} + B\bar{u}(t), t \in (0, \eta), \bar{y}(0) = S(T - \eta) y^0,$$

satisfies

(2.252) $$\bar{y}(\eta) = S(\eta) y^1.$$

Let $u \in L^2(0, T)$ be defined by

$$u(t) = 0, t \in (0, T - \eta),$$
$$u(t) = \bar{u}(t - T + \eta), t \in (T - \eta, T).$$

Let $y \in C^0([0, T]; H)$ be the solution of the Cauchy problem

$$\dot{y} = Ay + Bu(t), t \in (0, T), y(0) = y^0.$$

Then

(2.253) $$y(t) = S(t) y^0, \forall t \in [0, T - \eta],$$

(2.254) $$y(t) = \bar{y}(t - T + \eta), \forall t \in [T - \eta, T].$$

From (2.251), (2.252) and (2.254), one gets

$$\|y(T) - y^1\|_H \leqslant \varepsilon.$$

This concludes the proof of Theorem 2.45 on page 57. ∎

REMARK 2.50. In contrast to Theorem 2.45, note that, for a given $T > 0$, the null controllability in time T does not imply the approximate controllability in time T. For example, let $L > 0$ and let us take $H := L^2(0, L)$ and $U := \{0\}$. We consider the linear control system

(2.255) $$y_t + y_x = 0,\ t \in (0, T),\ x \in (0, L),$$
(2.256) $$y(t, 0) = u(t) = 0,\ t \in (0, T).$$

In Section 2.3.3.1, we shall see how to put this control system in the abstract framework $\dot{y} = Ay + Bu$; see also Section 2.1. It follows from (2.29), that, whatever $y^0 \in L^2(0, L)$ is, the solution to the Cauchy problem (see Definition 2.1 on page 25 and Section 2.3.3.1)

$$y_t + y_x = 0,\ t \in (0, T),\ x \in (0, L),$$
$$y(t, 0) = u(t) = 0,\ t \in (0, T),$$
$$y(0, x) = y^0(x),\ x \in (0, L),$$

satisfies
$$y(T, \cdot) = 0,\ \text{if}\ T \geqslant L.$$

In particular, if $T \geqslant L$, the linear control system (2.255)-(2.256) is null controllable but is not approximately controllable.

2.3.3. Examples. In this section we show how the abstract setting just presented can be applied to the two partial differential control systems we have encountered, namely a transport linear equation and a linear Korteweg-de Vries equation.

2.3.3.1. *The transport equation revisited.* We return to the transport control system that we have already considered in Section 2.1. Let $L > 0$. The linear control system we study is

(2.257) $$y_t + y_x = 0,\ t \in (0, T),\ x \in (0, L),$$
(2.258) $$y(t, 0) = u(t),\ t \in (0, T),$$

where, at time t, the control is $u(t) \in \mathbb{R}$ and the state is $y(t, \cdot) : (0, L) \to \mathbb{R}$.

For the Hilbert space H, we take $H := L^2(0, L)$. For the operator $A : D(A) \to H$ we take (as in Section 2.1; see (2.32) and (2.33))

$$\mathcal{D}(A) := \{f \in H^1(0, L);\ f(0) = 0\},$$
$$Af := -f_x,\ \forall f \in \mathcal{D}(A).$$

Then $\mathcal{D}(A)$ is dense in $L^2(0, L)$, A is closed, A is dissipative (see (2.36)). The adjoint A^* of A is defined by

$$\mathcal{D}(A^*) := \{f \in H^1(0, L);\ f(L) = 0\},$$
$$A^* f := f_x,\ \forall f \in \mathcal{D}(A^*).$$

By (2.37), the operator A^* is also dissipative. Hence, by Theorem A.10 on page 375, the operator A is the infinitesimal generator of a strongly continuous semigroup $S(t)$, $t \in [0, +\infty)$, of continuous linear operators on H.

For the Hilbert space U, we take $U := \mathbb{R}$. The operator $B : \mathbb{R} \to D(A^*)'$ is defined by

(2.259) $$(Bu)z = uz(0),\ \forall u \in \mathbb{R}, \forall z \in D(A^*).$$

Note that $B^* : D(A^*) \to \mathbb{R}$ is defined by
$$B^* z = z(0), \forall z \in D(A^*).$$

Let us check the regularity property (2.199). Let $z^0 \in D(A^*)$. Let
$$z \in C^0([0,T]; D(A^*)) \cap C^1([0,T]; L^2(0,L))$$
be defined by $z(t, \cdot) = S(t)^* z^0$. Inequality (2.199) is equivalent to

(2.260) $$\int_0^T z(t,0)^2 dt \leqslant C_T \int_0^L z^0(x)^2 dx.$$

Let us prove this inequality for $C_T := 1$. We have

(2.261) $$z_t = z_x, \, t \in (0,T), \, x \in (0,L),$$
(2.262) $$z(t,L) = 0, \, t \in (0,T),$$
(2.263) $$z(0,x) = z^0(x), \, x \in (0,L).$$

We multiply (2.261) by z and integrate on $[0,T] \times [0,L]$. Using (2.262), (2.263) and integrations by parts, we get

(2.264) $$\int_0^T z(t,0)^2 dt = \int_0^L z^0(x)^2 dx - \int_0^L z(T,x)^2 dx \leqslant \int_0^L z^0(x)^2 dx,$$

which shows that (2.260) holds for $C_T := 1$.

Note that there is a point which needs to be clarified: we have now two definitions of a solution to the Cauchy problem

(2.265) $$y_t + y_x = 0, \, t \in (0,T), \, x \in (0,L),$$
(2.266) $$y(t,0) = u(t),$$
(2.267) $$y(0,x) = y^0(x), \, x \in (0,L),$$

where T is given in $(0, +\infty)$, y^0 is given in $L^2(0,L)$ and u is given in $L^2(0,T)$. The first definition is the one given in Definition 2.1 on page 25. The second one is the one given in Definition 2.36 on page 53. (In fact, for the moment, there is no evidence that there exists any connection between "y is a solution in the sense of Definition 2.36" and y satisfies, in some "reasonable sense", (2.265), (2.266) and (2.267).) Let us prove that these two definitions actually lead to the same solution.

Let T be given in $(0, +\infty)$, y^0 be given in $L^2(0,L)$ and u be given in $L^2(0,T)$. Let $y \in C([0,T]; L^2(0,L))$ be a solution in the sense of Definition 2.1 on page 25. In order to prove that y is a solution in the sense of Definition 2.36 on page 53, it suffices to check that, for every $\tau \in [0,T]$ and every $z^\tau \in L^2(0,L)$,

(2.268) $$\int_0^L y(\tau,x) z^\tau(x) dx - \int_0^L y^0(x) z(\tau,x) dx = \int_0^\tau u(t) z(\tau - t, 0) dt,$$

where $z(t, \cdot) := S(t)^* z^\tau$. By density of $D(A^*)$ in $L^2(0,L)$ and the continuity of the left and the right hand sides of (2.268) with respect to z^τ for the $L^2(0,L)$-topology, it suffices to check that (2.268) holds for every $\tau \in [0,T]$ and every $z^\tau \in D(A^*)$.

Let $\tau \in [0, T]$ and $z^\tau \in D(A^*)$. Then we have

(2.269) $\quad z \in C^0([0, T]; H^1(0, L)) \cap C^1([0, T]; L^2(0, L)),$

(2.270) $\quad z_t - z_x = 0,\, t \in (0, T),\, x \in (0, L),$

(2.271) $\quad z(t, L) = 0,\, t \in (0, T),$

(2.272) $\quad z(0, x) = z^\tau(x),\, x \in (0, L).$

Let us point out that, by density and continuity arguments, (2.8) holds for every $\phi \in C^0([0, \tau]; H^1(0, L)) \cap C^1([0, \tau]; L^2(0, L))$ satisfying (2.7). We take $\phi \in C^0([0, \tau]; H^1(0, L)) \cap C^1([0, \tau]; L^2(0, L))$ defined by

(2.273) $\quad \phi(t, x) = z(\tau - t, x),\, t \in (0, \tau),\, x \in (0, L).$

Note that, by (2.271) and (2.273), (2.7) holds. Hence we have (2.8). Moreover, from (2.270) and (2.273), one gets

(2.274) $\quad \phi_t + \phi_x = 0,\, t \in (0, \tau),\, x \in (0, L).$

From (2.8) and (2.274), we have

$$-\int_0^\tau u(t)\phi(t, 0)dt + \int_0^L y(\tau, x)\phi(\tau, x)dx - \int_0^L y^0(x)\phi(0, x)dx = 0,$$

which, together with (2.272) and (2.273), gives (2.268).

Let us deal with the converse, that is: Prove that every solution in the sense of Definition 2.36 on page 53 is a solution in the sense of Definition 2.1 on page 25. Note that in fact this follows from what we have just proved (a solution in the sense of Definition 2.1 is a solution in the sense of Definition 2.36), the uniqueness of the solution to the Cauchy problem in the sense of Definition 2.36 and the existence of the solution to the Cauchy problem in the sense of Definition 2.1. Let us give a direct proof. Let $T > 0$, $u \in L^2(0, T)$, $y^0 \in L^2(0, L)$. Let $y \in C^0([0, T]; L^2(0, L))$ be a solution of the Cauchy problem (2.265)-(2.266)-(2.267) in the sense of Definition 2.36. Let $\tau \in [0, T]$. Let $\phi \in C^1([0, \tau] \times [0, L])$ be such that (2.7) holds. We want to prove that (2.8) holds. By density and continuity arguments we may assume that $\phi \in C^2([0, \tau] \times [0, L])$. Let $f \in C^1([0, \tau] \times [0, L])$ be defined by

(2.275) $\quad f(t, x) = -\phi_t(\tau - t, x) - \phi_x(\tau - t, x),\, \forall t \in [0, \tau],\, \forall x \in [0, L].$

Let $a \in C^1([0, \tau]; L^2(0, L)) \cap C^0([0, \tau]; D(A^*))$ be defined by

(2.276) $\quad a(t, \cdot) := \int_0^t S(t - s)^* f(s, \cdot)ds.$

Let $b \in C^1([0, \tau]; L^2(0, L)) \cap C^0([0, \tau]; D(A^*))$ be defined by

(2.277) $\quad b(t, x) = \phi(\tau - t, x) - a(t, x),\, \forall t \in [0, \tau],\, \forall x \in [0, L].$

One easily checks that

(2.278) $\quad b(t, \cdot) = S(t)^* \phi(\tau, \cdot),\, \forall t \in [0, \tau].$

From (2.203) and (2.278), we have

(2.279) $\quad \int_0^L y(\tau, x)\phi(\tau, x)dx - \int_0^L y^0(x)b(\tau, x)dx = \int_0^\tau u(t)b(\tau - t, 0)dt.$

Let $f(t) := f(t, \cdot)$, $t \in [0, \tau]$. From (2.203) we have, for every $t \in [0, \tau]$,

$$\text{(2.280)} \quad \int_0^L y(t,x) f(\tau - t, x) dx = \int_0^L y^0 S(t)^* f(\tau - t) dx$$
$$+ \int_0^t u(s) B^* S(t-s)^* f(\tau - t) ds.$$

Integrating (2.280) on $[0, \tau]$ and using Fubini's theorem together with (2.276), we get

$$\text{(2.281)} \quad \int_0^\tau \int_0^L y(t,x) f(\tau - t, x) dx dt = \int_0^L y^0 a(\tau, x) dx + \int_0^\tau u(t) a(\tau - t, 0) dt.$$

From (2.275), (2.277), (2.279) and (2.281), we get (2.8). This concludes the proof of the equivalence of Definition 2.1 and Definition 2.36. ∎

Let us now turn to the observability inequality (2.221) for exact controllability (a controllability type which was simply called "controllability" in Section 2.1). This inequality, with the above notation, reads

$$\text{(2.282)} \quad \int_0^T z(t,0)^2 dt \geqslant c \int_0^L z^0(x)^2 dx,$$

where $c > 0$ is independent of $z^0 \in D(A^*)$ and where $z(t, \cdot) = S(t)^* z^0$. The change of function $\tilde{z}(t, x) := z(T - t, x)$ shows that inequality (2.282) is equivalent to inequality (2.87).

REMARK 2.51. Note that the regularity property (2.260) was also already proved in Section 2.1.1; see equality (2.43), which, with the change of function $z(t, x) := y(t, L - x)$, gives (2.264) and therefore (2.260). The key estimates (regularity property and observability inequality) remain the same with the two approaches.

2.3.3.2. *The Korteweg-de Vries equation revisited.* We return to the Korteweg-de Vries control system that we have already considered in Section 2.2. Let $L > 0$. The linear control system we study is

$$\text{(2.283)} \quad y_t + y_x + y_{xxx} = 0, \, t \in (0, T), \, x \in (0, L),$$
$$\text{(2.284)} \quad y(t, 0) = y(t, L) = 0, \, y_x(t, L) = u(t), \, t \in (0, T),$$

where, at time t, the control is $u(t) \in \mathbb{R}$ and the state is $y(t, \cdot) : (0, L) \to \mathbb{R}$.

For the Hilbert space H, we take $H = L^2(0, L)$. For the operator $A : D(A) \to H$, we take (as in Section 2.1; see (2.106) and (2.107))

$$\mathcal{D}(A) := \{f \in H^3(0, L); \, f(0) = f(L) = f_x(L) = 0\},$$
$$Af := -f_x - f_{xxx}, \, \forall f \in \mathcal{D}(A).$$

Then $\mathcal{D}(A)$ is dense in $L^2(0, L)$, A is closed, A is dissipative (see (2.110)). The adjoint A^* of A is defined by

$$\mathcal{D}(A^*) := \{f \in H^3(0, L); \, f(0) = f(L) = f_x(0) = 0\},$$
$$A^* f := f_x + f_{xxx}, \, \forall f \in \mathcal{D}(A^*).$$

By (2.111), the operator A^* is also dissipative. Hence, by Theorem A.4 on page 374, the operator A is the infinitesimal generator of a strongly continuous semigroup $S(t)$, $t \in [0, +\infty)$, of continuous linear operators on $L^2(0, L)$.

For the Hilbert space U, we take $U := \mathbb{R}$. The operator $B : \mathbb{R} \to D(A^*)'$ is defined by

(2.285) $$(Bu)z = uz_x(L), \, \forall u \in \mathbb{R}, \forall z \in D(A^*).$$

Note that $B^* : D(A^*) \to \mathbb{R}$ is defined by

$$B^*z = z_x(L), \, \forall z \in D(A^*).$$

Let us check the regularity property (also called admissibility condition) (2.199). Let $z^0 \in D(A^*)$. Let

$$z \in C^0([0,T]; D(A^*)) \cap C^1([0,T]; L^2(0,L))$$

be defined by

(2.286) $$z(t, \cdot) = S(t)^* z^0.$$

The regularity property (2.199) is equivalent to

(2.287) $$\int_0^T |z_x(t,L)|^2 dt \leqslant C_T \int_0^L |z^0(x)|^2 dx.$$

From (2.286), one has

(2.288) $$z_t - z_x - z_{xxx} = 0 \text{ in } C^0([0, +\infty); L^2(0,L)),$$
(2.289) $$z(t,0) = z_x(t,0) = z(t,L) = 0, \, t \in [0, +\infty),$$
(2.290) $$z(0,x) = z^0(x), \, x \in [0,L].$$

Let $y \in C^0([0,T]; H^3(0,L)) \cap C^1([0,T]; L^2(0,L))$ be defined by

(2.291) $$y(t,x) = z(t, L-x), \, (t,x) \in [0,T] \times [0,L].$$

From (2.288), (2.289), (2.290) and (2.291), one has

(2.292) $$y_t + y_x + y_{xxx} = 0 \text{ in } C^0([0,T]; L^2(0,L)),$$
(2.293) $$y(t,L) = y_x(t,L) = y(t,0) = 0, \, t \in [0,T].$$

Hence, for every $\tau \in [0,T]$, we have (2.119) with $u = 0$. In particular, taking $\tau = T$, one has

(2.294) $$\int_0^T |y_x(t,0)|^2 dt = \int_0^L |y^0(x)|^2 dx - \int_0^L |y(T,x)|^2 dx \leqslant \int_0^L |y^0(x)|^2 dx.$$

From (2.291) and (2.294), one gets (2.287) with $C_T := 1$.

Let us now turn to the observability inequality (2.221) for the exact controllability (a controllability type which was simply called "controllability" in Section 2.2). In the case of our Korteweg-de Vries control system, this observability inequality reads

(2.295) $$\int_0^T |z_x(t,L)|^2 dx \geqslant c \int_0^L |z(0,x)|^2 dx,$$

for every $z \in C^0([0,T]; H^3(0,L)) \cap C^1([0,T]; L^2(0,L))$ satisfying (2.288) and (2.289). Defining $y \in C^0([0,T]; H^3(0,L)) \cap C^1([0,T]; L^2(0,L))$ by (2.291), one easily sees that (2.295) is indeed equivalent to (2.156).

Hence, again, the key estimates (regularity property and observability inequality) remain the same with the two approaches.

It remains to check that the two notions of solutions to the Cauchy problem

$$y_t + y_x + y_{xxx} = 0,\ t \in (0,T),\ x \in (0,L),$$
$$y(t,0) = y(t,L) = 0,\ y_x(t,L) = u(t),\ t \in (0,T),$$
$$y(0,x) = y^0(x),\ x \in (0,L),$$

i.e., the one given by Definition 2.36 on page 53 and the other given in Definition 2.21 on page 38, lead to the same solutions. This fact can be checked by proceeding as for the transport equation (see page 63).

2.4. Wave equation

Let $L > 0$, $T > 0$ and $a \in L^\infty((0,T) \times (0,L))$. We consider the linear control system

(2.296) $$y_{tt} - y_{xx} + a(t,x)y = 0,\ t \in (0,T),\ x \in (0,L),$$
(2.297) $$y(t,0) = 0,\ y_x(t,L) = u(t),\ t \in (0,T),$$

where, at time $t \in (0,T)$, the control is $u(t) \in \mathbb{R}$ and the state is

$$(y(t,\cdot), y_t(t,\cdot)) : (0,L) \to \mathbb{R}^2.$$

2.4.1. Well-posedness of the Cauchy problem. Let us first give a natural definition of solutions of the Cauchy problem

(2.298) $$y_{tt} - y_{xx} + a(t,x)y = 0,\ t \in (0,T),\ x \in (0,L),$$
(2.299) $$y(t,0) = 0, y_x(t,L) = u(t),\ t \in (0,T),$$
(2.300) $$y(0,x) = \alpha^0(x),\ y_t(0,x) = \beta^0(x),\ x \in (0,L),$$

where $\alpha^0 \in H^1_{(0)}(0,L) := \{\alpha \in H^1(0,L);\ \alpha(0) = 0\}$, $\beta^0 \in L^2(0,L)$ and $u \in L^2(0,T)$ are given. The vector space $H^1_{(0)}(0,L)$ is equipped with the scalar product

$$(\alpha_1, \alpha_2)_{H^1_{(0)}(0,L)} := \int_0^L \alpha_{1x}\alpha_{2x}dx.$$

With this scalar product, $H^1_{(0)}(0,L)$ is a Hilbert space. In order to motivate Definition 2.52 below, let us first assume that there exists a function y of class C^2 on $[0,T] \times [0,L]$ satisfying (2.298) to (2.300) in the usual sense. Let $\phi \in C^1([0,T] \times [0,L])$ be such that

(2.301) $$\phi(t,0) = 0,\ \forall t \in [0,T].$$

We multiply (2.298) by ϕ and integrate the obtained equality on $[0,\tau] \times [0,L]$, with $\tau \in [0,T]$. Using (2.299), (2.300), (2.301) and integrations by parts, one gets

(2.302) $$\int_0^L y_t(\tau,x)\phi(\tau,x)dx - \int_0^L \beta^0(x)\phi(0,x)dx$$
$$- \int_0^\tau \int_0^L (\phi_t y_t - \phi_x y_x - a\phi y)dxdt - \int_0^\tau u(t)\phi(t,L)dt = 0.$$

This equality leads to the following definition.

DEFINITION 2.52. Let $T > 0$, $\alpha^0 \in H^1_{(0)}(0,L)$, $\beta^0 \in L^2(0,L)$ and $u \in L^2(0,T)$ be given. A *solution of the Cauchy problem (2.298)-(2.299)-(2.300)* is a function $y \in L^\infty((0,T); H^1_{(0)}(0,L))$ such that $y_t \in L^\infty((0,T); L^2(0,L))$, satisfying

(2.303) $$y(0,\cdot) = \alpha^0,$$
(2.304) $$y(t,0) = 0, \, \forall t \in [0,T],$$

and such that, for every $\phi \in C^1([0,T] \times [0,L])$ satisfying (2.301), one has (2.302) for almost every $\tau \in (0,T)$.

With this definition, one has the following theorem.

THEOREM 2.53. *Let $T > 0$, $L > 0$ and $R > 0$. There exists $C := C(T,L,R) > 0$ such that, for every $a \in L^\infty((0,T) \times (0,L))$ satisfying*

(2.305) $$\|a\|_{L^\infty((0,T)\times(0,L))} \leqslant R,$$

for every $\alpha^0 \in H^1_{(0)}(0,L)$, for every $\beta^0 \in L^2(0,L)$ and for every $u \in L^2(0,T)$, the Cauchy problem (2.298)-(2.299)-(2.300) has a unique solution and this solution satisfies

(2.306) $\|y\|_{L^\infty((0,T); H^1_{(0)}(0,L))} + \|y_t\|_{L^\infty((0,T); L^2(0,L))}$
$$\leqslant C(\|\alpha^0\|_{H^1_{(0)}(0,L)} + \|\beta^0\|_{L^2(0,L)} + \|u\|_{L^2(0,T)}).$$

Moreover, $y(\cdot, L) \in L^\infty(0,T)$ is in fact in $H^1(0,T)$ and

(2.307) $$\|y_t(\cdot, L)\|_{L^2(0,T)} \leqslant C(\|\alpha^0\|_{H^1_{(0)}(0,L)} + \|\beta^0\|_{L^2(0,L)} + \|u\|_{L^2(0,T)}).$$

Finally

(2.308) $$y \in C^0([0,T]; H^1_{(0)}(0,L)) \cap C^1([0,T]; L^2(0,L)).$$

Inequality (2.307) is a hidden regularity property; see Remark 2.10 on page 32.

Proof of Theorem 2.53. We only prove the existence statement together with (2.306) and (2.307). For the uniqueness statement, one can proceed, for example, as in the proof of the uniqueness statement of Theorem 2.4 on page 27. One can also use the method explained in [**329**, (the proof of) Théorème 8.1, Chapitre 3, page 287] or [**291**, Theorem 3.1, Chapter IV, page 157]. For (2.308), see, for example, [**329**, (the proof of) Théorème 8.2, Chapitre III, page 296].
Let
$$H := \left\{ Y := \begin{pmatrix} \alpha \\ \beta \end{pmatrix}; \alpha \in H^1_{(0)}(0,L), \beta \in L^2(0,L) \right\}.$$
This vector space is equipped with the scalar product
$$\left(\begin{pmatrix} \alpha_1 \\ \beta_1 \end{pmatrix}, \begin{pmatrix} \alpha_2 \\ \beta_2 \end{pmatrix} \right)_H := (\alpha_1, \alpha_2)_{H^1_{(0)}(0,L)} + (\beta_1, \beta_2)_{L^2(0,L)}.$$
As usual we denote by $\|\cdot\|_H$ the norm in H associated to this scalar product. Let $A : \mathcal{D}(A) \subset H \to H$ be the linear operator defined by
$$\mathcal{D}(A) := \left\{ \begin{pmatrix} \alpha \\ \beta \end{pmatrix} \in H; \alpha \in H^2(0,L), \beta \in H^1_{(0)}(0,L), \alpha_x(L) = 0 \right\},$$
$$A\begin{pmatrix} \alpha \\ \beta \end{pmatrix} := \begin{pmatrix} \beta \\ \alpha_{xx} \end{pmatrix}, \, \forall \begin{pmatrix} \alpha \\ \beta \end{pmatrix} \in \mathcal{D}(A).$$

Note that

(2.309) $\quad\mathcal{D}(A)$ is dense in H,

(2.310) $\quad A$ is closed.

Moreover,

(2.311) $$(AY, Y)_H = \int_0^L (\beta_x \alpha_x + \alpha_{xx}\beta)dx = 0, \forall Y := \begin{pmatrix}\alpha\\\beta\end{pmatrix} \in \mathcal{D}(A).$$

Let A^* be the adjoint of A. One easily checks that

(2.312) $$A^* = -A.$$

From the Lumer-Phillips theorem (Theorem A.4 on page 374), (2.309), (2.310), and (2.312), A is the infinitesimal generator of a strongly continuous group of isometries $S(t)$, $t \in \mathbb{R}$, on H.

Let $T > 0$. Let us first treat the case where $a \in C^1([0,T] \times [0,L])$ and $u \in C^3([0,T])$ is such that

(2.313) $$u(0) = \dot{u}(0) = 0,$$

and

$$Y^0 := \begin{pmatrix}\alpha^0\\\beta^0\end{pmatrix} \in \mathcal{D}(A).$$

Then the map

$$f: [0,T] \times H \to H$$
$$\left(t, \begin{pmatrix}\alpha\\\beta\end{pmatrix}\right) \mapsto \left(x \mapsto \begin{pmatrix}0\\-a(t,x)\alpha(x) - \ddot{u}(t)x - a(t,x)u(t)x\end{pmatrix}\right)$$

is continuous in $t \in [0,T]$, uniformly globally Lipschitz in $(\alpha, \beta)^{\text{tr}} \in H$, and continuously differentiable from $[0,T] \times H$ into H. By a classical result on perturbations of linear evolution equations (see, for example, [**381**, Theorem 1.5, Chapter 6, page 187]) there exists $Z := (\alpha, \beta)^{\text{tr}} \in C^1([0,T]; H) \cap C^0([0,T]; \mathcal{D}(A))$ such that

(2.314) $$\frac{dZ}{dt} = AZ + \begin{pmatrix}0\\-a(t,x)\alpha - \ddot{u}(t)x - a(t,x)u(t)x\end{pmatrix},$$

(2.315) $$Z(0,\cdot) = Y^0.$$

Let $Y \in C^1([0,T]; H) \cap C^0([0,T]; H^2(0,L) \times H^1(0,L))$ be defined by

(2.316) $$Y(t,x) := Z(t,x) + \begin{pmatrix}u(t)x\\0\end{pmatrix}, \forall (t,x) \in [0,T] \times [0,L].$$

Let $\phi \in C^1([0,T] \times [0,L])$ satisfy (2.301) and let $\tau \in [0,T]$. Let y be the first component of Y. Then y is in $C^1([0,T]; H^1(0,L)) \cap C^0([0,T]; H^2(0,L))$ and satisfies (2.298), (2.299) and (2.300) in the usual sense. In particular, it is a solution of the Cauchy problem (2.298)-(2.299)-(2.300) in the sense of Definition 2.52. We multiply (2.298) by ϕ and integrate the obtained equality on $[0,\tau] \times [0,L]$. Using (2.299), (2.300), (2.301) and integrations by parts, one gets (2.302). We multiply (2.298) by y_t and integrate on $[0,L]$. Then, using an integration by parts together with (2.299), we get

(2.317) $$\dot{E}(\tau) = -\int_0^L ayy_t dx + u(\tau)y_t(\tau,L), \forall \tau \in [0,T],$$

with

$$(2.318) \quad E(\tau) := \frac{1}{2}\int_0^L (y_t^2(\tau,x) + y_x^2(\tau,x))dx.$$

Let $\tau \in [0,T]$. Let us now take $\phi := xy_x$ in (2.302) (which, by (2.308), holds for every $\tau \in [0,T]$) and integrate on $[0,\tau] \times [0,L]$. One gets, using (2.299) and integrations by parts,

$$(2.319) \quad \int_0^\tau E(t)dt + \int_0^L xy_t(\tau,x)y_x(\tau,x)dx - \int_0^L x\alpha_x^0\beta^0 dx$$
$$= -\int_0^\tau\int_0^L axy_xydxdt + \frac{L}{2}\int_0^\tau (u^2(t) + y_t^2(t,L))dt.$$

We have

$$(2.320) \quad 2|u(t)y_t(t,L)| \leqslant \varepsilon y_t(t,L)^2 + \frac{1}{\varepsilon}u(t)^2, \ \forall \varepsilon \in (0,+\infty).$$

From (2.305), (2.317), (2.318), (2.319), (2.320), Gronwall's lemma and Sobolev inequalities,

$$(2.321) \quad E(\tau) \leqslant CE(0) + C\int_0^\tau |u(t)||y_t(t,L)|dt$$
$$\leqslant \frac{1}{2}E(\tau) + CE(0) + C\int_0^\tau E(t)dt + C\int_0^\tau u^2(t)dt.$$

In (2.321) and until the end of this section, C denotes various constants which may vary from line to line and depend only on $T > 0$, $L > 0$ and $R \geqslant 0$. In particular, these constants C in (2.321) do not depend on $a \in L^\infty((0,T) \times (0,L))$ satisfying (2.305). From (2.321), we have

$$(2.322) \quad E(\tau) \leqslant CE(0) + C\int_0^\tau E(t)dt + C\int_0^\tau u^2(t)dt.$$

From Gronwall's lemma, (2.318), and (2.322), one gets

$$(2.323) \quad \int_0^L (y_x^2(\tau,x) + y_t^2(\tau,x))dx \leqslant C\int_0^L (|\alpha_x^0(x)|^2 + |\beta^0(x)|^2)dx$$
$$+ C\int_0^\tau u^2(t)dt, \ \forall \tau \in [0,T].$$

Let us point out that from (2.319) and (2.323) one gets

$$(2.324) \quad \|y_t(\cdot,L)\|_{L^2(0,T)} \leqslant C(\|\alpha^0\|_{H^1_{(0)}(0,L)} + \|\beta^0\|_{L^2(0,L)} + \|u\|_{L^2(0,T)}).$$

Now let $a \in L^\infty((0,T) \times (0,L))$ satisfy (2.305), let $u \in L^2(0,T)$ and let $(\alpha^0, \beta^0) \in H^1_{(0)}(0,L) \times L^2(0,L)$. Let $a_n \in C^1([0,T] \times [0,L])$, $u^n \in C^3([0,T])$ satisfy (2.313), $(\alpha_n^0, \beta_n^0)^{\text{tr}} \in \mathcal{D}(A)$ be such that, as $n \to \infty$,

$$(2.325) \quad a_n \to a \text{ in } L^2((0,T) \times (0,L)),$$

$$(2.326) \quad \|a_n\|_{L^\infty((0,T)\times(0,L))} \leqslant R,$$

$$(2.327) \quad u_n \to u \text{ in } L^2(0,T),$$

$$(2.328) \quad \alpha_n^0 \to \alpha^0 \text{ in } H^1(0,L) \text{ and } \beta_n^0 \to \beta^0 \text{ in } L^2(0,T).$$

Let $y_n \in C^1([0,T]; H^1(0,L)) \cap C^0([0,T]; H^2(0,L))$ be such that, in the usual sense (see above for the existence of y_n),

(2.329) $\quad y_{ntt} - y_{nxx} + a(t,x)y_n = 0, \, t \in (0,T), \, x \in (0,L),$

(2.330) $\quad y_n(t,0) = 0, \, y_{nx}(t,L) = u_n(t), \, t \in (0,T),$

(2.331) $\quad y_n(0,x) = \alpha_n^0(x), \, y_{nt}(0,x) = \beta_n^0(x), \, x \in (0,L).$

Let $\tau \in [0,T]$ and let $\phi \in C^1([0,T] \times [0,L])$ be such that (2.301) holds. From (2.329), (2.330), (2.331) and (2.301), we get (see the proof of (2.302))

(2.332) $\quad \displaystyle\int_0^L y_{nt}(\tau,x)\phi(\tau,x)dx - \int_0^L \beta_n^0(x)\phi(0,x)dx$
$$- \int_0^\tau \int_0^L (\phi_t y_{nt} - \phi_x y_{nx} - a_n \phi y_n)dxdt - \int_0^\tau u_n(t)\phi(t,L)dt = 0.$$

By (2.323),

(2.333) $\quad \displaystyle\int_0^L (y_{nx}^2(\tau,x) + y_{nt}^2(\tau,x))dx \leqslant C\int_0^L (|\alpha_{nx}^0(x)|^2 + |\beta_n^0(x)|^2)dx$
$$+ C\int_0^\tau u_n^2(t)dt, \, \forall \tau \in [0,T].$$

By (2.324),

(2.334) $\quad \|y_{nt}(\cdot,L)\|_{L^2(0,T)} \leqslant C(\|\alpha_n^0\|_{H^1_{(0)}(0,L)} + \|\beta_n^0\|_{L^2(0,L)} + \|u_n\|_{L^2(0,T)}).$

By (2.327), (2.328), (2.333), and (2.334),

(2.335) $\quad (y_n)_{n\in\mathbb{N}}$ is bounded in $L^\infty((0,T); H^1_{(0)}(0,L))$,

(2.336) $\quad (y_{nt})_{n\in\mathbb{N}}$ is bounded in $L^\infty((0,T); L^2(0,L))$,

(2.337) $\quad (y_{nt}(\cdot,L))_{n\in\mathbb{N}}$ is bounded in $L^2(0,T)$.

Hence, extracting subsequences if necessary, we may assume that there exists $y \in L^\infty((0,T); H^1_{(0)}(0,L))$, with $y_t \in L^\infty((0,T); L^2(0,L))$ and $y_t(\cdot,L) \in L^2(0,T)$, satisfying (2.306) and (2.307) such that

(2.338) $\quad y_{nx} \rightharpoonup y_x$ weakly in $L^2((0,T)\times(0,L))$ as $n\to\infty$,

(2.339) $\quad y_{nt} \rightharpoonup y_t$ weakly in $L^2((0,T)\times(0,L))$ as $n\to\infty$.

Letting $n \to \infty$ in (2.332) and using (2.325), (2.327), (2.328), (2.338) and (2.339), we get (2.302). This concludes the proof of the existence part of Theorem 2.53, together with the inequalities (2.306) and (2.307). ∎

2.4.2. Controllability. Let us now turn to the controllability of the control system (2.296)-(2.297). We start with a natural definition of controllability.

DEFINITION 2.54. Let $T > 0$. The control system (2.296)-(2.297) is *controllable in time T* if, for every $(\alpha^0, \beta^0) \in H^1_{(0)}(0,L) \times L^2(0,L)$ and every $(\alpha^1, \beta^1) \in H^1_{(0)}(0,L) \times L^2(0,L)$, there exists $u \in L^2(0,T)$ such that the solution y of the Cauchy problem (2.298)-(2.299)-(2.300) satisfies $(y(T,\cdot), y_t(T,\cdot)) = (\alpha^1, \beta^1)$.

With this definition we have the following theorem, which is due to Enrique Zuazua [**515, 516**].

THEOREM 2.55. *Let $T > 2L$. Then the control system (2.296)-(2.297) is controllable in time T. More precisely, there exists $C := C(T, L, R)$ such that, for every $a \in L^\infty((0,T) \times (0,L))$ satisfying*

(2.340) $$\|a\|_{L^\infty((0,T)\times(0,L))} \leqslant R,$$

for every $\alpha^0 \in H^1_{(0)}(0,L)$, for every $\beta^0 \in L^2(0,L)$, for every $\alpha^1 \in H^1_{(0)}(0,L)$ and for every $\beta^1 \in L^2(0,L)$, there exists $u \in L^2(0,T)$ satisfying

(2.341) $\|u\|_{L^2(0,T)} \leqslant C(\|\alpha^0\|_{H^1_{(0)}(0,L)} + \|\beta^0\|_{L^2(0,L)} + \|\alpha^1\|_{H^1_{(0)}(0,L)} + \|\beta^1\|_{L^2(0,L)})$

and such that the solution y of the Cauchy problem (2.298)-(2.299)-(2.300) satisfies $(y(T, \cdot), y_t(T \cdot)) = (\alpha^1, \beta^1)$.

In order to prove Theorem 2.55, we use the duality between controllability and observability (as in Section 2.1.2.3 for the transport equation and as in Section 2.2.2 for the Korteweg-de Vries equation).

Let $T > 0$. Let us define a linear map $\mathcal{F}_T : L^2(0,T) \to H$ in the following way. Let $u \in L^2(0,T)$. Let $y \in C^1([0,T]; L^2(0,L)) \cap C^0([0,T]; H^1(0,L))$ be the solution of the Cauchy problem

(2.342) $$y_{tt} - y_{xx} + ay = 0, \; t \in (0,T), \; x \in (0,L),$$

(2.343) $$y(t,0) = 0, \; y_x(t,L) = u(t), \; t \in (0,T),$$

(2.344) $$y(0,x) = 0, \; y_t(0,x) = 0, \; x \in (0,L).$$

Then
$$\mathcal{F}_T(u) := \begin{pmatrix} y(T, \cdot) \\ y_t(T, \cdot) \end{pmatrix}.$$

One has the following lemma, which is proved in the same way as Lemma 2.15 on page 34.

LEMMA 2.56. *The control system (2.94)-(2.95) is controllable in time T if and only if $\mathcal{F}_T$ is onto.*

Again, in order to prove that $\mathcal{F}_T$ is onto, one uses Proposition 2.16 on page 35. In order to compute the adjoint of $\mathcal{F}_T$, let us introduce some notations. Let Δ^{-1} be defined by

$$\Delta^{-1}: \begin{array}{ccl} L^2(0,L) & \to & H^2_{(0)} := \{z \in H^2(0,L); z(0) = z_x(L) = 0\} \\ f & \mapsto & z, \end{array}$$

where z is the solution of

$$z_{xx} = f, \; z(0) = z_x(L) = 0.$$

Let us consider the Cauchy problem

(2.345) $$\frac{d}{dt}\begin{pmatrix} P \\ Q \end{pmatrix} = \begin{pmatrix} Q - \Delta^{-1}(aQ) \\ P_{xx} \end{pmatrix},$$

(2.346) $$P_x(t,L) = 0, \; Q(t,0) = 0, \; t \in (0,T),$$

(2.347) $$P(T,x) = P^T(x), Q(T,x) = Q^T(x), \; x \in (0,L),$$

where $(P^T, Q^T)^{\text{tr}}$ is given in H. Let us recall that S denotes the group associated to A (see page 69). Then we adopt the following definition.

2.4. WAVE EQUATION

DEFINITION 2.57. A *solution of the Cauchy problem (2.345)-(2.346)-(2.347)* is a function $Z : [0, T] \to H$, $t \mapsto (P(t), Q(t))^{\mathrm{tr}}$ such that

$$Z \in C^0([0, T]; H),$$

$$Z(\tau_2) = S(\tau_2 - \tau_1)Z(\tau_1) - \int_{\tau_1}^{\tau_2} S(\tau_2 - t)(\Delta^{-1}(aQ), 0)^{\mathrm{tr}} dt, \; \forall (\tau_1, \tau_2) \in [0, T]^2,$$

$$Z(T) = (P^T, Q^T)^{\mathrm{tr}}.$$

Then arguing as in the proof of Theorem 2.53 on page 68 gives the following theorem.

THEOREM 2.58. *For every $(P^T, Q^T)^{\mathrm{tr}} \in H$, the Cauchy problem (2.345)-(2.346)-(2.347) has a unique solution. For such a solution, $Q(\cdot, L) \in H^{-1}(0, T)$ is in fact in $L^2(0, T)$. Moreover, there exists a constant $C := C(T, L, R)$ such that, for every $a \in L^\infty((0, T) \times (0, L))$ such that*

(2.348) $$\|a\|_{L^\infty((0,T)\times(0,L))} \leqslant R,$$

for every $(P^T, Q^T)^{\mathrm{tr}} \in H$, the solution $Z := (P, Q)^{\mathrm{tr}}$ of the Cauchy problem (2.345)-(2.346)-(2.347) satisfies

$$\|Z\|_{C^0([0,T];H)} \leqslant C(\|(P^T, Q^T)^{\mathrm{tr}}\|_H,$$

$$\|Q(\cdot, L)\|_{L^2(0,T)} \leqslant C(\|(P^T, Q^T)^{\mathrm{tr}}\|_H.$$

Now we make $\mathcal{F}_T^*$ explicit by the following lemma.

LEMMA 2.59. *Let*

$$Z^T := (P^T, Q^T)^{tr} \in H.$$

Let $Z := (P, Q)^{tr} \in C^0([0, T]; H)$ be the solution of the Cauchy problem (2.345)-(2.346)-(2.347). Then

(2.349) $$(\mathcal{F}_T^* Z^T)(t) = P_t(t, L) + (\Delta^{-1}(aQ(t, \cdot)))(L), \; t \in (0, T).$$

Proof of Lemma 2.59. Using the same approximation technique as in our proof of Theorem 2.53 on page 68, we may assume that

$$a \in C^1([0, T] \times [0, L]),$$
$$Z^T \in \mathcal{D}(A),$$
$$u \in C^3([0, T]), \; u(0) = 0, \; \dot{u}(0) = 0.$$

Let $y \in C^2([0, T]; L^2(0, L)) \cap C^1([0, T]; H^1(0, L)) \cap C^0([0, T]; H^2(0, L))$ be such that

(2.350) $$y_{tt} - y_{xx} + ay = 0,$$
(2.351) $$y(t, 0) = 0, \; y_x(t, L) = u(t), \; \forall t \in [0, T],$$
(2.352) $$y(0, \cdot) = 0, \; y_t(0, \cdot) = 0.$$

(See page 69, for the existence of such a y.) By the definitions of $\mathcal{F}_T$ and of y,

(2.353) $$\mathcal{F}_T(u) = \begin{pmatrix} y(T, \cdot) \\ y_t(T, \cdot) \end{pmatrix}.$$

Then, from (2.345), (2.346), (2.347), (2.350), (2.351) and (2.352), we get, with integrations by parts,

$$
\begin{aligned}
(Z^T, \mathcal{F}_T(u))_H &= \int_0^L P_x(T,x)y_x(T,x) + Q(T,x)y_t(T,x)dx \\
&= \int_0^T \int_0^L (P_x y_x + Q y_t)_t dx dt \\
&= \int_0^T \int_0^L (P_{tx} y_x + P_x y_{tx} + P_{xx} y_t + Q y_{xx} - aQy) dx dt \\
&= \int_0^T P_t(t,L) u(t) dt \\
&\quad + \int_0^T \int_0^L (-P_t y_{xx} - P_{xx} y_t + P_{xx} y_t + Q y_{xx} - aQy) dx dt \\
&= \int_0^T P_t(t,L) u(t) dt + \int_0^T \int_0^L ((\Delta^{-1}(aQ)) y_{xx} - aQy) dx dt \\
&= \int_0^T \left[P_t(t,L) + (\Delta^{-1}(aQ(t,\cdot)))(L) \right] u(t) dt,
\end{aligned}
$$

which concludes the proof of Lemma 2.59. ∎

The observability inequality is given in the following proposition due to Enrique Zuazua [**516**, Section 3].

PROPOSITION 2.60 (Observability inequality). *Let us assume that $T > 2L$. Then there exists $C := C(T, L, R) > 0$ such that, for every $a \in L^\infty((0,T) \times (0,L))$ such that (2.348) holds, and for every $(P^T, Q^T)^{tr} \in H$,*

(2.354) $$\|(P^T, Q^T)^{tr}\|_H \leqslant C \|\mathcal{F}_T^*(P^T, Q^T)^{tr}\|_{L^2(0,T)}.$$

Theorem 2.55 on page 72 is a corollary of Proposition 2.16 on page 35 and Proposition 2.60.

Let us prove only, instead of Proposition 2.60, the following weaker proposition (for a proof of Proposition 2.60, see [**516**, Section 3]).

PROPOSITION 2.61 (Weaker observability inequality). *Let $L > 0$ and $T > 2L$. Then there exist $R := R(T, L) > 0$ and $C^* = C^*(T, L) > 0$ such that, for every $a \in L^\infty((0,T) \times (0,L))$ such that*

(2.355) $$\|a\|_{L^\infty((0,T) \times (0,L))} \leqslant R,$$

and for every $(P^T, Q^T)^{tr} \in H$,

(2.356) $$\|(P^T, Q^T)^{tr}\|_H \leqslant C^* \|\mathcal{F}_T^*(P^T, Q^T)^{tr}\|_{L^2(0,T)}.$$

Proof of Proposition 2.61. By easy density arguments, we may assume that $a \in C^1([0,T] \times [0,L])$ and that $(P^T, Q^T)^{tr} \in \mathcal{D}(A)$. Then, if $Z := (P, Q)^{tr} \in$

2.4. WAVE EQUATION

$C^0([0,T];H)$ is the solution of the Cauchy problem (2.345)-(2.346)-(2.347),

(2.357) $\quad P \in C^0([0,T];H^2(0,L)) \cap C^1([0,T];H^1(0,L)),$

(2.358) $\quad Q \in C^0([0,T];H^1(0,L)) \cap C^1([0,T];L^2(0,L)),$

(2.359) $\quad P_t = Q - \Delta^{-1}(aQ) \text{ in } C^0([0,T];H^1(0,L)),$

(2.360) $\quad Q_t - P_{xx} = 0 \text{ in } C^0([0,T];L^2(0,L)),$

(2.361) $\quad P_x(t,L) = Q(t,0) = 0, \forall t \in [0,T].$

We now use the multiplier method (see page 36, and Remark 2.19 above for some comments on this method). We multiply (2.359) by $2xP_{xt}$ and integrate on $(0,T) \times (0,L)$. Using integrations by parts together with (2.361) we get

(2.362) $\quad 0 = L \int_0^T P_t^2(t,L)dt - \int_0^T \int_0^L (P_t^2 + P_x^2)dxdt - 2\int_0^L [QxP_x]_{t=0}^{t=T}dx$
$\quad - 2\int_0^T \int_0^L (x\Delta^{-1}(aQ))_x P_t dxdt + 2L\int_0^T (\Delta^{-1}(aQ(t,\cdot)))(L)P_t(t,L)dt.$

Let $E:[0,T] \to \mathbb{R}$ be defined by

(2.363) $\quad E(t) := \int_0^L (P_x^2(t,x) + Q^2(t,x))dx, \forall t \in [0,T].$

From (2.359), (2.360), (2.361) and (2.363), one has

(2.364) $\quad \dot{E} = 2\int_0^L aPQdx + 2P(t,0)(\Delta^{-1}(aQ)(t,\cdot))_x(0).$

From now on we assume that

(2.365) $\quad \|a\|_{L^\infty((0,T)\times(0,L))} \leqslant 1.$

We denote by C various positive constants which may depend only on T and L. In particular, these constants are independent of $a \in L^\infty((0,T) \times (0,L))$ satisfying (2.365), of (P^T, Q^T) and of $t \in [0,T]$. For $f \in L^2(0,L)$, one has the following expression of $\Delta^{-1}f$:

(2.366) $\quad (\Delta^{-1}f)(x) = -\int_0^x \int_{s_2}^L f(s_1)ds_1 ds_2.$

From (2.366), one gets the following standard estimate:

(2.367) $\quad \|\Delta^{-1}f\|_{H^2(0,L)} \leqslant C\|f\|_{L^2(0,L)}, \forall f \in L^2(0,L).$

From (2.364), (2.367), and the Poincaré inequality, one has

(2.368) $\quad |\dot{E}| \leqslant C\|a\|_{L^\infty((0,T)\times(0,L))}E.$

From (2.365), (2.368), and the Gronwall lemma, one has

(2.369) $\quad |E(t) - E(T)| \leqslant C\|a\|_{L^\infty((0,T)\times(0,L))}E(T), \forall t \in [0,T].$

From (2.359) and (2.363), one gets

(2.370)
$\quad |\int_0^L (P_t^2(t,x) + P_x^2(t,x))dx - E(t)| \leqslant C\|a\|_{L^\infty((0,T)\times(0,L))}E(t), \forall t \in [0,T].$

From (2.363), we get

(2.371) $$|2\int_0^L [QxP_x]_{t=0}^{t=T} dx| \leqslant L(E(0)+E(T)).$$

From (2.363) and (2.367), one easily checks that

(2.372) $\quad |\int_0^L (x\Delta^{-1}(aQ))_x(t,x)P_t(t,x)dx| \leqslant C\|a\|_{L^\infty((0,T)\times(0,L))}E(t), \forall t \in [0,T],$

(2.373) $\quad |(\Delta^{-1}(aQ(t,\cdot)))(L)| \leqslant C\|a\|_{L^\infty((0,T)\times(0,L))}E(t)^{1/2}, \forall t \in [0,T].$

From (2.362), (2.363), (2.365), (2.369), (2.370), (2.371), (2.372) and (2.373), one gets

$$(T-2L-C\|a\|_{L^\infty((0,T)\times(0,L))})E(T)$$
$$\leqslant L\int_0^T \left[P_t(t,L)+(\Delta^{-1}(aQ(t,\cdot)))(L)\right]^2 dt,$$

which, using Lemma 2.59 on page 73, concludes the proof of Proposition 2.61. ∎

2.4.3. Comments. There is a huge literature on the controllability of linear wave equations for any space dimension. One of the best results on this subject has been obtained by Claude Bardos, Gilles Lebeau and Jeffrey Rauch in [**34**]. See also the paper [**77**] by Nicolas Burq and Patrick Gérard and the paper [**76**] by Nicolas Burq for improvements or a simpler proof. See also the paper [**211**] by Robert Gulliver and Walter Littman for hyperbolic equations on Riemannian manifolds. Let us mention also the survey paper [**426**, Sections 3, 4 and 5] by David Russell and the books [**186**, Chapter 4] by Andrei Fursikov and Oleg Imanuvilov, [**325**] by Jacques-Louis Lions and [**277**] by Vilmos Komornik, where one can find plenty of results and useful references.

2.5. Heat equation

In this section Ω is a nonempty bounded open set of $\mathbb{R}^l$, and ω is a nonempty open subset of Ω. We consider the following linear control system:

(2.374) $\quad y_t - \Delta y = u(t,x), t \in (0,T), x \in \Omega,$

(2.375) $\quad y = 0$ on $(0,T) \times \partial\Omega,$

where, at time $t \in [0,T]$, the state is $y(t,\cdot) \in L^2(\Omega)$ and the control is $u(t,\cdot) \in L^2(\Omega)$. We require that

(2.376) $\quad u(\cdot,x) = 0, x \in \Omega \setminus \omega.$

Hence we consider the case of internal control; see Section 2.5.3 and Section 2.7 below for cases with boundary controls.

2.5.1. Well-posedness of the Cauchy problem. Again we start by giving a natural definition of a (weak) solution to the Cauchy problem associated to our control system (2.374)-(2.375), i.e., the Cauchy problem

(2.377) $\quad y_t - \Delta y = u(t,x), (t,x) \in (0,T) \times \Omega,$

(2.378) $\quad y = 0$ on $(0,T) \times \partial\Omega,$

(2.379) $\quad y(0,x) = y^0(x), x \in \Omega,$

2.5. HEAT EQUATION

where $T > 0$, $y^0 \in L^2(\Omega)$ and $u \in L^2((0,T) \times \Omega)$ are given. In order to motivate this definition, let us first assume that there exists a function y of class C^2 on $[0,T] \times \overline{\Omega}$ satisfying (2.377) to (2.379) in the usual sense. Let $\tau \in [0,T]$ and let $\phi \in C^2([0,\tau] \times \overline{\Omega})$ be such that

(2.380) $$\phi(t,x) = 0, \ \forall (t,x) \in [0,\tau] \times \partial\Omega.$$

We multiply (2.377) by ϕ and integrate the obtained equality on $[0,\tau] \times \Omega$. Using (2.378), (2.379), (2.380) and integrations by parts, one gets (at least if Ω is smooth enough)

$$-\int_0^\tau \int_\Omega (\phi_t + \Delta\phi) y \, dx \, dt - \int_0^\tau \int_\Omega u\phi \, dx \, dt$$
$$+ \int_\Omega y(\tau,x)\phi(\tau,x) \, dx - \int_\Omega y^0(x)\phi(0,x) \, dx = 0.$$

This equality leads to the following definition.

DEFINITION 2.62. Let $T > 0$, let $y^0 \in L^2(\Omega)$, and let $u \in L^2((0,T) \times \Omega)$ be given. A *solution of the Cauchy problem (2.377)-(2.378)-(2.379)* is a function $y \in C^0([0,T]; L^2(\Omega))$ such that, for every $\tau \in [0,T]$ and for every $\phi \in C^0([0,\tau]; H^1(\Omega))$ such that

(2.381) $$\phi(t,\cdot) \in H_0^1(\Omega), \ \forall t \in [0,\tau],$$

(2.382) $$\phi_t \in L^2((0,T) \times \Omega), \ \Delta\phi \in L^2((0,T) \times \Omega),$$

one has

(2.383) $$-\int_0^\tau \int_\Omega (\phi_t + \Delta\phi) y \, dx \, dt - \int_0^\tau \int_\Omega u\phi \, dx \, dt$$
$$+ \int_\Omega y(\tau,x)\phi(\tau,x) \, dx - \int_\Omega y^0(x)\phi(0,x) \, dx = 0.$$

In (2.381) and in the following, $H_0^1(\Omega)$ denotes the closure in $H^1(\Omega)$ of the set of functions $\varphi \in C^\infty(\Omega)$ with compact support. Let us recall that, if Ω is of class C^1, then

$$H_0^1(\Omega) = \{\varphi \in H^1(\Omega); \ \varphi = 0 \text{ on } \partial\Omega\}$$

(see, e.g., [3, Theorem 5.37, page 165]). With Definition (2.62), one has the following theorem.

THEOREM 2.63. *Let $T > 0$, $y^0 \in L^2(\Omega)$ and $u \in L^2((0,T) \times \Omega)$ be given. Then the Cauchy problem (2.377)-(2.378)-(2.379) has a unique solution. This solution satisfies*

(2.384) $$\|y(\tau,\cdot)\|_{L^2(\Omega)} \leqslant C(\|y^0\|_{L^2(\Omega)} + \|u\|_{L^2((0,T)\times\Omega)}), \ \forall \tau \in [0,T],$$

for some $C > 0$ which does not depend on $(y^0, u) \in L^2(\Omega) \times L^2((0,T) \times \Omega)$.

Proof of Theorem 2.63. Let

$$H := L^2(\Omega),$$

equipped with the usual scalar product. Let $A : \mathcal{D}(A) \subset H \to H$ be the linear operator defined by

$$\mathcal{D}(A) := \left\{ y \in H_0^1(\Omega); \ \Delta y \in L^2(\Omega) \right\},$$
$$Ay := \Delta y \in H.$$

Note that, if Ω is smooth enough (for example of class C^2), then
$$\mathcal{D}(A) = H_0^1(\Omega) \cap H^2(\Omega). \tag{2.385}$$
However, without any regularity assumption on Ω, (2.385) is wrong in general (see in particular [**204**, Theorem 2.4.3, page 57] by Pierre Grisvard). One easily checks that
$$\mathcal{D}(A) \text{ is dense in } L^2(\Omega), \tag{2.386}$$
$$A \text{ is closed.} \tag{2.387}$$
Moreover,
$$(Ay, y)_H = -\int_\Omega |\nabla y|^2 dx, \, \forall y \in D(A). \tag{2.388}$$
Let A^* be the adjoint of A. One easily checks that
$$A^* = A. \tag{2.389}$$
From the Lumer-Phillips theorem (Theorem A.4 on page 374), (2.386), (2.387), (2.388) and (2.389), A is the infinitesimal generator of a strongly continuous semigroup of linear contractions $S(t)$, $t \in [0, +\infty)$, on H.

There are now two strategies to prove Theorem 2.63.

(i) For the existence statement of Theorem 2.63, one can proceed as in the second proof of the existence statement of Theorem 2.4 on page 27; see also the proofs of the existence statement of Theorem 2.23 on page 39 or Theorem 2.53 on page 68. For the uniqueness statement of Theorem 2.63 one can proceed as for the uniqueness statement of Theorem 2.23 on page 39.

(ii) One can use Theorem 2.37 on page 53 on the Cauchy problem for abstract linear control systems. For the Hilbert space U we take $L^2(\omega)$. The linear map $B \in \mathcal{L}(U; D(A^*)')$ is the map which is defined by
$$(Bu)\varphi = \int_\omega u\varphi dx.$$
Note that $B \in \mathcal{L}(U; H)$. Hence the regularity property (2.199) is automatically satisfied. Therefore Theorem 2.63 follows from Theorem 2.37 on page 53 provided that one proves that the two notions of solutions to the Cauchy problem (2.377)-(2.378)-(2.379), the one given by Definition 2.36 on page 53, and the one given in Definition 2.62 on the preceding page, lead to the same solutions. This fact can be checked by proceeding as for the transport equation (see page 63).
∎

2.5.2. Controllability. Due to the smoothing effect of the heat equation, whatever $y^0 \in L^2(\Omega)$ and $u \in L^2((0,T) \times \Omega)$ satisfying (2.376) are, the solution $y : (0,T) \times \Omega \to \mathbb{R}$ of the Cauchy problem (2.377)-(2.378)-(2.379) is such that y is of class C^∞ in $(0, T] \times (\Omega \setminus \overline{\omega})$. Hence, if $\Omega \not\subset \overline{\omega}$, one cannot expect to have the controllability that we have found for the transport equation, the Korteweg-de Vries equation or the wave equation. More precisely, if $\Omega \not\subset \overline{\omega}$, there are $y^1 \in L^2(\Omega)$ such that, for every $y^0 \in L^2(\Omega)$, for every $T > 0$ and for every $u \in L^2((0,T) \times \Omega)$ satisfying (2.376), $y(T, \cdot) \neq y^1$. As proposed by Andrei Fursikov and Oleg Imanuvilov in [**185, 187**], for the heat equation, as for many equations where there is a regularizing effect, the good notion of controllability is not to go from a given state to another given state in a fixed time but to go from a given state to a given trajectory. (A related notion of controllability, in the framework of finite-dimensional control systems, has been also proposed by Jan Willems [**503**] for a different purpose.) This leads to the following definition.

DEFINITION 2.64. *The control system (2.374)-(2.375)-(2.376) is* controllable *if, for every T, for every $y^0 \in L^2(\Omega)$, for every $\hat{y} \in C^0([0,T]; L^2(\Omega))$, and for every $\hat{u} \in L^2((0,T) \times \Omega)$ such that*

(2.390) $$\hat{u}(t,x) = 0, \, (t,x) \in (0,T) \times \Omega \setminus \overline{\omega},$$

and such that $\hat{y}$ is the solution of the Cauchy problem

(2.391) $$\hat{y}_t - \Delta \hat{y} = \hat{u}(t,x), \, (t,x) \in (0,T) \times \Omega,$$
(2.392) $$\hat{y} = 0 \text{ on } (0,T) \times \partial\Omega,$$
(2.393) $$\hat{y}(0,x) = \hat{y}(0,\cdot)(x), \, x \in \Omega,$$

there exists $u \in L^2((0,T) \times \Omega)$ satisfying (2.376) such that the solution y of the Cauchy problem (2.377)-(2.378)-(2.379) satisfies

(2.394) $$y(T, \cdot) = \hat{y}(T, \cdot).$$

REMARK 2.65. Using the linearity of the control system (2.374)-(2.375)-(2.376), it is not hard to check that the notion of controllability introduced in Definition 2.64 is equivalent to what is called null controllability in Section 2.3.2; see Definition 2.39 on page 55.

One has the following theorem, which is due to Hector Fattorini and David Russell [**162**, Theorem 3.3] if $l = 1$, to Oleg Imanuvilov [**242, 243**] (see also the book [**186**] by Andrei Fursikov and Oleg Imanuvilov) and to Gilles Lebeau and Luc Robbiano [**307**] for $l > 1$.

THEOREM 2.66. *Let us assume that Ω is of class C^2 and connected. Then the control system (2.374)-(2.375)-(2.376) is controllable.*

Proof of Theorem 2.66. From Theorem 2.44 on page 56, one has the following proposition, due to Szymon Dolecki and David Russell [**146**].

PROPOSITION 2.67. *Assume that, for every $T > 0$, there exists $M > 0$ such that, for every $y^0 \in L^2(\Omega)$, the solution y of the Cauchy problem*

(2.395) $$y_t - \Delta y = 0, \, (t,x) \in (0,T) \times \Omega,$$
(2.396) $$y = 0 \text{ on } (0,T) \times \partial\Omega,$$
(2.397) $$y(0,x) = y^0(x), \, x \in \Omega,$$

satisfies

(2.398) $$\|y(T,\cdot)\|^2_{L^2(\Omega)} \leqslant M^2 \int_0^T \int_\omega y^2 dx dt.$$

Then the control system (2.374)-(2.375)-(2.376) is controllable.

Indeed, with the definitions of A, B, U and H given in Section 2.5.1, inequality (2.223) reads as inequality (2.398) with $M := c^{-1/2}$. (Inequality (2.398) is the observability inequality for the control system (2.374)-(2.375)-(2.376).)

Proof of the observability inequality (2.398). We follow [**186**, Chapter 1] by Andrei Fursikov and Oleg Imanuvilov; we establish a Carleman inequality [**83**], which implies the observability inequality (2.398). Without loss of generality, we may assume that $y^0 \in H_0^1(\Omega)$. Let y be the solution of the Cauchy problem (2.395)-(2.396)-(2.397). Let ω_0 be a nonempty open subset of ω such that the closure $\overline{\omega}_0$ of ω_0 in $\mathbb{R}^l$ is a subset of ω. The first step is the following lemma, due to Oleg Imanuvilov [**243**, Lemma 1.2] (see also [**186**, Lemma 1.1 page 4]).

LEMMA 2.68. *There exists $\psi \in C^2(\overline{\Omega})$ such that*

(2.399) $$\psi > 0 \text{ in } \Omega, \, \psi = 0 \text{ on } \partial\Omega,$$

(2.400) $$|\nabla \psi(x)| > 0, \, \forall x \in \overline{\Omega} \setminus \omega_0.$$

Proof of Lemma 2.68. Lemma 2.68 is obviously true if $l = 1$. From now on we assume that $l \geqslant 2$. Let $g \in C^2(\overline{\Omega})$ be such that:

- The set of $x \in \overline{\Omega}$ such that $\nabla g(x) = 0$ is finite and does not meet $\partial\Omega$.
- $g > 0$ on Ω and $g = 0$ on $\partial\Omega$.

The existence of such a g follows from classical arguments of Morse theory; see [**186**, pages 20–21]. Let us denote by a_i, $i \in \{1, \ldots, k\}$ the $x \in \Omega$ such that $\nabla g(x) = 0$. Let $\gamma_i \in C^\infty([0,1]; \Omega)$ be such that

(2.401) $$\gamma_i \text{ is one to one for every } i \in \{1, \ldots, k\},$$

(2.402) $$\gamma_i([0,1]) \cap \gamma_j([0,1]) = \emptyset, \, \forall (i,j) \in \{1, \ldots, k\}^2 \text{ such that } i \neq j,$$

(2.403) $$\gamma_i(0) = a_i, \, \forall i \in \{1, \ldots, k\},$$

(2.404) $$\gamma_i(1) \in \omega_0, \, \forall i \in \{1, \ldots, k\}.$$

The existence of such γ_i's follows from the connectedness of Ω. It relies on easy transversality arguments if $l \geqslant 3$ (if $l \geqslant 3$, two embedded curves which intersect can be perturbed a little bit so that they do not intersect anymore). If $l = 2$, one proceeds by induction on k and by noticing that $\Omega \setminus \Gamma$, where Γ is a finite number of disjoint embedded paths in Ω, is connected. Now let $X \in C^\infty(\mathbb{R}^n; \mathbb{R}^n)$ be such that

(2.405) $$\overline{\{x \in \mathbb{R}^n; X(x) \neq 0\}} \subset \Omega,$$

(2.406) $$X(\gamma_i(t)) = \gamma_i'(t), \, \forall i \in \{1, \ldots, k\}.$$

Let Φ be the flow associated to the vector field X, i.e., $\Phi : \mathbb{R} \times \mathbb{R}^n \to \mathbb{R}^n$, $(t,x) \mapsto \Phi(t,x)$ satisfies

$$\frac{\partial \Phi}{\partial t} = X(\Phi), \, \Phi(0,x) = x, \, \forall x \in \mathbb{R}^n.$$

From (2.406), one has

(2.407) $$\Phi(t, a_i) = \gamma_i(t), \, \forall t \in [0,1].$$

From (2.404) and (2.407), one has $\Phi(1, a_i) \in \omega_0$. Note that, for every $\tau \in \mathbb{R}$, $\Phi(\tau, \cdot)$ is a diffeomorphism of $\mathbb{R}^n$ (its inverse map is $\Phi(-\tau, \cdot)$). By (2.405), for every $\tau \in \mathbb{R}$, $\Phi(\tau, \Omega) = \Omega$ and $\Phi(\tau, \cdot)$ is equal to the identity map on a neighborhood of $\partial\Omega$. Then one easily checks that $\psi : \overline{\Omega} \to \mathbb{R}$ defined by

$$\psi(x) := g(\Phi(-1, x)), \, \forall x \in \overline{\Omega},$$

satisfies all the required properties. This concludes the proof of Lemma 2.68. ∎

2.5. HEAT EQUATION

Let us fix ψ as in Lemma 2.68. Let $\alpha : (0,T) \times \overline{\Omega} \to (0, +\infty)$ and $\phi : (0,T) \times \overline{\Omega} \to (0, +\infty)$ be defined by

$$\alpha(t,x) = \frac{e^{2\lambda \|\psi\|_{C^0(\overline{\Omega})}} - e^{\lambda \psi(x)}}{t(T-t)}, \ \forall (t,x) \in (0,T) \times \overline{\Omega}, \tag{2.408}$$

$$\phi(t,x) = \frac{e^{\lambda \psi(x)}}{t(T-t)}, \ \forall (t,x) \in (0,T) \times \overline{\Omega}, \tag{2.409}$$

where $\lambda \in [1, +\infty)$ will be chosen later on. Let $z : [0,T] \times \overline{\Omega} \to \mathbb{R}$ be defined by

$$z(t,x) := e^{-s\alpha(t,x)} y(t,x), \ (t,x) \in (0,T) \times \overline{\Omega}, \tag{2.410}$$

$$z(0,x) = z(T,x) = 0, \ x \in \overline{\Omega}, \tag{2.411}$$

where $s \in [1, +\infty)$ will be chosen later on. From (2.395), (2.408), (2.409) and (2.410), we have

$$P_1 + P_2 = P_3 \tag{2.412}$$

with

$$P_1 := -\Delta z - s^2 \lambda^2 \phi^2 |\nabla \psi|^2 z + s\alpha_t z, \tag{2.413}$$

$$P_2 := z_t + 2s\lambda \phi \nabla \psi \nabla z + 2s\lambda^2 \phi |\nabla \psi|^2 z, \tag{2.414}$$

$$P_3 := -s\lambda \phi (\Delta \psi) z + s\lambda^2 \phi |\nabla \psi|^2 z. \tag{2.415}$$

Let $Q := (0,T) \times \Omega$. From (2.412), we have

$$2 \iint_Q P_1 P_2 \, dxdt \leqslant \iint_Q P_3^2 \, dxdt. \tag{2.416}$$

Let n denote the outward unit normal vector field on $\partial \Omega$. Note that z vanishes on $[0,T] \times \partial \Omega$ (see (2.396) and (2.410)) and on $\{0,T\} \times \overline{\Omega}$ (see (2.411)). Then straightforward computations using integrations by parts lead to

$$2 \iint_Q P_1 P_2 \, dxdt = I_1 + I_2 \tag{2.417}$$

with

$$I_1 := \iint_Q (2s^3 \lambda^4 \phi^3 |\nabla \psi|^4 |z|^2 + 4s\lambda^2 \phi |\nabla \psi|^2 |\nabla z|^2) \, dxdt \tag{2.418}$$

$$- \int_0^T \int_{\partial \Omega} 2s\lambda \phi \frac{\partial \psi}{\partial n} \left(\frac{\partial z}{\partial n} \right)^2 d\sigma dt,$$

$$I_2 := \iint_Q \Big(4s\lambda (\phi \psi_i)_j z_i z_j - 2s\lambda (\phi \psi_i)_i |\nabla z|^2 \tag{2.419}$$

$$+ 2s^3 \lambda^3 \phi^3 (|\nabla \psi|^2 \psi_i)_i z^2 - 2s\lambda^2 (\phi |\nabla \psi|^2)_{ii} z^2$$

$$- s\alpha_{tt} z^2 - 2s^2 \lambda (\phi \psi_i \alpha_t)_i z^2 + 4s^2 \lambda^2 \phi \alpha_t |\nabla \psi|^2 z^2 + 2s^2 \lambda^2 \phi \phi_t |\nabla \psi|^2 z^2 \Big) dxdt.$$

In (2.419) and until the end of the proof of Theorem 2.66 on page 79, we use the usual repeated-index sum convention. By (2.400) and (2.409), there exists Λ such that, for every $\lambda \geqslant \Lambda$, we have, on $(0,T) \times (\Omega \setminus \omega_0)$,

(2.420)
$$-4\lambda^2 \phi |\nabla \psi|^2 |a|^2 \leqslant 4\lambda (\phi \psi_i)_j a_i a_j - 2\lambda (\phi \psi_i)_i |a|^2, \, \forall a = (a_1, \ldots, a_n)^{\mathrm{tr}} \in \mathbb{R}^n,$$
(2.421)
$$-\lambda^4 \phi^3 |\nabla \psi|^4 \leqslant 2\lambda^3 \phi^3 (|\nabla \psi|^2 \psi_i)_i.$$

We take $\lambda := \Lambda$. Note that, by (2.399),

(2.422)
$$\frac{\partial \psi}{\partial n} \leqslant 0 \text{ on } \partial \Omega.$$

Moreover, using (2.408) and (2.409), one gets, the existence of $C > 0$, such that, for every $(t,x) \in (0,T) \times \Omega$,

(2.423) $\quad |\alpha_{tt}| + |(\phi \psi_i \alpha_t)_i| + |\phi \alpha_t |\nabla \psi|^2| + |\phi \phi_t |\nabla \psi|^2|$
$$+ |\phi^3(|\nabla \psi|^2 \psi_i)_i| + |\phi^3 |\nabla \psi|^4| \leqslant \frac{C}{t^3(T-t)^3},$$

(2.424) $\quad |\phi(\Delta \psi)| + |\phi |\nabla \psi|^2| + |(\phi \psi_i)_i| + |(\phi \psi_i)_i| + |(\phi |\nabla \psi|^2)_{ii}| \leqslant \dfrac{C}{t(T-t)},$

(2.425) $\quad |(\phi \psi_i)_j| \leqslant \dfrac{C}{t(T-t)}, \, \forall (i,j) \in \{1,\ldots,l\}^2.$

From (2.400) and (2.409), one gets the existence of $C > 0$ such that

(2.426)
$$\frac{1}{t^3(T-t)^3} \leqslant C \phi^3 |\nabla \psi|^4 (t,x), \, \forall (t,x) \in (0,T) \times (\Omega \setminus \omega_0).$$

Using (2.416) to (2.426), we get the existence of $C > 0$ such that, for every $s \geqslant 1$ and for every y^0,

(2.427) $\quad s^3 \displaystyle\int_{(0,T)} \int_{\Omega \setminus \omega_0} \dfrac{|z|^2}{t^3(T-t)^3} dx dt \leqslant C s^2 \iint_Q \dfrac{|z|^2}{t^3(T-t)^3} dx dt$
$$+ C s^3 \int_{(0,T)} \int_{\omega_0} \frac{|\nabla z|^2 + |z|^2}{t^3(T-t)^3} dx dt.$$

Hence, for $s \geqslant 1$ large enough, there exists $c_0 > 0$ independent of y^0 such that

(2.428)
$$\int_{T/3}^{2T/3} \int_\Omega |z|^2 dx dt \leqslant c_0 \int_0^T \int_{\omega_0} \frac{|\nabla z|^2 + |z|^2}{t^3(T-t)^3} dx dt.$$

We choose such an s and such a c_0. Coming back to y using (2.408) and (2.410), we deduce from (2.428) the existence of $c_1 > 0$ independent of y^0 such that

(2.429)
$$\int_{T/3}^{2T/3} \int_\Omega |y|^2 dx dt \leqslant c_1 \int_0^T \int_{\omega_0} t(T-t)(|\nabla y|^2 + |y|^2) dx dt.$$

Let $\rho \in C^\infty(\overline{\Omega})$ be such that

$$\rho = 1 \text{ in } \omega_0,$$
$$\rho = 0 \text{ in } \overline{\Omega} \setminus \omega.$$

We multiply (2.395) by $t(T-t)\rho y$ and integrate on Q. Using (2.396) and integrations by parts, we get the existence of $c_2 > 0$ independent of y^0 such that

$$\text{(2.430)} \qquad \int_0^T \int_{\omega_0} t(T-t)(|\nabla y|^2 + |y|^2)dxdt \leqslant c_2 \int_0^T \int_\omega |y|^2 dxdt.$$

From (2.429) and (2.430), we get

$$\text{(2.431)} \qquad \int_{T/3}^{2T/3} \int_\Omega |y|^2 dxdt \leqslant c_1 c_2 \int_0^T \int_\omega |y|^2 dxdt.$$

Let us now multiply (2.395) by y and integrate on Ω. Using integrations by parts together with (2.396), we get

$$\text{(2.432)} \qquad \frac{d}{dt} \int_\Omega |y(t,x)|^2 dx \leqslant 0.$$

From (2.431) and (2.432), one gets that (2.398) holds with

$$M := \sqrt{\frac{3c_1 c_2}{T}}.$$

This concludes the proof of Theorem 2.66 on page 79. ∎

REMARK 2.69. Let us emphasize that c_1 and c_2 depend on ω, Ω, and T. Sharp estimates depending on ω, Ω, and T on the minimum of the set of constants $M > 0$ satisfying (2.398) have been obtained by Enrique Fernández-Cara and Enrique Zuazua in [**168**] and by Luc Miller in [**357**].

The interest in the minimum of the set of constants $M > 0$ satisfying (2.398) is justified in the following exercise.

EXERCISE 2.70. *Let $T > 0$ be given. Let K_1 be the minimum of the constants $M > 0$ such that inequality (2.398) holds for the solution y of the Cauchy problem (2.395)-(2.396)-(2.397) for every $y^0 \in L^2(\Omega)$. For $y^0 \in L^2(\Omega)$, let $U(y^0)$ be the set of $u \in L^2((0,T) \times \Omega)$ such that u vanishes on $(0,T) \times (\Omega \setminus \omega)$ and such that this control u steers the control system (2.374)-(2.375) from y^0 to 0 in time T, which means that the solution y of the Cauchy problem (2.377)-(2.378)-(2.379) satisfies $y(T, \cdot) = 0$. Let*

$$K_2 := \sup_{\|y^0\|_{L^2(\Omega)} \leqslant 1} \{\min\{\|u\|_{L^2(0,T)}; u \in U(y^0)\}\}.$$

*Prove that $K_1 = K_2$. (**Hint.** Just apply Theorem 2.44 on page 56).*

2.5.3. Motion planning for the one-dimensional heat equation. This section is borrowed from the paper [**295**] by Béatrice Laroche, Philippe Martin and Pierre Rouchon. Let us consider the following control heat equation:

$$\text{(2.433)} \qquad y_t - y_{xx} = 0, \; t \in (0,T), \; x \in (0,1),$$
$$\text{(2.434)} \qquad y_x(t,1) = u(t), \; t \in (0,T),$$
$$\text{(2.435)} \qquad y_x(t,0) = 0, \; t \in (0,T).$$

It models a one-dimensional rod $(0,1)$. Heat is added from a steam chest at the boundary $x = 1$, while the boundary $x = 0$ is assumed to be perfectly insulated. For the control system (2.433)-(2.434)-(2.435), the state at time t is $y(t, \cdot) : [0, 1] \to \mathbb{R}$ and the control is $u(t) \in \mathbb{R}$.

Proceeding as for Definition 2.1 on page 25, Definition 2.21 on page 38 or Definition 2.52 on page 68, one is led to adopt the following definition.

DEFINITION 2.71. Let $y^0 \in L^2(0,1)$, $T > 0$ and $u \in H^1(0,T)$. A solution of the Cauchy problem (2.433), (2.434) and (2.435) is a function $y \in C^0([0,T]; L^2(0,L))$ such that, for every $\tau \in [0,T]$ and for every

$$\varphi \in C^1([0,\tau]; L^2(0,1)) \cap C^0([0,\tau]; H^2(0,1))$$

such that

$$\varphi_x(t,0) = \varphi_x(t,1) = 0, \, \forall t \in [0,\tau],$$

one has

$$\int_0^1 y(\tau,x)\varphi(\tau,x)dx - \int_0^1 y^0(x)\varphi(0,x)dx - \int_0^\tau \int_0^1 y(\varphi_t + \varphi_{xx})dxdt$$
$$- \int_0^\tau u(t)\varphi(t,1)dt = 0.$$

One has the following theorem.

THEOREM 2.72. Let $T > 0$, $y^0 \in L^2(0,1)$ and $u \in H^1(0,T)$. Then the Cauchy problem (2.433), (2.434) and (2.435) has a unique solution y. This solution satisfies

(2.436) $\qquad \|y\|_{C^0([0,T]; L^2(0,L))} \leqslant C(T)(\|y^0\|_{L^2(0,1)} + \|u\|_{H^1(0,T)}),$

where $C(T) > 0$ is independent on y^0 and u.

Proof of Theorem 2.72. It is very similar to the proofs of Theorem 2.4 on page 27, Theorem 2.23 on page 39 or Theorem 2.53 on page 68. We briefly sketch it. Let $H := L^2(0,1)$ and let $A : \mathcal{D}(A) \to H$ be the linear operator defined by

$$\mathcal{D}(A) := \{f \in H^2(0,1); f_x(0) = f_x(1) = 0\},$$
$$Af = f_{xx}, \, \forall f \in H.$$

One easily checks that

$$\mathcal{D}(A) \text{ is dense in } H,$$
$$A \text{ is closed and dissipative,}$$
$$A = A^*.$$

Hence, by Theorem A.10 on page 375, the operator A is the infinitesimal generator of a strongly continuous semigroup $S(t)$, $t \in [0, +\infty)$ of continuous linear operators on H. Then one easily checks that $y \in C^0([0,T]; L^2(0,1))$ is a solution of the Cauchy problem (2.433), (2.434) and (2.435) if and only if

(2.437)
$$y(t, \cdot) = u(t)\theta + S(t)(y^0 - u(0)\theta) - \int_0^t S(t-s)(\dot{u}(s)\theta - u(s)\theta_{xx})ds, \, \forall t \in [0,T],$$

with

(2.438) $\qquad \theta(x) := \dfrac{x^2}{2}, \, \forall x \in [0,1].$

Inequality (2.436) follows from (2.437) and (2.438). This concludes the proof of Theorem 2.72. ∎

It has been proved by Hector Fattorini and David Russell in [**162**, Theorem 3.3] that, for every $T > 0$ the linear control system (2.433)-(2.434)-(2.435) is null controllable in time T: for every $y^0 \in L^2(0,1)$, every $\tilde{y}^0 \in L^2(0,1)$ and every $T > 0$, there exists $u \in L^2(0,T)$ such that the solution y to the Cauchy problem (2.433), (2.434) and (2.435) satisfies $y(T, \cdot) = S(T)\tilde{y}^0$. (See Definition 2.39 on page 55.) Hence, by Theorem 2.45 on page 57, the linear control system (2.433)-(2.434)-(2.435) is approximately controllable in time T, for every $T > 0$. That is, for every $\varepsilon > 0$, for every $T > 0$, for every $y^0 \in L^2(0,1)$ and for every $y^1 \in L^2(0,1)$, there exists $u \in L^2(0,T)$ such that the solution of the Cauchy problem

(2.439) $$y_t - y_{xx} = 0,\ t \in (0,T),\ x \in (0,1),$$

(2.440) $$y_x(t,1) = u(t),\ t \in (0,T),$$

(2.441) $$y_x(t,0) = 0,\ t \in (0,T),$$

(2.442) $$y(0,x) = y^0(x),\ x \in (0,1),$$

satisfies

$$\|y(T,\cdot) - y^1\|_{L^2(0,1)} \leqslant \varepsilon.$$

(See Definition 2.40 on page 55.) In fact the approximate controllability is easier to get than the null controllability and has been proved earlier than the null controllability. It goes back to Hector Fattorini [**159**].

In this section we present a method to construct u explicitly. This method is based on the notion of differential flatness due to Michel Fliess, Jean Lévine, Pierre Rouchon and Philippe Martin [**175**]. Roughly speaking a control system is differentially flat if every trajectory can be expressed in terms of a function z (called a flat output) and its derivatives. The control system (2.433)-(2.434)-(2.435) is flat and a flat output is

$$z(t) := y(t,0).$$

To see this, let us consider the following system

(2.443) $$y_{xx} = y_t,\ t \in (0,T),\ x \in (0,1),$$

(2.444) $$y_x(t,0) = 0,\ t \in (0,T),$$

(2.445) $$y(t,0) = z(t),\ t \in (0,T).$$

This system is in the Cauchy-Kovalevsky form (see e.g. [**89**, Theorem 1.1]). If one seeks formal solutions

$$y(t,x) := \sum_{i=0}^{\infty} a_i(t) \frac{x^i}{i!}$$

of (2.443)-(2.444)-(2.445), then one gets, for every $i \in \mathbb{N}$,

$$a_{2i}(t) = z^{(i)}(t),$$
$$a_{2i+1}(t) = 0.$$

Therefore, formally again,

(2.446) $$y(t,x) = \sum_{i=0}^{\infty} z^{(i)}(t) \frac{x^{2i}}{(2i)!},$$

(2.447) $$u(t) = \sum_{i=1}^{\infty} \frac{z^{(i)}(t)}{(2i-1)!}.$$

If one wants to get, instead of formal solutions, a true solution, one needs to have the convergence of the series in (2.446)-(2.447) in a suitable sense. A sufficient condition for the convergence of these series can be given in terms of the Gevrey order of z. Let us recall the following definition, due to Maurice Gevrey [**191**, page 132] (see also [**89**, Definition 1.1]).

DEFINITION 2.73. Let $z : t \in [0,T] \mapsto z(t) \in \mathbb{R}$ be of class C^∞. Then z is *Gevrey of order* $s \in [1,+\infty)$ if there exist $M > 0$ and $R > 0$ such that
$$|z^{(m)}(t)| \leqslant M \frac{(m!)^s}{R^m}, \, \forall t \in [0,T], \, \forall m \in \mathbb{N}.$$

Clearly a Gevrey function of order s is also of order s' for every $s' > s$. Note also that a Gevrey function of order 1 is analytic. In particular, every Gevrey function f of order 1 such that

(2.448) $$f^{(i)}(T) = 0 \text{ for every } i \in \mathbb{N}$$

is identically equal to 0. This is no longer the case if one considers Gevrey functions of order strictly larger than 1. For example, let $\gamma \in (0,+\infty)$, $T > 0$ and $\varphi_\gamma : [0,T] \to \mathbb{R}$ be defined by

$$\varphi_\gamma(t) := \begin{cases} 0 & \text{if } t \in \{0,T\}, \\ \exp\left(\frac{-1}{((T-t)t)^\gamma}\right) & \text{if } t \in (0,T). \end{cases}$$

Then φ_γ satisfies (2.448) and is Gevrey of order $1 + (1/\gamma)$ (see, for example, [**89**, page 16]). This implies that $\Phi_\gamma : [0,T] \to \mathbb{R}$ defined by

(2.449) $$\Phi_\gamma(t) := \frac{\int_0^t \varphi_\gamma(\tau)d\tau}{\int_0^T \varphi_\gamma(\tau)d\tau}, \, \forall t \in [0,T],$$

is also Gevrey of order $1 + (1/\gamma)$. This function Φ_γ will be used later on.

Definition 2.73 of Gevrey functions of order s on $[0,T]$ can be generalized to functions of two variables in the following way.

DEFINITION 2.74. Let $y : (t,x) \in [0,T] \times [0,1] \mapsto y(t,x) \in \mathbb{R}$ be of class C^∞. Then y is *Gevrey of order* $s_1 \in [1,+\infty)$ *in* t *and of order* $s_2 \in [1,+\infty)$ *in* x if there exist $M > 0$, $R_1 > 0$ and $R_2 > 0$ such that
$$\left|\frac{\partial^{m+n} y}{\partial t^m \partial x^n}(t,x)\right| \leqslant M \frac{(m!)^{s_1}(n!)^{s_2}}{R_1^m R_2^n}, \, \forall(t,x) \in [0,T] \times [0,1], \, \forall(m,n) \in \mathbb{N}^2.$$

Now, concerning the convergence of the series in (2.446)-(2.447), one has the following theorem.

THEOREM 2.75 ([**295**, Theorem 1]). *Assume that* $z \in C^\infty([0,T])$ *is a Gevrey function of order* $s \in [1,2)$. *Then, for every* $t \in [0,T]$, *the series in (2.446)-(2.447) are convergent. The function* y *defined by (2.446) is of class* C^∞ *and is a Gevrey function of order* s *in* t *and of order* 1 *in* x. *In particular, the control* u *defined by (2.447) is of class* C^∞ *and is a Gevrey function of order* s. *Moreover,* (y,u) *is a trajectory (i.e., a solution) of the control system (2.433)-(2.434)-(2.435).*

We omit the proof of this theorem. This theorem allows us to construct many trajectories of the control system (2.433)-(2.434)-(2.435) and in fact all the trajectories such that y is of class C^∞ and is a Gevrey function of order s in t and of order 1 in x for any given s in $[1,2)$.

2.5. HEAT EQUATION

Let us explain how to use this construction in order to move from a given state y^0 into a given neighborhood (for the $L^2(0,1)$-topology) of another given state y^1. Let y^0 and y^1 be given in $L^2(0,1)$. By the Müntz-Szasz theorem (see, for example, Theorem [**420**, Theorem 15.26, page 336]; one can alternatively use the Stone-Weierstrass theorem, see for example [**421**, 7.32 Theorem, page 162]) the set of polynomials of even degree is dense in $C^0([0,1])$ and therefore in $L^2(0,1)$. Hence, if $\varepsilon > 0$ is given, there exist two polynomials P^0 and P^1,

$$P^0(x) = \sum_{i=0}^{n} p_i^0 \frac{x^{2i}}{(2i)!}, \ p_i^0 \in \mathbb{R}, \ i \in \{0,\ldots,n\},$$

$$P^1(x) = \sum_{i=0}^{n} p_i^1 \frac{x^{2i}}{(2i)!}, \ p_i^1 \in \mathbb{R}, \ i \in \{0,\ldots,n\},$$

such that

$$\|y^0 - P^0\|_{L^2(0,1)} \leqslant \varepsilon, \ \|y^1 - P^1\|_{L^2(0,1)} \leqslant \varepsilon.$$

Let $\gamma \in (1,\infty)$ and let $T > 0$. Let $z \in C^\infty([0,T])$ be defined by

$$z(t) := \sum_{i=0}^{n} p_i^0 \frac{t^i}{i!}(1 - \Phi_\gamma(t)) + p_i^1 \frac{(t-T)^i}{i!}\Phi_\gamma(t).$$

Then z is a Gevrey function of order $1 + (1/\gamma) \in (1,2)$. Note that

(2.450) $\qquad z^{(i)}(0) = p_i^0$ and $z^{(i)}(T) = p_i^1, \ \forall i \in \{0,\ldots,n\},$

(2.451) $\qquad z^{(i)}(0) = 0$ and $z^{(i)}(T) = 0, \ \forall i \in \mathbb{N} \setminus \{0,\ldots,n\}.$

Following (2.447), let

$$u(t) := \sum_{i=1}^{\infty} \frac{z^{(i)}(t)}{(2i-1)!}, \ t \in [0,T].$$

By Theorem 2.75 on the previous page, $u(t)$ is well defined and the function u is of class C^∞ on $[0,T]$. Let $\bar{u} \in C^1([0,T])$ be any approximation of u in the following sense:

$$\|\bar{u} - u\|_{C^1([0,T])} \leqslant \varepsilon.$$

For example, one can take

$$\bar{u}(t) := \sum_{i=1}^{N} \frac{z^{(i)}(t)}{(2i-1)!}, \ t \in [0,T],$$

for a large enough N. Then one has the following theorem, which is proved in [**295**].

THEOREM 2.76. *The control $t \in [0,T] \mapsto u(t) \in \mathbb{R}$ steers the control system (2.433)-(2.434)-(2.435) from the state P^0 to the state P^1 during the time interval $[0,T]$. Moreover, there exists a constant K, which is independent of $\varepsilon \in (0,+\infty)$, $T > 0$, $y^0 \in L^2(0,1)$, $y^1 \in L^2(0,1)$, the choices of P^0, P^1 and $\bar{u}$, such that the* *solution of the Cauchy problem*

$$y_t - y_{xx} = 0,\ t \in (0,T),\ x \in (0,1),$$
$$y_x(t,0) = 0,\ t \in (0,T),$$
$$y_x(t,1) = \bar{u}(t),\ t \in (0,T),$$
$$y(0,x) = y^0(x),\ x \in (0,1),$$

satisfies

$$\|y(T,\cdot) - y^1\|_{L^2(0,1)} \leqslant K\varepsilon.$$

We omit the proof of this theorem.

REMARK 2.77. One can find further generalizations of the flatness approach to more general one-dimensional heat equations in [**295**]. This approach has also been used for many other partial differential equations. Let us mention, for example, the works

- [**150**] by François Dubois, Nicolas Petit and Pierre Rouchon and [**385**] by Nicolas Petit and Pierre Rouchon on water tanks problem (see also Section 6.3 below),
- [**152**] by William Dunbar, Nicolas Petit, Pierre Rouchon and Philippe Martin on the Stefan problem,
- [**371**] for the control of a vibrating string with an interior mass by Hugues Mounier, Joachim Rudolph, Michel Fliess and Pierre Rouchon,
- [**384**] for the control of a heavy chain by Nicolas Petit and Pierre Rouchon,
- [**416, 417**] by Pierre Rouchon for the control of a quantum particle in a potential well (see also Chapter 9 below).

REMARK 2.78. When the series (2.447) diverges, one can try to use the "smallest term summation". Such a summation technique is explained by Jean-Pierre Ramis in [**397**]. This summation has been applied in [**295**]. Numerically it leads to quite good steering controls, which turn out to be much softer. The theoretical justification remains to be done.

2.5.4. Heat control system with a finite number of controls. Let m be a positive integer and let $(f_1, \ldots f_m) \in L^2(\Omega)^m$, where Ω is again a nonempty bounded open subset of $\mathbb{R}^l$. Let us consider the control system

(2.452) $$\begin{cases} y_t(t,x) - \Delta y(t,x) = \sum_{i=1}^m u_i(t) f_i(x),\ (t,x) \in (0,T) \times \Omega, \\ y(t,x) = 0,\ (t,x) \in (0,T) \times \partial\Omega, \\ y(0,x) = y^0(x),\ x \in \Omega, \end{cases}$$

where, at time t, the state is $y(t,\cdot)$ and the control is $u(t) = (u_1(t), \ldots, u_m(t))^{\mathrm{tr}} \in \mathbb{R}^m$.

The null controllability problem consists of finding, for every $y^0 \in L^2(\Omega)$, m control functions $u_1, \ldots, u_m$ in $L^2(0,T)$ steering the control system (2.452) from y^0 to 0 in time T, i.e., such that the solution y of the Cauchy problem (2.452) satisfies $y(T,\cdot) = 0$. (For definition, existence and uniqueness of a solution to the Cauchy problem (2.452), see Section 2.5.1.) The main result of this section is that one never has null controllability if $l \geqslant 2$. More precisely, the following theorem, due to Sergei Avdonin and Sergei Ivanov [**22**, Theorem IV.1.3, page 178] holds.

THEOREM 2.79. *Let us assume that $l \geqslant 2$. Let m be a positive integer and let $f_1, \ldots, f_m \in L^2(\Omega)$. Then there exists $y^0 \in H_0^1(\Omega)$ with $\Delta y^0 \in L^2(\Omega)$ such that,*

for every $T > 0$ and for every $u_1, \ldots, u_m \in L^2(0,T)$, the solution y of the Cauchy problem (2.452) satisfies
$$y(T, \cdot) \neq 0.$$

We propose in this section a proof which is a joint work with Yacine Chitour and slightly differs from the proof given in [**22**]. We first translate the question of existence of controls $u_1, \ldots, u_m$ in $L^2(0,T)$ steering the control system (2.452) from y^0 to 0 in time T into the existence of m entire functions (i.e., holomorphic functions on $\mathbb{C}$) $\hat{u}_1, \ldots, \hat{u}_m$ solutions of an interpolation problem (2.459) described below.

Let $y^0 \in H_0^1(\Omega)$ be such that $\Delta y^0 \in L^2(\Omega)$. Assume that there exist m functions $u_1, \ldots, u_m$ in $L^2(0,T)$ such that the solution y to (2.452) satisfies $y(T, \cdot) = 0$. Let $\theta : \mathbb{R} \to [0,1]$ be a function of class C^∞ such that

(2.453)
$$\begin{cases} \theta(t) = 0, & \text{if } t \geq T, \\ \theta(t) = 1, & \text{if } t \leq 0. \end{cases}$$

Applying the change of the unknown

(2.454)
$$y(t,x) = \theta(t) y^0(x) + \tilde{y}(t,x),$$

we get that

(2.455)
$$\tilde{y}_t(t,x) - \Delta \tilde{y}(t,x) = \sum_{i=1}^m u_i(t) f_i(x) - \dot{\theta}(t) y^0(x)$$
$$+ \theta(t) \Delta y^0(x), (t,x) \in (0,T) \times \Omega,$$

(2.456) $\quad\quad\quad\quad\quad \tilde{y}(t,x) = 0, (t,x) \in (0,T) \times \partial\Omega,$

(2.457) $\quad\quad\quad\quad\quad \tilde{y}(0,x) = \tilde{y}(T,x) = 0, x \in \Omega.$

For $f \in L^1(0,T)$, the Laplace transform $\hat{f}$ of f is defined by $\hat{f}(s) := \int_0^T f(t) e^{-st} dt$, $s \in \mathbb{C}$. In this section, we adopt a slightly different definition and set $\hat{f}(s) := \int_0^T f(t) e^{st} dt$ for subsequent computational simplifications.

Consider the following Laplace transforms (with respect to the time t).

$$\hat{\theta}(s) := \int_0^T \theta(t) e^{st} dt, \quad \zeta(s,x) := \int_0^T \tilde{y}(t,x) e^{st} dt,$$
$$\hat{u}_i(s) := \int_0^T u_i(t) e^{st} dt, \ 1 \leqslant i \leqslant m.$$

Clearly $\hat{\theta}$ and $\hat{u}_i$, $i = 1, \ldots, m$ are holomorphic functions from $\mathbb{C}$ into $\mathbb{C}$ and $s \mapsto \zeta(s, \cdot)$ is a holomorphic function from $\mathbb{C}$ into $H_0^1(\Omega)$.

From (2.455), (2.456) and (2.457), we readily get that, for every $s \in \mathbb{C}$,
(2.458)
$$\begin{cases} \zeta(s, \cdot) \in H_0^1(\Omega), \\ -s\zeta(s, \cdot) - \Delta \zeta(s, \cdot) = \left(\sum_{i=1}^m \hat{u}_i(s) f_i\right) + y^0 + \hat{\theta}(s) \left(\Delta y^0 + s y^0\right) \text{ in } H^{-1}(\Omega). \end{cases}$$

As on page 77, let us denote by A the Laplace–Dirichlet operator:
$$\mathcal{D}(A) := \{v \in H_0^1(\Omega); \Delta v \in L^2(\Omega)\},$$
$$Av = \Delta v, \, \forall v \in \mathcal{D}(A).$$

It is well known that
$$A^* = A,$$
A is onto and A^{-1} is compact from $L^2(\Omega)$ into $L^2(\Omega)$.

Then the eigenvalues of A are real and the Hilbert space $L^2(\Omega)$ has a complete orthonormal system of eigenfunctions for the operator A (see, for example, [**71**, Théorème VI.11, page 97] or [**141**, Théorème 3, page 31–32] or [**267**, page 277]).

If s is an eigenvalue of $-A$, then, by the Fredholm alternative theorem, equation (2.458), where $\zeta(s,\cdot) \in H_0^1(\Omega)$ is the unknown, has a solution (if and) only if

$$(2.459) \qquad \sum_{i=1}^m \hat{u}_i(s) \int_\Omega f_i w \, dx + \int_\Omega y^0 w \, dx = 0,$$

for every eigenfunction $w \in \mathcal{D}(A)$ associated to the eigenvalue s.

We are now able to express the initial null controllability of system (2.452) in terms of a complex analysis problem. We need the following notation.

Let $\mathcal{A}_{0,B}$ be the class of entire functions g such that there exists $K_g > 0$ for which

$$(2.460) \qquad |g(s)| \leqslant K_g e^{B \max\{0, \operatorname{Re}(s)\}}, \; \forall s \in \mathbb{C}.$$

By the (easy part of the) Paley-Wiener theorem, the Laplace transform of any function in $L^2(0,T)$ belongs to $\mathcal{A}_{0,T}$. Also, let $\mathcal{A}$ be the set of holomorphic functions $f : \mathbb{C} \to \mathbb{C}$ such that, for some positive constants K and a depending on f,

$$(2.461) \qquad |f(s)| \leq K e^{a|\operatorname{Re} s|}, \; \forall s \in \mathbb{C}.$$

Clearly $\mathcal{A}_{0,T} \subset \mathcal{A}$.

Let us recall a classical result providing a sufficient condition for the nonexistence of a nonzero holomorphic function f in $\mathcal{A}$ depending on the distribution of the zeros of f.

LEMMA 2.80. *Let $f : \mathbb{C} \to \mathbb{C}$ be in $\mathcal{A}$. Let us assume that there exists a sequence $(r_k)_{k \geqslant 1}$ of distinct positive real numbers such that*

$$(2.462) \qquad \sum_{k=1}^\infty \frac{1}{r_k} = \infty,$$

$$(2.463) \qquad f(r_k) = 0.$$

Then f is identically equal to 0.

Lemma 2.80 is a consequence of a much more general theorem due to Mary Cartwright and Norman Levinson; see [**315**, Theorem 1, page 127]. It will be used repeatedly in the sequel, together with the following definition.

DEFINITION 2.81. A subset S of positive real numbers is said to be *admissible* if it contains a sequence of distinct positive real numbers $(r_k)_{k \geqslant 1}$ such that (2.462) holds.

We argue by contradiction: until the end of the proof of Theorem 2.79 we assume that

$$(2.464) \qquad \begin{cases} \forall y^0 \in \mathcal{D}(A), \exists T > 0, \exists (u_1, \ldots, u_m) \in L^2(0,T)^m \text{ such that} \\ \text{the solution } y \text{ of } (2.452) \text{ satisfies } y(T,\cdot) = 0. \end{cases}$$

The proof of Theorem 2.79 is now decomposed in two steps: the case $m = 1$ and the case $m \geqslant 2$.

2.5.4.1. *The case $m = 1$.* By (2.464), for every $y^0 \in \mathcal{D}(A)$, there exists $u \in L^2(0,T)$ such that the solution y of the Cauchy problem (2.452) satisfies

$$y(T, \cdot) = 0. \tag{2.465}$$

Using (2.459) and (2.465), we get that, for every eigenvalue s of the Laplace–Dirichlet operator $-A$ and for every eigenfunction $w : \Omega \to \mathbb{R}$ associated to s,

$$\hat{u}(s) \int_\Omega f_1 w\, dx + \int_\Omega y^0 w\, dx = 0. \tag{2.466}$$

We first claim that, for every nonzero eigenfunction w of $-A$,

$$\int_\Omega f_1 w\, dx \neq 0. \tag{2.467}$$

Indeed, if there exists a nonzero eigenfunction w of $-A$ such that

$$\int_\Omega f_1 w\, dx = 0,$$

then, from (2.466), it follows that

$$\int_\Omega y^0 w\, dx = 0, \tag{2.468}$$

but there exists $y^0 \in \mathcal{D}(A)$ which does not satisfy (2.468) (e.g., $y^0 := w$). Hence one has (2.467). Let $0 < \lambda_1 < \lambda_2 < \ldots < \lambda_j < \lambda_{j+1} < \ldots$ be the ordered sequence of the eigenvalues of the Laplace-Dirichlet operator $-A$. Let y^0 be a nonzero eigenfunction for the eigenvalue λ_1. Clearly, $y^0 \in \mathcal{D}(A)$. Moreover, if s is an eigenvalue of $-A$ different from λ_1, then

$$\int_\Omega y^0 w\, dx = 0$$

for every eigenfunction w associated to the eigenvalue s. Therefore (2.466) and (2.467) imply that $\hat{u}(\lambda_i) = 0$ for every $i \in \mathbb{N} \setminus \{0, 1\}$. Let us assume, for the moment, that the following lemma holds. (This lemma is a classical Weyl estimate; for a much more precise estimate when Ω is smooth; see the paper [**248**] by Victor Ivriĭ.)

LEMMA 2.82. *Assume that (2.464) holds. Then there exists $A > 0$ (depending on Ω) such that*

$$\lambda_j \leqslant A j^{2/l}, \ \forall j \in \mathbb{N} \setminus \{0\}. \tag{2.469}$$

By applying Lemma 2.80 and Lemma 2.82, we conclude that $\hat{u} = 0$ and hence $u = 0$. Therefore, for such a y^0, system (2.452) reduces to

$$\begin{cases} y_t(t,x) - \Delta y(t,x) = 0, & \text{if } (t,x) \in (0,T) \times \Omega, \\ y(t,x) = 0, & \text{if } (t,x) \in (0,T) \times \partial\Omega, \\ y(0,x) = y^0(x), & \text{if } x \in \Omega. \end{cases} \tag{2.470}$$

The solution to (2.470) is given by

$$y(t,x) := e^{-\lambda_1 t} y^0(x),$$

and so $y(T, \cdot) \neq 0$. This is a contradiction with the fact that $y(T, \cdot) = 0$. ∎

It remains to prove Lemma 2.82 on the previous page. This lemma holds for every $m \in \mathbb{N}$ and since we will need it to treat the case $m \geqslant 2$, we give a proof of Lemma 2.82 which works for every $m \in \mathbb{N}$.

We start with a very classical lemma on the multiplicity of the eigenvalues.

LEMMA 2.83. *Assume that (2.464) holds. Then, for every eigenvalue λ of $-A$, the multiplicity of λ is at most m.*

Indeed, if the multiplicity of the eigenvalue λ of $-A$ is strictly larger than m, there exists $w \in \mathcal{D}(A)$ such that

(2.471) $$-Aw = \lambda w,$$

(2.472) $$\int_\Omega w f_i dx = 0, \forall i \in \{1, \ldots, m\},$$

(2.473) $$w \neq 0.$$

Let y^0 be in $\mathcal{D}(A)$. Let $T > 0$ and let $u_1, \ldots, u_m$ be in $L^2(0,T)$. Let us denote by y the solution of the Cauchy problem (2.452). From (2.452), (2.471) and (2.472), one first gets

$$\frac{d}{dt} \int_\Omega y(t,x) w(x) dx = -\lambda \int_\Omega y(t,x) w(x) dx$$

and then

(2.474) $$\int_\Omega y(T,x) w(x) dx = e^{-\lambda T} \int_\Omega y^0 w dx.$$

Let us take $y^0 = w$. From (2.473) and (2.474), one gets that $y(T, \cdot) \neq 0$, in contradiction with our controllability assumption (2.464). This concludes the proof of Lemma 2.83. ∎

Let us now go back to the proof of Lemma 2.82 on the preceding page. For a bounded nonempty open subset U of $\mathbb{R}^l$, let $A_U : \mathcal{D}(A_U) \subset L^2(U) \to L^2(U)$ be the linear operator defined by

$$\mathcal{D}(A_U) := \{y \in H_0^1(U); \Delta y \in L^2(U)\},$$
$$A_U y := \Delta y \in L^2(U).$$

Let us denote by $0 < \mu_1(U) \leqslant \mu_2(U) \leqslant \ldots \leqslant \mu_j(U) \leqslant \mu_{j+1}(U) \leqslant \ldots$ the ordered sequence of the eigenvalues of the operator $-A_U$ *repeated according to their multiplicity*. Let U and U' be bounded nonempty open subsets of $\mathbb{R}^l$ such that $U \subset U'$. Extending by 0 on $U' \setminus U$ every function in $H_0^1(U)$ we have $H_0^1(U) \subset H_0^1(U')$. Then, using the classical min-max characterization of the μ_j's (see, for instance, [**401**, Theorem XIII.1, pages 76–77]) one gets

(2.475) $$\mu_j(U') \leqslant \mu_j(U), \ \forall j \in \mathbb{N} \setminus \{0\}.$$

For $\varepsilon > 0$, let $Q_\varepsilon := (0,\varepsilon)^l$. Let $\varepsilon > 0$ and $c := (c_1, \ldots, c_l) \in \mathbb{R}^n$ be such that $c + Q_\varepsilon := \{x+y; y \in Q_\varepsilon\} \subset \Omega$. By (2.475),

(2.476) $$\mu_j(\Omega) \leqslant \mu_j(c + Q_\varepsilon), \ \forall j \in \mathbb{N} \setminus \{0\}.$$

But, for every $(j_1, \ldots, j_l) \in (\mathbb{N} \setminus \{0\})^l$, the function
$$x := (x_1, \ldots, x_l) \in c + Q_\varepsilon \mapsto \prod_{k=1}^l \sin\left(\frac{\pi j_k(x_k - c_k)}{\varepsilon}\right)$$
is an eigenfunction of A_{c+Q_ε} and the corresponding eigenvalue is
$$\sum_{k=1}^l \frac{\pi^2 j_k^2}{\varepsilon^2}.$$
Hence there exists $A' > 0$ (depending on $\varepsilon > 0$) such that
(2.477) $\quad\quad\quad\quad \mu_j(c + Q_\varepsilon) \leqslant A' j^{2/l}, \ \forall j \in \mathbb{N} \setminus \{0\}.$

By Lemma 2.83 on the preceding page,
$$\lambda_j \leqslant \mu_{mj}(\Omega), \ \forall j \in \mathbb{N} \setminus \{0\},$$
which, together with (2.476) and (2.477), gives (2.469) if $A := A' m^{2/l}$. Thus Lemma 2.82 is proved. This concludes the proof of Theorem 2.79 if $m = 1$. ∎

2.5.4.2. *The case* $m \geqslant 2$. For $j > 0$, let w_j be an eigenfunction of the Laplace–Dirichlet operator $-\Delta_\Omega$ corresponding to the eigenvalue λ_j and satisfying
$$\int_\Omega w_j^2 dx = 1.$$

For every $a \in \mathcal{D}(A)$, let $(a_j)_{j>0}$ be the sequence defined by $a_j := \int_\Omega aw_j dx$. For such a function a, we use $\aleph(a)$ to denote the set given by
$$\aleph(a) := \{j \in \mathbb{N} \setminus \{0\}; \ a_j = 0\}.$$
It is clear that, for every subset S of $\mathbb{N} \setminus \{0\}$ which is not equal to $\mathbb{N} \setminus \{0\}$, there exists a nonzero element $a \in \mathcal{D}(A)$ such that $\aleph(a) = S$.

For every $a \in \mathcal{D}(A)$, let $\hat{u}_1^a, \ldots, \hat{u}_m^a$ be the Laplace transforms of controls $u_1^a, \ldots, u_m^a$ in $L^2(0, T)$ steering the control system (2.452) from a to 0 in time T. Let $U(a)$ be the column vector of complex functions of coordinates $\hat{u}_1^a, \ldots, \hat{u}_m^a$: $U(a) := (\hat{u}_1^a, \ldots, \hat{u}_m^a)^{\mathrm{tr}}$.

For $j > 0$, we define the column vector F_j of coordinates $\int_\Omega f_i w_j dx$, for $1 \leqslant i \leqslant m$. Then equation (2.459) becomes
(2.478) $\quad\quad\quad\quad\quad U(a)^{\mathrm{tr}}(\lambda_j) F_j = -a_j,$
for every $j > 0$ and $a \in \mathcal{D}(A)$.

Adapting the argument of the proof of (2.467) to the case $m \geqslant 2$, it is clear that, for every $j > 0$, F_j is a nonzero vector. Let us define m nonzero functions $a^1, \ldots, a^m$ in $\mathcal{D}(A)$ and $\aleph_l := \aleph(a^l)$, $1 \leq l \leq m$, such that
(2.479) $\quad\quad\quad\quad\quad\quad\quad \aleph_l \subset \aleph_{l-1},$

(2.480) $\ \forall l \in \{1, \ldots, m\}, \exists \delta > 0$ such that
$$\#\{j; j \leqslant k, j \in \aleph_l\} \geqslant \delta k, \ \forall k \in \mathbb{N} \text{ with } k \geqslant 1/\delta,$$

(2.481) $\ \forall l \in \{1, \ldots, m\}, \exists \delta > 0$ such that
$$\#\{j; j \leqslant k, j \in \aleph_{l-1} \setminus \aleph_l\} \geqslant \delta k, \ \forall k \in \mathbb{N} \text{ with } k \geqslant 1/\delta.$$
Here, $\aleph_0 = \mathbb{N} \setminus \{0\}$. A possible choice is $\aleph_l = (2^l \mathbb{N}) \setminus \{0\}$, for $1 \leqslant l \leqslant m$.

Let us define the $m \times m$ matrix of complex functions V with rows $V_l := U(a^l)^{\mathrm{tr}}$, for $1 \leqslant l \leqslant m$. For $j \in \aleph_m$, the equations (2.478) obtained for each a^l, $1 \leqslant l \leqslant m$, can be written at once by using V, namely

(2.482) $$V(\lambda_j) F_j = 0.$$

Since F_j is nonzero, this implies that det $V(\lambda_j) = 0$. From (2.480) for $l = m$ and Lemma 2.82 on page 91 we get that $\{\lambda_j; j \in \aleph_m\}$ is admissible (see Definition 2.81 on page 90). Moreover, det V, as the $\hat{u}_1^a, \ldots, \hat{u}_m^a$, is in $\mathcal{A}$, where $\mathcal{A}$ is defined on page 90. Then the holomorphic function det V verifies the assumptions of Lemma 2.80 on page 90, and therefore is identically zero.

We obtain that, for every $j \in \aleph_{m-1} \setminus \aleph_m$, the rows $U(a^l)^{\mathrm{tr}}(\lambda_j)$, for $1 \leqslant l \leqslant m$, are linearly dependent. This translates into the existence of $\alpha_j^1, \ldots, \alpha_j^m$, not all equal to zero, such that $R_m(\lambda_j) := \sum_{l=1}^{m} \alpha_j^l U(a^l)^{\mathrm{tr}}(\lambda_j) = 0$. Using (2.478), $j \in \aleph_{m-1}$ and $R_m(\lambda_j) \cdot F_j = 0$, we get that $0 = \alpha_j^m U(a^m)^{\mathrm{tr}}(\lambda_j) F_j = -\alpha_j^m a_j^m$. This last equality implies that $\alpha_j^m = 0$ (since $j \notin \aleph_m$). Therefore, for every $j \in \aleph_{m-1} \setminus \aleph_m$, the rows $U(a^l)^{\mathrm{tr}}(\lambda_j)$, for $1 \leqslant l \leqslant m - 1$, are linearly dependent. This fact can be expressed as follows. Let V_m be the $(m-1) \times m$ matrix of complex functions with rows given by $U(a^l)^{\mathrm{tr}}$, $1 \leqslant l \leqslant m-1$. Then, for every $j \in \aleph_{m-1} \setminus \aleph_m$, the rank of $V_m(\lambda_j)$ is less than $m - 1$. Equivalently, that means that, for every $j \in \aleph_{m-1} \setminus \aleph_m$ and every minor M of V_m of order $m - 1$ (viewed as a holomorphic function), $M(\lambda_j) = 0$. But, using (2.481) for $l = m - 1$ and Lemma 2.82 on page 91, it is then easy to see that every such minor M verifies the assumptions of Lemma 2.80 on page 90. Therefore, every minor M of V_m of order $m - 1$ is equal to zero (as a holomorphic function), i.e., for every $s \in \mathbb{C}$, the rank of $V_m(s)$ is less than $m - 1$. This implies that, for every $j \in \aleph_{m-2} \setminus \aleph_{m-1}$, the rows $U(a^l)^{\mathrm{tr}}(\lambda_j)$, $1 \leqslant l \leqslant m - 1$, are linearly dependent.

Repeating the previous construction, we arrive, in $m - 1$ steps, at the fact that the rank of the row vector $U(a^1)$ is less than one, i.e., $U(a^1)$ is the zero function. This implies, by (2.478), that $a_j^1 = 0$ for every $j \geqslant 0$, contradicting the fact that a^1 is not equal to zero. The proof of Theorem 2.79 is complete. ∎

REMARK 2.84. The above method can also be used to get other obstructions to controllability. For example, it allows us to prove in [**91**] that, for generic bounded open subset $\Omega \in \mathbb{R}^l$ with $l \geqslant 2$, the steady-state controllability for the heat equation with boundary controls dependent only on time does not hold.

REMARK 2.85. The noncontrollability result stated in Theorem 2.79 on page 88 is not directly due to the dimension l but to the growth of the eigenvalues of the Laplace-Dirichlet operator. It turns out that the growth of the eigenvalues is linked to the dimension l. But, even in dimension 1, there are partial differential equations for which the growth of the eigenvalues is again too slow to have controllability with a finite number of controls. This is, for example, the case for suitable fractional powers of the Laplacian in dimension 1 on a finite interval or for the Laplacian on the half line. For these cases, one gets again obstruction to null controllability with a finite number of controls. These results are due to Sorin Micu and Enrique Zuazua. In fact, they have gotten stronger negative results, showing that, for very few states the control system can be steered from these states to zero in finite time. See [**354**] for the case where the domain Ω is a half line, [**355**] for the case of a half space, and [**356**] for a fractional order parabolic equation.

2.6. A one-dimensional Schrödinger equation

Let I be the open interval $(-1,1)$. For $\gamma \in \mathbb{R}$, let $A_\gamma : D(A_\gamma) \subset L^2(I;\mathbb{C}) \to L^2(I;\mathbb{C})$ be the operator defined on

(2.483) $$D(A_\gamma) := H^2(I;\mathbb{C}) \cap H_0^1(I;\mathbb{C})$$

by

(2.484) $$A_\gamma \varphi := -\varphi_{xx} - \gamma x \varphi.$$

In (2.483) and in the following,

$$H_0^1(I;\mathbb{C}) := \{\varphi \in H^1((0,L);\mathbb{C}); \varphi(0) = \varphi(L)\},$$

as usual. We denote by $\langle \cdot, \cdot \rangle$ the usual Hermitian scalar product in the Hilbert space $L^2(I;\mathbb{C})$:

(2.485) $$\langle \varphi, \psi \rangle := \int_I \varphi(x)\overline{\psi(x)} dx.$$

Let us recall that, in (2.485) and throughout the whole book, $\overline{z}$ denotes the complex conjugate of the complex number z. Note that

(2.486) $$\mathcal{D}(A_\gamma) \text{ is dense in } L^2(I;\mathbb{C}),$$

(2.487) $$A_\gamma \text{ is closed,}$$

(2.488) $$A_\gamma^* = A_\gamma, \text{ (i.e., } A_\gamma \text{ is self-adjoint),}$$

(2.489) $$A_\gamma \text{ has compact resolvent.}$$

Let us recall that (2.489) means that there exists a real α in the resolvent set of A_γ such that the operator $(\alpha \text{Id} - A_\gamma)^{-1}$ is compact from $L^2(I;\mathbb{C})$ into $L^2(I;\mathbb{C})$, where Id denotes the identity map on H (see, for example, [**267**, pages 36 and 187]). Then (see, for example, [**267**, page 277]), the Hilbert space $L^2(I;\mathbb{C})$ has a complete orthonormal system $(\varphi_{k,\gamma})_{k \in \mathbb{N} \setminus \{0\}}$ of eigenfunctions for the operator A_γ:

$$A_\gamma \varphi_{k,\gamma} = \lambda_{k,\gamma} \varphi_{k,\gamma},$$

where $(\lambda_{k,\gamma})_{k \in \mathbb{N} \setminus \{0\}}$ is an increasing sequence of positive real numbers. Let $\mathbb{S}$ be the unit sphere of $L^2(I;\mathbb{C})$:

(2.490) $$\mathbb{S} := \{\phi \in L^2(I;\mathbb{C}); \int_I |\phi(x)|^2 dx = 1\}$$

and, for $\phi \in \mathbb{S}$, let $T_\mathbb{S} \phi$ be the tangent space to $\mathbb{S}$ at ϕ:

(2.491) $$T_\mathbb{S} \phi := \{\Phi \in L^2(I;\mathbb{C}); \Re \langle \Phi, \phi \rangle = 0\},$$

where, as usual, $\Re z$ denotes the real part of the complex number z.

In this section we consider the following linear control system:

(2.492) $$\Psi_t = i\Psi_{xx} + i\gamma x \Psi + iux\psi_{1,\gamma}, \ (t,x) \in (0,T) \times I,$$

(2.493) $$\Psi(t,-1) = \Psi(t,1) = 0, \ t \in (0,T),$$

where

(2.494) $$\psi_{1,\gamma}(t,x) := e^{-i\lambda_{1,\gamma} t} \varphi_{1,\gamma}(x), \ (t,x) \in (0,T) \times I.$$

This is a control system where, at time $t \in [0,T]$:
- The state is $\Psi(t,\cdot) \in L^2(I;\mathbb{C})$ with $\Psi(t,\cdot) \in T_\mathbb{S}(\psi_{1,\gamma}(t,\cdot))$.
- The control is $u(t) \in \mathbb{R}$.

The physical interest of this linear control system is motivated in Section 4.2.2 (see in particular the linear control system (4.87)-(4.88)) and in Chapter 9 (see in particular the control system (Σ_γ^l) defined on page 261).

Let us first deal with the Cauchy problem

(2.495) $$\Psi_t = i\Psi_{xx} + i\gamma x\Psi + iux\psi_{1,\gamma},\, (t,x) \in (0,T) \times I,$$
(2.496) $$\Psi(t,-1) = \Psi(t,1) = 0,\, t \in (0,T),$$
(2.497) $$\Psi(0,x) = \Psi^0(x),$$

where $T > 0$, $u \in L^1(0,T)$ and $\Psi^0 \in L^2(I;\mathbb{C})$ are given. By (2.488),

$$(-iA_\gamma)^* = -(-iA_\gamma).$$

Therefore, by Theorem A.16 on page 377, $-iA_\gamma$ is the infinitesimal generator of a strongly continuous group of linear isometries on $L^2(I;\mathbb{C})$. We denote by $S_\gamma(t)$, $t \in \mathbb{R}$, this group.

Our notion of solution to the Cauchy problem (2.495)-(2.496)-(2.497) is given in the following definition (see Theorem A.7 on page 375).

DEFINITION 2.86. Let $T > 0$, $u \in L^1(0,T)$ and $\Psi^0 \in L^2(I;\mathbb{C})$. A *solution* $\Psi : [0,T] \times I \to \mathbb{C}$ *to the Cauchy problem (2.495)-(2.496)-(2.497)* is the function $\Psi \in C^0([0,T]; L^2(I;\mathbb{C}))$ defined by

(2.498) $$\Psi(t) = S_\gamma(t)\Psi^0 + \int_0^t S_\gamma(t-\tau)iu(\tau)x\psi_{1,\gamma}(\tau,\cdot)d\tau.$$

Let us now turn to the controllability of our control system (2.492)-(2.493). Let

(2.499) $$H^3_{(0)}(I;\mathbb{C}) := \{\psi \in H^3(I;\mathbb{C});\, \psi(-1) = \psi_{xx}(-1) = \psi(1) = \psi_{xx}(1) = 0\}.$$

The goal of this section is to prove the following controllability result due to Karine Beauchard [**40**, Theorem 5, page 862].

THEOREM 2.87. *There exists $\gamma_0 > 0$ such that, for every $T > 0$, for every $\gamma \in (0,\gamma_0]$, for every $\Psi^0 \in T_\mathbb{S}\psi_{1,\gamma}(0,\cdot) \cap H^3_{(0)}(I;\mathbb{C})$ and for every $\Psi^1 \in T_\mathbb{S}\psi_{1,\gamma}(T,\cdot) \cap H^3_{(0)}(I;\mathbb{C})$, there exists $u \in L^2(0,T)$ such that the solution of the Cauchy problem*

(2.500) $$\Psi_t = i\Psi_{xx} + i\gamma x\Psi + iu(t)x\psi_{1,\gamma},\, t \in (0,T),\, x \in I,$$
(2.501) $$\Psi(t,-1) = \Psi(t,1) = 0,\, t \in (0,T),$$
(2.502) $$\Psi(0,x) = \Psi^0(x),\, x \in I,$$

satisfies

(2.503) $$\Psi(T,x) = \Psi^1(x),\, x \in I.$$

We are also going to see that the conclusion of Theorem 2.87 does not hold for $\gamma = 0$ (as already noted by Pierre Rouchon in [**417**]).

Proof of Theorem 2.87. Let $T > 0$,

$$\Psi^0 \in T_\mathbb{S}(\psi_{1,\gamma}(0,\cdot)) \text{ and } \Psi^1 \in T_\mathbb{S}(\psi_{1,\gamma}(T,\cdot)).$$

Let $u \in L^2(0,T)$. Let Ψ be the solution of the Cauchy problem (2.500)-(2.501)-(2.502). Let us decompose $\Psi(t,\cdot)$ in the complete orthonormal system $(\varphi_{k,\gamma})_{k \in \mathbb{N}\setminus\{0\}}$

of eigenfunctions for the operator A_γ:

$$\Psi(t,\cdot) = \sum_{k=1}^{\infty} y_k(t)\varphi_{k,\gamma}.$$

Taking the Hermitian product of (2.500) with $\varphi_{k,\gamma}$, one readily gets, using (2.501) and integrations by parts,

(2.504) $$\dot{y}_k = -i\lambda_{k,\gamma} y_k + ib_{k,\gamma} u(t) e^{-i\lambda_{1,\gamma} t},$$

with

(2.505) $$b_{k,\gamma} := \langle \varphi_{k,\gamma}, x\varphi_{1,\gamma} \rangle \in \mathbb{R}.$$

Note that (2.502) is equivalent to

(2.506) $$y_k(0) = \langle \Psi^0, \varphi_{k,\gamma} \rangle, \forall k \in \mathbb{N} \setminus \{0\}.$$

From (2.505) and (2.506) one gets

(2.507) $$y_k(T) = e^{-i\lambda_{k,\gamma} T}\left(\langle \Psi^0, \varphi_{k,\gamma} \rangle + ib_{k,\gamma} \int_0^T u(t)e^{i(\lambda_{k,\gamma}-\lambda_{1,\gamma})t}dt\right).$$

By (2.507), (2.503) is equivalent to the following so-called moment problem on u:

(2.508) $$b_{k,\gamma} \int_0^T u(t)e^{i(\lambda_{k,\gamma}-\lambda_{1,\gamma})t}dt$$
$$= i\left(\langle \Psi^0, \varphi_{k,\gamma} \rangle - \langle \Psi^1, \varphi_{k,\gamma} \rangle e^{i\lambda_{k,\gamma} T}\right), \forall k \in \mathbb{N} \setminus \{0\}.$$

Let us now explain why for $\gamma = 0$ the conclusion of Theorem 2.87 does not hold. Indeed, one has

(2.509) $$\varphi_{n,0}(x) := \sin(n\pi x/2), n \in \mathbb{N} \setminus \{0\}, \text{ if } n \text{ is even,}$$
(2.510) $$\varphi_{n,0}(x) := \cos(n\pi x/2), n \in \mathbb{N} \setminus \{0\}, \text{ if } n \text{ is odd.}$$

In particular, $x\varphi_{1,0}\varphi_{k,0}$ is an odd function if k is odd. Therefore

$$b_{k,0} = 0 \text{ if } k \text{ is odd.}$$

Hence, by (2.508), if there exists k odd such that

$$\langle \Psi^0, \varphi_{k,0} \rangle - \langle \Psi^1, \varphi_{k,0} \rangle e^{i\lambda_{k,0} T} \neq 0,$$

there is no control $u \in L^2(0,T)$ such that the solution of the Cauchy problem (2.500)-(2.501)-(2.502) (with $\gamma = 0$) satisfies (2.503).

Let us now turn to the case where γ is small but not 0. Since Ψ^0 is in $T_\mathbb{S}(\psi_{1,\gamma}(0,\cdot))$,

(2.511) $$\Re\langle \Psi^0, \varphi_{1,\gamma} \rangle = 0.$$

Similarly, the fact that Ψ^1 is in $T_\mathbb{S}(\psi_{1,\gamma}(T,\cdot))$ tells us that

(2.512) $$\Re(\langle \Psi^1, \varphi_{1,\gamma} \rangle e^{i\lambda_{1,\gamma} T}) = 0.$$

The key ingredient to prove Theorem 2.87 is the following theorem.

THEOREM 2.88. *Let $(\mu_i)_{i\in\mathbb{N}\setminus\{0\}}$ be a sequence of real numbers such that*

(2.513) $$\mu_1 = 0,$$

(2.514) *there exists $\rho > 0$ such that $\mu_{i+1} - \mu_i \geqslant \rho$, $\forall i \in \mathbb{N} \setminus \{0\}$.*

Let $T > 0$ be such that

(2.515) $$\lim_{x \to +\infty} \frac{N(x)}{x} < \frac{T}{2\pi},$$

where, for every $x > 0$, $N(x)$ is the largest number of μ_j's contained in an interval of length x. Then there exists $C > 0$ such that, for every sequence $(c_k)_{k\in\mathbb{N}\setminus\{0\}}$ of complex numbers such that

(2.516) $$c_1 \in \mathbb{R},$$

(2.517) $$\sum_{k=1}^{\infty} |c_k|^2 < \infty,$$

there exists a (real-valued) function $u \in L^2(0,T)$ such that

(2.518) $$\int_0^T u(t) e^{i\mu_k t} dt = c_k, \forall k \in \mathbb{N} \setminus \{0\},$$

(2.519) $$\int_0^T u(t)^2 dt \leqslant C \sum_{k=1}^{\infty} |c_k|^2.$$

REMARK 2.89. Theorem 2.88 is due do Jean-Pierre Kahane [**262**, Theorem III.6.1, page 114]; see also [**53**, pages 341–365] by Arne Beurling. See also, in the context of control theory, [**422**, Section 3] by David Russell who uses prior works [**246**] by Albert Ingham, [**400**] by Ray Redheffer and [**439**] by Laurent Schwartz. For a proof of Theorem 2.88, see, for example, [**282**, Section 1.2.2], [**280**, Chapter 9] or [**22**, Chapter II, Section 4]. Improvements of Theorem 2.88 have been obtained by Stéphane Jaffard, Marius Tucsnak and Enrique Zuazua in [**252, 253**], by Stéphane Jaffard and Sorin Micu in [**251**], by Claudio Baiocchi, Vilmos Komornik and Paola Loreti in [**27**] and by Vilmos Komornik and Paola Loreti in [**279**] and in [**280**, Theorem 9.4, page 177].

Note that, by (2.511) and (2.512),

(2.520) $$i(\langle \Psi^0, \varphi_{1,\gamma}\rangle - \langle \Psi^1, \varphi_{1,\gamma}\rangle e^{i\lambda_{1,\gamma}T}) \in \mathbb{R}.$$

Hence, in order to apply Theorem 2.87 to our moment problem, it remains to estimate $\lambda_{k,\gamma}$ and $b_{k,\gamma}$. This is done in the following propositions, due to Karine Beauchard.

PROPOSITION 2.90 ([**40**, Proposition 41, pages 937–938]). *There exist $\gamma_0 > 0$ and $C_0 > 0$ such that, for every $\gamma \in [-\gamma_0, \gamma_0]$ and for every $k \in \mathbb{N} \setminus \{0\}$,*

$$\left|\lambda_{k,\gamma} - \frac{\pi^2 k^2}{4}\right| \leqslant C_0 \frac{\gamma^2}{k}.$$

PROPOSITION 2.91 ([**40**, Proposition 1, page 860]). *There exist $\gamma_1 > 0$ and $C > 0$ such that, for every $\gamma \in (0, \gamma_1]$ and for every even integer $k \geqslant 2$,*

$$\left|b_{k,\gamma} - \frac{(-1)^{\frac{k}{2}+1} 8k}{\pi^2 (k^2-1)^2}\right| < \frac{C\gamma}{k^3},$$

and for every odd integer $k \geqslant 3$,

$$\left| b_{k,\gamma} - \gamma \frac{2(-1)^{\frac{k-1}{2}}(k^2+1)}{\pi^4 k(k^2-1)^2} \right| < \frac{C\gamma^2}{k^3}.$$

It is a classical result that

$$\varphi := \sum_{k=1}^{+\infty} d_k \varphi_{k,\gamma} \in H^3_{(0)}(I;\mathbb{C})$$

if and only if

$$\sum_{k=1}^{+\infty} k^6 |d_k|^2 < +\infty.$$

Hence, Theorem 2.87 readily follows from Theorem 2.88 applied to the moment problem (2.508) with the help of Proposition 2.90 and Proposition 2.91. ∎

2.7. Singular optimal control: A linear 1-D parabolic-hyperbolic example

In this section, which is borrowed from our joint work [**124**] with Sergio Guerrero, we consider the problem of the null controllability of a family of 1-D linear parabolic control equations depending on two parameters, namely the viscosity and the coefficient of the transport term. We study the dependence, with respect to these parameters and the time of controllability, of the norm of the optimal controls. In particular, we give estimates on the optimal control as the viscosity tends to 0. Let $(\varepsilon, T, L, M) \in (0,+\infty)^3 \times \mathbb{R}$. We consider the following parabolic linear control system

(2.521) $\quad y_t - \varepsilon\, y_{xx} + M\, y_x = 0, \, t \in (0,T), \, x \in (0,L),$

(2.522) $\quad y(t,0) = u(t), \, y(t,L) = 0, \, t \in (0,T),$

where, at time t, the state is $y(t,\cdot) : (0,L) \to \mathbb{R}$ and the control is $u(t) \in \mathbb{R}$.

This section is organized as follows:

- In Section 2.7.1, we study the Cauchy problem associated to the control system (2.521)-(2.522). To achieve this goal, we use the abstract framework detailed in Section 2.3.
- In Section 2.7.2 we study the null controllability of system (2.521)-(2.522) and the dependence of the cost of the null controllability of system (2.521)-(2.522) with respect to the four parameters ε, T, L, M.

2.7.1. Well-posedness of the Cauchy problem.
This section concerns the following Cauchy problem:

(2.523) $\quad y_t - \varepsilon\, y_{xx} + M\, y_x = 0, \, t \in (0,T), \, x \in (0,L),$

(2.524) $\quad y(t,0) = u(t), \, y(t,L) = 0, \, t \in (0,T),$

(2.525) $\quad y(0,x) = y^0(x), \, x \in (0,L).$

Let us recall that

$$H^{-1}(0,L) := \{\xi_x; \, \xi \in L^2(0,L)\} \subset \mathcal{D}'(0,L),$$

where $\mathcal{D}'(0,L)$ denotes the set of distributions on $(0,L)$. Let us denote by J the following linear map

$$
\begin{aligned}
H^{-1}(0,L) &\to H_0^1(0,L) \\
f &\mapsto \alpha,
\end{aligned}
\tag{2.526}
$$

where $H_0^1(0,L) := \{\beta \in H^1(0,L); \beta(0) = \beta(L) = 0\}$ and $\alpha \in H_0^1(0,L)$ is such that

$$-\alpha_{xx} = f \text{ in } \mathcal{D}'(0,L).$$

The vector space $H^{-1}(0,L)$ is equipped with the following scalar product

$$(f,g)_{H^{-1}(0,L)} = \int_0^L (Jf)_x (Jg)_x dx, \; \forall (f,g) \in H^{-1}(0,L) \times H^{-1}(0,L). \tag{2.527}$$

Equipped with this scalar product $H^{-1}(0,L)$ is a Hilbert space and J is a surjective isometry between $H^{-1}(0,L)$ and $H_0^1(0,L)$ provided that $H_0^1(0,L)$ is equipped with the scalar product

$$(\alpha,\beta)_{H_0^1(0,L)} = \int_0^L \alpha_x \beta_x dx, \; \forall (\alpha,\beta) \in H_0^1(0,L) \times H_0^1(0,L).$$

Note that, by an integration by parts, it follows from (2.527) that

$$(f,g)_{H^{-1}(0,L)} = \int_0^L (Jf) g \, dx, \; \forall (f,g) \in H^{-1}(0,L) \times L^2(0,L). \tag{2.528}$$

Let A be the unbounded linear operator on $H^{-1}(0,L)$ defined by

$$D(A) := H_0^1(0,L), \tag{2.529}$$
$$Af := \varepsilon f_{xx} - M f_x, \forall f \in D(A). \tag{2.530}$$

Then the operator is densely defined and closed. Let $f \in D(A) \cap H^2(0,L)$ and $\alpha = Jf$. One has, using (2.528) and integrations by parts,

$$
\begin{aligned}
(Af, f)_{H^{-1}(0,L)} &= \int_0^L (\varepsilon f_{xx} - M f_x) \alpha \, dx \\
&= -\varepsilon \int_0^L f^2 dx - M \int_0^L \alpha_{xx} \alpha_x dx \\
&= -\varepsilon \int_0^L f^2 dx - \frac{M}{2} (\alpha_x(L)^2 - \alpha_x(0)^2).
\end{aligned}
\tag{2.531}
$$

Since $\alpha \in H_0^1(0,L) \cap H^2(0,L)$ there exists $\xi \in [0,L]$ such that $\alpha_x(\xi) = 0$. Hence, for every $s \in [0,L]$,

$$\alpha_x(s)^2 \leqslant 2 \int_0^L |\alpha_x \alpha_{xx}| dx \leqslant \frac{\varepsilon}{|M|+1} \int_0^L f^2 dx + \frac{|M|+1}{\varepsilon} \int_0^L \alpha_x^2 dx. \tag{2.532}$$

From (2.531) and (2.532), one gets, for every $f \in D(A) \cap H^2(0,L)$,

$$(Af,f)_{H^{-1}(0,L)} \leqslant |M| \frac{|M|+1}{\varepsilon} \int_0^L \alpha_x^2 dx = |M| \frac{|M|+1}{\varepsilon} \|f\|_{H^{-1}(0,L)}^2. \tag{2.533}$$

Let $\mathrm{Id}_{H^{-1}(0,L)}$ be the identity map from $H^{-1}(0,L)$ into itself. By the density of $D(A) \cap H^2(0,L)$ in $D(A)$ for the graph norm of A (see page 374) and (2.533), the operator $A - C\mathrm{Id}_{H^{-1}(0,L)}$ is dissipative (see Definition A.2 on page 373) for $C > 0$ large enough.

Concerning the adjoint A^* of A, one easily checks that

(2.534)
$$D(A^*) = \{f \in H^1(0,L); -\varepsilon f(0) + M(Jf)_x(0) = -\varepsilon f(L) + M(Jf)_x(L) = 0\},$$

(2.535) $\qquad\qquad A^*f = J^{-1}(-\varepsilon f + M(Jf)_x),$

(2.536) $\qquad\qquad A^*$ is closed and densely defined.

Let $f \in D(A^*) \cap H^2(0,L)$ and $\alpha := Jf$. One has, using (2.528) and (2.535),

(2.537)
$$\begin{aligned}(A^*f, f)_{H^{-1}(0,L)} &= \int_0^L (-\varepsilon f + M(Jf)_x)f\,dx \\ &= -\varepsilon \int_0^L f^2\,dx - M\int_0^L \alpha_{xx}\alpha_x\,dx \\ &= -\varepsilon \int_0^L f^2\,dx - \frac{M}{2}(\alpha_x(L)^2 - \alpha_x(0)^2).\end{aligned}$$

From (2.532) and (2.537), one gets that $(A - C\text{Id}_{H^{-1}(0,L)})^*$, as $A - C\text{Id}_{H^{-1}(0,L)}$, is dissipative for $C > 0$ large enough. Let $C > 0$ be such that $A - C\text{Id}_{H^{-1}(0,L)}$ and $(A - C\text{Id}_{H^{-1}(0,L)})^*$ are dissipative. By Theorem A.10 on page 375, $A - C\text{Id}_{H^{-1}(0,L)}$ is the infinitesimal generator of a strongly continuous semigroup $S_C(t)$, $t \in [0,+\infty)$, of continuous linear operators on $H^{-1}(0,L)$. For $t \in [0,+\infty)$, let

$$\begin{aligned}S(t): H &\to H \\ f &\mapsto S(t)f := e^{Ct}S_C(t)f.\end{aligned}$$

One easily checks that $S(t)$, $t \in [0,+\infty)$, is a strongly continuous semigroup of continuous linear operators on $H^{-1}(0,L)$ and that the infinitesimal generator (see Definition A.9 on page 375) of this semigroup is A.

Let us now turn to the operator B. We first take $U := \mathbb{R}$. In order to motivate our definition of B, let us point out that, straightforward integrations by parts show that, if $f(L) = 0$, then

(2.538) $\quad (\varepsilon f_{xx} - Mf_x, g)_{H^{-1}(0,L)} = \varepsilon f(0)(Jg)_x(0)$
$$+ (f, A^*g)_{H^{-1}(0,L)}, \forall f \in H^1(0,L), \forall g \in D(A^*).$$

Looking at Definition 2.36 on page 53, one sees that if
$$y \in C^1([0,T]; H^{-1}(0,L)) \cap C^0([0,T]; H^1(0,L))$$
is a classical solution of (2.523) and (2.524) for some $u: [0,T] \to \mathbb{R}$, and if $z \in C^1([0,T]; H^{-1}) \cap C^0([0,T]; D(A))$ is a solution of
$$\dot{z} = A^*z,$$
we must have

(2.539) $\qquad (\varepsilon f_{xx} - Mf_x, g)_{H^{-1}(0,L)} - (f, A^*g)_{H^{-1}(0,L)} = f(0)Bg,$

with
$$f := y(0),\; g := z(0).$$

Hence, from (2.538) and (2.539), one sees that the definition of B must be the following:
$$\begin{aligned}\mathbb{R} &\to D(A^*)' \\ u &\mapsto (g \in D(A^*) \mapsto \varepsilon u(Jg)_x(0)).\end{aligned}$$

Clearly, the linear map B is well defined and continuous from $\mathbb{R}$ into $D(A^*)'$. Let us check the regularity property (2.199). Let $T > 0$ and $z^0 \in D(A^*)$ and let $z \in C^1([0,T]; H^{-1}(0,L)) \cap C^0([0,T]; D(A^*))$ be defined by

(2.540) $$z(t) := S(t)^* z^0, \, \forall t \in [0,T].$$

Let $\varphi \in C^1([0,T]; H_0^1(0,L)) \cap C^0([0,T]; H^3(0,L))$ be defined by

(2.541) $$\varphi(t) = Jz(t), \, \forall t \in [0,T].$$

Then, with $\varphi(t,x) := \varphi(t)(x)$,

(2.542) $$\varphi_t = \varepsilon \varphi_{xx} + M\varphi_x, \, t \in (0,T), \, x \in (0,L),$$
(2.543) $$\varphi(t,0) = \varphi(t,L) = 0, \, \forall t \in [0,T].$$

Moreover, the inequality in the regularity property (2.199) is equivalent to

(2.544) $$\int_0^T |\varphi_x(t,0)|^2 dt \leqslant \frac{C_T}{\varepsilon^2} \int_0^L |\varphi_x(0,x)|^2 dx.$$

Let us prove inequality (2.544). We multiply (2.542) by $-\varphi_{xx}$. Using (2.543) and integrations by parts, we get

(2.545) $$\frac{d}{dt} \int_0^L |\varphi_x(t,x)|^2 dx = -\varepsilon \int_0^L |\varphi_{xx}(t,x)|^2 dx$$
$$- \frac{M}{2}(|\varphi_x(t,L)|^2 - |\varphi_x(t,0)|^2), \, \forall t \in [0,T].$$

Using (2.532) with $\alpha := \varphi(t,\cdot)$, one has

(2.546) $$|\varphi_x(t,s)|^2 \leqslant \frac{\varepsilon}{|M|+1} \int_0^L |\varphi_{xx}(t,x)|^2 dx + \frac{|M|+1}{\varepsilon} \int_0^L |\varphi_x(t,x)|^2 dx.$$

Using (2.545) and (2.546), we have (discuss on the sign of M)

(2.547)
$$\frac{d}{dt} \int_0^L |\varphi_x(t,x)|^2 dx \leqslant -c_1 \int_0^L |\varphi_{xx}(t,x)|^2 dx + c_2 \int_0^L |\varphi_x(t,x)|^2 dx, \, \forall t \in [0,T],$$

with

(2.548) $$c_1 := \frac{\varepsilon}{2} > 0, \, c_2 := \frac{|M|(|M|+1)}{2\varepsilon} > 0.$$

From (2.547) and (2.548), one gets

(2.549) $$\int_0^L |\varphi_x(t,x)|^2 dx \leqslant e^{c_2 T} \int_0^L |\varphi_x(0,x)|^2 dx, \, \forall t \in [0,T],$$

(2.550) $$\int_0^T \int_0^L |\varphi_{xx}(t,x)|^2 dx dt \leqslant \frac{e^{c_2 T}}{c_1} \int_0^L |\varphi_x(0,x)|^2 dx.$$

From (2.546), (2.549) and (2.550), one gets (2.544) with

$$C_T := \varepsilon^2 \left(\frac{\varepsilon}{c_1(|M|+1)} + T \frac{|M|+1}{\varepsilon} \right) e^{c_2 T}.$$

Finally, using Definition 2.36 on page 53 together with Theorem 2.37 on page 53, one gets the definition of a solution to the Cauchy problem (2.523), (2.524) and (2.525), together with the existence and uniqueness of the solution to this problem.

Moreover, proceeding as in Section 2.3.3.1 and Section 2.3.3.2, one can easily check that, if $y^0 \in H^2(0,L)$ and $u \in H^2(0,T)$ are such that

$$y^0(0) = u(0) \text{ and } y(L) = 0,$$

then the solution y to the Cauchy problem (2.523), (2.524) and (2.525) is such that

$$y \in C^1([0,T];L^2(0,L)) \cap C^0([0,T];H^2(0,L))$$

and satisfies (2.523), (2.524) and (2.525) in the usual senses.

2.7.2. Null controllability and its cost. The control system (2.521)-(2.522) is null controllable for every time $T > 0$. Let us recall (see Definition 2.39 on page 55) that this means that, for every $y^0 \in H^{-1}(0,L)$ and for every $(\varepsilon, T, M) \in (0, +\infty)^2 \times \mathbb{R}$, there exists $u \in L^2(0,T)$ such that the solution of (2.523)-(2.524)-(2.525) satisfies $y(T,\cdot) = 0$. This controllability result is due to Hector Fattorini and David Russell [**162**, Theorem 3.3]; see also Oleg Imanuvilov [**242, 243**], Andrei Fursikov and Oleg Imanuvilov [**186**], and Gilles Lebeau and Luc Robbiano [**307**] for parabolic control systems in dimension larger than 1. (The last reference does not explicitly deal with transport terms; but the proof of [**307**] can perhaps be adapted to treat these terms.) The proof in [**186**] is the one we have given for Theorem 2.66 on page 79. This null controllability result also follows from Section 2.7.2.2 below if $T > 0$ is large enough.

For $y^0 \in H^{-1}(0,L)$, we denote by $U(\varepsilon, T, L, M, y^0)$ the set of controls $u \in L^2(0,T)$ such that the corresponding solution of (2.523)-(2.524)-(2.525) satisfies $y(T,\cdot) = 0$. Next, we define the quantity which measures the cost of the null controllability for system (2.521)-(2.522):

$$(2.551) \quad K(\varepsilon, T, L, M) := \sup_{\|y^0\|_{H^{-1}(0,L)} \leqslant 1} \{\inf\{\|u\|_{L^2(0,T)} : u \in U(\varepsilon, T, L, M, y^0)\}\}.$$

REMARK 2.92. One easily checks that $U(\varepsilon, T, L, M, y^0)$ is a closed affine subspace of $L^2(0,T)$. Hence the infimum in (2.551) is achieved.

REMARK 2.93. In [**124**], we have in fact considered, instead of K defined by (2.551), the quantity

$$(2.552) \quad K^*(\varepsilon, T, L, M) := \sup_{\|y^0\|_{L^2(0,L)} \leqslant 1} \{\inf\{\|u\|_{L^2(0,T)} : u \in U(\varepsilon, T, L, M, y^0)\}\}.$$

But the proofs given here are (essentially) the same as the ones given in [**124**]; see also Remark 2.98.

In this section our goal is to give estimates on $K(\varepsilon, T, L, M)$, in particular as $\varepsilon \to 0^+$. Let us point out that simple scaling arguments lead to the relations

$$(2.553) \quad K(\varepsilon, T, L, M) = a^{1/4} K\left(\varepsilon, aT, a^{1/2}L, \frac{M}{a^{1/2}}\right)$$

and

$$(2.554) \quad K(\varepsilon, T, L, M) = a^{3/4} K(a\varepsilon, T, a^{1/2}L, a^{1/2}M),$$

for every $(a, \varepsilon, T, L, M) \in (0, +\infty)^4 \times \mathbb{R}$.

In order to understand the behavior of $K(\varepsilon, T, L, M)$ as $\varepsilon \to 0^+$, it is natural to look at the limits of trajectories of the control system (2.521)-(2.522) as $\varepsilon \to 0^+$. This is done in the following proposition, proved in [**124**, Appendix].

PROPOSITION 2.94. Let (T, L, M) be given in $(0, +\infty)^2 \times \mathbb{R}^*$ and let $y^0 \in L^2(0, L)$. Let $(\varepsilon_n)_{n \in \mathbb{N}}$ be a sequence of positive real numbers which tends to 0 as $n \to +\infty$. Let $(u_n)_{n \in \mathbb{N}}$ be a sequence of functions in $L^2(0, T)$ such that, for some $u \in L^2(0, T)$,

(2.555) $\quad u_n$ converges weakly to u in $L^2(0, T)$ as $n \to +\infty$.

For $n \in \mathbb{N}$, let us denote by $y_n \in C^0([0, T]; H^{-1}(0, L))$ the solution of

(2.556) $\quad y_{nt} - \varepsilon_n y_{nxx} + M y_{nx} = 0, \ (t, x) \in (0, T) \times (0, L),$

(2.557) $\quad y_n(t, 0) = u_n(t), \ y_n(t, L) = 0, \ t \in (0, T),$

(2.558) $\quad y_n(0, x) = y^0(x), \ x \in (0, L).$

For $M > 0$, let $y \in C^0([0, T]; L^2(0, L))$ be the solution of

(2.559) $\quad \begin{cases} y_t + M y_x = 0, & (t, x) \in (0, T) \times (0, L), \\ y(t, 0) = u(t), & t \in (0, T), \\ y(0, x) = y^0(x), & x \in (0, L), \end{cases}$

and, for $M < 0$, let $y \in C^0([0, T]; L^2(0, L))$ be the solution of

(2.560) $\quad \begin{cases} y_t + M y_x = 0, & (t, x) \in (0, T) \times (0, L), \\ y(t, L) = 0, & t \in (0, T), \\ y(0, x) = y^0(x), & x \in (0, L). \end{cases}$

Then

$$y_n \rightharpoonup y \text{ weakly in } L^2((0, T) \times (0, L)) \text{ as } n \to \infty.$$

For the definition, existence and uniqueness of solutions to the Cauchy problems (2.559) and (2.560), see Definition 2.1 on page 25 and Theorem 2.4 on page 27. Definition 2.1 and Theorem 2.4, which deal with the case $M = 1$, can be easily adapted to the case $M > 0$ and $M < 0$.

It follows directly from Proposition 2.94 and the first proof of Theorem 2.6 on page 29 that, for every (T, L, M) with $T < L/|M|$, one has

(2.561) $\quad \lim_{\varepsilon \to 0^+} K(\varepsilon, T, L, M) = +\infty.$

Our first main result gives an estimate for the rate of convergence in (2.561).

THEOREM 2.95. There exists $C_0 > 0$ such that, for every $(\varepsilon, T, L) \in (0, +\infty)^3$, one has, for every $M > 0$,

(2.562) $\quad K(\varepsilon, T, L, M) \geqslant C_0 \dfrac{\varepsilon^{-1} T^{-1/2} L^{1/2}}{1 + \dfrac{L^{5/2} M^{5/2}}{\varepsilon^{5/2}}} \exp\left(\dfrac{M}{2\varepsilon}(L - TM) - \dfrac{\pi^2 \varepsilon T}{L^2}\right),$

and, for every $M < 0$,

(2.563) $\quad K(\varepsilon, T, L, M) \geqslant C_0 \dfrac{\varepsilon^{-1} T^{-1/2} L^{1/2}}{1 + \dfrac{L^{5/2} |M|^{5/2}}{\varepsilon^{5/2}}} \exp\left(\dfrac{|M|}{2\varepsilon}(2L - T|M|) - \dfrac{\pi^2 \varepsilon T}{L^2}\right).$

The proof of this theorem is given in Section 2.7.2.1. It relies on harmonic analysis.

Concerning upper bounds of $K(\varepsilon, T, L, M)$, let us point out that

- If $M > 0$ and $T > L/M$, the control $u := 0$ steers the control system (2.559) (where the state is $y(t, \cdot) \in L^2(0, L)$ and the control is $u(t) \in \mathbb{R}$) from any state to 0 in time T. (This means that, if $M > 0$, $T > L/M$ and $u = 0$, then, for the function y defined in Proposition 2.94, $y(T, \cdot) = 0$ whatever $y^0 \in L^2(0, L)$ is.)
- If $M < 0$ and $T > L/|M|$, then, for the function y defined in Proposition 2.94, $y(T, \cdot) = 0$ whatever $y^0 \in L^2(0, L)$ is.

This could lead us to hope that, for every $(T, L, M) \in (0, +\infty)^2 \times \mathbb{R}^*$ with $T > L/|M|$,

$$K(\varepsilon, T, L, M) \to 0 \text{ as } \varepsilon \to 0^+. \tag{2.564}$$

As shown by Theorem 2.95, this turns out to be false for $M < 0$ and $T \in (L/|M|, 2L/|M|)$. Our next theorem shows that (2.564) holds if $T|M|/L$ is large enough.

THEOREM 2.96. *Let a, A, b and B be the four positive constants*

$$a := 4.3, \ A := 2.61, \ b := 57.2, \ B := 18.1. \tag{2.565}$$

There exists $C_1 > 0$ such that, for every $(\varepsilon, T, L) \in (0, +\infty)^3$ and for every $M \in \mathbb{R}^*$ with

$$\frac{|M|L}{\varepsilon} \geqslant C_1: \tag{2.566}$$

- *If $M > 0$ and*

$$T \geqslant a\frac{L}{M}, \tag{2.567}$$

  then

$$K(\varepsilon, T, L, M) \leqslant C_1 \varepsilon^{-3/2} L^{1/2} M \exp\left(-\frac{L^2}{2\varepsilon T}\left(\frac{3}{4}\left(\frac{2TM}{3L} - 1\right)^2 - A\right)\right). \tag{2.568}$$

- *If $M < 0$ and*

$$T \geqslant b\frac{L}{|M|}, \tag{2.569}$$

  then

$$K(\varepsilon, T, L, M) \leqslant C_1 \varepsilon^{-1} M^{1/2} \tag{2.570}$$
$$\exp\left(-\frac{L^2}{2\varepsilon T}\left(\frac{3}{4}\left(\frac{2T|M|}{3L} - 1\right)^2 - B\frac{T|M|}{L}\right)\right).$$

The proof of this theorem is given in Section 2.7.2.2. It relies on a decay estimate for the solution of (2.521)-(2.522) when $u = 0$ and a Carleman estimate for the solutions of the adjoint system of (2.521)-(2.522).

REMARK 2.97. Theorem 2.95 and Theorem 2.96 have been generalized in part by Sergio Guerrero and Gilles Lebeau in [**209**] to higher dimensions and to the case where M may depend on t and x.

2.7.2.1. *Proof of Theorem 2.95.* Let $(\varepsilon, T, L, M) \in (0, +\infty)^3 \times \mathbb{R}^*$. For the sake of simplicity, throughout this section, K stands for $K(\varepsilon, T, L, M)$. Let us define $y^0 \in L^2(0, L)$ by

$$y^0(x) := \sin\left(\frac{\pi x}{L}\right) \exp\left(\frac{Mx}{2\varepsilon}\right), \, x \in (0, L).$$

Let us estimate $\|y^0\|_{H^{-1}(0,L)}$. Let $z^0 := Jy^0$. One easily gets

$$(2.571) \quad z^0(x) = -\frac{1}{2i} \frac{1}{\left(\frac{M}{2\varepsilon} + \frac{i\pi}{L}\right)^2} \exp\left(\left(\frac{M}{2\varepsilon} + \frac{i\pi}{L}\right)x\right)$$
$$+ \frac{1}{2i} \frac{1}{\left(\frac{M}{2\varepsilon} - \frac{i\pi}{L}\right)^2} \exp\left(\left(\frac{M}{2\varepsilon} - \frac{i\pi}{L}\right)x\right) + a_1 x + a_0,$$

with

$$(2.572) \qquad a_0 := \frac{\pi M}{\varepsilon L} \frac{1}{\left(\frac{M^2}{4\varepsilon^2} + \frac{\pi^2}{L^2}\right)^2},$$

$$(2.573) \qquad a_1 := -\frac{B}{L} - \frac{\pi M}{\varepsilon L} \frac{1}{\left(\frac{M^2}{4\varepsilon^2} + \frac{\pi^2}{L^2}\right)^2} \exp\left(\frac{ML}{2\varepsilon}\right).$$

From (2.571), (2.572), (2.573), one gets the existence of $C_2 > 0$ such that, for every $(\varepsilon, T, L, M) \in (0, +\infty)^3 \times \mathbb{R}^*$,

$$(2.574) \qquad \|y^0\|^2_{H^{-1}(0,L)} \leqslant C_2 \frac{\varepsilon^3}{M^3} \frac{\exp\left(\frac{LM}{\varepsilon}\right)}{1 + \frac{\varepsilon^3}{L^3 M^3}}, \text{ if } M > 0,$$

$$(2.575) \qquad \|y^0\|^2_{H^{-1}(0,L)} \leqslant C_2 \frac{\varepsilon^3}{M^3} \frac{1}{1 + \frac{\varepsilon^3}{L^3 |M|^3}}, \text{ if } M < 0.$$

Let $u \in U(\varepsilon, T, L, M, y^0)$ be such that (see Remark 2.92 on page 103)

$$\|u\|_{L^2(0,T)} = \min\{\|\tilde{u}\|_{L^2(0,T)} : \tilde{u} \in U(\varepsilon, T, L, M, y^0)\}.$$

In particular, the solution of (2.523)-(2.524)-(2.525) satisfies

$$(2.576) \qquad\qquad\qquad y(T, \cdot) = 0,$$

and we have

$$(2.577) \qquad\qquad \|u\|_{L^2(0,T)} \leqslant K \|y^0\|_{H^{-1}(0,L)}.$$

Let $\varphi \in C^2([0, T] \times [0, L])$ be such that

$$(2.578) \qquad \begin{cases} \varphi_t + \varepsilon \varphi_{xx} + M \varphi_x = 0, \, \forall (t, x) \in [0, T] \times [0, L], \\ \varphi(t, 0) = \varphi(t, L) = 0, \, \forall t \in [0, T]. \end{cases}$$

Let $z \in C^2([0,T]; L^2(0,L))$ and $z^T \in L^2(0,L)$ be defined by

(2.579) $$z(t) := J^{-1}(\varphi(t,\cdot)), \forall t \in [0,T],$$
(2.580) $$z^T := z(T).$$

Using (2.578), one gets

(2.581) $$z(t) = S^*(T-t)z^T, \forall t \in [0,T].$$

From Definition 2.36 on page 53 for $\tau := T$, (2.523), (2.524), (2.525), (2.580) and (2.581), one gets that

$$(y(T,\cdot), z^T)_{H^{-1}(0,L)} - (y^0, z(0))_{H^{-1}(0,L)} = \varepsilon \int_0^T u(t)(Jz(t))_x(t,0)dt,$$

which, together with (2.528), (2.576), (2.579) and (2.580), leads to

(2.582) $$-\int_0^L y^0(x)\varphi(0,x)dx = \varepsilon \int_0^T u(t)\varphi_x(t,0)dt.$$

Let $k \in \mathbb{N} \setminus \{0\}$. Let us define $\varphi : [0,T] \times [0,L] \to \mathbb{R}$ by requiring, for every $(t,x) \in [0,T] \times [0,L]$,

$$\varphi(t,x) := \exp\left(-\frac{Mx}{2\varepsilon}\right) \sin\left(\frac{k\pi x}{L}\right) \exp\left(\left(\frac{M^2}{4\varepsilon} + \frac{k^2\pi^2\varepsilon}{L^2}\right)t\right).$$

One easily checks that (2.578) holds. Hence, by (2.582), one has

(2.583)
$$\frac{k\pi\varepsilon}{L} \int_0^T u(t) \exp\left(\left(\frac{M^2}{4\varepsilon} + \frac{k^2\pi^2\varepsilon}{L^2}\right)t\right) dt = -\int_0^L \sin\left(\frac{\pi x}{L}\right) \sin\left(\frac{k\pi x}{L}\right) dx.$$

Let us now define a function $v : \mathbb{C} \to \mathbb{C}$ by

(2.584) $$v(s) := \int_{-T/2}^{T/2} u\left(t + \frac{T}{2}\right) e^{-ist} dt, s \in \mathbb{C}.$$

From (2.583) and (2.584), we get

(2.585) $$v\left(i\left(\frac{M^2}{4\varepsilon} + \frac{k^2\pi^2\varepsilon}{L^2}\right)\right) = 0 \quad \text{if } k \in \mathbb{N} \setminus \{0,1\}$$

and

(2.586) $$v\left(i\left(\frac{M^2}{4\varepsilon} + \frac{\pi^2\varepsilon}{L^2}\right)\right) = -\frac{L^2}{2\pi\varepsilon} \exp\left(-\left(\frac{M^2 T}{8\varepsilon} + \frac{\pi^2 \varepsilon T}{2L^2}\right)\right).$$

From (2.577) and (2.584), we get

$$|v(s)| \leqslant \exp\left(\frac{T|\mathrm{Im}(s)|}{2}\right) \int_0^T |u(t)|dt \leqslant K\, T^{1/2} \exp\left(\frac{T|\mathrm{Im}(s)|}{2}\right) \|y^0\|_{H^{-1}(0,L)},$$

which, together with (2.574) and (2.575), gives the existence of $C_3 > 0$ such that, for every $(\varepsilon, T, L, M) \in (0, +\infty)^3 \times \mathbb{R}^*$,

$$
(2.587) \quad |v(s)| \leqslant C_3 K \varepsilon^{3/2} T^{1/2} M^{-3/2} \frac{\exp\left(\dfrac{LM}{2\varepsilon} + \dfrac{T|\mathrm{Im}(s)|}{2}\right)}{1 + \dfrac{\varepsilon^{3/2}}{L^{3/2} M^{3/2}}}, \text{ if } M > 0,
$$

$$
(2.588) \quad |v(s)| \leqslant C_3 K \varepsilon^{3/2} T^{1/2} |M|^{-3/2} \frac{\exp\left(\dfrac{T|\mathrm{Im}(s)|}{2}\right)}{1 + \dfrac{\varepsilon^{3/2}}{L^{3/2} |M|^{3/2}}}, \text{ if } M < 0.
$$

Let us define a map $f : \mathbb{C} \to \mathbb{C}$ by

$$
f(s) := v\left(\frac{s - iM^2}{4\varepsilon}\right), \; s \in \mathbb{C}.
$$

We readily have that f is an entire function satisfying

$$
(2.589) \quad f(b_k) = 0, \; k \in \mathbb{N} \setminus \{0, 1\},
$$

with

$$
(2.590) \quad b_k := i\left(2M^2 + \frac{4k^2\pi^2\varepsilon^2}{L^2}\right), \; k \in \mathbb{N} \setminus \{0\}.
$$

Additionally, (2.587) and (2.588) translate into

(2.591)
$$
|f(s)| \leqslant C_3 K \varepsilon^{3/2} T^{1/2} M^{-3/2} \frac{\exp\left(\dfrac{LM}{2\varepsilon}\right) \exp\left(\dfrac{T|\mathrm{Im}(s) - M^2|}{8\varepsilon}\right)}{1 + \dfrac{\varepsilon^{3/2}}{L^{3/2} M^{3/2}}}, \text{ if } M > 0,
$$

$$
(2.592) \quad |f(s)| \leqslant C_3 K \varepsilon^{3/2} T^{1/2} M^{-3/2} \frac{\exp\left(\dfrac{T|\mathrm{Im}(s) - M^2|}{8\varepsilon}\right)}{1 + \dfrac{\varepsilon^{3/2}}{L^{3/2} |M|^{3/2}}}, \text{ if } M < 0.
$$

Let $(a_\ell)_{\ell \in \mathbb{N}}$ be the sequence of zeros of f in $\mathbb{C}_+ := \{s \in \mathbb{C} : \mathrm{Im}(s) > 0\}$, each zero being repeated according to its multiplicity. By the classical representation of entire functions of exponential type in $\mathbb{C}_+$ (see, for example, [281, Theorem page 56]), (2.591) and (2.592), one has for $s = x_1 + ix_2 \in \mathbb{C}_+$,

$$
(2.593) \quad \ln|f(s)| = \sum_{\ell=1}^{\infty} \ln\left|\frac{s - a_\ell}{s - \overline{a_\ell}}\right| + \sigma x_2 + \frac{x_2}{\pi} \int_{-\infty}^{+\infty} \frac{\ln|f(\tau)|}{|\tau - s|^2} d\tau,
$$

where σ is a real number independent of s such that

$$
(2.594) \quad \sigma \leqslant \frac{T}{8\varepsilon}.
$$

Note that, by (2.586) and (2.590),

$$
(2.595) \quad f(b_1) = v\left(i\left(\frac{M^2}{4\varepsilon} + \frac{\pi^2\varepsilon}{L^2}\right)\right) = -\frac{L^2}{2\pi\varepsilon}\exp\left(-T\left(\frac{M^2}{8\varepsilon} + \frac{\pi^2\varepsilon}{2L^2}\right)\right).
$$

By (2.589), $\{b_k;\, k \in \mathbb{N} \setminus \{0,1\}\} \subset \{a_\ell;\, \ell \in \mathbb{N}\}$. Hence

$$\text{(2.596)} \qquad \sum_{\ell=0}^{\infty} \ln\left|\frac{b_1 - a_\ell}{b_1 - \overline{a}_\ell}\right| \leqslant \sum_{k=2}^{\infty} \ln\left|\frac{b_1 - b_k}{b_1 - \overline{b}_k}\right| =: I_0$$

(recall that $\mathrm{Im}(a_\ell) > 0$ for every $\ell \in \mathbb{N}$).

Let us estimate I_0 from above. We have

$$I_0 = \sum_{k=2}^{\infty} \ln\left(\frac{(k^2-1)\pi^2\varepsilon^2/L^2}{M^2 + (k^2+1)\pi^2\varepsilon^2/L^2}\right) \leqslant J_0,$$

with

$$J_0 := \int_2^{\infty} \ln\left(\frac{\pi^2\varepsilon^2 x^2}{M^2 L^2 + \pi^2\varepsilon^2 x^2}\right) dx.$$

Using the change of variable

$$\tau = \frac{\pi\varepsilon}{L|M|} x$$

and an integration by parts, we get

$$J_0 = \frac{L|M|}{2\pi\varepsilon} \int_{\frac{2\pi\varepsilon}{L|M|}}^{\infty} \ln\left(\frac{\tau^2}{1+\tau^2}\right) d\tau$$

$$= -2\ln\left(\frac{1}{1+\left(\frac{LM}{2\pi\varepsilon}\right)^2}\right) - \frac{2L|M|}{\pi\varepsilon} \int_{\frac{2\pi\varepsilon}{L|M|}}^{\infty} \frac{1}{1+\tau^2} d\tau$$

$$= -2\ln\left(\frac{1}{1+\left(\frac{LM}{2\pi\varepsilon}\right)^2}\right) - \frac{2L|M|}{\pi\varepsilon}\left(\frac{\pi}{2} - \arctan\left(\frac{2\pi\varepsilon}{L|M|}\right)\right).$$

Hence, there exists $C_4 > 0$ such that, for every $(\varepsilon, L, M) \in (0,+\infty)^2 \times \mathbb{R}^*$,

$$\text{(2.597)} \qquad I_0 \leqslant -\frac{L|M|}{\varepsilon} + 4\ln\left(1 + \frac{L|M|}{\varepsilon}\right) + C_4.$$

Let us now get an upper bound for

$$J_1 := \frac{\mathrm{Im}(b_1)}{\pi} \int_{-\infty}^{+\infty} \frac{\ln|f(\tau)|}{|\tau - b_1|^2} d\tau.$$

Using (2.591) and (2.592), we get, after straightforward computations, the existence of $C_5 > 0$ such that, for every $(\varepsilon, T, L, M) \in (0,+\infty)^3 \times \mathbb{R}^*$,

$$\text{(2.598)} \qquad J_1 \leqslant \frac{LM}{2\varepsilon} + \frac{TM^2}{8\varepsilon} + \frac{1}{2}\ln\left(\frac{K^2\varepsilon^3 TM^{-3}}{1+\frac{\varepsilon^3}{L^3 M^3}}\right) + C_5, \text{ if } M > 0,$$

$$\text{(2.599)} \qquad J_1 \leqslant \frac{TM^2}{8\varepsilon} + \frac{1}{2}\ln\left(\frac{K^2\varepsilon^3 T|M|^{-3}}{1+\frac{\varepsilon^3}{L^3|M|^3}}\right) + C_5, \text{ if } M < 0.$$

Theorem 2.95 then follows from (2.590), (2.593) applied with $s = b_1$, (2.594), (2.595), (2.596), (2.597), (2.598) and (2.599). ∎

2.7.2.2. *Proof of Theorem 2.96 on page 105.* We only treat the case where

(2.600) $$M > 0.$$

(The case $M < 0$ can be treated in a similar way; see [**124**].) Let $(\varepsilon, T, L, M) \in (0 + \infty)^4$. Considering Theorem 2.44 on page 56, together with (2.540), (2.541), (2.542) and (2.542), we consider the following partial differential equation:

(2.601) $$\varphi_t - \varepsilon\varphi_{xx} - M\varphi_x = 0,\ (t,x) \in (0,T) \times (0,L),$$

(2.602) $$\varphi(t,0) = \varphi(t,L) = 0,\ t \in (0,T),$$

(2.603) $$\varphi(0,x) = \varphi^0(x),\ x \in (0,L).$$

By Theorem 2.44, in order to prove (2.568), it suffices to prove the following observability inequality. Assume that (2.566) and (2.567) hold. Then, for every $\varphi^0 \in H_0^1(0,L)$, the solution φ of (2.601)-(2.602)-(2.603) satisfies

(2.604) $$\int_0^L |\varphi(T,x)|^2 dx \leqslant C_1^2 \varepsilon^{-1} L M^2$$

$$\times \exp\left(-\frac{L^2}{\varepsilon T}\left(\frac{3}{4}\left(\frac{2TM}{3L} - 1\right)^2 - A\right)\right) \int_0^T |\varphi_x(t,0)|^2 dt.$$

REMARK 2.98. Concerning K^* defined by (2.552), one can prove that it is the infimum of the set of $C > 0$ such that

(2.605) $$\int_0^L |\varphi(T,x)|^2 dx \leqslant C\varepsilon^2 \int_0^T |\varphi_x(t,0)|^2 dt, \forall \varphi^0 \in H_0^1(0,L) \cap H^2(0,L).$$

This result does not follow directly from Theorem 2.44 on page 56 since the linear map $\mathcal{F}_T \colon u \mapsto y(T,\cdot)$ where $y : (0,T) \times (0,L) \to \mathbb{R}$ is the solution of the Cauchy problem (2.523), (2.524) and (2.525) with $y^0 := 0$ is an *unbounded* linear operator (from a subspace of) $L^2(0,T)$ into $L^2(0,L)$. However, one can adapt the proof of Theorem 2.44 on page 56 to treat this unbounded case by noticing that in Lemma 2.48 on page 58 one does not require C_3 to be a continuous linear operator: It suffices that C_3 is a closed and densely defined linear operator.

In order to prove the observability inequality (2.604), we first prove decay estimates for the solution of (2.601)-(2.602)-(2.603).

LEMMA 2.99. *Let $\tau \in (L/M, +\infty)$. Then, for every $\varphi^0 \in L^2(0,L)$, the weak solution of (2.601)-(2.602)-(2.603) satisfies*

(2.606) $$\|\varphi(\tau,\cdot)\|_{L^2(0,L)}^2 \leqslant \frac{L^2}{4\varepsilon\pi\tau} \exp\left(-\frac{(M\tau - L)^2}{2\varepsilon\tau}\right) \|\varphi^0\|_{L^2(0,L)}^2.$$

Proof of Lemma 2.99. Let us first consider the solution

$$\widetilde{\varphi} \in C^0([0,\tau]; L^2(\mathbb{R}))$$

of the following system:

$$\begin{cases} \widetilde{\varphi}_t - \varepsilon\widetilde{\varphi}_{xx} - M\widetilde{\varphi}_x = 0,\ (t,x) \in (0,\tau) \times (-\infty, +\infty), \\ \widetilde{\varphi}(0,x) = \begin{cases} |\varphi(0,x)| & \text{if } x \in (0,L), \\ 0 & \text{otherwise}. \end{cases} \end{cases}$$

Then, by the maximum principle for parabolic equations (see for example [**292**, Chapter 1, Section 2]), we know that

(2.607) $$|\varphi(t,x)| \leqslant \widetilde{\varphi}(t,x), \ \forall (t,x) \in (0,\tau] \times (0,L).$$

It is well known that the solution $\widetilde{\varphi}$ is given in terms of the fundamental solution of the heat equation by

$$\widetilde{\varphi}(t,x) = \frac{1}{2(\varepsilon \pi t)^{1/2}} \int_0^L \exp\left(-\frac{(x+Mt-y)^2}{4\varepsilon t}\right) |\varphi(0,y)| dy.$$

(Write $\varphi(t,x) = \psi(\varepsilon t, x + Mt) = \psi(\tau, \xi)$, with $\psi_\tau - \psi_{\xi\xi} = 0$ and see, for example, [**475**, (5.9)-(5.10), page 217].) Now, we readily estimate the exponential term by its L^∞-norm and we deduce, provided that $\tau > L/M$, that

$$|\widetilde{\varphi}(\tau,x)| \leqslant \frac{1}{2(\varepsilon \pi \tau)^{1/2}} \exp\left(-\frac{(M\tau - L)^2}{4\varepsilon\tau}\right) \|\varphi(0,\cdot)\|_{L^1(0,L)}, \ \forall x \in (0,L),$$

which, together with (2.607), implies Lemma 2.99. ∎

We now establish a similar decay estimate but for the H_0^1-norm. Our goal is to deduce from Lemma 2.99 the following lemma.

LEMMA 2.100. *Let $\eta > 0$. Then there exists $C > 0$ such that, for every $(\varepsilon, \tau, L, M) \in (0, +\infty)^4$ such that*

(2.608) $$\frac{\tau M}{L} \geqslant 1 + \eta,$$

(2.609) $$LM \geqslant \varepsilon,$$

and for every $\varphi^0 \in L^2(0,L)$, the weak solution of (2.601)-(2.602)-(2.603) verifies

(2.610) $$\|\varphi_x(\tau,\cdot)\|_{L^2(0,L)}^2 \leqslant C \frac{LM^3}{\varepsilon^3} \exp\left(-\frac{(M\tau - L)^2}{2\varepsilon\tau}\right) \|\varphi^0\|_{L^2(0,L)}^2.$$

Proof of Lemma 2.100. For the moment being, let us assume that the following lemma holds.

LEMMA 2.101. *For every $(\varepsilon, \tau, L, M) \in (0, +\infty)^4$ and for every $\varphi^0 \in L^2(0,L)$, the weak solution of (2.601)-(2.602)-(2.603) verifies*

(2.611) $$\|\varphi_x(\tau,\cdot)\|_{L^2(0,L)}^2 \leqslant \left(\frac{1}{\varepsilon\tau} + \frac{\tau M^4}{\varepsilon^3}\right) \|\varphi^0\|_{L^2(0,L)}^2.$$

Let

(2.612) $$\tau_1 := \tau - \frac{\eta\varepsilon}{2M^2}.$$

From (2.608), (2.609) and (2.612),

$$\tau_1 > \frac{L}{M}.$$

This inequality allows to apply Lemma 2.99 with $\tau := \tau_1$. One gets

(2.613) $$\|\varphi(\tau_1,\cdot)\|_{L^2(0,L)}^2 \leqslant \frac{L^2}{4\varepsilon\pi\tau_1} \exp\left(-\frac{(M\tau_1 - L)^2}{2\varepsilon\tau_1}\right) \|\varphi^0\|_{L^2(0,L)}^2.$$

Let us now apply Lemma 2.101 with $\varphi^0 := \varphi(\tau_1, \cdot)$ and $\tau := \tau - \tau_1$. One gets

(2.614) $\quad \|\varphi_x(\tau, \cdot)\|_{L^2(0,L)}^2 \leqslant \left(\dfrac{1}{\varepsilon(\tau - \tau_1)} + \dfrac{(\tau - \tau_1)M^4}{\varepsilon^3} \right) \|\varphi(\tau_1, \cdot)\|_{L^2(0,L)}^2.$

From (2.613) and (2.614), we have

(2.615) $\quad \|\varphi_x(\tau, \cdot)\|_{L^2(0,L)}^2 \leqslant R \|\varphi(\tau_1, \cdot)\|_{L^2(0,L)}^2,$

with

(2.616) $\quad R := \left(\dfrac{1}{\varepsilon(\tau - \tau_1)} + \dfrac{(\tau - \tau_1)M^4}{\varepsilon^3} \right) \dfrac{L^2}{4\varepsilon\pi\tau_1} \exp\left(-\dfrac{(M\tau_1 - L)^2}{2\varepsilon\tau} \right).$

Note that

$$\dfrac{(M\tau - L)^2}{\varepsilon\tau} - \dfrac{(M\tau_1 - L)^2}{\varepsilon\tau_1} \leqslant \dfrac{\eta}{2},$$

$$\left(\dfrac{1}{\varepsilon(\tau - \tau_1)} + \dfrac{(\tau - \tau_1)M^4}{\varepsilon^3} \right) \dfrac{L^2}{4\varepsilon\pi\tau_1} \leqslant \dfrac{1}{4\pi} \left(\dfrac{2}{\eta} + \dfrac{\eta}{2} \right) \dfrac{LM^3}{\varepsilon^3},$$

which, together with (2.615) and (2.616), gives (2.610) for $C > 0$ large enough.

It remains to prove Lemma 2.101. Without loss of generality, we may assume $\varphi^0 \in H_0^1(0, L) \cap H^2(0, L)$. Then

$$\varphi \in C^0([0, \tau]; H^2(0, L)) \cap C^1([0, \tau]; L^2(0, L)),$$

which is a sufficient regularity to perform the following computations. We first multiply (2.601) by 2φ and integrate on $[0, L]$. Using an integration by parts together with (2.602), one gets that

(2.617) $\quad \dfrac{\mathrm{d}}{\mathrm{d}t} \displaystyle\int_0^L \varphi^2 dx + 2\varepsilon \int_0^L \varphi_x^2 dx = 0.$

We then multiply (2.601) by $-2\varphi_{xx}$ and integrate on $[0, L]$. Using an integration by parts together with (2.602), one gets that

(2.618) $\quad \dfrac{\mathrm{d}}{\mathrm{d}t} \displaystyle\int_0^L \varphi_x^2 dx + 2\varepsilon \int_0^L \varphi_{xx}^2 dx = M\varphi_x(t,L)^2 - M\varphi_x(t,0)^2.$

Proceeding as in the proof of (2.532), we get

(2.619) $\quad \|\varphi_x(t, \cdot)\|_{L^\infty(0,L)}^2 \leqslant \dfrac{\varepsilon}{M} \displaystyle\int_0^L \varphi_{xx}^2 dx + \dfrac{M}{\varepsilon} \int_0^L \varphi_x^2 dx.$

Using (2.602), we get

(2.620) $\quad \displaystyle\int_0^L \varphi_x^2 dx = -\int_0^L \varphi_{xx} \varphi \, dx \leqslant \dfrac{\varepsilon^2}{M^2} \int_0^L \varphi_{xx}^2 dx + \dfrac{M^2}{4\varepsilon^2} \int_0^L \varphi^2 dx.$

From (2.617), (2.618), (2.619) and (2.620), we have

(2.621) $\quad \dfrac{\mathrm{d}}{\mathrm{d}t} \displaystyle\int_0^L \varphi_x^2 dx \leqslant \dfrac{M^4}{4\varepsilon^3} \int_0^L \varphi^2 dx \leqslant \dfrac{M^4}{4\varepsilon^3} \|\varphi^0\|_{L^2(0,L)}^2.$

Integrating this inequality on $[t, \tau]$, we get

(2.622) $\quad \displaystyle\int_0^L \varphi_x(\tau, x)^2 dx \leqslant \int_0^L \varphi_x(t,x)^2 dx + (\tau - t) \dfrac{M^4}{4\varepsilon^3} \|\varphi^0\|_{L^2(0,L)}^2.$

Integrating (2.617) on $[0,\tau]$, we have

$$2\varepsilon \int_0^\tau \int_0^L \varphi_x^2 dx dt \leqslant \|\varphi^0\|_{L^2(0,L)}^2. \tag{2.623}$$

Finally, integrating (2.622) on $[0,\tau]$ and using (2.623), we get

$$\int_0^L \varphi_x(\tau, x)^2 dx \leqslant \left(\frac{1}{2\varepsilon\tau} + \frac{\tau M^4}{8\varepsilon^3}\right) \|\varphi^0\|_{L^2(0,L)}^2.$$

This concludes the proof of Lemma 2.101 and therefore also the proof of Lemma 2.100. ∎

We now establish, as in Section 2.5.2, a Carleman inequality. Let us first perform a change of variables in order to restrict ourselves to the case where $\varepsilon = L = 1$:

$$\begin{cases} \widetilde{t} = \varepsilon t/L^2, \\ \widetilde{x} = x/L. \end{cases} \tag{2.624}$$

In the new variables, we have, with $\widetilde{\varphi}(\widetilde{t}, \widetilde{x}) := \varphi(t, x)$,

$$\begin{cases} \widetilde{\varphi}_{\widetilde{t}} - \widetilde{\varphi}_{\widetilde{x}\widetilde{x}} - \varepsilon^{-1} M L \widetilde{\varphi}_{\widetilde{x}} = 0, & (\widetilde{t}, \widetilde{x}) \in (0, \varepsilon T/L^2) \times (0,1), \\ \widetilde{\varphi}(\widetilde{t}, 0) = \widetilde{\varphi}(\widetilde{t}, 1) = 0, & \widetilde{t} \in (0, \varepsilon T/L^2). \end{cases} \tag{2.625}$$

Let

$$\widetilde{M} := \frac{ML}{\varepsilon}, \tag{2.626}$$

$$\widetilde{T} := \frac{\varepsilon T}{L^2}. \tag{2.627}$$

Then (2.566) and (2.567) become, respectively,

$$\widetilde{M} \geqslant C_1, \tag{2.628}$$

$$\widetilde{M}\widetilde{T} \geqslant a. \tag{2.629}$$

Let us define a weight function (compare with (2.408) coming from [**186**]),

$$\alpha(\widetilde{t}, \widetilde{x}) := \frac{\beta(\widetilde{x})}{\widetilde{t}}, \quad (\widetilde{t}, \widetilde{x}) \in (0, \widetilde{T}) \times (0,1), \tag{2.630}$$

where $0 \leqslant \beta \in C^2([0,1])$ will be chosen below. We also introduce the function

$$\psi := e^{-\alpha}\widetilde{\varphi}.$$

This function verifies that

$$P_1 + P_2 = P_3, \tag{2.631}$$

with

$$P_1 := \psi_{\widetilde{x}\widetilde{x}} + \alpha_{\widetilde{x}}^2 \psi + \widetilde{M}\alpha_{\widetilde{x}}\psi - \alpha_{\widetilde{t}}\psi,$$
$$P_2 := -\psi_{\widetilde{t}} + 2\alpha_{\widetilde{x}}\psi_{\widetilde{x}} + \widetilde{M}\psi_{\widetilde{x}},$$
$$P_3 := -\alpha_{\widetilde{x}\widetilde{x}}\psi;$$

compare with (2.413), (2.414) and (2.415). As in Section 2.5.2, we take the L^2-norm in identity (2.631) and then expand the double product

$$\|P_1\|^2_{L^2(Q)} + \|P_2\|^2_{L^2(Q)} + 2(P_1, P_2)_{L^2(Q)} = \|P_3\|^2_{L^2(Q)}, \tag{2.632}$$

where Q stands for the open set $(0, \widetilde{T}) \times (0, 1)$.

Let us compute $2(P_1, P_2)_{L^2(Q)}$. Let us first compute the terms concerning $\psi_{\widetilde{x}\widetilde{x}}$. We have

$$-(\psi_{\widetilde{x}\widetilde{x}}, \psi_{\widetilde{t}})_{L^2(Q)} = \frac{1}{2} \int_0^1 |\psi_{\widetilde{x}}(\widetilde{T}, \widetilde{x})|^2 \, d\widetilde{x}.$$

Moreover,

$$2(\psi_{\widetilde{x}\widetilde{x}}, \alpha_{\widetilde{x}}\psi_{\widetilde{x}})_{L^2(Q)} = \int_0^{\widetilde{T}} (\alpha_{\widetilde{x}}(\widetilde{t}, 1)|\psi_{\widetilde{x}}(\widetilde{t}, 1)|^2 - \alpha_{\widetilde{x}}(\widetilde{t}, 0)|\psi_{\widetilde{x}}(\widetilde{t}, 0)|^2) \, d\widetilde{t}$$
$$- \iint_Q \alpha_{\widetilde{x}\widetilde{x}} |\psi_{\widetilde{x}}|^2 \, d\widetilde{x} \, d\widetilde{t}.$$

Finally,

$$\widetilde{M}(\psi_{\widetilde{x}\widetilde{x}}, \psi_{\widetilde{x}})_{L^2(Q)} = (\widetilde{M}/2) \int_0^{\widetilde{T}} (|\psi_{\widetilde{x}}(\widetilde{t}, 1)|^2 - |\psi_{\widetilde{x}}(\widetilde{t}, 0)|^2) \, d\widetilde{t}.$$

As long as the term $\alpha_{\widetilde{x}}^2 \psi$ is concerned, we first have

$$-(\alpha_{\widetilde{x}}^2 \psi, \psi_{\widetilde{t}})_{L^2(Q)} = \iint_Q \alpha_{\widetilde{x}} \alpha_{\widetilde{x}\widetilde{t}} |\psi|^2 \, d\widetilde{x} \, d\widetilde{t} - \frac{1}{2} \int_0^1 \alpha_{\widetilde{x}}^2(T, \widetilde{x}) |\psi(0, \widetilde{x})|^2 \, d\widetilde{x}.$$

Next,

$$2(\alpha_{\widetilde{x}}^2 \psi, \alpha_{\widetilde{x}} \psi_{\widetilde{x}})_{L^2(Q)} = -3 \iint_Q \alpha_{\widetilde{x}\widetilde{x}} \alpha_{\widetilde{x}}^2 |\psi|^2 \, d\widetilde{x} \, d\widetilde{t}.$$

Finally,

$$\widetilde{M}(\alpha_{\widetilde{x}}^2 \psi, \psi_{\widetilde{x}})_{L^2(Q)} = -\widetilde{M} \iint_Q \alpha_{\widetilde{x}\widetilde{x}} \alpha_{\widetilde{x}} |\psi|^2 \, d\widetilde{x} \, d\widetilde{t}.$$

Let us next look at the terms concerning $\widetilde{M} \alpha_{\widetilde{x}} \psi$ in $2(P_1, P_2)_{L^2(Q)}$. First, we have

$$-\widetilde{M}(\alpha_{\widetilde{x}} \psi, \psi_{\widetilde{t}})_{L^2(Q)} = (\widetilde{M}/2) \iint_Q \alpha_{\widetilde{x}\widetilde{t}} |\psi|^2 \, d\widetilde{x} \, d\widetilde{t} - (\widetilde{M}/2) \int_0^1 \alpha_{\widetilde{x}}(T, \widetilde{x}) |\psi(0, \widetilde{x})|^2 \, d\widetilde{x}.$$

Then we find

$$2\widetilde{M}(\alpha_{\widetilde{x}} \psi, \alpha_{\widetilde{x}} \psi_{\widetilde{x}})_{L^2(Q)} = -2\widetilde{M} \iint_Q \alpha_{\widetilde{x}} \alpha_{\widetilde{x}\widetilde{x}} |\psi|^2 \, d\widetilde{x} \, d\widetilde{t}.$$

The last term provides

$$\widetilde{M}^2 (\alpha_{\widetilde{x}} \psi, \psi_{\widetilde{x}})_{L^2(Q)} = -(\widetilde{M}^2/2) \iint_Q \alpha_{\widetilde{x}\widetilde{x}} |\psi|^2 \, d\widetilde{x} \, d\widetilde{t}.$$

Finally, we deal with the computations of the term $\alpha_{\widetilde{t}} \psi$. First, we obtain

$$(\alpha_{\widetilde{t}} \psi, \psi_{\widetilde{t}})_{L^2(Q)} = -(1/2) \iint_Q \alpha_{\widetilde{t}\widetilde{t}} |\psi|^2 \, d\widetilde{x} \, d\widetilde{t} + (1/2) \int_0^1 \alpha_{\widetilde{t}}(T, \widetilde{x}) |\psi(T, \widetilde{x})|^2 \, d\widetilde{x}.$$

Additionally, we find

$$-(\alpha_{\widetilde{t}} \psi, 2\alpha_{\widetilde{x}} \psi_{\widetilde{x}})_{L^2(Q)} = \iint_Q (\alpha_{\widetilde{t}} \alpha_{\widetilde{x}\widetilde{x}} + \alpha_{\widetilde{t}\widetilde{x}} \alpha_{\widetilde{x}}) |\psi|^2 \, d\widetilde{x} \, d\widetilde{t}.$$

Finally,
$$-(\alpha_{\widetilde{t}}\psi, \widetilde{M}\psi_{\widetilde{x}})_{L^2(Q)} = (\widetilde{M}/2)\iint_Q \alpha_{\widetilde{t}\widetilde{x}}|\psi|^2\,d\widetilde{x}\,d\widetilde{t}.$$

Putting all these computations together, we conclude that the double product term is
(2.633)
$$2(P_1, P_2)_{L^2(Q)} = \int_0^1 |\psi_{\widetilde{x}}(T, \widetilde{x})|^2\,d\widetilde{x}$$
$$+ \int_0^{\widetilde{T}} ((2\alpha_{\widetilde{x}}(\widetilde{t}, 1) + \widetilde{M})|\psi_{\widetilde{x}}(\widetilde{t}, 1)|^2 - (2\alpha_{\widetilde{x}}(\widetilde{t}, 0) + \widetilde{M})|\psi_{\widetilde{x}}(\widetilde{t}, 0)|^2)\,d\widetilde{t}$$
$$-2\iint_Q \alpha_{\widetilde{x}\widetilde{x}}|\psi_{\widetilde{x}}|^2\,d\widetilde{x}\,d\widetilde{t} + 4\iint_Q \alpha_{\widetilde{x}}\alpha_{\widetilde{x}\widetilde{t}}|\psi|^2\,d\widetilde{x}\,d\widetilde{t} - \int_0^1 \alpha_{\widetilde{x}}^2(T, \widetilde{x})|\psi(T, \widetilde{x})|^2\,d\widetilde{x}$$
$$-6\iint_Q \alpha_{\widetilde{x}\widetilde{x}}\alpha_{\widetilde{x}}^2|\psi|^2\,d\widetilde{x}\,d\widetilde{t} + 2\widetilde{M}\iint_Q \alpha_{\widetilde{x}\widetilde{t}}|\psi|^2\,d\widetilde{x}\,d\widetilde{t} - \widetilde{M}\int_0^1 \alpha_{\widetilde{x}}(T, \widetilde{x})|\psi(T, \widetilde{x})|^2\,d\widetilde{x}$$
$$-6\widetilde{M}\iint_Q \alpha_{\widetilde{x}}\alpha_{\widetilde{x}\widetilde{x}}|\psi|^2\,d\widetilde{x}\,d\widetilde{t} - \widetilde{M}^2\iint_Q \alpha_{\widetilde{x}\widetilde{x}}|\psi|^2\,d\widetilde{x}\,d\widetilde{t}$$
$$-\iint_Q \alpha_{\widetilde{t}\widetilde{t}}|\psi|^2\,d\widetilde{x}\,d\widetilde{t} + \int_0^1 \alpha_{\widetilde{t}}(T, \widetilde{x})|\psi(T, \widetilde{x})|^2\,d\widetilde{x} + 2\iint_Q \alpha_{\widetilde{t}}\alpha_{\widetilde{x}\widetilde{x}}|\psi|^2\,d\widetilde{x}\,d\widetilde{t}.$$

On the other hand, we have the following for the right hand side term:

(2.634)
$$\|P_3\|^2_{L^2(Q)} = \iint_Q \alpha_{\widetilde{x}\widetilde{x}}^2|\psi|^2\,d\widetilde{x}\,d\widetilde{t}.$$

Combining (2.633) and (2.634) with (2.632), we obtain

$$\int_0^{\widetilde{T}} (2\alpha_{\widetilde{x}}(\widetilde{t}, 1) + \widetilde{M})|\psi_{\widetilde{x}}(\widetilde{t}, 1)|^2\,d\widetilde{t} - 2\iint_Q \alpha_{\widetilde{x}\widetilde{x}}|\psi_{\widetilde{x}}|^2\,d\widetilde{x}\,d\widetilde{t}$$
$$-6\iint_Q \alpha_{\widetilde{x}\widetilde{x}}\alpha_{\widetilde{x}}^2|\psi|^2\,d\widetilde{x}\,d\widetilde{t} + 2\iint_Q \alpha_{\widetilde{t}}\alpha_{\widetilde{x}\widetilde{x}}|\psi|^2\,d\widetilde{x}\,d\widetilde{t} - 6\widetilde{M}\iint_Q \alpha_{\widetilde{x}}\alpha_{\widetilde{x}\widetilde{x}}|\psi|^2\,d\widetilde{x}\,d\widetilde{t}$$
$$-\widetilde{M}^2\iint_Q \alpha_{\widetilde{x}\widetilde{x}}|\psi|^2\,d\widetilde{x}\,d\widetilde{t}$$
$$\leqslant \iint_Q \alpha_{\widetilde{x}\widetilde{x}}^2|\psi|^2\,d\widetilde{x}\,d\widetilde{t} + \int_0^{\widetilde{T}} (2\alpha_{\widetilde{x}}(\widetilde{t}, 0) + \widetilde{M})|\psi_{\widetilde{x}}(\widetilde{t}, 0)|^2\,d\widetilde{t}$$
$$+4\iint_Q \alpha_{\widetilde{x}}\alpha_{\widetilde{x}\widetilde{t}}|\psi|^2\,d\widetilde{x}\,d\widetilde{t} - 2\widetilde{M}\iint_Q \alpha_{\widetilde{x}\widetilde{t}}|\psi|^2\,d\widetilde{x}\,d\widetilde{t} + \iint_Q \alpha_{\widetilde{t}\widetilde{t}}|\psi|^2\,d\widetilde{x}\,d\widetilde{t}$$
$$+\int_0^1 \alpha_{\widetilde{x}}^2(T, \widetilde{x})|\psi(T, \widetilde{x})|^2\,d\widetilde{x} + \widetilde{M}\int_0^1 \alpha_{\widetilde{x}}(T, \widetilde{x})|\psi(T, \widetilde{x})|^2\,d\widetilde{x}$$
$$-\int_0^1 \alpha_{\widetilde{t}}(T, \widetilde{x})|\psi(T, \widetilde{x})|^2\,d\widetilde{x}.$$

From the definition of α (given in (2.630)), we find

(2.635)
$$\int_0^{\widetilde{T}} (2\frac{\beta'(1)}{\widetilde{t}} + \widetilde{M})|\psi_{\widetilde{x}}(\widetilde{t},1)|^2 \, d\widetilde{t} - 2\iint_Q \frac{\beta''(\widetilde{x})}{\widetilde{t}}|\psi_{\widetilde{x}}|^2 \, d\widetilde{x} \, d\widetilde{t}$$
$$-6\iint_Q \frac{\beta''(\widetilde{x})(\beta'(\widetilde{x}))^2}{\widetilde{t}^3}|\psi|^2 \, d\widetilde{x} \, d\widetilde{t} - 2\iint_Q \frac{\beta(\widetilde{x})\beta''(\widetilde{x})}{(\widetilde{T}-\widetilde{t})^3}|\psi|^2 \, d\widetilde{x} \, d\widetilde{t}$$
$$-6\widetilde{M}\iint_Q \frac{\beta'(\widetilde{x})\beta''(\widetilde{x})}{(\widetilde{T}-\widetilde{t})^2}|\psi|^2 \, d\widetilde{x} \, d\widetilde{t} - \widetilde{M}^2 \iint_Q \frac{\beta''(\widetilde{x})}{\widetilde{t}}|\psi|^2 \, d\widetilde{x} \, d\widetilde{t}$$
$$\leqslant \iint_Q \frac{(\beta''(\widetilde{x}))^2}{(\widetilde{T}-\widetilde{t})^2}|\psi|^2 \, d\widetilde{x} \, d\widetilde{t} + \int_0^{\widetilde{T}} \left(2\frac{\beta'(0)}{\widetilde{T}-\widetilde{t}} + \widetilde{M}\right)|\psi_{\widetilde{x}}(\widetilde{t},0)|^2 \, d\widetilde{t}$$
$$+4\iint_Q \frac{(\beta'(\widetilde{x}))^2}{(\widetilde{T}-\widetilde{t})^3}|\psi|^2 \, d\widetilde{x} \, d\widetilde{t} + 2\widetilde{M}\iint_Q \frac{\beta'(\widetilde{x})}{(\widetilde{T}-\widetilde{t})^2}|\psi|^2 \, d\widetilde{x} \, d\widetilde{t}$$
$$+2\iint_Q \frac{\beta(\widetilde{x})}{(\widetilde{T}-\widetilde{t})^3}|\psi|^2 \, d\widetilde{x} \, d\widetilde{t} + \frac{1}{\widetilde{T}^2}\int_0^1 (\beta'(\widetilde{x}))^2|\psi(T,\widetilde{x})|^2 \, d\widetilde{x}$$
$$+\frac{\widetilde{M}}{\widetilde{T}}\int_0^1 \beta'(\widetilde{x})|\psi(T,\widetilde{x})|^2 \, d\widetilde{x} + \frac{1}{\widetilde{T}^2}\int_0^1 \beta(\widetilde{x})|\psi(T,\widetilde{x})|^2 \, d\widetilde{x}.$$

Let us now define the function $\beta : [0,1] \to \mathbb{R}$. Let $\delta \in (0,1)$. We take the function β satisfying

(2.636)
$$\beta''(\widetilde{x}) = -\frac{1}{1-\delta}\frac{2(\beta'(\widetilde{x}))^2 + \beta(\widetilde{x})}{3(\beta'(\widetilde{x}))^2 + \beta(\widetilde{x})}, \quad \widetilde{x} \in [0,1],$$

together with the initial conditions $\beta(0) = \delta$ and $\beta'(0) = 0.807$. One can check, that, for $\delta > 0$ small enough,

(2.637) $\qquad\qquad \beta > 0, \ \beta' > 0 \text{ and } \beta'' < 0 \text{ on } [0,1],$

(2.638) $\qquad\qquad\qquad\qquad \beta(1) < 0.435.$

We choose $\delta > 0$ such that (2.637) and (2.638) hold. From (2.635), (2.636) and (2.637), we have

(2.639)
$$-2\iint_Q \frac{\beta''(\widetilde{x})}{\widetilde{t}}|\psi_{\widetilde{x}}|^2 \, d\widetilde{x} \, d\widetilde{t} - 2\delta \iint_Q \beta''(\widetilde{x})\frac{\beta(\widetilde{x}) + 3(\beta'(\widetilde{x}))^2}{\widetilde{t}^3}|\psi|^2 \, d\widetilde{x} \, d\widetilde{t}$$
$$-6\widetilde{M}\iint_Q \frac{\beta'(\widetilde{x})\beta''(\widetilde{x})}{\widetilde{t}^2}|\psi|^2 \, d\widetilde{x} \, d\widetilde{t} - \widetilde{M}^2 \iint_Q \frac{\beta''(\widetilde{x})}{\widetilde{t}}|\psi|^2 \, d\widetilde{x} \, d\widetilde{t}$$
$$\leqslant \iint_Q \frac{(\beta''(\widetilde{x}))^2}{\widetilde{t}^2}|\psi|^2 \, d\widetilde{x} \, d\widetilde{t} + \int_0^{\widetilde{T}} (2\frac{\beta'(0)}{\widetilde{t}} + \widetilde{M})|\psi_{\widetilde{x}}(\widetilde{t},0)|^2 \, d\widetilde{t}$$
$$+2\widetilde{M}\iint_Q \frac{\beta'(\widetilde{x})}{\widetilde{t}^2}|\psi|^2 \, d\widetilde{x} \, d\widetilde{t} + \frac{1}{\widetilde{T}^2}\int_0^1 (\beta'(\widetilde{x}))^2|\psi(T,\widetilde{x})|^2 \, d\widetilde{x}$$
$$+\frac{\widetilde{M}}{\widetilde{T}}\int_0^1 \beta'(\widetilde{x})|\psi(T,\widetilde{x})|^2 \, d\widetilde{x} + \frac{1}{\widetilde{T}^2}\int_0^1 \beta(\widetilde{x})|\psi(T,\widetilde{x})|^2 \, d\widetilde{x}.$$

Additionally, using (2.637) and the fact that $\beta''(\widetilde{x}) \leqslant -2/((1-\delta)3)$, we get the existence of a universal constant $m_1 > 0$ (in particular, independent of ε, T, L, M, φ^0) such that, if

(2.640) $\qquad\qquad\qquad\qquad \widetilde{M} \geqslant m_1,$

then

$$(2.641) \quad -3\widetilde{M} \iint_Q \frac{\beta'(\widetilde{x})\beta''(\widetilde{x})}{\widetilde{t}^2}|\psi|^2 \, d\widetilde{x} \, d\widetilde{t}$$
$$\geqslant \iint_Q \frac{(\beta''(\widetilde{x}))^2}{\widetilde{t}^2}|\psi|^2 \, d\widetilde{x} \, d\widetilde{t} + 2\widetilde{M} \iint_Q \frac{\beta'(\widetilde{x})}{\widetilde{t}^2}|\psi|^2 \, d\widetilde{x} \, d\widetilde{t}.$$

From (2.637), (2.639) and (2.641), we have

$$(2.642) \quad -\widetilde{M}^2 \iint_Q \frac{\beta''(\widetilde{x})}{\widetilde{t}}|\psi|^2 \, d\widetilde{x} \, d\widetilde{t} \leqslant \int_0^{\widetilde{T}} (2\frac{\beta'(0)}{\widetilde{t}} + \widetilde{M})|\psi_{\widetilde{x}}(\widetilde{t},0)|^2 \, d\widetilde{t}$$
$$+ \frac{1}{\widetilde{T}^2}\int_0^1 (\beta'(\widetilde{x}))^2|\psi(T,\widetilde{x})|^2 \, d\widetilde{x} + \frac{\widetilde{M}}{\widetilde{T}}\int_0^1 \beta'(\widetilde{x})|\psi(T,\widetilde{x})|^2 \, d\widetilde{x}$$
$$+ \frac{1}{\widetilde{T}^2}\int_0^1 \beta(\widetilde{x})|\psi(T,\widetilde{x})|^2 \, d\widetilde{x}.$$

Let us recall that $\psi := e^{-\alpha}\widetilde{\varphi}$. Then, from (2.602), (2.629), (2.630), (2.637), (2.640) and (2.642), we deduce that

$$(2.643) \quad \widetilde{M}\iint_Q \frac{1}{\widetilde{t}}e^{-2\alpha}|\widetilde{\varphi}|^2 \, d\widetilde{x} \, d\widetilde{t} \leqslant C\left(\int_0^{\widetilde{T}}|\widetilde{\varphi}_{\widetilde{x}}(\widetilde{t},0)|^2 \, d\widetilde{t} + \frac{1}{\widetilde{T}}\int_0^1|\widetilde{\varphi}(0,\widetilde{x})|^2 \, d\widetilde{x}\right).$$

In (2.643) and from here on, C will stand for generic positive constants independent of ε, M, T, L and φ^0.

Let $0 < \gamma < 1/3$ be fixed (one can take for example $\gamma := 1/6$). By (2.630) and (2.637), $e^{-2\alpha}$ reaches its minimum in the region $[(1-3\gamma)\widetilde{T}/3, \widetilde{T}/3] \times [0,1]$ at $(\widetilde{t},\widetilde{x}) = (\widetilde{T}/3, 1)$. Hence

$$(2.644) \quad \frac{\widetilde{M}}{\widetilde{T}}e^{-2\alpha(\widetilde{T}/3,1)}\int_0^1\int_{(1-3\gamma)\widetilde{T}/3}^{\widetilde{T}/3}|\widetilde{\varphi}|^2 \, d\widetilde{t} \, d\widetilde{x}$$
$$\leqslant C\left(\int_0^{\widetilde{T}}|\widetilde{\varphi}_{\widetilde{x}}(\widetilde{t},0)|^2 \, d\widetilde{t} + \frac{1}{\widetilde{T}}\int_0^1|\widetilde{\varphi}(0,\widetilde{x})|^2 \, d\widetilde{x}\right).$$

From (2.630), we deduce, using (2.565) and (2.638), that

$$(2.645) \quad \exp\{-2\alpha(\widetilde{T}/3, 1)\} = \exp\{-6\beta(1)/\widetilde{T}\} > \exp\{-A/\widetilde{T}\}.$$

Coming back to our original variables (see (2.624), (2.626) and (2.627)), we get from (2.644) and (2.645),

$$(2.646) \quad \int_0^L\int_{(1-3\gamma)T/3}^{T/3}|\varphi|^2 dt dx$$
$$\leqslant Ce^{AL^2/(\varepsilon T)}\left(\frac{\varepsilon^2 T}{M}\int_0^T|\varphi_x(t,0)|^2 dt + \frac{L}{M}\int_0^L|\varphi(T,x)|^2 dx\right).$$

From now on, we assume that (2.609) holds. Let us apply the dissipativity result stated in Lemma 2.100 on page 111 to $\varphi^0 := \varphi(t,\cdot)$, with $t \in [(1-3\gamma)T/3, T/3]$

and $\tau := T - t$. This is possible, since, by (2.565) and (2.567), $(2T/3) \geqslant 3(L/M)$. We obtain

(2.647) $$\int_0^L \int_{(1-3\gamma)T/3}^{T/3} |\varphi|^2 dt dx \geqslant \frac{\varepsilon^3 T}{CLM^3} e^{3((2T/3)M-L)^2/(4\varepsilon T)} \int_0^L |\varphi_x(T,x)|^2 dx.$$

Note that the Poincaré inequality, together with (2.602), tells us that

(2.648) $$\int_0^L \varphi(T,x)^2 dx \leqslant CL^2 \int_0^L \varphi_x(T,x)^2 dx.$$

Let us also point out that (2.565) and (2.567) imply that

$$\frac{3}{4}\left(\frac{2TM}{3L} - 1\right)^2 - A \geqslant \frac{1}{C}\frac{M^2 T^2}{L^2},$$

which combined with (2.567), (2.646), (2.647) and (2.648), yields the existence of $C_1 > \max\{m_1, 1\}$ such that $(LM/\varepsilon) \geqslant C_1$ and (2.567) imply (2.604). This concludes the proof of Theorem 2.96 on page 105. ∎

2.7.2.3. *An open problem.* Of course, as one can already see in [**124**], the constants 4.3 and 57.2 in Theorem 2.96 are not optimal. Our open problem in this section concern improvements of these constants. In particular, it is natural to ask

OPEN PROBLEM 2.102. *Assume that*

$$M > 0, \ TM > L.$$

Is it true that

$$K(\varepsilon, T, L, M) \to 0 \ as \ \varepsilon \to 0?$$

Assume that

$$M < 0, \ T|M| > 2L.$$

Is it true that

$$K(\varepsilon, T, L, M) \to 0 \ as \ \varepsilon \to 0?$$

2.8. Bibliographical complements

In this chapter we have already given references to books and papers. But there are of course many other references which must also be mentioned. If one restricts to books or surveys we would like to add in particular (but this is a very incomplete list):

- The book [**47**] by Alain Bensoussan on stochastic control.
- The books [**49, 50**] by Alain Bensoussan, Giuseppe Da Prato, Michel Delfour and Sanjoy Mitter, which deal, in particular, with differential control systems with delays and partial differential control systems with specific emphasis on controllability, stabilizability and the Riccati equations.
- The book [**139**] by Ruth Curtain and Hans Zwart which deals with general infinite-dimensional linear control systems theory. It includes the usual classical topics in linear control theory such as controllability, observability, stabilizability, and the linear-quadratic optimal problem. For a more advanced level on this general approach, one can look at the book [**462**] by Olof Staffans.

- The book [**140**] by René Dáger and Enrique Zuazua on partial differential equations on planar graphs modeling networked flexible mechanical structures (with extensions to the heat, beam and Schrödinger equations on planar graphs).
- The book [**154**] by Abdelhaq El Jaï and Anthony Pritchard on the input-output map and the importance of the location of the actuators/sensors for a better controllability/observability.
- The books [**160, 161**] by Hector Fattorini on optimal control for infinite-dimensional control problems (linear or nonlinear, including partial differential equations).
- The book [**182**] by Andrei Fursikov on study of optimal control problems for infinite-dimensional control systems with many examples coming from physical systems governed by partial differential equations (including the Navier-Stokes equations).
- The book [**280**] by Vilmos Komornik and Paola Loreti on harmonic (and nonharmonic) analysis methods with many applications to the controllability of various time-reversible systems.
- The book [**294**] by John Lagnese and Günter Leugering on optimal control on networked domains for elliptic and hyperbolic equations, with a special emphasis on domain decomposition methods.
- The books [**301, 302**] by Irena Lasiecka and Roberto Triggiani which deal with finite horizon quadratic regulator problems and related differential Riccati equations for general parabolic and hyperbolic equations with numerous important specific examples.
- The book [**487**] by Marius Tucsnak and George Weiss on passive and conservative linear systems, with a detailed chapter on the controllability of these systems.
- The survey [**521**] by Enrique Zuazua on recent results on the controllability of linear partial differential equations. It includes the study of the controllability of wave equations, heat equations, in particular with low regularity coefficients, which is important to treat semi-linear equations (see Section 4.3), fluid-structure interaction models.
- The survey [**519**] by Enrique Zuazua on numerical methods to get optimal controls for linear partial differential equations.

Part 2

Controllability of nonlinear control systems

A major method to study the local controllability around an equilibrium is to look at the controllability of the linearized control system around this equilibrium. Indeed, using the inverse function theorem, the controllability of this linearized control system implies the local controllability of the nonlinear control system, in any cases in finite dimension (see Section 3.1, and in particular Theorem 3.8 on page 128) and in many cases in infinite dimension (see Section 4.1 for an application to a nonlinear Korteweg-de Vries equation). In infinite dimension the situation can be more complicated due to some problems of "loss of derivatives" as explained in Section 4.2. However, suitable iterative schemes (in particular the Nash-Moser process) can allow to handle these cases; see Section 4.2.1 and Section 4.2.2.

When the linearized control system around the equilibrium is not controllable, the situation is more complicated. However, for finite-dimensional systems, one knows powerful tools to handle this situation. These tools rely on iterated Lie brackets. They lead to many sufficient or necessary conditions for local controllability of a nonlinear control system. We recall some of these conditions in Section 3.2, Section 3.3 and in Section 3.4.

In infinite dimension, iterated Lie brackets give some interesting results as we will see in Chapter 5. However, we will also see in the same chapter that these iterated Lie brackets do not work so well in many interesting cases. We present three methods to get in some cases controllability results for some control systems modeled by partial differential equations even if the linearized control system around the equilibrium is not controllable. These methods are:

1. the return method (Chapter 6),
2. quasi-static deformations (Chapter 7),
3. power series expansion (Chapter 8).

Let us briefly describe them.

Return method. The idea of the return method goes as follows. Let us assume that one can find a trajectory of the nonlinear control system such that:

- it starts and ends at the equilibrium,
- the linearized control system around this trajectory is controllable.

Then, in general, the implicit function theorem implies that one can go from every state close to the equilibrium to every other state close to the equilibrium. In Chapter 6, we sketch some results in flow control which have been obtained by this method, namely

1. global controllability results for the Euler equations of incompressible fluids (Section 6.2.1),
2. global controllability results for the Navier-Stokes equations of incompressible fluids (Section 6.2.2.2),
3. local controllability of a 1-D tank containing a fluid modeled by the shallow water equations (Section 6.3).

Quasi-static deformations. The quasi-static deformation method allows one to prove in some cases that one can move from a given equilibrium to another given equilibrium if these two equilibria are connected in the set of equilibria. The idea is just to move very slowly the control (quasi-static deformation) so that at each time the state is close to the curve of equilibria connecting the two given equilibria. If some of these equilibria are unstable, one also uses suitable feedback laws in order to stabilize them: without these feedback laws the quasi-static deformation method would not work in general. We present an application to a semilinear heat equation (Section 7.2).

Power series expansion. In this method one makes some power series expansion in order to decide whether the nonlinearity allows to move in the (oriented) directions which are not controllable for the linearized control system around the equilibrium. We present an application to the Korteweg-de Vries equation (Section 8.2).

These three methods can be used together. We present an example for a Schrödinger control system in Chapter 9.

CHAPTER 3

Controllability of nonlinear systems in finite dimension

Throughout this chapter, we denote by (C) the nonlinear control system

$$(C) \qquad \dot{x} = f(x, u),$$

where $x \in \mathbb{R}^n$ is the state, $u \in \mathbb{R}^m$ is the control, with $(x, u) \in \mathcal{O}$ where $\mathcal{O}$ is a nonempty open subset of $\mathbb{R}^n \times \mathbb{R}^m$. Unless otherwise specified, we assume that $f \in C^\infty(\mathcal{O}; \mathbb{R}^n)$. Let us recall the definition of an equilibrium of the control system $\dot{x} = f(x, u)$.

DEFINITION 3.1. An *equilibrium* of the control system $\dot{x} = f(x, u)$ is a couple $(x_e, u_e) \in \mathcal{O}$ such that

$$(3.1) \qquad f(x_e, u_e) = 0.$$

Let us now give the definition we use in this book for small-time local controllability (we should in fact say small-time local controllability with controls close to u_e).

DEFINITION 3.2. Let $(x_e, u_e) \in \mathcal{O}$ be an equilibrium of the control system $\dot{x} = f(x, u)$. The control system $\dot{x} = f(x, u)$ is *small-time locally controllable at the equilibrium* (x_e, u_e) if, for every real number $\varepsilon > 0$, there exists a real number $\eta > 0$ such that, for every $x^0 \in B_\eta(x_e) := \{x \in \mathbb{R}^n; |x - x_e| < \eta\}$ and for every $x^1 \in B_\eta(x_e)$, there exists a measurable function $u : [0, \varepsilon] \to \mathbb{R}^m$ such that

$$|u(t) - u_e| \leqslant \varepsilon, \ \forall t \in [0, \varepsilon],$$
$$(\dot{x} = f(x, u(t)), x(0) = x^0) \Rightarrow (x(\varepsilon) = x^1).$$

One does not know any checkable necessary and sufficient condition for small-time local controllability for general control systems, even for analytic control systems. However, one knows powerful necessary conditions and powerful sufficient conditions. In this chapter we recall some of these conditions. In particular,

- If the linearized control system at (x_e, u_e) is controllable, then the nonlinear control system $\dot{x} = f(x, u)$ is small-time locally controllable at (x_e, u_e); see Section 3.1.
- We recall the necessary Lie algebra rank condition. It relies on iterated Lie brackets. We explain why iterated Lie brackets are natural for the problem of controllability; see Section 3.2.
- We study in detail the case of the driftless control affine systems, i.e, the case where $f(x, u) = \sum_{i=1}^{m} u_i f_i(x)$. For these systems the above Lie algebra rank condition turns out to be sufficient, even for global controllability; see Section 3.3.

- Among the iterated Lie brackets, we describe some of them which are "good" to give the small-time local controllability and some of them which lead to obstructions to small-time local controllability; see Section 3.4.

3.1. The linear test

In this section we prove that if a linearized control system at an equilibrium (resp. along a trajectory) is controllable, then the nonlinear control system is locally controllable at this equilibrium (resp. along this trajectory). We present an application of this useful result.

We assume that the map f is only of class C^1 on $\mathcal{O}$. We study in this section:

- the local controllability along a trajectory,
- the local controllability at an equilibrium point.

3.1.1. Local controllability along a trajectory. We start with the definition of a trajectory.

DEFINITION 3.3. A *trajectory of the control system* $\dot{x} = f(x, u)$ is a function $(\bar{x}, \bar{u}) : [T_0, T_1] \to \mathcal{O}$ such that

(3.2) $$T_0 < T_1,$$

(3.3) $$\bar{x} \in C^0([T_0, T_1]; \mathbb{R}^n),\ \bar{u} \in L^\infty((T_0, T_1); \mathbb{R}^m),$$

(3.4) $$\exists\ \text{a compact set}\ K \subset \mathcal{O}\ \text{such that}\ (\bar{x}(t), \bar{u}(t)) \in K$$
$$\text{for almost every } t \in (T_0, T_1),$$

(3.5) $$\bar{x}(t_2) = \bar{x}(t_1) + \int_{t_1}^{t_2} f(\bar{x}(t), \bar{u}(t))dt,\ \forall (t_1, t_2) \in [T_0, T_1].$$

Let us make some comments on this definition:

- If $\bar{u}$ is also continuous on $[T_0, T_1]$, then (3.4) is equivalent to
$$(\bar{x}(t), \bar{u}(t)) \in \mathcal{O},\ \forall t \in [T_0, T_1].$$
Moreover, in this case, $\bar{x}$ is of class C^1 and (3.5) is equivalent to
$$\dot{\bar{x}}(t) = f(\bar{x}(t), \bar{u}(t)),\ \forall t \in [T_0, T_1].$$

- Taking into account (3.3) and (3.4), (3.5) is equivalent to

(3.6) $$\dot{\bar{x}} = f(\bar{x}, \bar{u})\ \text{in}\ \mathcal{D}'(T_0, T_1)^n.$$

In (3.6), $\mathcal{D}'(T_0, T_1)$ denotes the set of distributions on (T_0, T_1). Hence (3.6) just means that, for every $\varphi : (T_0, T_1) \to \mathbb{R}^n$ of class C^∞ and with compact support,
$$\int_{T_0}^{T_1} (\bar{x}(t) \cdot \dot{\varphi}(t) + f(\bar{x}(t), \bar{u}(t)) \cdot \varphi(t))dt = 0.$$

For simplicity, when no confusion is possible, we write simply "$\dot{\bar{x}} = f(\bar{x}, \bar{u})$" instead of "$\dot{\bar{x}} = f(\bar{x}, \bar{u})$ in $\mathcal{D}'(T_0, T_1)^n$".

Let us now define the notion of "local controllability along a trajectory".

DEFINITION 3.4. Let $(\bar{x}, \bar{u}) : [T_0, T_1] \to \mathcal{O}$ be a trajectory of the control system $\dot{x} = f(x, u)$. The control system $\dot{x} = f(x, u)$ is *locally controllable along the trajectory* $(\bar{x}, \bar{u})$ if, for every $\varepsilon > 0$, there exists $\eta > 0$ such that, for every $(a, b) \in \mathbb{R}^n \times \mathbb{R}^n$

with $|a-\bar{x}(T_0)| < \eta$ and $|b-\bar{x}(T_1)| < \eta$, there exists a trajectory $(x, u) : [T_0, T_1] \to \mathcal{O}$ such that
$$x(T_0) = a, \ x(T_1) = b,$$
$$|u(t) - \bar{u}(t)| \leqslant \varepsilon, \ t \in [T_0, T_1].$$

Let us also introduce the definition of the linearized control system along a trajectory.

DEFINITION 3.5. The *linearized control system along the trajectory* $(\bar{x}, \bar{u}) : [T_0, T_1] \to \mathcal{O}$ is the linear time-varying control system
$$\dot{x} = \frac{\partial f}{\partial x}(\bar{x}(t), \bar{u}(t))x + \frac{\partial f}{\partial u}(\bar{x}(t), \bar{u}(t))u, \ t \in [T_0, T_1],$$
where, at time $t \in [T_0, T_1]$, the state is $x(t) \in \mathbb{R}^n$ and the control is $u(t) \in \mathbb{R}^m$.

With these definitions, one can state the following classical and useful theorem.

THEOREM 3.6. *Let $(\bar{x}, \bar{u}) : [T_0, T_1] \to \mathcal{O}$ be a trajectory of the control system $\dot{x} = f(x, u)$. Let us assume that the linearized control system along the trajectory $(\bar{x}, \bar{u})$ is controllable (see Definition 1.2 on page 4). Then the nonlinear control system $\dot{x} = f(x, u)$ is locally controllable along the trajectory $(\bar{x}, \bar{u})$.*

Proof of Theorem 3.6. Let $\mathcal{F}$ be the map
$$\mathbb{R}^n \times L^\infty((T_0, T_1); \mathbb{R}^m) \to \mathbb{R}^n \times \mathbb{R}^n$$
$$(a, u) \mapsto \mathcal{F}(a, u) := (a, x(T_1)),$$
where $x : [T_0, T_1] \to \mathbb{R}^n$ is the solution of the Cauchy problem
$$\dot{x} = f(x, u(t)), \ x(T_0) = a.$$
It is a well known result (see, e.g., [**458**, Theorem 1, page 57]) that this map $\mathcal{F}$ is well defined and of class C^1 on a neighborhood of $(\bar{x}(T_0), \bar{u})$ in $\mathbb{R}^n \times L^\infty((T_0, T_1); \mathbb{R}^m)$. Its differential $\mathcal{F}'(\bar{x}(T_0), \bar{u})$ is the following linear map:
$$\mathbb{R}^n \times L^\infty((T_0, T_1); \mathbb{R}^m) \to \mathbb{R}^n \times \mathbb{R}^n$$
$$(a, u) \mapsto (a, x(T_1)),$$
where $x : [T_0, T_1] \to \mathbb{R}^n$ is the solution of the Cauchy problem
$$\dot{x} = \frac{\partial f}{\partial x}(\bar{x}(t), \bar{u}(t))x + \frac{\partial f}{\partial u}(\bar{x}(t), \bar{u}(t))u, \ x(T_0) = a.$$
It follows from the hypothesis of Theorem 3.6 (i.e., the controllability of the linearized control system along the trajectory $(\bar{x}, \bar{u})$) that $\mathcal{F}'(\bar{x}(T_0), \bar{u})$ is onto. Since $\mathbb{R}^n \times \mathbb{R}^n$ is of dimension $2n$, this gives the existence of a linear subspace E of dimension $2n$ such that
$$(3.7) \qquad \mathcal{F}'(\bar{x}(T_0), \bar{u})E = \mathbb{R}^n \times \mathbb{R}^n.$$
We apply the usual inverse function theorem to the restriction of $\mathcal{F}$ to E. We get the existence of a $\nu > 0$ and of a map $\mathcal{G}$ of class C^1 from $B_\nu(\bar{x}(T_0)) \times B_\nu(\bar{x}(T_1))$ into E, with $B_\nu(z) := \{x \in \mathbb{R}^n; |x - z| < \nu\}$ such that
$$(3.8) \qquad \mathcal{G}(\bar{x}(T_0), \bar{x}(T_1)) = (\bar{x}(T_0), \bar{u}) \in E \subset \mathbb{R}^n \times L^\infty((T_0, T_1); \mathbb{R}^m),$$
$$(3.9) \qquad \mathcal{F} \circ \mathcal{G}(a, b) = (a, b), \ \forall (a, b) \in B_\nu(\bar{x}(T_0)) \times B_\nu(\bar{x}(T_1)).$$
Let us write $\mathcal{G}(a, b) = (\mathcal{G}_1(a, b), \mathcal{G}_2(a, b)) \in \mathbb{R}^n \times L^\infty((T_0, T_1); \mathbb{R}^m)$. Let us take
$$(a, b) \in B_\nu(\bar{x}(T_0)) \times B_\nu(\bar{x}(T_1)).$$

From (3.9), we get that $\mathcal{G}_1(a,b) = a$ and that, if $x : [T_0, T_1] \to \mathbb{R}^n$ is the solution of the Cauchy problem
$$\dot{x} = f(x, \mathcal{G}_2(a,b)(t)), \ x(T_0) = a,$$
then $x(T_1) = b$. Finally, since $\mathcal{G}$ is of class C^1, it follows from (3.8) that there exists $C > 0$ such that, $\forall (a,b) \in B_\nu(\bar{x}(T_0)) \times B_\nu(\bar{x}(T_1))$,
$$\|\mathcal{G}_2(a,b) - \bar{u}\|_{L^\infty((T_0, T_1); \mathbb{R}^m)} \leqslant C|(a,b) - (\bar{x}(T_0), \bar{x}(T_1))|$$
$$\leqslant C(|a - \bar{x}(T_0)| + |b - \bar{x}(T_1)|).$$
This concludes the proof of Theorem 3.6. ∎

3.1.2. Local controllability at an equilibrium point. In this section the trajectory $(\bar{x}, \bar{u})$ is constant: there exists $(x_e, u_e) \in \mathcal{O}$ such that
$$(\bar{x}(t), \bar{u}(t)) = (x_e, u_e), \ t \in [T_0, T_1].$$
In particular,
$$(3.10) \qquad f(x_e, u_e) = 0,$$
i.e., (x_e, u_e) is an equilibrium of the control system $\dot{x} = f(x, u)$ (see Definition 3.1 on page 125). Following Definition 3.5, we introduce the following definition.

DEFINITION 3.7. Let (x_e, u_e) be an equilibrium of the control system $\dot{x} = f(x, u)$. The *linearized control system at* (x_e, u_e) *of the control system* $\dot{x} = f(x, u)$ is the linear control system
$$(3.11) \qquad \dot{x} = \frac{\partial f}{\partial x}(x_e, u_e) x + \frac{\partial f}{\partial u}(x_e, u_e) u$$
where, at time t, the state is $x(t) \in \mathbb{R}^n$ and the control is $u(t) \in \mathbb{R}^m$.

Then, from Theorem 3.6 on the previous page, we get the following classical theorem, which is a special case of a theorem due to Lawrence Markus [**343**, Theorem 3] (see also [**309**, Theorem 1, page 366]).

THEOREM 3.8. *Let (x_e, u_e) be an equilibrium point of the control system $\dot{x} = f(x, u)$. Let us assume that the linearized control system of the control system $\dot{x} = f(x, u)$ at (x_e, u_e) is controllable. Then the nonlinear control system $\dot{x} = f(x, u)$ is small-time locally controllable at (x_e, u_e).*

Theorem 3.8, which is easy and natural, is very useful. Let us recall that one can easily check whether the controllability of the linearized control system of the control system $\dot{x} = f(x, u)$ at (x_e, u_e) holds by using the Kalman rank condition (Theorem 1.16 on page 9).

Let us give an application in the following example.

EXAMPLE 3.9. *This example deals with the control of the attitude of a rigid spacecraft with control torques provided by thruster jets. Let $\eta = (\phi, \theta, \psi)$ be the Euler angles of a frame attached to the spacecraft representing rotations with respect to a fixed reference frame. Let $\omega = (\omega_1, \omega_2, \omega_3)$ be the angular velocity of the frame attached to the spacecraft with respect to the reference frame, expressed in the frame attached to the spacecraft, and let J be the inertia matrix of the satellite. The evolution of the spacecraft is governed by the equations*
$$(3.12) \qquad J\dot{\omega} = S(\omega)J\omega + \sum_{i=1}^m u_i b_i, \ \dot{\eta} = A(\eta)\omega,$$

where the $u_i \in \mathbb{R}$, $1 \leqslant i \leqslant m$, are the controls. The $b_1, \ldots, b_m$ are m fixed vectors in $\mathbb{R}^3$ ($u_i b_i \in \mathbb{R}^3$, $1 \leqslant i \leqslant m$ are the torques applied to the spacecraft), $S(\omega)$ is the matrix representation of the wedge-product, i.e.,

$$(3.13) \qquad S(\omega) = \begin{pmatrix} 0 & \omega_3 & -\omega_2 \\ -\omega_3 & 0 & \omega_1 \\ \omega_2 & -\omega_1 & 0 \end{pmatrix},$$

and

$$(3.14) \qquad A(\eta) = \begin{pmatrix} \cos\theta & 0 & \sin\theta \\ \sin\theta\tan\phi & 1 & -\cos\theta\tan\phi \\ -\sin\theta/\cos\phi & 0 & \cos\theta/\cos\phi \end{pmatrix}.$$

The state of our control system is $(\eta_1, \eta_2, \eta_3, \omega_1, \omega_2, \omega_3)^{tr} \in \mathbb{R}^6$ and the control is $(u_1, \ldots, u_m)^{tr} \in \mathbb{R}^m$. Concerning the problem of controllability of (3.14), without loss of generality, we assume that the vectors $b_1, \ldots, b_m$ are linearly independent and that $1 \leqslant m \leqslant 3$. We consider the equilibrium $(0,0) \in \mathbb{R}^{6+m}$. The linearized control system at this equilibrium is

$$(3.15) \qquad \dot{\eta} = \omega, \; \dot{\omega} = \sum_{i=1}^{i=m} u_i J^{-1} b_i$$

where the state is $(\eta_1, \eta_2, \eta_3, \omega_1, \omega_2, \omega_3)^{tr} \in \mathbb{R}^6$ and the control is $(u_1, \ldots, u_m)^{tr} \in \mathbb{R}^m$. Using the Kalman rank condition (Theorem 1.16 on page 9), one easily checks that the linear control system (3.15) is controllable if and only if $m = 3$. Hence, from Theorem 3.8 on the previous page, we get that, if $m = 3$, the control system (3.12) is small-time locally controllable at the equilibrium $(0,0) \in \mathbb{R}^6 \times \mathbb{R}^3$. Of course for $m \leqslant 2$, one cannot deduce anything about the nonlocal controllability of (3.12) from the nonlocal controllability of the linearized control system (3.15). We will study the small-time local controllability of (3.12) for $m \leqslant 2$ in Example 3.35 on page 144.

3.2. Iterated Lie brackets and the Lie algebra rank condition

Let us first recall the definition of the Lie bracket of two vector fields.

DEFINITION 3.10. Let Ω be a nonempty open subset of $\mathbb{R}^n$. Let
$$X := (X^1, \ldots, X^n)^{\text{tr}} \in C^1(\Omega; \mathbb{R}^n),$$
$$Y := (Y^1, \ldots, Y^n)^{\text{tr}} \in C^1(\Omega; \mathbb{R}^n).$$
Then the *Lie bracket* $[X, Y] := ([X,Y]^1, \ldots, [X,Y]^n)^{\text{tr}}$ of X and Y is the element in $C^0(\Omega; \mathbb{R}^n)$ defined by

$$(3.16) \qquad [X, Y](x) := Y'(x)X(x) - X'(x)Y(x), \; \forall x \in \Omega.$$

In other words, the components of $[X, Y](x)$ are

$$(3.17) \quad [X,Y]^j(x) = \sum_{k=1}^n X^k(x) \frac{\partial Y^j}{\partial x_k}(x) - Y^k(x) \frac{\partial X^j}{\partial x_k}(x), \; \forall j \in \{1, \ldots, n\}, \; \forall x \in \Omega.$$

For $X \in C^\infty(\Omega; \mathbb{R}^n)$, $Y \in C^\infty(\Omega; \mathbb{R}^n)$ and $Z \in C^\infty(\Omega; \mathbb{R}^n)$, one has the Jacobi identity

$$(3.18) \qquad [X, [Y, Z]] + [Y, [Z, X]] + [Z, [X, Y]] = 0.$$

Note also that, if $X \in C^\infty(\mathbb{R}^n; \mathbb{R}^n)$ and $Y \in C^\infty(\mathbb{R}^n; \mathbb{R}^n)$ are defined by
$$X(x) := Ax, \; Y(x) := Bx, \; \forall x \in \mathbb{R}^n,$$
where $A \in \mathcal{M}_{n,n}(\mathbb{R})$ and $B \in \mathcal{M}_{n,n}(\mathbb{R})$ are given, then straightforward computations show that

(3.19) $$[X, Y](x) = (BA - AB)x, \; \forall x \in \mathbb{R}^n.$$

The following definition will also be useful.

DEFINITION 3.11. Let Ω be a nonempty open subset of $\mathbb{R}^n$, $X \in C^\infty(\Omega; \mathbb{R}^n)$ and $Y \in C^\infty(\Omega; \mathbb{R}^n)$. One defines, by induction on $k \in \mathbb{N}$, $\mathrm{ad}_X^k Y \in C^\infty(\Omega; \mathbb{R}^n)$ by
$$\mathrm{ad}_X^0 Y = Y,$$
$$\mathrm{ad}_X^{k+1} Y = [X, \mathrm{ad}_X^k Y], \; \forall k \in \mathbb{N}.$$

Before giving the Lie algebra rank condition, let us first explain why it is quite natural that Lie brackets appear for controllability conditions:
- for necessary conditions of controllability because of the Frobenius theorem (Theorem 3.25 on page 141),
- for sufficient conditions because of the following observation. Let us assume that $m = 2$ and that $f(x, u) = u_1 f_1(x) + u_2 f_2(x)$ with $f_1 \in C^\infty(\Omega)$, $f_2 \in C^\infty(\Omega)$, where Ω is a nonempty open subset of $\mathbb{R}^n$. Let us start from a given initial state $a \in \Omega$. Then one knows how to move in the (oriented) directions $\pm f_1(a)$ even with small control: just take $u_1 = \eta \neq 0$ and $u_2 = 0$. Similarly, one knows how to move in the (oriented) directions $\pm f_2(a)$. Let us explain how to move in the direction of the Lie bracket $[f_1, f_2](a)$.

Let $\Psi_i : \mathbb{R} \times \mathbb{R}^n \to \mathbb{R}$, $(t, x) \to \Psi_i^t(x)$ be the flow associated to f_i (for $i \in \{1, 2\}$), that is

(3.20) $$\frac{\partial \Psi_i}{\partial t} = f_i(\Psi_i), \; \Psi_i^0(x) = x.$$

Let $\eta_1 \in \mathbb{R}$ and $\eta_2 \in \mathbb{R}$ be given. Then straightforward computations show that

(3.21) $$\lim_{\varepsilon \to 0} \frac{(\Psi_2^{-\eta_2 \varepsilon}(\Psi_1^{-\eta_1 \varepsilon}(\Psi_2^{\eta_2 \varepsilon}(\Psi_1^{\eta_1 \varepsilon}(a))))) - a}{\varepsilon^2} = \eta_1 \eta_2 [f_1, f_2](a).$$

Hence, if one starts at a, that is, if $x(0) = a$, and if one takes
$$(u_1(t), u_2(t)) = (\eta_1, 0), \text{ for } t \in (0, \varepsilon),$$
$$(u_1(t), u_2(t)) = (0, \eta_2), \text{ for } t \in (\varepsilon, 2\varepsilon),$$
$$(u_1(t), u_2(t)) = (-\eta_1, 0), \text{ for } t \in (2\varepsilon, 3\varepsilon),$$
$$(u_1(t), u_2(t)) = (0, -\eta_2), \text{ for } t \in (3\varepsilon, 4\varepsilon),$$
then, as $\varepsilon \to 0$,

(3.22) $$x(4\varepsilon) = a + \eta_1 \eta_2 \varepsilon^2 [f_1, f_2](a) + o(\varepsilon^2).$$

Hence we have indeed succeeded to move in the (oriented) directions $\pm[f_1, f_2](a)$.

Even in the case of a control system with a drift term, Lie brackets also appear naturally. The drift term is the vector field $x \mapsto f(x, u_e)$. One says that one has a driftless control system if this vector field is identically equal to 0. Let us consider for example the case where $m = 1$ and

(3.23) $$f(x, u) = f_0(x) + u f_1(x),$$

with

(3.24) $$f_0(a) = 0.$$

(Hence, if $u_e = 0$, the drift is f_0.) We start again at time 0 at the state a. One can clearly move in the (oriented) directions $\pm f_1(a)$. Let $\eta \in \mathbb{R}$ be fixed and let us consider, for $\varepsilon > 0$, the control defined on $[0, 2\varepsilon]$ by

$$u(t) = -\eta \text{ for } t \in (0, \varepsilon),$$
$$u(t) = \eta \text{ for } t \in (\varepsilon, 2\varepsilon).$$

Let $x : [0, 2\varepsilon] \to \mathbb{R}^n$ be the solution of

$$\dot{x} = f_0(x) + u(t) f_1(x), \ x(0) = a.$$

Then, straightforward computations show that, as $\varepsilon \to 0$,

(3.25) $$x(2\varepsilon) = a + \varepsilon^2 \eta [f_0, f_1](a) + o(\varepsilon^2).$$

Hence we have succeeded to move in the (oriented) directions $\pm[f_0, f_1](a)$. Of course the case of a system with a drift term is more complicated than the case without drift. For example, in the previous driftless case, one can keep going and for example move in the direction of $[[f_1, f_2], f_1](a)$. However, in the case with the drift term, one cannot always move in the direction of $[[f_1, f_0], f_1](a)$. Indeed, let us consider, for example, the following control system:

(3.26) $$\dot{x}_1 = x_2^2, \ \dot{x}_2 = u,$$

where the state is $x = (x_1, x_2)^{\text{tr}} \in \mathbb{R}^2$ and the control is $u \in \mathbb{R}$. Hence

$$f_0(x) = (x_2^2, 0)^{\text{tr}}, \ f_1(x) = (0, 1)^{\text{tr}}.$$

One easily checks that

$$[[f_1, f_0], f_1](x) = -(2, 0)^{\text{tr}}.$$

However, for a trajectory $t \to (x(t), u(t))$ of (3.26), $t \to x_1(t)$ is a nondecreasing function. Hence one cannot move in the direction of $[[f_1, f_0], f_1](0)$ for the control system (3.26).

But there are "iterated" Lie brackets which are indeed "good" for systems with a drift term. Let us give the simplest one. We start with a definition.

DEFINITION 3.12. The control system $\dot{x} = f(x, u)$ is said to be *control affine* if there are $m + 1$ maps f_i, $i \in \{0, \ldots, m\}$, such that

$$f(x, u) := f_0(x) + \sum_{i=1}^{m} u_i f_i(x), \ \forall (x, u) \in \mathcal{O}.$$

Let us assume that the control system $\dot{x} = f(x, u)$ is control affine and that $\mathcal{O} = \Omega \times \mathbb{R}^m$, where Ω is a nonempty open subset of $\mathbb{R}^n$. Hence

$$\dot{x} = f(x, u) := f_0(x) + \sum_{i=1}^{m} u_i f_i(x)$$

with $f_i \in C^\infty(\Omega; \mathbb{R}^n)$, $i \in \{0, \ldots, m\}$. Let $a \in \Omega$ be such that

$$f_0(a) = 0.$$

We explain how to move in the (oriented) directions $\pm \mathrm{ad}_{f_0}^k f_j(a)$, for every $k \in \mathbb{N}$ and for every $j \in \{1, \ldots, m\}$. Let us fix $j \in \{1, \ldots, m\}$ and $k \in \mathbb{N}$. Let $\phi \in C^k([0,1])$ be such that

$$\phi^{(l)}(0) = \phi^{(l)}(1) = 0, \ \forall l \in \{0, \ldots, k-1\}. \tag{3.27}$$

Let $\eta \in (0,1]$ and let $\varepsilon \in [0,1]$. Let us define $u : [0, \eta] \to \mathbb{R}^m$, $t \mapsto (u_1(t), \ldots, u_m(t))^{\mathrm{tr}}$ by

$$u_i(t) = 0 \text{ if } i \neq j, \ \forall t \in [0, \eta], \tag{3.28}$$

$$u_j(t) = \varepsilon \phi^{(k)}(t/\eta), \ \forall t \in [0, \eta]. \tag{3.29}$$

Let $x : [0, \eta] \to \mathbb{R}^n$ be the solution of the Cauchy problem

$$\dot{x} = f_0(x) + \sum_{i=1}^m u_i(t) f_i(x), \ x(0) = a.$$

Let us check that

$$x(\eta) = a + \varepsilon \eta^{k+1} \left(\int_0^1 \phi(\tau) d\tau \right) \mathrm{ad}_{f_0}^k f_j(a) + O(\varepsilon^2 + \eta^{k+2}) \text{ as } |\varepsilon| + |\eta| \to 0. \tag{3.30}$$

Let $A \in \mathcal{L}(\mathbb{R}^n; \mathbb{R}^n)$ and $B \in \mathcal{L}(\mathbb{R}^m; \mathbb{R}^n)$ be defined by

$$A := \frac{\partial f_0}{\partial x}(a), \ B := \frac{\partial f}{\partial u}(a, 0) = (f_1(a), \ldots, f_m(a)).$$

Let $y : [0, \eta] \to \mathbb{R}^n$ be the solution of the Cauchy problem

$$\dot{y} = Ay + Bu(t), \ y(0) = 0.$$

Using Gronwall's lemma and the change of time $\tau := t/\eta$, one easily checks that there exist $C > 0$ and $\varepsilon_0 > 0$ such that, for every $\varepsilon \in [0, \varepsilon_0]$ and for every $\eta \in (0, 1]$,

$$|x(t) - (a + y(t))| \leqslant C\varepsilon^2, \ \forall t \in [0, \eta]. \tag{3.31}$$

Let us estimate $y(\eta)$. One has, using (3.27), (3.28), (3.29), integrations by parts and the change of variable $\tau := t/\eta$,

$$\begin{aligned} y(\eta) &= \varepsilon \int_0^\eta \phi^{(k)}(t/\eta) e^{(\eta-t)A} f_j(a) dt \\ &= -\eta \varepsilon \int_0^\eta \phi^{(k-1)}(t/\eta) e^{(\eta-t)A} A f_j(a) dt \\ &\vdots \\ &= (-1)^k \eta^k \varepsilon \int_0^\eta \phi(t/\eta) e^{(\eta-t)A} A^k f_j(a) dt \\ &= (-1)^k \eta^{k+1} \varepsilon \int_0^1 \phi(\tau) e^{\eta(1-\tau)A} A^k f_j(a) d\tau. \end{aligned}$$

Hence, for some $C > 0$ independent of $\varepsilon \in [0, +\infty)$ and of $\eta \in (0, 1]$,

$$\left| y(\eta) - (-1)^k \eta^{k+1} \varepsilon \left(\int_0^1 \phi(\tau) d\tau \right) A^k f_j(a) \right| \leqslant C \varepsilon \eta^{k+2}. \tag{3.32}$$

But, as one easily checks,

$$\mathrm{ad}_{f_0}^k f_j(a) = (-1)^k A^k f_j(a). \tag{3.33}$$

Equality (3.30) follows from (3.31), (3.32) and (3.33). Note that by taking $\eta = \varepsilon^{1/(2k+2)}$ in (3.30), one gets

$$(3.34) \quad x(\eta) = a + \varepsilon^{3/2}\left(\int_0^1 \phi(\tau)d\tau\right)\mathrm{ad}_{f_0}^k f_j(a) + O(\varepsilon^{(3/2)+(1/(2k+2))}) \text{ as } \varepsilon \to 0,$$

which shows that one can move in the (oriented) directions $\pm\mathrm{ad}_{f_0}^k f_j(a)$. From this fact one gets the small-time local controllability at the equilibrium $(a, 0)$ of the control system $\dot{x} = f_0(x) + \sum_{i=1}^m u_i(t) f_i(x)$ if

$$(3.35) \quad \mathrm{Span}\{\mathrm{ad}_{f_0}^k f_j(a); k \in \mathbb{N}, j \in \{1, \ldots, m\}\} = \mathbb{R}^n.$$

This can be done, for example, by using the Brouwer fixed-point theorem, as does Halina Frankowska in [**176**, **177**] (see also Section 3.5.2) or a constructive iterative scheme as Matthias Kawski does in [**270**].

Note that (3.35) is equivalent to the fact that the linearized control system of $\dot{x} = f(x, u) = f_0(x) + \sum_{i=1}^m u_i f_i(x)$ around the equilibrium point $(x_e, u_e) = (a, 0)$ is controllable. This indeed follows from (3.33) and the Kalman rank condition (Theorem 1.16 on page 9). (Note that the equality

$$\mathrm{Span}\{\mathrm{ad}_{f_0}^k f_j(a); k \in \mathbb{N}, j \in \{1, \ldots, m\}\}$$
$$= \mathrm{Span}\{\mathrm{ad}_{f_0}^k f_j(a); k \in \{0, \ldots, n-1\}, j \in \{1, \ldots, m\}\}$$

follows from the Cayley-Hamilton theorem (see (1.41)) and (3.33)). Hence, with these iterated Lie brackets $\mathrm{ad}_{f_0}^k f_j(a)$, we just recover the linear test, i.e., Theorem 3.8 on page 128.

Let us now give a necessary condition for local controllability. We start with some new definitions.

DEFINITION 3.13. Let Ω be a nonempty open subset of $\mathbb{R}^n$ and let $\mathcal{F}$ be a family of vector fields of class C^∞ in Ω. We denote by $\mathrm{Lie}(\mathcal{F})$ *the Lie algebra generated by the vector fields in $\mathcal{F}$*, i.e., the smallest linear subspace E of $C^\infty(\Omega; \mathbb{R}^n)$ satisfying the following two conditions:

$$(3.36) \quad \mathcal{F} \subset E,$$
$$(3.37) \quad (X \in E \text{ and } Y \in E) \Rightarrow ([X, Y] \in E).$$

(Such a smallest subspace exists: just note that $E := C^\infty(\Omega; \mathbb{R}^n)$ satisfies (3.36) and (3.37) and take the intersection of all the linear subspaces of $C^\infty(\Omega; \mathbb{R}^n)$ satisfying (3.36) and (3.37).)

For the next definition, let Ω be the open subset of $\mathbb{R}^n$ defined by

$$\Omega := \{x \in \mathbb{R}^n; (x, u_e) \in \mathcal{O}\}.$$

For every $\alpha \in \mathbb{N}^m$,

$$\frac{\partial^{|\alpha|} f}{\partial u^\alpha}(\cdot, u_e) \in C^\infty(\Omega; \mathbb{R}^n).$$

DEFINITION 3.14 ([**106**]). The *strong jet accessibility subspace of the control system* $\dot{x} = f(x, u)$ at an equilibrium point (x_e, u_e) is the linear subspace of $\mathbb{R}^n$, denoted by $\mathcal{A}(x_e, u_e)$, defined by

$$(3.38) \quad \mathcal{A}(x_e, u_e) := \left\{g(x_e); g \in \mathrm{Lie}\left(\frac{\partial^{|\alpha|} f}{\partial u^\alpha}(\cdot, u_e), \alpha \in \mathbb{N}^m\right)\right\}.$$

EXAMPLE 3.15. Let us assume that, for some open neighborhood ω of x_e in $\mathbb{R}^n$, there exist $r > 0$, and f_0, f_1, ..., f_m in $C^\infty(\omega; \mathbb{R}^n)$ such that, for every $x \in \omega$ and for every $u = (u_1, \ldots, u_m)^{tr} \in \mathbb{R}^m$ such that $|u - u_e| < r$,
$$(x, u) \in \mathcal{O},$$
$$f(x, u) = f_0(x) + \sum_{i=1}^{m} u_i f_i(x).$$

Then
$$\mathcal{A}(x_e, u_e) = \{g(x_e); g \in \mathrm{Lie}(\{f_0, f_1, \ldots, f_m\})\}.$$

Let us now define the Lie algebra rank condition.

DEFINITION 3.16. The control system $\dot{x} = f(x, u)$ satisfies the *Lie algebra rank condition* at an equilibrium (x_e, u_e) if
(3.39) $$\mathcal{A}(x_e, u_e) = \mathbb{R}^n.$$

With these definitions, one has the following well-known necessary condition for small-time local controllability of analytic control systems due to Robert Hermann [221] and Tadashi Nagano [372] (see also [465] by Héctor Sussmann).

THEOREM 3.17. *Assume that the control system $\dot{x} = f(x, u)$ is small-time locally controllable at the equilibrium point (x_e, u_e) and that f is analytic. Then the control system $\dot{x} = f(x, u)$ satisfies the Lie algebra rank condition at (x_e, u_e).*

Note that the assumption on analyticity cannot be removed. Indeed, consider the following case
$$n = m = 1, \mathcal{O} = \mathbb{R} \times \mathbb{R}, (x_e, u_e) = (0, 0) \in \mathbb{R} \times \mathbb{R},$$
$$f(x, u) = u e^{-1/u^2} \text{ if } u \neq 0, f(x, 0) = 0.$$

Then $\mathcal{A}(0, 0) = 0$. However, the control system $\dot{x} = f(x, u)$ is small-time controllable at the equilibrium point (x_e, u_e).

The necessary condition for small-time controllability given in Theorem 3.17 is sufficient for important control systems such as, for example, the linear control systems $\dot{x} = Ax + Bu$. This follows from the Kalman rank condition (Theorem 1.16 on page 9) and the fact that
$$\mathcal{A}(x_e, u_e) = \mathrm{Span} \{A^i Bu; i \in \{0, \ldots, n-1\}, u \in \mathbb{R}^m\}$$
(see (3.33)). It is also a sufficient condition for driftless control affine systems, that is, in the case $\dot{x} = \sum_{i=1}^{m} u_i f_i(x)$; this is the classical Rashevski-Chow theorem [398, 94] that we recall and prove in the following section.

3.3. Controllability of driftless control affine systems

In fact, for driftless control affine systems one also has a global controllability result as shown by the following theorem proved independently by Petr Rashevski in [398] and by Wei-Liang Chow in [94].

THEOREM 3.18. *Let Ω be a connected nonempty open subset of $\mathbb{R}^n$. Let us assume that $\mathcal{O} = \Omega \times \mathbb{R}^m$ and that for some $f_1, \ldots, f_m \in C^\infty(\Omega; \mathbb{R}^n)$,*
$$f(x, u) = \sum_{i=1}^{m} u_i f_i(x), \forall (x, u) \in \Omega \times \mathbb{R}^m.$$

Let us also assume that

(3.40) $$\mathcal{A}(x,0) = \mathbb{R}^n, \forall x \in \Omega.$$

Then, for every $(x^0, x^1) \in \Omega \times \Omega$ and for every $T > 0$, there exists u belonging to $L^\infty((0,T); \mathbb{R}^m)$ such that the solution of the Cauchy problem

$$\dot{x} = f(x, u(t)), \ x(0) = x^0,$$

is defined on $[0,T]$ and satisfies $x(T) = x^1$.

Theorem 3.18 is a global controllability result. In general such a controllability result does not imply small-time local controllability. However, the proof of Theorem 3.18 given below also gives the small-time local controllability. More precisely, one also has the following theorem.

THEOREM 3.19. *Let us assume that Ω is a nonempty open subset of $\mathbb{R}^n$, that $\Omega \times \{0\} \subset \mathcal{O}$ and that, for some $f_1, \ldots, f_m \in C^\infty(\Omega; \mathbb{R}^n)$,*

$$f(x, u) = \sum_{i=1}^m u_i f_i(x), \ \forall (x, u) \in \mathcal{O}.$$

Let $x_e \in \Omega$ be such that

$$\mathcal{A}(x_e, 0) = \mathbb{R}^n.$$

Then the control system $\dot{x} = f(x, u)$ is small-time locally controllable at $(x_e, 0) \in \mathcal{O}$.

EXAMPLE 3.20. *Let us consider the following control system (usually called the nonholonomic integrator)*

(3.41) $$\dot{x}_1 = u_1, \ \dot{x}_2 = u_2, \ \dot{x}_3 = x_1 u_2 - x_2 u_1.$$

Thus $n = 3$, $m = 2$, $\Omega = \mathbb{R}^3$ and, for every $x = (x_1, x_2, x_3)^{tr} \in \mathbb{R}^3$,

(3.42) $$f_1(x) := \begin{pmatrix} 1 \\ 0 \\ -x_2 \end{pmatrix}, \ f_2(x) := \begin{pmatrix} 0 \\ 1 \\ x_1 \end{pmatrix}.$$

One easily checks that, for every $x = (x_1, x_2, x_3)^{tr} \in \mathbb{R}^3$,

(3.43) $$[f_1, f_2](x) = \begin{pmatrix} 0 \\ 0 \\ 2 \end{pmatrix}.$$

From (3.42) and (3.43),

$$\text{Span}\{f_1(x), f_2(x), [f_1, f_2](x)\} = \mathbb{R}^3, \forall x \in \mathbb{R}^3.$$

Hence the control system (3.41) satisfies the Lie algebra rank condition (3.40) at every point $x \in \mathbb{R}^3$. Therefore, by Theorem 3.18, this control system is globally controllable and, by Theorem 3.19, it is small-time locally controllable at every equilibrium $(x_e, u_e) = (x_e, 0) \in \mathbb{R}^3 \times \mathbb{R}^2$.

Proof of Theorem 3.18. Let us give a proof of this theorem which does not require any knowledge of submanifolds or geometry. Just in order to simplify the notations, we assume that the vector fields $f_1, \ldots, f_m$ are complete, which means that the associated flow maps $(t, x) \mapsto \Psi_i^t(x)$ (see (3.20)) are defined on the whole $\mathbb{R} \times \Omega$. In fact, this can always be assumed by replacing if necessary $f_i(x)$ by

$\tilde{f}_i(x) = \rho(x)(1+|f_i(x)|^2)^{-1}f_i(x)$ where $\rho \in C^\infty(\Omega, (0,1])$ is such that, if $\Omega \neq \mathbb{R}^n$, there exists $C > 0$ such that

$$(3.44) \qquad \rho(x) \leqslant C \mathrm{dist}(x, \mathbb{R}^n \setminus \Omega), \forall x \in \Omega.$$

(The existence of such a ρ can be easily proved by means of a partition of unity argument; see, for example, [**464**, Theorem 2, Chapter 6, page 171] for a much more precise existence result.) Indeed:

- The vector fields $\tilde{f}_i$ are complete.
- The condition (3.40) implies the same condition for the function $f(x,u) := \sum_1^m u_i \tilde{f}_i(x)$.
- The conclusion of Theorem 3.18 for $f(x,u) := \sum_1^m u_i \tilde{f}_i(x)$ implies the conclusion of Theorem 3.18 for $f(x,u) := \sum_1^m u_i f_i(x)$.

Let us first point out that, for every $t < 0$,

$$(3.45) \qquad (\dot{y} = -f_i(y),\ y(0) = x) \Rightarrow (\Psi_i^t(x) = y(|t|)).$$

Let $\mathfrak{A}(a) \subset \Omega$ be the set consisting of the $\Psi_{i_k}^{t_k} \circ \Psi_{i_{k-1}}^{t_{k-1}} \circ \ldots \circ \Psi_{i_1}^{t_1}(a)$ where

$$(3.46) \qquad k \in \mathbb{N} \setminus \{0\},$$
$$(3.47) \qquad i_j \in \{1,\ldots,m\},\ \forall j \in \{1,\ldots,k\},$$
$$(3.48) \qquad t_j \in \mathbb{R},\ \forall j \in \{1,\ldots,k\}.$$

Let us point out that it suffices to prove that

$$(3.49) \qquad \mathfrak{A}(a) = \Omega,\ \forall a \in \Omega.$$

Indeed, let k, $(i_j)_{j \in \{1,\ldots,k\}}$ and $(t_j)_{j \in \{1,\ldots,k\}}$ be such that (3.46), (3.47) and (3.48) hold. Let $a \in \Omega$ and let

$$(3.50) \qquad b := \Psi_{i_k}^{t_k} \circ \Psi_{i_{k-1}}^{t_{k-1}} \circ \ldots \circ \Psi_{i_1}^{t_1}(a) \in \Omega.$$

Let $\tau_0 = 0$ and, for $j \in \{1,\ldots,k\}$, let

$$\tau_j := \sum_{i=1}^j |t_i|.$$

Let $\bar{u} \in L^\infty((0,\tau_k);\mathbb{R}^m)$ be defined by

$$\bar{u}(t) = \varepsilon_j e_{i_j},\ \forall t \in (\tau_{j-1}, \tau_j),\ \forall j \in \{1,\ldots,k\},$$

where $(e_1,\ldots,e_m)$ is the canonical basis of $\mathbb{R}^m$ and $\varepsilon_j := \mathrm{Sign}(t_j),\ j \in \{1,\ldots,k\}$. Let $\bar{x} \in C^0([0,\tau_k];\Omega)$ be the solution of the Cauchy problem

$$\dot{\bar{x}} = f(\bar{x},\bar{u}(t)),\ \bar{x}(0) = a.$$

Clearly (see also (3.45)),

$$\bar{x}(t) = \Psi_{i_1}^{\varepsilon_1 t}(a),\ \forall t \in [0,\tau_1].$$

Similarly, by induction on j, one easily checks that

$$(3.51) \quad \bar{x}(t) = \Psi_{i_j}^{\varepsilon_j(t-\tau_{j-1})} \circ \Psi_{i_{j-1}}^{t_{j-1}} \circ \ldots \circ \Psi_{i_1}^{t_1}(a),\ \forall t \in (\tau_{j-1},\tau_j),\ \forall j \in \{2,\ldots,k\}.$$

From (3.50) and (3.51), one has

$$\bar{x}(\tau_k) = b.$$

Let $T > 0$ and let $(\tilde{x}, \tilde{u}) \in C^0([0,T]; \Omega) \times L^\infty((0,T); \mathbb{R}^m)$ be defined by

$$\tilde{x}(t) = \bar{x}\left(\frac{\tau_k t}{T}\right), \forall t \in [0,T], \tilde{u}(t) = \frac{\tau_k}{T} \bar{u}\left(\frac{\tau_k t}{T}\right), \forall t \in (0,T).$$

Then $(\tilde{x}, \tilde{u})$ is a trajectory of the control system $\dot{x} = f(x, u)$ such that $\tilde{x}(0) = a$ and $\tilde{x}(T) = b$. Hence property (3.49) implies Theorem 3.18.

Let us now prove that property (3.49) holds. From the definition of $\mathfrak{A}(d)$ for $d \in \Omega$, one easily gets that

(3.52) $\quad (b \in \mathfrak{A}(a) \text{ and } c \in \mathfrak{A}(b)) \Rightarrow (c \in \mathfrak{A}(a)), \quad \forall (a,b,c) \in \Omega \times \Omega \times \Omega.$

Moreover, since

$$\Psi_i^t \circ \Psi_i^{-t}(x) = x, \quad \forall x \in \Omega, \quad \forall t \in \mathbb{R}, \quad \forall i \in \{1, \ldots, m\},$$

one has

$$b = \Psi_{i_k}^{t_k} \circ \Psi_{i_{k-1}}^{t_{k-1}} \circ \ldots \circ \Psi_{i_1}^{t_1}(a) \Rightarrow a = \Psi_{i_1}^{-t_1} \circ \ldots \circ \Psi_{i_{k-1}}^{-t_{k-1}} \circ \Psi_{i_k}^{-t_k}(b)$$

and therefore

(3.53) $\quad (b \in \mathfrak{A}(a) \Rightarrow a \in \mathfrak{A}(b)), \quad \forall (a,b) \in \Omega \times \Omega.$

Since $a \in \mathfrak{A}(a)$, it follows from (3.52) and (3.53) that the relation

(3.54) $\quad b$ is related to a if and only if $b \in \mathfrak{A}(a)$

defines an equivalence relation on Ω. Since Ω is connected, we get that, for every $a \in \Omega$, $\mathfrak{A}(a) = \Omega$ if one has

(3.55) $\quad \mathfrak{A}(a)$ is an open subset of $\Omega, \forall a \in \Omega.$

(Note that Ω is the disjoint union of the equivalence classes of the relation (3.54) and that the equivalent class of a is $\mathfrak{A}(a)$.) Let us assume, for the moment, that the following lemma holds.

LEMMA 3.21. *Under assumption (3.40) of Theorem 3.18,*

$$\mathfrak{A}(a) \text{ has a nonempty interior}, \forall a \in \Omega.$$

We assume (3.40). Let us prove (3.55) within the following two steps
Step 1. Let us prove that

(3.56) $\quad \mathfrak{A}(a)$ is a neighborhood of $a, \forall a \in \Omega.$

From Lemma 3.21, there exists $b \in \mathfrak{A}(a)$ such that $\mathfrak{A}(a)$ is a neighborhood of b. Since $b \in \mathfrak{A}(a)$, there exist $k \in \mathbb{N} \setminus \{0\}$, k indices $i_1, \ldots, i_k$ in $\{1, \ldots, m\}$ and k real numbers $t_1, \ldots, t_k$ such that $b = \psi(a)$ with $\psi = \Psi_{i_k}^{t_k} \circ \Psi_{i_{k-1}}^{t_{k-1}} \circ \ldots \circ \Psi_{i_1}^{t_1}$. Since ψ is continuous, $\psi^{-1}(\mathfrak{A}(a))$ is a neighborhood of a. Moreover, ψ is a homeomorphism of Ω: one has $\psi^{-1} = \Psi_{i_1}^{-t_1} \circ \ldots \circ \Psi_{i_{k-1}}^{-t_{k-1}} \circ \Psi_{i_k}^{-t_k}$. This expression of ψ^{-1} and (3.52) show that $\psi^{-1}(\mathfrak{A}(a)) \subset \mathfrak{A}(a)$, which concludes the proof of (3.56).

Step 2. $\mathfrak{A}(a)$ is an open subset of Ω. Indeed, let $b \in \mathfrak{A}(a)$. Then $\mathfrak{A}(b)$ is a neighborhood of b. However, from (3.52), $\mathfrak{A}(b) \subset \mathfrak{A}(a)$. Hence, by (3.56) applied to b, $\mathfrak{A}(a)$ is a neighborhood of b.

It only remains to prove Lemma 3.21. A key tool for its proof is the following classical lemma.

LEMMA 3.22. *Let k be in $\{1, \ldots, n-1\}$. Let $\varphi : \mathbb{R}^k \to \Omega$, $y \mapsto \varphi(y)$ be a map of class C^∞ in a neighborhood of $\overline{y} \in \mathbb{R}^k$. Assume that $\varphi'(\overline{y})$ is one-to-one. Then there exist an open subset ω of $\mathbb{R}^k$ containing $\overline{y}$, an open subset U of Ω containing $\varphi(\omega)$ and a map $V \in C^\infty(U; \mathbb{R}^{n-k})$ such that*

$$V(\varphi(y)) = 0, \; \forall y \in \omega,$$

$$\ker V'(\varphi(y)) = \varphi'(y)(\mathbb{R}^k), \; \forall y \in \omega.$$

Proof of Lemma 3.22. Let ℓ be a linear map from $\mathbb{R}^{n-k}$ to $\mathbb{R}^n$ such that

(3.57) $$\ell(\mathbb{R}^{n-k}) + \varphi'(\overline{y})(\mathbb{R}^k) = \mathbb{R}^n.$$

Let

$$\psi : \mathbb{R}^k \times \mathbb{R}^{n-k} \to \mathbb{R}^n, \; (y, z) \mapsto \psi(y, z) = \varphi(y) + \ell(z).$$

From (3.57), $\psi'(\overline{y}, 0)$ is onto and therefore invertible. By the inverse function theorem, there exist an open subset U_1 of $\mathbb{R}^k \times \mathbb{R}^{n-k}$ containing $(\overline{y}, 0)$, an open subset U of $\mathbb{R}^n$ containing $\varphi(\overline{y})$ $(= \psi(\overline{y}, 0))$ and $\mathcal{T} \in C^\infty(U; U_1)$ such that

$$(\psi(y, z) = x \text{ and } (y, z) \in U_1) \Rightarrow (x \in U \text{ and } (y, z) = \mathcal{T}(x)).$$

Let $P_2 : \mathbb{R}^k \times \mathbb{R}^{n-k} \to \mathbb{R}^{n-k}$, $(y, z) \mapsto z$. One easily sees that it suffices to take

(i) for ω, an open subset of $\mathbb{R}^k$ containing $\overline{y}$ and such that $\omega \times \{0\} \subset U_1$,
(ii) $V = P_2 \circ \mathcal{T}$.

This concludes the proof of Lemma 3.22. ∎

Let us now prove Lemma 3.21. Let $a \in \Omega$. Let us first point out that the vectors $f_1(a), \ldots, f_m(a)$ cannot all be equal to 0. Indeed, if all these vectors are equal to 0, then $\mathcal{A}(a, 0) = 0$ (note that, if $g \in C^\infty(\Omega; \mathbb{R}^n)$ and $h \in C^\infty(\Omega; \mathbb{R}^n)$ are such that $g(a) = 0$ and $h(a) = 0$, then $[g, h](a) = 0$), in contradiction with (3.40). Let $i_1 \in \{1, \ldots, m\}$ be such that

(3.58) $$f_{i_1}(a) \neq 0.$$

Let $\varphi : \mathbb{R} \to \mathbb{R}^n$ be defined by $\varphi(t) := \Psi_{i_1}^t(a)$. We have $\varphi'(0) = f_{i_1}(a)$. Hence, by (3.58), $\varphi'(0) \neq 0$. If $n = 1$, $\{\Psi_{i_1}^t(a); t \in \mathbb{R}\}$ is a neighborhood of a. Hence we may assume that $n \geqslant 2$. We apply Lemma 3.22 with $k = 1$, and $\overline{y} = 0$. We get the existence of $\varepsilon_1 > 0$, of an open subset Ω_1 of Ω containing $\{\Psi_{i_1}^t(a) \, ; \, t \in (-\varepsilon_1, \varepsilon_1)\}$ and of $V \in C^\infty(\Omega; \mathbb{R}^{n-1})$ such that

(3.59) $$V(\Psi_{i_1}^t(a)) = 0, \; \forall t \in (-\varepsilon_1, \varepsilon_1),$$
(3.60) $$\ker V'(\Psi_{i_1}^t(a)) = \mathbb{R} f_{i_1}(\Psi_{i_1}^t(a)), \; \forall t \in (-\varepsilon_1, \varepsilon_1).$$

Then there exist $t_1^1 \in (-\varepsilon_1, \varepsilon_1)$ and $i_2 \in \{1, \ldots, m\}$ such that

$$V'(\Psi_{i_1}^{t_1^1}(a)) f_{i_2}(\Psi_{i_1}^{t_1^1}(a)) \neq 0.$$

Indeed, if this is not the case, then

$$E := \{g \in C^\infty(\Omega; \mathbb{R}^n) \, ; \, V'(\Psi_{i_1}^t(a)) g(\Psi_{i_1}^t(a)) = 0, \, \forall t \in (-\varepsilon_1, \varepsilon_1)\}$$

is a vector subspace of $C^\infty(\Omega; \mathbb{R}^n)$ which contains the set $\{f_1, \ldots, f_m\}$. Let us assume, for the moment, that

(3.61) $$(g \in E \text{ and } h \in E) \Rightarrow ([g, h] \in E).$$

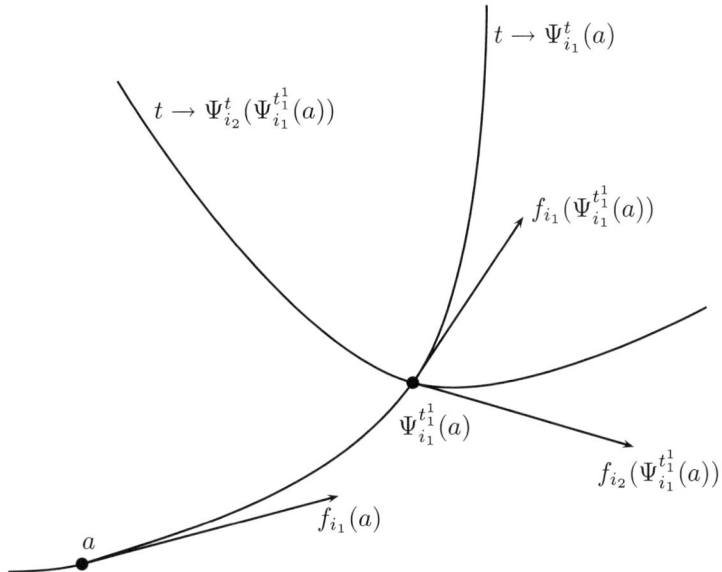

FIGURE 1. $f_{i_1}(\Psi_{i_1}^{t_1^1}(a))$, $f_{i_2}(\Psi_{i_1}^{t_1^1}(a))$ are linearly independent

Then
$$\text{Lie}\,(f_1,\ldots,f_m) \subset E$$
and therefore
$$\mathcal{A}(a,0) \subset \text{Ker}\,V'(a),$$
in contradiction with (3.40) for $x=a$ and (3.60) for $t=0$.

Let us check (3.61). This is a consequence of the following lemma.

LEMMA 3.23. *Let $\theta \in C^\infty(\Omega;\mathbb{R}^{n-1})$ be such that*
$$\theta(\Psi_{i_1}^t(a)) = 0,\, \forall t \in (-\varepsilon_1,\varepsilon_1) \tag{3.62}$$
and $F \in C^\infty(\Omega;\mathbb{R}^n)$ be such that
$$V'(\Psi_{i_1}^t(a))\,F(\Psi_{i_1}^t(a)) = 0,\, \forall t \in (-\varepsilon_1,\varepsilon_1). \tag{3.63}$$
Then
$$\theta'(\Psi_{i_1}^t(a))\,F(\Psi_{i_1}^t(a)) = 0,\, \forall t \in (-\varepsilon_1,\varepsilon_1). \tag{3.64}$$

Indeed, it suffices to apply this lemma to $(\theta, F) := (\bar\theta, \bar F) := (\nabla V \cdot g, h)$ and to $(\theta, F) := (\tilde\theta, \tilde F) := (\nabla V \cdot h, g)$ and to check that
$$\nabla \tilde\theta \cdot \tilde F - \nabla \bar\theta \cdot \bar F = \nabla V \cdot [g,h]. \tag{3.65}$$

Let us now prove Lemma 3.23. Differentiating (3.62) with respect to t we get
$$\theta'(\Psi_{i_1}^t(a))\,f_{i_1}(\Psi_{i_1}^t(a)) = 0,\, \forall t \in (-\varepsilon_1,\varepsilon_1). \tag{3.66}$$

But (3.60) and (3.63) imply that $F(\Psi_{i_1}^t(a)) = \lambda(t) f_{i_1}(\Psi_{i_1}^t(a))$, with $\lambda(t) \in \mathbb{R}$ for $t \in (-\varepsilon_1,\varepsilon_1)$. This implies, together with (3.66), that (3.64) holds. Thus Lemma 3.23 is proved.

Therefore, there exist $t_1^1 \in (-\varepsilon_1, \varepsilon_1)$ and $i_2 \in \{1, \ldots, m\}$ such that

(3.67) $$V'(\Psi_{i_1}^{t_1^1}(a)) f_{i_2}(\Psi_{i_1}^{t_1^1}(a)) \neq 0,$$

which, together with (3.60), implies that

(3.68) $\quad f_{i_1}(\Psi_{i_1}^{t_1^1}(a))$ and $f_{i_2}(\Psi_{i_1}^{t_1^1}(a))$ are linearly independent

(see figure 1).

Let $\varphi : \mathbb{R}^2 \to \mathbb{R}^n$, $(t_1, t_2) \mapsto \Psi_{i_2}^{t_2} \circ \Psi_{i_1}^{t_1}(a)$. One has $\varphi(\mathbb{R}^2) \subset \mathfrak{A}(a)$ and

$$\frac{\partial \varphi}{\partial t_2}(t_1^1, 0) = f_{i_2}(\Psi_{i_1}^{t_1^1}(a)),$$
$$\frac{\partial \varphi}{\partial t_1}(t_1^1, 0) = f_{i_1}(\Psi_{i_1}^{t_1^1}(a)).$$

In particular, with (3.68), $\varphi'(t_1^1, 0)$ is one-to-one. If $n = 2$, $\varphi'(t_1^1, 0)$ is then invertible and the inverse function theorem gives us that $\varphi(\mathbb{R}^2)$ (and therefore $\mathfrak{A}(a)$) is a neighborhood of $\varphi(t_1^1, 0)$. Hence we may assume that $n \geqslant 3$ and one proceeds as above. For convenience let us briefly sketch how it is done. We apply Lemma 3.22 on page 138 with $k = 2$, $\overline{y} = (t_1^1, 0)$. One gets the existence of an open subset ω of $\mathbb{R}^2$ containing $\overline{y}$, an open subset U of Ω containing $\varphi(\omega)$ and a map $V \in C^\infty(U; \mathbb{R}^{n-2})$ such that

$$V(\varphi(t_1, t_2)) = 0, \, \forall (t_1, t_2) \in \omega,$$
(3.69) $\quad \ker V'(\varphi(t_1, t_2)) = \varphi'(t_1, t_2)(\mathbb{R}^2), \, \forall (t_1, t_2) \in \omega.$

Once more, one points out that there exist $(t_1^2, t_2^2) \in \omega$ and $i_3 \in \{1, \ldots, m\}$ such that

(3.70) $$V'(\varphi(t_1^2, t_2^2)) f_{i_3}(\varphi(t_1^2, t_2^2)) \neq 0.$$

Indeed, if this is not the case, then

$$E := \{f \in C^\infty(\Omega; \mathbb{R}^n); \, V'(\varphi(y)) f(\varphi(y)) = 0, \, \forall y \in \omega\}$$

is a vector subspace of $C^\infty(\Omega; \mathbb{R}^n)$ which contains the set $\{f_1, \ldots, f_m\}$. As above, one checks that (3.61) holds and therefore

$$\text{Lie}\,(f_1, \ldots, f_m) \subset E.$$

Hence
$$\mathcal{A}(\varphi(\overline{y}), 0) \subset \text{Ker}\, V'(\varphi(\overline{y})),$$

which leads to a contradiction with (3.40). Let

$$\psi : \mathbb{R}^3 \to \Omega, \, (t_1, t_2, t_3) \mapsto \Psi_{i_3}^{t_3} \circ \Psi_{i_2}^{t_2} \circ \Psi_{i_1}^{t_1}(a).$$

Then $\psi(\mathbb{R}^3) \subset \mathfrak{A}(a)$. From (3.69) and (3.70), we get that

$$\psi'(t_1^2, t_2^2, 0)(\mathbb{R}^3) = \mathbb{R} f_{i_3}(\varphi(t_1^2, t_2^2)) \oplus \ker V'(\varphi(t_1^2, t_2^2)).$$

Therefore $\psi'(t_1^2, t_2^2, 0)$ is one-to-one (note that $\ker V'(x)$ is of dimension at least 2 for every $x \in \varphi(\omega)$). If $n = 3$, $\psi'(t_1^2, t_2^2, 0)$ is then invertible and the inverse function theorem gives us that $\psi(\mathbb{R}^3)$, and therefore also $\mathfrak{A}(a)$, is a neighborhood of $\psi(t_1^2, t_2^2, 0)$. If $n > 3$, one keeps going (one can easily proceed by induction). This concludes the proof of Theorem 3.18. ∎

REMARK 3.24. Héctor Sussmann and Wensheng Liu have given in [**472**] an explicit method to produce trajectories joining two given points for driftless control affine systems satisfying the Lie algebra rank condition at every $(x, 0)$.

Let us end this section by recalling the Frobenius theorem.

THEOREM 3.25. *Let us assume again that $f_1, \ldots, f_m \in C^\infty(\Omega; \mathbb{R}^n)$. Let us also assume that there exists an integer $k \in \{0, \ldots, n\}$ such that, at every point $x \in \Omega$,*

the dimension of $\mathcal{A}(x, 0)$ is k, $\forall x \in \Omega$.

Then, for every $a \in \Omega$, $\mathfrak{A}(a)$ is a submanifold of Ω of dimension k.

For a proof of this version of the Frobenius theorem, see, e.g., [**258**, Theorem 4] (for a more classical local version, see, e.g., [**496**, Theorem 2.20, page 48]).

Since a trajectory of the control system

$$\dot{x} = f(x, u) := f_0(x) + \sum_{i=1}^{m} u_i f_i(x),$$

where the state is $x \in \Omega$ and the control is $(u_1, \ldots, u_m)^{\text{tr}} \in \mathbb{R}^m$, is a trajectory of the control system

$$\dot{x} = u_0 f_0(x) + \sum_{i=1}^{m} u_i f_i(x),$$

where the state is $x \in \Omega$ and the control is $(u_0, u_1, \ldots, u_m)^{\text{tr}} \in \mathbb{R}^{m+1}$, we get as a corollary of the Frobenius Theorem 3.25:

COROLLARY 3.26. *Let us assume again that $f_0, f_1, \ldots, f_m \in C^\infty(\Omega; \mathbb{R}^n)$. Let us also assume that there exists an integer $k \in \{1, \ldots, n\}$ such that, at every point $x \in \Omega$,*

Span$\{h(x); h \in \text{Lie}\{f_0, f_1, \ldots, f_m\}\}$ is of dimension k.

Then, for every $x^0 \in \Omega$, the set of x^1 which can be reached from x^0 (i.e., the set of $x^1 \in \Omega$ such that, for some $T > 0$ and some $u \in L^\infty((0, T); \mathbb{R}^m)$, the solution of the Cauchy problem $\dot{x} = f(x, u(t))$, $x(0) = x^0$, satisfies $x(T) = x^1$) is contained in a submanifold of Ω of dimension k.

3.4. Bad and good iterated Lie brackets

The necessary condition for controllability in the analytic case given in Theorem 3.17 on page 134 is not sufficient in general, as the two following simple control systems show

(3.71) $\quad n = 1,\ m = 1,\ \mathcal{O} = \mathbb{R} \times \mathbb{R},\ \dot{x} = u^2,$

(3.72) $\quad n = 2,\ m = 1,\ \mathcal{O} = \mathbb{R}^2 \times \mathbb{R},\ \dot{x}_1 = x_2^2, \dot{x}_2 = u.$

(System (3.72) has already been studied on page 131.) The control system (3.71) (resp. (3.72)) satisfies the Lie algebra rank condition at $(0, 0) \in \mathbb{R} \times \mathbb{R}$ (resp. $(0, 0) \in \mathbb{R}^2 \times \mathbb{R}$). However, this control system is not small-time locally controllable at $(0, 0) \in \mathbb{R} \times \mathbb{R}$ (resp. $(0, 0) \in \mathbb{R}^2 \times \mathbb{R}$). Indeed, whatever the trajectory $(x, u) : [0, T] \to \mathbb{R} \times \mathbb{R}$ (resp. $(x, u) : [0, T] \to \mathbb{R}^2 \times \mathbb{R}$) of the control system (3.71) (resp. (3.72)) is, one has $x(0) \leqslant x(T)$ (resp. $x_1(0) \leqslant x_1(T)$ with $(x_1, x_2) = x^{\text{tr}}$).

Let us now give sufficient conditions for small-time local controllability. To slightly simplify the notations, we assume that

(3.73) $$x_e = 0, \ u_e = 0.$$

(In any case, this assumption is not restrictive; just make a translation on the state and on the control.) We simply say "small-time locally controllable" instead of "small-time locally controllable at $(0,0)$". Our assumption (3.10) is now

(3.74) $$f(0,0) = 0.$$

Let us start with the case of a control affine system, i.e.,

$$\mathcal{O} = \Omega \times \mathbb{R}^m, \ f(x, u) = f_0(x) + \sum_{i=1}^{m} u_i f_i(x),$$

where Ω is an open subset of $\mathbb{R}^n$ containing 0 and the f_i's are in $C^\infty(\Omega; \mathbb{R}^n)$. In this case (3.74) is equivalent to $f_0(0) = 0$. Let $L(f_0, \ldots, f_m)$ be the free Lie algebra generated by $f_0, \ldots, f_m$ and let us denote by $\mathrm{Br}(f) \subset L(f_0, \ldots, f_m)$ the set of *formal iterated* Lie brackets of $\{f_0, f_1, \ldots, f_m\}$; see [**469**] for more details and precise definitions. For example,

(3.75) $$h := [[[f_0, [f_1, f_0]], f_1], f_0] \in \mathrm{Br}(f),$$
(3.76) $$h := [[[[f_0, [f_1, f_0]], f_1], f_0], [f_1, [f_2, f_1]]] \in \mathrm{Br}(f).$$

For $h \in L(f_0, \ldots, f_m)$, let $h(0) \in \mathbb{R}^n$ be the "evaluation" of h at 0. For h in $\mathrm{Br}(f)$ and $i \in \{0, \ldots, m\}$, let $\delta_i(h)$ be the number of times that f_i appears in h. For example,

1. with h given by (3.75), $\delta_0(h) = 3$, $\delta_1(h) = 2$ and $\delta_i(h) = 0$, for every $i \in \{2, \ldots, m\}$,
2. with h given by (3.76), $\delta_0(h) = 3$, $\delta_1(h) = 4$, $\delta_2(h) = 1$ and $\delta_i(h) = 0$, for every $i \in \{3, \ldots, m\}$.

EXERCISE 3.27. *Let*

(3.77) $$\mathfrak{L} := \mathrm{Span}\left\{ h(0); h \in \mathrm{Br}(f) \ \text{with} \ \sum_{i=1}^{m} \delta_i(h) = 1 \right\}.$$

Prove the following equalities:

(3.78) $$\mathfrak{L} = \mathrm{Span}\{\mathrm{ad}_{f_0}^k f_i(0); i \in \{1, \ldots, m\}, k \in \mathbb{N}\},$$
(3.79) $$\mathfrak{L} = \mathrm{Span}\{\mathrm{ad}_{f_0}^k f_i(0); i \in \{1, \ldots, m\}, k \in \{0, 1, \ldots, n-1\}\}.$$

(**Hint for (3.78)**. *Use the Jacobi identity (3.18).*)

Let $\mathfrak{S}_m$ be the group of permutations of $\{1, \ldots, m\}$. For π in $\mathfrak{S}_m$, let $\tilde{\pi}$ be the automorphism of $L(f_0, \ldots, f_m)$ which sends f_0 to f_0 and f_i to $f_{\pi(i)}$ for $i \in \{1, \ldots, m\}$. For $h \in \mathrm{Br}(f)$, we let

$$\sigma(h) := \sum_{\pi \in \mathfrak{S}_m} \tilde{\pi}(h) \in L(f_0, \ldots, f_m).$$

For example, if h is given by (3.75) and $m = 2$, one has

$$\sigma(h) = [[[f_0, [f_1, f_0]], f_1], f_0] + [[[f_0, [f_2, f_0]], f_2], f_0].$$

Let us introduce a definition.

DEFINITION 3.28 ([**469**, Section 7]). For $\theta \in [0, +\infty]$, the control system $\dot{x} = f_0(x) + \sum_{i=1}^{m} u_i f_i(x)$ satisfies the *Sussmann condition* $S(\theta)$ if it satisfies the Lie algebra rank condition (3.39) at $(0,0)$ and if, for every $h \in \text{Br}(f)$ with $\delta_0(h)$ odd and $\delta_i(h)$ even for every i in $\{1, \ldots, m\}$, $\sigma(h)(0)$ is in the span of the $g(0)$'s where the g's are in $\text{Br}(f)$ and satisfy

$$\tag{3.80} \theta \delta_0(g) + \sum_{i=1}^{m} \delta_i(g) < \theta \delta_0(h) + \sum_{i=1}^{m} \delta_i(h).$$

with the convention that, when $\theta = +\infty$, (3.80) is replaced by $\delta_0(g) < \delta_0(h)$.

Héctor Sussmann has proved in [**469**] the following theorem.

THEOREM 3.29. [**469**, Theorem 7.3] *If, for some θ in $[0,1]$, the control system $\dot{x} = f_0(x) + \sum_{i=1}^{m} u_i f_i(x)$ satisfies the condition $S(\theta)$, then it is small-time locally controllable.*

The following result is easy to check.

PROPOSITION 3.30. *Let θ be in $[0,1]$. Then the control system $\dot{x} = f_0(x) + \sum_{i=1}^{m} u_i f_i(x)$ satisfies the condition $S(\theta)$ if and only if the control system*

$$\dot{x} = f_0(x) + \sum_{i=1}^{m} y_i f_i(x), \ \dot{y} = u,$$

where the state is $(x, y) \in \mathbb{R}^n \times \mathbb{R}^m$ and the control is $u \in \mathbb{R}^m$, satisfies the condition $S(\theta/(1-\theta))$ (with the convention $1/0 = +\infty$).

This proposition allows us to extend the condition $S(\theta)$ to $\dot{x} = f(x, u)$ in the following way.

DEFINITION 3.31. Let $\theta \in [0,1]$. The control system $\dot{x} = f(x, u)$ satisfies the condition $S(\theta)$ if the control system $\dot{x} = f(x, y), \ \dot{y} = u$ satisfies $S(\theta/(1-\theta))$.

What is called the Hermes condition is $S(0)$, a condition which was conjectured to be sufficient for small-time local controllability by Henry Hermes, who has proved his conjecture in the plane in [**225**] and has got partial results for more general systems in [**228**]. Héctor Sussmann has treated the case of scalar-input (i.e., $m = 1$) control affine systems in [**468**].

It follows from [**469**] that we have:

THEOREM 3.32. *If, for some θ in $[0,1]$, the control system $\dot{x} = f(x, u)$ satisfies the condition $S(\theta)$, then it is small-time locally controllable.*

Proof of Theorem 3.32. Apply [**469**] to $\dot{x} = f(x, y), \ \dot{y} = u$ with the constraint $\int_0^t |u(s)| \, ds \leqslant r$ (instead of $|u| \leqslant 1$). ■

One can find other sufficient conditions for small-time local controllability in [**5**] by Andrei Agrachev, in [**6, 7**] by Andrei Agrachev and Revaz Gamkrelidze, in [**55**] by Rosa Maria Bianchini, in [**56**] by Rosa Maria Bianchini and Gianna Stefani, in [**270**] by Matthias Kawski, in [**482**] by Alexander Tret'yak, and in references of these papers.

EXAMPLE 3.33. If $f(x,u) = \sum_{i=1}^{m} u_i f_i(x)$ or if $f(x,u) = Ax + Bu$ (i.e., for driftless control affine systems and linear control systems), the control system $\dot{x} = f(x,u)$ satisfies the Hermes condition if and only if it satisfies the Lie algebra rank condition at $(0,0)$. Hence, by Sussmann's Theorem 3.29 on the preceding page, one recovers the Rashevski-Chow theorem, i.e., that for driftless control affine systems, small-time local controllability is implied by the Lie algebra rank condition at $(0,0)$ (Theorem 3.19 on page 135) and that the Kalman rank condition

$$\text{Span } \{A^i B u; \, i \in [0, n-1], \, u \in \mathbb{R}^m\} = \mathbb{R}^n$$

implies the controllability of the linear control system $\dot{x} = Ax + Bu$ (Theorem 1.16 on page 9).

EXAMPLE 3.34. We take $n = 2$, $m = 1$ and

$$\dot{x}_1 = x_2^3, \, \dot{x}_2 = u.$$

In other words, our system is

$$\dot{x} = f_0(x) + u f_1(x),$$

with

$$f_0 \begin{pmatrix} x_1 \\ x_2 \end{pmatrix} = \begin{pmatrix} x_2^3 \\ 0 \end{pmatrix}, \, f_1 \begin{pmatrix} x_1 \\ x_2 \end{pmatrix} = \begin{pmatrix} 0 \\ 1 \end{pmatrix}.$$

Then

$$f_1 \begin{pmatrix} 0 \\ 0 \end{pmatrix} = \begin{pmatrix} 0 \\ 1 \end{pmatrix}, \, [f_1, [f_1, [f_1, f_0]]] \begin{pmatrix} 0 \\ 0 \end{pmatrix} = \begin{pmatrix} 6 \\ 0 \end{pmatrix}$$

span $\mathbb{R}^2$. Hence our system satisfies the Lie algebra rank condition at $(0,0) \in \mathbb{R}^2 \times \mathbb{R}$. Moreover, if $h \in Br(f)$ with $\delta_0(h)$ odd and $\delta_1(h)$ even, then $h(0) = 0$. Hence our system satisfies the Hermes condition and therefore is small-time locally controllable.

EXAMPLE 3.35. Let us go back to the problem of the control of the attitude of a rigid spacecraft. In this example our control system is (3.12) with $b_1, \ldots, b_m$ independent and $1 \leqslant m \leqslant 3$. Then the following results hold.

- For $m = 3$, as we have seen in Section 3.1 (page 129), the control system (3.12) is small-time locally controllable. Bernard Bonnard in [61] (see also the paper [137] by Peter Crouch) has also proved that the control system (3.12) is globally controllable in large time (that is, given two states, there exist a time $T > 0$ and an open-loop control $u \in L^\infty(0,T)$ steering the control system (3.12) from the first state to the second one). His proof relies on the Poincaré recurrence theorem, as previously introduced by Lawrence Markus and George Sell in [344], and by Claude Lobry in [334]. For a recent presentation of this result, we refer to [189, Th.II.23, Chapitre II, pages 139–140].
- For $m = 2$, the control system (3.12) satisfies the Lie algebra rank condition at $(0,0) \in \mathbb{R}^6 \times \mathbb{R}^2$ if and only if (see [61, 137])

(3.81) $\quad \text{Span } \{b_1, b_2, S(\omega) J^{-1} \omega; \, \omega \in \text{ Span } \{b_1, b_2\}\} = \mathbb{R}^3.$

  Moreover, if (3.81) holds, then:
  - The control system (3.12) satisfies Sussmann's condition $S(1)$, and so is small-time locally controllable; see [272] by El-Yazid Keraï.

- The control system (3.12) is globally controllable in large time; this result is due to Bernard Bonnard [**61**], see also [**137**].
- If $m = 1$, the control system (3.12) satisfies the Lie algebra rank condition at $(0,0) \in \mathbb{R}^6 \times \mathbb{R}$ if and only if (see [**61, 137**])

(3.82) $\quad$ Span $\{b_1, S(b_1)J^{-1}b_1 S(\omega)J^{-1}\omega;\ \omega \in$ Span $\{b_1, S(b_1)J^{-1}b_1\}\} = \mathbb{R}^3$.

Moreover,
- The control system (3.12) does not satisfy a necessary condition for small-time local controllability due to Héctor Sussmann [**468**, Proposition 6.3] (see the next theorem) and so it is not small-time locally controllable. This result is due to El-Yazid Keraï [**272**].
- If (3.82) holds, the control system (3.12) is globally controllable in large time. This result is due to Bernard Bonnard [**61**]; see also [**137**] by Peter Crouch.

For analysis of the small-time local controllability of other underactuated mechanical systems, let us mention, in particular, [**402**] by Mahmut Reyhanoglu, Arjan van der Schaft and Harris McClamroch and Ilya Kolmanovsky.

With the notations we have introduced we can also state the following necessary condition for small-time local controllability due to Gianna Stefani [**463**, Theorem 1], which improves a prior result due to Héctor Sussmann [**468**, Proposition 6.3].

THEOREM 3.36. *Let us assume that f_0 and f_1 are analytic in an open neighborhood Ω of $0 \in \mathbb{R}^n$ and that*

(3.83) $\quad\quad\quad\quad\quad\quad\quad\quad f_0(0) = 0.$

We consider the control system

(3.84) $\quad\quad\quad\quad\quad\quad\quad \dot{x} = f_0(x) + u f_1(x),$

where the state is $x \in \Omega$ and the control is $u \in \mathbb{R}$. Let us also assume that the control system (3.84) is small-time locally controllable. Then

(3.85) $\quad \operatorname{ad}_{f_1}^{2k} f_0(0) \in$ Span $\{h(0);\ h \in Br(f);\ \delta_1(h) \leqslant 2k-1\},\ \forall k \in \mathbb{N} \setminus \{0\}.$

(Let us recall that $\operatorname{ad}_{f_1}^p f_0$, $p \in \mathbb{N}$, are defined in Definition 3.11 on page 130.) The prior result [**468**, Proposition 6.3] corresponds to the case $k = 1$.

Sketch of the proof of Theorem 3.36. We follow Héctor Sussmann's proof of [**468**, Proposition 6.3]. Let us assume that (3.85) does not hold for $k = 1$. One first proves the existence of an analytic function ϕ on a neighborhood $\mathcal{V}$ of 0 such that

(3.86) $\quad\quad\quad\quad\quad\quad\quad\quad \phi(0) = 0,$

(3.87) $\quad\quad \phi'(0)(h(0)) = 0, \forall h \in Br(f)$ such that $\delta_1(h) \leqslant 1,$

(3.88) $\quad\quad\quad\quad\quad\quad \phi'(0)([f_1, [f_1, f_0]]) = 1,$

(3.89) $\quad\quad\quad\quad\quad\quad \phi'(x)(f_1(x)) = 0,\ \forall x \in \mathcal{V}.$

To prove the existence of such a ϕ, let $g_1, \ldots, g_n$ be analytic vector fields defined on a neighborhood of 0 such that

$$(g_1(0), \ldots, g_n(0)) \text{ is a basis of } \mathbb{R}^n,$$
$$g_1 = f_1$$

and, for some $k \in \{1, \ldots, n-1\}$,

$$(g_1(0), \ldots, g_k(0)) \text{ is a basis of Span } \{h(0); h \in \mathrm{Br}(f); \delta_1(h) \leqslant 1\},$$
$$g_{k+1} = [f_1, [f_1, f_0]].$$

Let us denote by Ψ_i the flow associated to g_i (see (3.20)). Then there exists a neighborhood $\mathcal{V}$ of 0 such that every $x \in \mathcal{V}$ has a unique expression of the following form:

$$x = \Psi_1^{t_1} \circ \Psi_2^{t_2} \circ \ldots \circ \Psi_{n-1}^{t_{n-1}} \circ \Psi_n^{t_n}(0).$$

Then it is not hard to check that, if we define ϕ by $\phi(x) := t_{k+1}$, one has (3.86), (3.87), (3.88) and (3.89).

For the next step, let us introduce some notations. Let $r \in \mathbb{N} \setminus \{0\}$. For a multi-index $I = (i_1, \ldots, i_r) \in \{0,1\}^r$, let f_I be the partial differential operator defined by

(3.90) $$f_I(\theta) = L_{f_{i_1}} L_{f_{i_2}} \ldots L_{f_{i_{r-1}}} L_{f_{i_r}} \theta.$$

In (3.90) and in the following, for $X = (X_1, \ldots, X_n)^{\mathrm{tr}} : \mathbb{R}^n \to \mathbb{R}^n$ and $V : \mathbb{R}^n \to \mathbb{R}$, $L_X V : \mathbb{R}^n \to \mathbb{R}$ denotes the (Lie) derivative of V in the direction of X:

(3.91) $$L_X V := \sum_{i=1}^{i=n} X_i \frac{\partial V}{\partial x_i} = V'(x) X(x).$$

For $u \in L^\infty(0,T)$, $r \in \mathbb{N} \setminus \{0\}$ and multi-index $I = (i_1, \ldots, i_r) \in \{0,1\}^r$, we define $u_I \in L^\infty(0,T)$ by

$$u_I(t_r) = \int_0^{t_r} \int_0^{t_{r-1}} \int_0^{t_{r-2}} \ldots \int_0^{t_2} \int_0^{t_1} u_{i_r}(t_{r-1}) u_{i_{r-1}}(t_{r-2})$$
$$\ldots u_{i_2}(t_1) u_{i_1}(t_0) dt_0 dt_1 \ldots dt_{r-2} dt_{r-1}, \forall t_r \in [0,T],$$

with

$$u_0(t) = 1 \text{ and } u_1(t) = u(t), t \in (0,T).$$

(We have slightly modified the notations of [**468**]: what we call here $u_I(t)$ is denoted $\int_0^t u_I$ in [**468**]; see [**468**, (3.3)].) Let $\mathcal{E}$ be the set of all multi-indices $I = (i_1, \ldots, i_r) \in \{0,1\}^r$ with $r \in \mathbb{N} \setminus \{0\}$. The next step is the following proposition.

PROPOSITION 3.37 ([**468**, Proposition 4.3]). *Let $\phi : \mathbb{R}^n \to \mathbb{R}$ be an analytic function on a neighborhood of 0 in $\mathbb{R}^n$ such that $\phi(0) = 0$. Then there exists $T > 0$ such that, for every $u \in L^\infty(0,T)$ with $\|u\|_{L^\infty(0,T)} \leqslant 1$, one has*

(3.92) $$\sum_{I \in \mathcal{E}} |u_I(t)| |(f_I(\phi))(0)| < +\infty, \forall t \in [0,T],$$

(3.93) $$\phi(x(t)) = \sum_{I \in \mathcal{E}} u_I(t) (f_I(\phi))(0), \forall t \in [0,T],$$

where $x : [0,T] \to \mathbb{R}^n$ is the solution of the Cauchy problem

$$\dot{x} = f_0(x) + u(t) f_1(x), x(0) = 0.$$

We omit the proof of this proposition, which is also explicitly or implicitly contained in [**173, 174**] by Michel Fliess, [**286**] by Arthur Krener, [**73**] by Roger Brockett (see also [**75**]) and [**193**] by Gilbert Elmer. It can now be found in classical books in nonlinear control theory; see e.g. [**247**, Section 3.1]. Let us emphasize that the right hand side of (3.93) is the Chen-Fliess series, introduced by Kuo-Tsai Chen in [**90**] for geometrical purposes, and by Michel Fliess in [**174**] for control theory.

Let us analyze the right hand side of (3.93). Using (3.83) (which is always assumed and is in any case implied by the small-time local controllability of the control system (3.84)), we get

(3.94) $$(f_I(\phi))(0) = 0, \forall I = (i_1, \ldots, i_r) \in \mathcal{E} \text{ with } i_1 = 0.$$

By (3.89),

(3.95) $$f_I(\phi) = 0, \forall I = (i_1, \ldots, i_r) \in \mathcal{E} \text{ with } i_r = 1.$$

Let us now assume that

(3.96) $$I = (i_1, \ldots, i_r) \in \mathcal{E} \text{ with } i_1 = 1 \text{ and } i_k = 0, \forall k \in \{2, \ldots, r\}.$$

Let us check that we then have

(3.97) $$(f_I(\phi))(0) = 0.$$

For $k \in \mathbb{N} \setminus \{0\}$, let 0_k be the multi-index $(j_1, \ldots, j_k)$ with $j_l = 0$ for every $l \in \{1, \ldots, k\}$. Then

(3.98) $$f_I(\phi) = L_{f_1} f_{0_{r-1}}(\phi) = L_{[f_1, f_0]} f_{0_{r-2}}(\phi) + L_{f_0} L_{f_1} f_{0_{r-2}}(\phi).$$

But (3.83) implies that

(3.99) $$L_{f_0} L_{f_1} f_{0_{r-2}}(\phi)(0) = 0.$$

Similarly to (3.98), we have

(3.100) $$L_{[f_1, f_0]} f_{0_{r-2}}(\phi) = L_{[[f_1, f_0], f_0]} f_{0_{r-3}}(\phi) + L_{f_0} L_{[f_1, f_0]} f_{0_{r-3}}(\phi).$$

As for (3.99), (3.83) implies that

$$L_{f_0} L_{[f_1, f_0]} f_{0_{r-3}}(\phi)(0) = 0.$$

Continuing on, we end up with

$$f_I(\phi)(0) = (-1)^{r-1} (L_{\text{ad}_{f_0}^{r-1} f_1} \phi)(0),$$

which, together with (3.87), implies that

(3.101) $$f_I(\phi)(0) = 0.$$

Let us now study the case $I = (1, 1, 0)$. Let $v \in C^0([0, T])$ be defined by

$$v(t) := \int_0^t u(\tau) d\tau, \forall t \in [0, T].$$

We have

$$u_{(1,1,0)}(t_3) = \int_0^{t_3}\int_0^{t_2}\int_0^{t_1} u(t_1)u(t_0) dt_0 dt_1 dt_2$$
(3.102)
$$= \int_0^{t_3}\int_0^{t_2} \dot{v}(t_1)v(t_1) dt_1 dt_2$$
$$= \frac{1}{2}\int_0^{t_3} v(t_2)^2 dt_2.$$

From (3.88) and (3.102), we get

(3.103) $$u_{(1,1,0)}(t)(f_{(1,1,0)}(\phi))(0) = \frac{1}{2}\int_0^t v(\tau)^2 d\tau.$$

Finally, let us assume that $I = (i_1, \ldots, i_r) \in \mathcal{E}$ is such that

(3.104) there are at least two $l \in \{1, \ldots, r\}$ such that $i_l = 1$,

(3.105) $i_1 = 1$ and $i_r = 0$,

(3.106) $r \geqslant 4$.

Let us write $I = (1, 0_k, 1, J)$ with $J := (j_1, \ldots, j_l) \in \mathcal{E}$, $k \in \mathbb{N}$ and the convention $(1, 0_0, 1, J) := (1, 1, J)$. We have $l = r - 2 - k \geqslant 1$, $l + k \geqslant 2$ and $j_l = 0$. Then after computations which are detailed in [**468**, pages 708–710], one gets the existence of $C > 0$ independent of I such that, for every $T > 0$, for every $t \in [0, T]$, and for every $u \in L^\infty((0,T); [-1,1])$,

(3.107) $$|u_I(t)||(f_I(\phi))(0)| \leqslant C^{l+k+2} t^{l+k-1} \frac{(l+k)!(l+k+2)^5}{l!k!} \int_0^t v(\tau)^2 d\tau.$$

Finally, from (3.93), (3.94), (3.95), (3.97), (3.103) and (3.107), one gets the existence of $T > 0$ such that, for every $t \in [0, T]$ and for every $u \in L^\infty((0,T); [-1,1])$,

(3.108) $$\phi(x(t)) \geqslant \frac{1}{3}\int_0^t v(\tau)^2 d\tau$$

(see [**468**, page 710] for the details). Hence the control system (3.84) is not small-time locally controllable. This concludes the sketch proof of Theorem 3.36 for the case $k = 1$. ∎

Let us give an example of application of Theorem 3.36.

EXAMPLE 3.38. *In this example $n = 4$ and $m = 2$ and the control system we consider is given by*

(3.109) $$\dot{x}_1 = x_2, \; \dot{x}_2 = -x_1 + u, \; \dot{x}_3 = x_4, \; \dot{x}_4 = -x_3 + 2x_1 x_2,$$

where the state is $(x_1, x_2, x_3, x_4)^{tr} \in \mathbb{R}^4$ and the control is $u \in \mathbb{R}$. Hence

(3.110) $$f_0(x) = (x_2, -x_1, x_4, -x_3 + 2x_1 x_2)^{tr},$$
(3.111) $$f_1(x) = (0, 1, 0, 0)^{tr}.$$

Straightforward computations lead to

(3.112) $$[f_1, f_0](x) = (1, 0, 0, 2x_1)^{tr}, \forall x \in \mathbb{R}^4,$$

(3.113) $$ad_{f_1}^k f_0(x) = (0, 0, 0, 0)^{tr}, \forall k \in \mathbb{N} \setminus \{0, 1\}, \forall x \in \mathbb{R}^4,$$

(3.114) $$ad_{f_0}^{2k} f_1(0) = (-1)^k (0, 1, 0, 0)^{tr}, \forall k \in \mathbb{N},$$

(3.115) $$ad_{f_0}^{2k+1} f_1(0) = (-1)^{k+1} (1, 0, 0, 0)^{tr}, \forall k \in \mathbb{N},$$

(3.116) $$ad_{f_0}^2 f_1(x) = (0, -1, 2x_1, 0)^{tr}, \forall x = (x_1, x_2, x_3, x_4)^{tr} \in \mathbb{R}^4,$$

(3.117) $$[[f_1, f_0], ad_{f_0}^2 f_1](0) = (0, 0, 2, 0)^{tr},$$

(3.118) $$[f_0, [[f_1, f_0], ad_{f_0}^2 f_1]](0) = (0, 0, 0, 2)^{tr}.$$

Note that, by (3.111), (3.112), (3.117) and (3.118),

$$\text{Span } \{f_1(0), [f_1, f_0](0), [[f_1, f_0], ad_{f_0}^2 f_1](0), [f_0, [[f_1, f_0], ad_{f_0}^2 f_1]](0)\} = \mathbb{R}^4,$$

which implies that the control system (3.109) satisfies the Lie algebra rank condition at $(0,0) \in \mathbb{R}^4 \times \mathbb{R}$. Equation (3.113) shows that assumption (3.85) is satisfied. Hence it seems that one cannot apply Theorem 3.36. However, letting

$$y_1 := x_1, \ y_2 := x_3, \ y_3 := x_4, \ v := x_2,$$

one easily checks that the small-time local controllability of the control system (3.109) at $(0,0) \in \mathbb{R}^4 \times \mathbb{R}$ implies the small-time local controllability at $(0,0) \in \mathbb{R}^3 \times \mathbb{R}$ of the control system

(3.119) $$\dot{y}_1 = v, \ \dot{y}_2 = y_3, \ \dot{y}_3 = -y_2 + 2vy_1,$$

where the state is $y = (y_1, y_2, y_3)^{tr} \in \mathbb{R}^3$ and the control is $v \in \mathbb{R}$. Let $g_0 \in C^\infty(\mathbb{R}^3; \mathbb{R}^3)$, $g_1 \in C^\infty(\mathbb{R}^3; \mathbb{R}^3)$ and $g \in C^\infty(\mathbb{R}^3 \times \mathbb{R}; \mathbb{R}^3)$ be defined by

$$g_0(y) := (0, y_3, -y_2)^{tr}, \ g_1(y) := (1, 0, 2y_1)^{tr},$$
$$g(y, v) := g_0(y) + vg_1(y), \ \forall (y, v) \in \mathbb{R}^3 \times \mathbb{R}.$$

Hence the control system (3.119) can be written

$$\dot{y} = g_0(y) + vg_1(y) = g(y, v).$$

Now, straightforward computations give

(3.120) $$[g_1, g_0](y) = (0, 2y_1, 0)^{tr},$$

(3.121) $$ad_{g_1}^2 g_0(0) = (0, 2, 0)^{tr},$$

(3.122) $$ad_{g_0}^k g_1(0) = (0, 0, 0)^{tr}, \forall k \in \mathbb{N} \setminus \{0\}.$$

Since $g_1(0) = (1, 0, 0)^{tr}$, (3.122) gives us (see also Exercise 3.27 on page 142)

(3.123) $$\text{Span } \{h(0); \ h \in Br(g); \delta_1(h) \leqslant 1\} = \mathbb{R}(1, 0, 0)^{tr}.$$

Equation (3.121), together with (3.123), shows that (3.85) does not hold for $k = 1$ and for the control system (3.119). Hence, by Theorem 3.36 the control system (3.119) is not small-time locally controllable at $(0,0) \in \mathbb{R}^3 \times \mathbb{R}$. This implies that the original control system (3.109) is not small-time locally controllable at $(0,0) \in \mathbb{R}^4 \times \mathbb{R}$.

In Example 6.4 on page 190 we will see that our control system (3.109) is in fact controllable in large time.

REMARK 3.39. One can find other necessary conditions for local controllability in [**226**] by Henry Hermes, in [**231**] by Henry Hermes and Matthias Kawski, in [**270**] by Matthias Kawski, in [**463**] by Gianni Stefani, in [**468**] by Héctor Sussmann, and the references therein.

3.5. Global results

To get global controllability results is usually much more complicated than to get local controllability results. There are very few theorems leading to global controllability and they all require very restrictive assumptions on the control system. The most famous global controllability result is the Rashevski-Chow theorem (Theorem 3.18 on page 134) which deals with driftless control affine systems. Let us mention two other important methods.

1. A general enlarging technique, due to Velimir Jurdjevic and Ivan Kupka [**259, 260**], which allows us to replace the initial control system by a control system having more controls but the same "controllability"; see also [**153**] by Rachida El Assoudi, Jean-Paul Gauthier and Ivan Kupka as well as [**258**, Chapter 6, Section 3] by Velimir Jurdjevic, and [**62**, Section 5.5] by Bernard Bonnard and Monique Chyba.
2. A method based on the Poincaré recurrence theorem. It was introduced by Lawrence Markus and George Sell in [**344**], and by Claude Lobry in [**334**]. For a recent presentation of this method, we refer to [**189**, Th.II.23, Chapitre II, pages 139–140] and to [**258**, Theorem 5, Chapter 4, p. 114].

In this section, we present two more methods which allow us to get global controllability. These methods are less general than the methods previously mentioned, but they turn out to be useful also for infinite-dimensional control systems.

- The first one deals with perturbations of linear control systems. It relies on a method which has been introduced by Enrique Zuazua in [**515, 516**] to get global controllability results for semilinear wave equations. We detail it in Section 3.5.1.
- The second one relies on homogeneity arguments. For example, it shows that if a "quadratic control system" is locally controllable, then any control system, which is a perturbation of this quadratic control system by a linear term, is globally controllable. This has been used in various papers to get global controllability results for the Navier-Stokes equations from the controllability of the Euler equations (see Remark 3.47 on page 156 as well as Section 6.2.2.2—in particular, Remark 6.18 on page 201—and the references therein). We detail this second method in Section 3.5.2.

3.5.1. Lipschitz perturbations of controllable linear control systems.
We consider the control system

(3.124) $$\dot{x} = A(t,x)x + B(t,x)u + f(t,x),\ t \in [T_0, T_1],$$

where, at time t, the state is $x(t) \in \mathbb{R}^n$ and the control is $u(t) \in \mathbb{R}^m$. We assume that $A : [T_0, T_1] \times \mathbb{R}^n \to \mathcal{L}(\mathbb{R}^n; \mathbb{R}^n)$, $B : [T_0, T_1] \times \mathbb{R}^n \to \mathcal{L}(\mathbb{R}^m; \mathbb{R}^n)$ and $f : [T_0, T_1] \times \mathbb{R}^n \to \mathbb{R}^n$ are smooth enough (for example, assuming that these maps are continuous with respect to (t, x) and locally Lipschitz with respect to x is sufficient).

3.5. GLOBAL RESULTS

We also assume that

(3.125) $$A \in L^\infty((T_0, T_1) \times \mathbb{R}^n; \mathcal{L}(\mathbb{R}^n; \mathbb{R}^n)),$$
(3.126) $$B \in L^\infty((T_0, T_1) \times \mathbb{R}^n; \mathcal{L}(\mathbb{R}^m; \mathbb{R}^n)),$$
(3.127) $$f \in L^\infty((T_0, T_1) \times \mathbb{R}^n; \mathbb{R}^n).$$

For $z \in C^0([T_0, T_1]; \mathbb{R}^n)$, let us denote by $\mathfrak{C}_z \in \mathcal{M}_{n,n}(\mathbb{R})$ the controllability Gramian of the time-varying linear control system

$$\dot{x} = A(t, z(t))x + B(t, z(t))u, \ t \in (T_0, T_1),$$

(see Definition 1.10 on page 6).

The goal of this section is to prove the following theorem.

THEOREM 3.40. *Let us also assume that for every $z \in C^0([T_0, T_1]; \mathbb{R}^n)$, $\mathfrak{C}_z$ is invertible and that there exists $M > 0$ such that*

(3.128) $$|\mathfrak{C}_z^{-1}|_{\mathcal{M}_{n,n}(\mathbb{R})} \leqslant M, \ \forall z \in C^0([T_0, T_1]; \mathbb{R}^n).$$

Then the control system (3.124) is globally controllable (on $[T_0, T_1]$): for every $x^0 \in \mathbb{R}^n$ and for every $x^1 \in \mathbb{R}^n$, there exists $u \in L^\infty((T_0, T_1); \mathbb{R}^m)$ such that the solution of the Cauchy problem

$$\dot{x} = A(t, x)x + B(t, x)u(t) + f(t, x), \ x(T_0) = x^0$$

satisfies

$$x(T_1) = x^1.$$

Proof of Theorem 3.40. As we have said, we follow a method introduced by Enrique Zuazua in [**515, 516**]. Let $x^0 \in \mathbb{R}^n$ and $x^1 \in \mathbb{R}^n$. Let us define

$$\mathcal{F} : z \in C^0([T_0, T_1]; \mathbb{R}^n) \mapsto \mathcal{F}(z) \in C^0([T_0, T_1]; \mathbb{R}^n)$$

in the following way. Let $u_z \in L^2((T_0, T_1); \mathbb{R}^m)$ be the control of minimal L^2-norm which steers the time-varying linear control system with a remainder term

$$\dot{x} = A(t, z(t))x + B(t, z(t))u + f(t, z(t)), \ t \in [T_0, T_1],$$

from the state x^0 to the state x^1. Then $\mathcal{F}(z) := x$ is the solution of the Cauchy problem

(3.129) $$\dot{x} = A(t, z(t))x + B(t, z(t))u_z(t) + f(t, z(t)), \ x(T_0) = x^0.$$

Let us point out that such an u_z exists and its expression is the following one:

(3.130) $$u_z(t) = B(t)^{\text{tr}} R(T_1, t)^{\text{tr}} \mathfrak{C}_z^{-1}$$
$$(x^1 - \int_{T_0}^{T_1} R_z(T_1, \tau) f(\tau, z(\tau)) d\tau - R_z(T_1, T_0) x^0), \ t \in (T_0, T_1),$$

where R_z is the resolvent of the time-varying linear system

$$\dot{x} = A(t, z(t))x, \ t \in [T_0, T_1],$$

(see Definition 1.4 on page 4). Equality (3.130) can be checked by considering the solution $\tilde{x} \in C^1([T_0, T_1]; \mathbb{R}^n)$ of the Cauchy problem

$$\dot{\tilde{x}} = A(t, z(t))\tilde{x} + f(t, z(t)), \ x(T_0) = 0.$$

One has (see Proposition 1.9 on page 6)

$$\tilde{x}(T_1) = \int_{T_0}^{T_1} R_z(T_1, \tau) f(\tau, z(\tau)) d\tau. \tag{3.131}$$

Clearly u_z is the control of minimal L^2 norm which steers the control system

$$\dot{x} = A(t, z(t))x + B(t, z(t))u$$

from the state x^0 to the state $x^1 - \tilde{x}(T_1)$. Hence (3.130) follows from Proposition 1.13 on page 8 and (3.131).

Note that, by the definition of u_z,

$$x(T_1) = x^1,$$

where $x : [T_0, T_1] \to \mathbb{R}^n$ is the solution of the Cauchy problem (3.129). Clearly

(3.132) $\quad \mathcal{F}$ is a continuous map from $C^0([T_0, T_1]; \mathbb{R}^n)$ into $C^0([T_0, T_1]; \mathbb{R}^n)$.

Moreover, if z is a fixed point of $\mathcal{F}$, then u_z steers the control system (3.124) from x^0 to x^1 during the time interval $[T_0, T_1]$. By (3.125), (3.126), (3.127), (3.128) and (3.130), there exists $M_1 > 0$ such that

$$\|u_z\|_{L^\infty(T_0, T_1)} \leqslant M_1, \; \forall z \in C^0([T_0, T_1]; \mathbb{R}^n). \tag{3.133}$$

From the definition of $\mathcal{F}$, (3.125), (3.126), (3.127), (3.129) and (3.133), there exists $M_2 > 0$ such that

$$\|\mathcal{F}(z)\|_{C^0([T_0, T_1]; \mathbb{R}^n)} \leqslant M_2, \; \forall z \in C^0([T_0, T_1]; \mathbb{R}^n), \tag{3.134}$$

$$|\mathcal{F}(z)(t_2) - \mathcal{F}(z)(t_1)| \tag{3.135}$$
$$\leqslant M_2 |t_2 - t_1|, \; \forall z \in C^0([T_0, T_1]; \mathbb{R}^n), \; \forall (t_1, t_2) \in [T_0, T_1]^2.$$

From (3.134), (3.135) and the Ascoli theorem (see, for example, [**419**, A 5 Ascoli's theorem, page 369])

(3.136) $\quad \overline{\mathcal{F}(C^0([T_0, T_1]; \mathbb{R}^n))}$ is a compact subset of $C^0([T_0, T_1]; \mathbb{R}^n)$.

From (3.132), (3.136) and the Schauder fixed-point theorem (Theorem B.17 on page 391), $\mathcal{F}$ has a fixed point. This concludes the proof of Theorem 3.40. ∎

Applying Theorem 3.40 to the case where A and B do not depend on time we get the following corollary, which is a special case of a result due to Dahlard Lukes ([**337**, Theorem 2.1], [**338**]).

COROLLARY 3.41. *Let $A \in \mathcal{L}(\mathbb{R}^n; \mathbb{R}^n)$ and $B \in \mathcal{L}(\mathbb{R}^m; \mathbb{R}^n)$ be such that the linear control*

$$\dot{x} = Ax + Bu, \; x \in \mathbb{R}^n, \; u \in \mathbb{R}^m,$$

is controllable. Let $f \in C^1(\mathbb{R}^n; \mathbb{R}^n)$. Let us assume that f is bounded on $\mathbb{R}^n$. Then, for every $T > 0$, the nonlinear control system

$$\dot{x} = Ax + Bu + f(x), \; x \in \mathbb{R}^n, \; u \in \mathbb{R}^m,$$

is globally controllable in time T: for every $x^0 \in \mathbb{R}^n$ and for every $x^1 \in \mathbb{R}^n$, there exists $u \in L^\infty((T_0, T_1); \mathbb{R}^n)$ such that the solution of the Cauchy problem

$$\dot{x} = Ax + Bu + f(x), \; x(T_0) = x^0$$

satisfies

$$x(T_1) = x^1.$$

EXERCISE 3.42. *Prove that the assumption that f is bounded cannot be removed.* (**Hint.** *Take $n = 2$, $m = 1$, and consider simply the following control system $\dot{x}_1 = x_2 - f(x), \dot{x}_2 = u$ with $f(x) = x_2$.*)

EXERCISE 3.43. *Prove that the conclusion of Corollary 3.41 holds if the assumption "f is bounded on $\mathbb{R}^n$" is replaced by the assumption*

(3.137) $$|f(x^2) - f(x^1)| \leqslant \varepsilon |x^2 - x^1|, \forall (x^2, x^1) \in \mathbb{R}^n \times \mathbb{R}^n,$$

provided that $\varepsilon > 0$ is small enough (the smallness depending on A and B, but not on x^0 and x^1); see also [**337**, *Theorem 2.2*].

EXERCISE 3.44 (See [**223**, Theorem 1.2] by Henry Hermes). *Let $T > 0$. Let us consider the following control system*

(3.138) $$\dot{x} = f(t, x(t)) + B(t)u, \, t \in (0, T),$$

where the state is $x \in \mathbb{R}^n$ and the control is $u \in \mathbb{R}^m$. Let us assume that $f \in C^1([0, T] \times \mathbb{R}^n; \mathbb{R}^n)$ and $B \in L^\infty((0, T); \mathcal{L}(\mathbb{R}^m; \mathbb{R}^n))$. Let us also assume that there exists $C > 0$ such that

$$|f(t, x)| \leqslant C, \, \forall (t, x) \in [0, T] \times \mathbb{R}^n,$$

$$\int_0^T x^{tr} B(t) B(t)^{tr} x \, dt \geqslant C^{-1} |x|^2, \, \forall x \in \mathbb{R}^n.$$

Prove that the control system (3.138) is globally controllable in time $T > 0$.

3.5.2. Global controllability and homogeneity.
In this section we show how some simple homogeneity arguments can be used to get global controllability.

We do not try to present the most general result; our goal is just to present the method on a simple case.

Let $A \in \mathcal{L}(\mathbb{R}^n; \mathbb{R}^n)$, $B \in \mathcal{L}(\mathbb{R}^m; \mathbb{R}^n)$ and $F \in C^1(\mathbb{R}^n; \mathbb{R}^n)$. We assume that there exists $p \in \mathbb{N}$ such that

(3.139) $$p > 1,$$
(3.140) $$F(\lambda x) = \lambda^p F(x), \, \forall \lambda \in (0, +\infty), \forall x \in \mathbb{R}^n.$$

Our control system is

(3.141) $$\dot{x} = Ax + F(x) + Bu,$$

where the state is $x \in \mathbb{R}^n$ and the control is $u \in \mathbb{R}^m$.

Our idea is to get a global controllability result for the control system (3.141) from a local controllability property of the control system

(3.142) $$\dot{x} = F(x) + Bu.$$

For the control system (3.142), the state is again $x \in \mathbb{R}^n$ and the control is $u \in \mathbb{R}^m$. This local controllability property, which will be assumed in this section, is the following one: there exist $\bar{T} > 0$, $\rho > 0$ and a continuous function $\bar{u} : \bar{B}_\rho \to L^1((0, \bar{T})); \mathbb{R}^m)$ such that

(3.143) $$((\dot{x} = F(x) + B\bar{u}(a)(t), x(0) = a) \Rightarrow (x(\bar{T}) = 0)), \forall a \in \bar{B}_\rho.$$

Here and everywhere in this section, for every $\mu > 0$,

$$B_\mu := \{x \in \mathbb{R}^n; |x| < \mu\}, \, \bar{B}_\mu := \{x \in \mathbb{R}^n; |x| \leqslant \mu\}.$$

Note that, as we will see later on (Proposition 11.24 on page 301), for every $\theta \in [0,1]$, this controllability property is implied by the Sussmann condition $S(\theta)$ (see Definition 3.31 on page 143)); see also Open Problem 11.25 on page 301.

REMARK 3.45. From (3.139) and (3.140), one gets

(3.144) $$F(0) = 0,$$
(3.145) $$F'(0) = 0.$$

Hence $(0,0) \in \mathbb{R}^n \times \mathbb{R}^m$ is an equilibrium of the control system (3.142) and the linearized control system around $(0,0) \in \mathbb{R}^n \times \mathbb{R}^m$ of this control system is the linear control system

(3.146) $$\dot{x} = Bu.$$

Therefore, if $B(\mathbb{R}^m) \neq \mathbb{R}^n$ (in particular, if $m < n$), this linearized control system is not controllable.

Our goal is to prove the following theorem.

THEOREM 3.46. *Under the above assumptions, the control system (3.141) is globally controllable in small time. More precisely, for every $x_0 \in \mathbb{R}^n$, for every $x_1 \in \mathbb{R}^n$, and for every $T > 0$, there exists $u \in L^1((0,T); \mathbb{R}^m)$ such that*

$$(\dot{x} = Ax + F(x) + Bu(t), \ x(0) = x^0) \Rightarrow (x(T) = x^1).$$

Proof of Theorem 3.46. Let $x_0 \in \mathbb{R}^n$, $x_1 \in \mathbb{R}^n$ and $\eta > 0$. Let Φ be the map defined by

$$\begin{array}{rcl} [0,\eta] \times \bar{B}_\rho & \to & \mathbb{R}^n \\ (\varepsilon, a) & \mapsto & \bar{x}(0), \end{array}$$

where $\bar{x} : [0, \bar{T}] :\to \mathbb{R}^n$ is the solution of the Cauchy problem

$$\dot{\bar{x}} = \varepsilon^{p-1} A\bar{x} + F(\bar{x}) + B\bar{u}(a)(t), \ \bar{x}(\bar{T}) = \varepsilon x^1.$$

By (3.139) and (3.143), if $\eta > 0$ is small enough, Φ is well defined. We choose such an η. Then Φ is continuous on $[0, \eta] \times \bar{B}_\rho$ and

(3.147) $$\Phi(0, a) = a, \ \forall a \in B_\rho.$$

In particular,

(3.148) $$\text{degree}\,(\Phi(0, \cdot), B_\rho, 0) = 1.$$

For the definition of the degree, see Appendix B. Thus there exists $\eta_1 \in (0, \eta]$ such that

(3.149) $$\text{degree}\,(\Phi(\varepsilon, \cdot), B_\rho, \varepsilon x^0) = 1, \ \varepsilon \in [0, \eta_1].$$

Let

(3.150) $$\bar{\tau} = \eta_1^{p-1} \bar{T},$$
(3.151) $$\tau \in (0, \bar{\tau}],$$
(3.152) $$\varepsilon = \left(\frac{\tau}{\bar{T}}\right)^{1/(p-1)}.$$

By (3.150), (3.151) and (3.152),

(3.153) $$\varepsilon \in (0, \eta_1].$$

From (3.149), (3.153) and the homotopy invariance of the degree (see Proposition B.8 on page 387), there exists $a \in \bar{B}_\rho$ such that

(3.154) $$\Phi(\varepsilon, a) = \varepsilon x^0.$$

Let $\bar{x} : [0, \bar{T}] \to \mathbb{R}^n$ be the solution of the Cauchy problem

(3.155) $$\dot{\bar{x}} = \varepsilon^{p-1} A\bar{x} + F(\bar{x}) + B\bar{u}(a)(t),$$

(3.156) $$\bar{x}(\bar{T}) = \varepsilon x^1.$$

By (3.154), one has

(3.157) $$\bar{x}(0) = \varepsilon x^0.$$

Let $x : [0, \tau] \to \mathbb{R}^n$ and $u : [0, \tau] \to \mathbb{R}^m$ be defined by

(3.158) $$x(t) = \frac{1}{\varepsilon} \bar{x}\left(\frac{t}{\varepsilon^{p-1}}\right), \, u(t) = \frac{1}{\varepsilon^p} \bar{u}(a)\left(\frac{t}{\varepsilon^{p-1}}\right), \, \forall t \in [0, \tau].$$

From (3.140), (3.155) and (3.158), we get
$$\dot{x} = Ax + F(x) + Bu(t).$$

From (3.152), (3.156) and (3.158), one has
$$x(\tau) = x^1.$$

From (3.157) and (3.158), one gets
$$x(0) = x^0.$$

This shows that one can steer the control system (3.141) from x^0 to x^1 in an arbitrary *small* time. It remains to prove that one can make this motion in an arbitrary time. For that, let $T > 0$. Applying the previous result with $x^1 = 0$, one gets the existence of $T_0 \leqslant T/2$ and of $u_0 \in L^1((0, T_0); \mathbb{R}^m)$ such that the solution of the Cauchy problem
$$\dot{x} = Ax + F(x) + Bu_0(t), \, x(0) = x^0,$$
satisfies
$$x(T_0) = 0.$$
Similarly, applying the previous result with $x^0 = 0$, one gets the existence of $T_1 \leqslant T/2$ and of $u_1 \in L^1((0, T_1); \mathbb{R}^m)$ such that the solution of the Cauchy problem
$$\dot{x} = Ax + F(x) + Bu_1(t), \, x(0) = 0,$$
satisfies
$$x(T_1) = x^1.$$
Finally, let $u \in L^1((0, T); \mathbb{R}^m)$ be defined by
$$u(t) = u_0(t), \, t \in (0, T_0),$$
$$u(t) = 0, \, t \in (T_0, T - T_1),$$
$$u(t) = u_1(t - T + T_1), \, t \in (T - T_1, T).$$
(Note that, since $T_0 \leqslant T/2$ and $T_1 \leqslant T/2$, one has $T_0 \leqslant T - T_1$.) Then (let us recall (3.144)) the solution of the Cauchy problem
$$\dot{x} = Ax + F(x) + Bu(t), x(0) = x^0,$$
satisfies
$$x(T) = x^1.$$

This concludes the proof of Theorem 3.46. ∎

REMARK 3.47. The above argument is similar to the one introduced in [**111**] in the framework of control of partial differential equations to get a global controllability result for the Navier-Stokes from a local controllability result for the Euler equations; see also [**123, 188, 9, 446**] as well as Section 6.2.2.2.

3.6. Bibliographical complements

There are many books on the subject treated in this chapter. Let us, for example, mention the following classical books, where one can also find more advanced results: [**8**] by Andrei Agrachev and Yuri Sachkov, [**247**] by Alberto Isidori, [**258**] by Velimir Jurdjevic, [**361**] by Richard Montgomery and [**376**] by Henk Nijmeijer and Arjan van der Schaft. In these books, one can find many results of controllability relying on the Lie brackets approach. For books dealing with more specific problems that one encounters in robotics, let us mention, in particular, [**18**] by Brigitte d'Andréa-Novel, [**82**] edited by Carlos Canudas-de-Wit, [**303**] edited by Jean-Paul Laumond, [**436**] by Claude Samson, Michel Le Borgne and Bernard Espiau.

There is a subject that we have not at all discussed, namely the problem of optimal control. Now one wants to go from x^0 to x^1 in an optimal way, that is, the way that achieves some minimizing criterion. This if of course an important problem both from the mathematical and practical points of view. There is a huge literature on this subject. For the Pontryagin [**390**] maximum principle side (open-loop approach), let us mention, in particular, the following books or recent papers:

- [**62**] by Bernard Bonnard and Monique Chyba, on singular trajectories and their role in optimal control theory.
- [**63**] by Bernard Bonnard, Ludovic Faubourg and Emmanuel Trélat on the control of space vehicle.
- [**100**] by Francis Clarke, Yuri Ledyaev, Ronald Stern and Peter Wolenski as well as [**96**] by Francis Clarke, for the use of nonsmooth analysis to get generalizations of the classical maximum principle.
- [**290**] by Huibert Kwakernaak and Raphael Sivan, a classical book on linear optimal control systems.
- [**309**] by Ernest Lee and Lawrence Markus, one of the most classical books on this subject.
- [**440**] by Atle Seierstad and Knut Sydsæter, with a special emphasis on applications of optimal control to economics.
- [**470**] by Héctor Sussmann, where one can find numerous generalizations of the classical maximum principle.
- [**481**] by Emmanuel Trélat, which also deals with classical materials of control theory.
- [**495**] by Richard Vinter, which deals with optimal control viewed from the nonsmooth analysis side.
- [**509**] by Jerzy Zabczyk, which deals to finite and infinite-dimensional control systems.

For the Hamilton-Jacobi side (closed-loop approach), let us mention, besides some of the above references, the following books:

- [**35**] by Guy Barles and [**331**] by Pierre-Louis Lions, both on viscosity solutions of the Hamilton-Jacobi equation (a notion introduced by Michael Crandall and Pierre-Louis Lions in [**135**]).
- [**64**] by Ugo Boscain and Benedetto Piccoli, with a detailed study of the optimal synthesis problem in dimension 2.
- [**81**] by Piermarco Cannarsa and Carlo Sinestrari, which shows the importance of semiconcavity for the Hamilton-Jacobi equation.
- [**510**] by Jerzy Zabczyk, which deals with stochastic control in discrete time.

CHAPTER 4

Linearized control systems and fixed-point methods

In this chapter, we first consider (Section 4.1 and Section 4.2) the problem of the controllability around an equilibrium of a nonlinear partial differential equation such that the linearized control system around this equilibrium is controllable. In finite dimension, we have already seen (Theorem 3.8 on page 128) that, in such a situation, the nonlinear control system is locally controllable around the equilibrium. Of course in infinite dimension one expects that a similar result holds. We prove that this is indeed the case for the following equations.

- The nonlinear Korteweg-de Vries equation. In this case one can prove the local controllability of the nonlinear control equation (Theorem 4.3 on page 161) by means of a standard fixed-point method.
- A hyperbolic equation. In this case there is a problem of loss of derivatives which prevents the use of a standard fixed-point method. One uses instead an ad-hoc fixed-point method, which is specific to hyperbolic systems.
- A nonlinear one-dimensional Schrödinger equation. There is again a problem of loss of derivatives. This problem is overcome by the use of a Nash-Moser method.

Of course these nonlinear equations have to be considered only as examples given here to illustrate possible methods which can be used when the linearized control system around the equilibrium is controllable: these methods actually can be applied to many other equations.

Sometimes these methods, which lead to *local* controllability results, can be adapted to give *global* controllability results if the nonlinearity is not too strong at infinity. We have already seen this for finite-dimensional control systems (see Section 3.5). In Section 4.3 we show how to handle the case of some nonlinear partial differential equations on the example of a nonlinear one-dimensional wave equation.

4.1. The Linear test: The regular case

In this section, we consider nonlinear control systems modeled by means of partial differential equations such that the linearized control system around some equilibrium is controllable. We want to deduce from this property the local controllability of the nonlinear control system around this equilibrium. On a nonlinear Korteweg-de Vries equation, we explain how to get that local controllability by means of a natural fixed-point strategy, which can be used for many other equations.

We consider the following nonlinear control system

(4.1) $\quad y_t + y_x + y_{xxx} + yy_x = 0,\ t \in (0,T),\ x \in (0,L),$

(4.2) $\quad y(t,0) = y(t,L) = 0,\ y_x(t,L) = u(t),\ t \in (0,T),$

where, at time $t \in [0,T]$, the control is $u(t) \in \mathbb{R}$ and the state is $y(t,\cdot) : (0,L) \mapsto \mathbb{R}$. Equation (4.1) is a Korteweg-de Vries equation, which serves to model various physical phenomena, for example, the propagation of small amplitude long water waves in a uniform channel (see, e.g., [**144**, Section 4.4, pages 155–157] or [**502**, Section 13.11]). Let us recall that Jerry Bona and Ragnar Winther pointed out in [**60**] that the term y_x in (4.1) has to be added to model the water waves when x denotes the spatial coordinate in a *fixed* frame. The nonlinear control system (4.1)-(4.2) is called the nonlinear KdV equation.

The linearized control system around $(y,u) = (0,0)$ is the control system

(4.3) $\quad y_t + y_x + y_{xxx} = 0,\ x \in (0,L),\ t \in (0,T),$

(4.4) $\quad y(t,0) = y(t,L) = 0,\ y_x(t,L) = u(t),\ t \in (0,T),$

where, at time t, the control is $u(t) \in \mathbb{R}$ and the state is $y(t,\cdot) : (0,L) \to \mathbb{R}$. We have previously seen (see Theorem 2.25 on page 42) that if

(4.5) $\quad L \notin \mathcal{N} := \left\{ 2\pi\sqrt{\dfrac{j^2 + l^2 + jl}{3}};\ j, l \in \mathbb{N} \setminus \{0\} \right\},$

then, for every time $T > 0$, the control system (4.3)-(4.4) is controllable in time T. Hence one may expect that the nonlinear control system (4.1)-(4.2) is at least locally controllable if (4.5) holds. The goal of this section is to prove that this is indeed true, a result due to Lionel Rosier [**407**, Theorem 1.3].

4.1.1. Well-posedness of the Cauchy problem. Let us first define the notion of solutions for the Cauchy problem associated to (4.1)-(4.2). The same procedure motivating the definition (Definition 2.21 on page 38) of solutions to the Cauchy problem for the linearized control system (4.3)-(4.4), leads to the following definition.

DEFINITION 4.1. Let $T > 0$, $f \in L^1((0,T); L^2(0,L))$, $y^0 \in L^2(0,L)$ and $u \in L^2(0,T)$ be given. A *solution of the Cauchy problem*

(4.6) $\quad y_t + y_x + y_{xxx} + yy_x = f,\ x \in [0,L],\ t \in [0,T],$

(4.7) $\quad y(t,0) = y(t,L) = 0,\ y_x(t,L) = u(t),\ t \in [0,T],$

(4.8) $\quad y(0,x) = y^0(x),\ x \in [0,L],$

is a function $y \in C^0([0,T]; L^2(0,L)) \cap L^2((0,T); H^1(0,L))$ such that, for every $\tau \in [0,T]$ and for every $\phi \in C^3([0,\tau] \times [0,L])$ such that

(4.9) $\quad \phi(t,0) = \phi(t,L) = \phi_x(t,0) = 0,\ \forall t \in [0,\tau],$

one has

(4.10) $\quad -\displaystyle\int_0^\tau \int_0^L (\phi_t + \phi_x + \phi_{xxx}) y\, dx\, dt - \int_0^\tau u(t)\phi_x(t,L)\, dt + \int_0^\tau \int_0^L \phi y y_x\, dx\, dt$
$\quad + \displaystyle\int_0^L y(\tau, x)\phi(\tau, x)\, dx - \int_0^L y^0(x)\phi(0,x)\, dx = \int_0^\tau \int_0^L f y\, dx\, dt.$

Then one has the following theorem which is proved in [**122**, Appendix A].

THEOREM 4.2. *Let $T > 0$. Then there exists $\varepsilon > 0$ such that, for every $f \in L^1((0,T); L^2(0,L))$, $y^0 \in L^2(0,L)$ and $u \in L^2(0,T)$ satisfying*

$$\|f\|_{L^1((0,T);L^2(0,L))} + \|y^0\|_{L^2(0,L)} + \|u\|_{L^2(0,T)} \leqslant \varepsilon,$$

the Cauchy problem (4.6)-(4.7)-(4.8) has a unique solution.

The proof is rather lengthy and technical. We omit it.

4.1.2. Local controllability of the nonlinear KdV equation. The goal of this section is to prove the following local controllability result due to Lionel Rosier.

THEOREM 4.3 ([**407**, Theorem 1.3]). *Let $T > 0$, and let us assume that*

(4.11) $$L \notin \mathcal{N},$$

with

(4.12) $$\mathcal{N} := \left\{ 2\pi \sqrt{\frac{j^2 + l^2 + jl}{3}}; j, l \in \mathbb{N} \setminus \{0\} \right\}.$$

Then there exist $C > 0$ and $r_0 > 0$ such that for every $y^0, y^1 \in L^2(0,L)$, with $\|y^0\|_{L^2(0,L)} < r_0$ and $\|y^1\|_{L^2(0,L)} < r_0$, there exist

$$y \in C^0([0,T], L^2(0,L)) \cap L^2((0,T); H^1(0,L))$$

and $u \in L^2(0,T)$ satisfying (4.1)-(4.2) such that

(4.13) $$y(0, \cdot) = y^0,$$
(4.14) $$y(T, \cdot) = y^1,$$
(4.15) $$\|u\|_{L^2(0,T)} \leqslant C(\|y^0\|_{L^2(0,L)} + \|y^1\|_{L^2(0,L)}).$$

Proof of Theorem 4.3. Let us first recall (see Remark 2.33 on page 48) that, by the controllability of the linearized controlled KdV system (4.3)-(4.4) in time T, there exists a continuous linear map

$$\begin{array}{rcl} \Gamma: L^2(0,L) & \to & L^2(0,T) \\ y^1 & \mapsto & u \end{array}$$

such that the solution $y \in C^0([0,T]; L^2(0,L))$ of the Cauchy problem

$$y_t + y_x + y_{xxx} = 0, (t,x) \in (0,T) \times (0,L),$$
$$y(t,0) = y(t,L) = 0, y_x(t,L) = u(t), t \in (0,T),$$
$$y(0,x) = 0, x \in (0,L),$$

satisfies

$$y(T,x) = y^1(x), x \in (0,L).$$

Let

$$B := C^0([0,T]; L^2(0,L)) \cap L^2((0,T); H^1(0,L))$$

endowed with the norm

$$\|y\|_B = \text{Max}\left\{ \|y(t,\cdot)\|_{L^2(0,L)}; t \in [0,T] \right\} + \left(\int_0^T \|y(t,\cdot)\|_{H^1(0,L)}^2 dt \right)^{\frac{1}{2}}.$$

Let us denote by $\mathcal{Y}$ the continuous linear map
$$\begin{array}{rcl} L^2(0,T) & \to & C^0([0,T]; L^2(0,L)) \\ u & \mapsto & y, \end{array}$$
where y is the solution of
$$y_t + y_x + y_{xxx} = 0, \ (t,x) \in (0,T) \times (0,L),$$
$$y(t,0) = y(t,L) = 0, \ y_x(t,L) = u(t), \ t \in (0,T),$$
$$y(0,x) = 0, \ x \in (0,L)$$
(see Theorem 2.23 on page 39.) Proceeding as in the proof of Proposition 2.26 on page 42, one can check that $y \in B$ and that the linear map $\mathcal{Y}$ is continuous from $L^2(0,T)$ into B; see [**407**, Proof of Proposition 3.7] for the complete details).

Let S be the continuous linear map
$$\begin{array}{rcl} L^2(0,L) & \to & B \\ y^0 & \mapsto & y, \end{array}$$
where y is the solution of
$$y_t + y_x + y_{xxx} = 0, \ (t,x) \in (0,T) \times (0,L),$$
$$y(t,0) = y(t,L) = 0, \ y_x(t,L) = 0, \ t \in (0,T),$$
$$y(0,x) = y^0(x), \ x \in (0,L);$$
see Proposition 2.26 on page 42.

Let Ψ be the continuous linear map
$$\begin{array}{rcl} L^1((0,T); L^2(0,L)) & \to & B \\ f & \mapsto & y, \end{array}$$
where y is the solution of

(4.16) $$y_t + y_x + y_{xxx} = f, \ (t,x) \in (0,T) \times (0,L),$$
(4.17) $$y(t,0) = y(t,L) = 0, \ y_x(t,L) = 0, \ t \in (0,T),$$
(4.18) $$y(0,x) = 0, \ x \in (0,L).$$

Of course the definition of "y is a solution of (4.16)-(4.17)-(4.18)" is obtained by forgetting the term yy_x in Definition 4.1. Then the uniqueness of y follows from Theorem 2.23 on page 39. For the existence part:

1. One first treats the case where $f \in C^1([0,T]; L^2(0,L))$ by applying Theorem A.7 on page 375.
2. Then one gets the estimate $\|y\|_B \leqslant C\|f\|_{L^1((0,T);L^2(0,L))}$ by proceeding as in the proof of (2.141) and (2.142) (see Proposition 2.26 on page 42).
3. One finally treats the general case by using the density of $C^1([0,T]; L^2(0,L))$ in $L^1((0,T); L^2(0,L))$.

Let $\mathfrak{F}$ be the nonlinear map
$$\begin{array}{rcl} L^2((0,T); H^1(0,L)) & \to & L^2((0,T); H^1(0,L)) \\ y & \mapsto & z \end{array}$$
where
$$z := Sy^0 + \mathcal{Y} \circ \Gamma(y^1 - (Sy^0)(T,\cdot) + \Psi(yy_x)(T,\cdot)) - \Psi(yy_x).$$

Then, after some lengthy but straightforward estimates, it is not too hard to check that there exists $K > 0$ independent of $y \in L^2((0,T); H^1(0,L))$, of $y^0 \in L^2(0,L)$ and of $y^1 \in L^2(0,L)$ such that

(4.19) $\|\mathfrak{F}(y)\|_{L^2((0,T);H^1(0,L))}$
$$\leqslant K(\|y\|^2_{L^2((0,T);H^1(0,L))} + \|y^0\|_{L^2(0,L)} + \|y^1\|_{L^2(0,L)}),$$

(4.20) $\|\mathfrak{F}(z) - \mathfrak{F}(y)\|_{L^2((0,T);H^1(0,L))}$
$$\leqslant K\|z - y\|_{L^2((0,T);H^1(0,L))}(\|z\|_{L^2((0,T);H^1(0,L))} + \|y\|_{L^2((0,T);H^1(0,L))}).$$

We take

(4.21) $$R = \frac{1}{4K}, \quad r = \frac{3R^2}{2}.$$

Let $y^0 \in L^2(0,L)$ and $y^1 \in L^2(0,L)$ be such that

(4.22) $$\|y^0\|_{L^2(0,L)} \leqslant r \text{ and } \|y^1\|_{L^2(0,L)} \leqslant r.$$

Let
$$B_R := \left\{ y \in L^2((0,T); H^1(0,L)); \|y\|_{L^2((0,T);H^1(0,L))} \leqslant R \right\}.$$

The set B_R is a closed subset of the Hilbert space $L^2((0,T); H^1(0,L))$. From (4.19), (4.20), (4.21) and (4.22), we get that

$$\mathfrak{F}(B_R) \subset B_R,$$
$$\|\mathfrak{F}(z) - \mathfrak{F}(y)\|_{L^2((0,T);H^1(0,L))} \leqslant \frac{1}{2}\|z - y\|_{L^2((0,T);H^1(0,L))}, \forall (y,z) \in B_R^2.$$

Hence, by the Banach fixed-point theorem, $\mathfrak{F}$ has a (unique) fixed point in B_R. But, since y is a fixed point of $\mathfrak{F}$, the solution of the Cauchy problem

$$y_t + y_x + y_{xxx} + yy_x = 0, (t,x) \in (0,T) \times (0,L),$$
$$y(t,0) = y(t,L) = 0, y_x(t,L) = u(t), t \in (0,T),$$
$$y(0,x) = y^0(x), x \in (0,L),$$

with

(4.23) $$u := \Gamma(y^1 - (Sy^0)(T,\cdot) + \Psi(yy_x)(T,\cdot)) \in L^2(0,T),$$

satisfies
$$y(T,\cdot) = y^1.$$

Moreover, by (4.19), (4.21) and the fact that $y \in B_R$,

(4.24) $$\|y\|_{L^2((0,T);H^1(0,L))} \leqslant \frac{4}{3}K(\|y^0\|_{L^2(0,L)} + \|y^1\|_{L^2(0,L)}).$$

From (4.23) and (4.24), one gets the existence of $C > 0$, independent of y^0 and y^1 satisfying (4.22), such that

$$\|u\|_{L^2(0,T)} \leqslant C(\|y^0\|_{L^2(0,L)} + \|y^1\|_{L^2(0,L)}).$$

This concludes the proof of Theorem 4.3. ∎

EXERCISE 4.4. *This exercise is the continuation of Exercise 2.34 on page 49. Let $L > 0$ and $T > 0$. We consider the following nonlinear control system:*

$$(4.25) \qquad y_t + y_x + \int_0^x y(t,s)ds + \left(\int_0^x y(t,s)ds\right)^2 = 0,\ t \in (0,T),\ x \in (0,L),$$

$$(4.26) \qquad y(t,0) = u(t),\ t \in (0,T),$$

where, at time $t \in [0,T]$, the state is $y(t,\cdot) \in L^2(0,L)$ and the control is $u(t) \in \mathbb{R}$. For $y^0 \in L^2(0,L)$ and $u \in L^2(0,T)$, one says that $y : (0,T) \times (0,L) \to \mathbb{R}$ is a solution of the Cauchy problem

$$(4.27) \qquad y_t + y_x + \int_0^x y(t,s)ds + \left(\int_0^x y(t,s)ds\right)^2 = 0,\ t \in (0,T),\ x \in (0,L),$$

$$(4.28) \qquad y(t,0) = u(t),\ t \in (0,T),$$

$$(4.29) \qquad y(0,x) = y^0(x),\ x \in (0,L),$$

if $y \in C^0([0,T]; L^2(0,L))$ and if

$$-\int_0^\tau \int_0^L (\phi_t + \phi_x - \int_x^L \phi(t,s)ds) y\, dx\, dt$$

$$-\int_0^\tau u(t)\phi(t,0)dt + \int_0^\tau \int_0^L \phi(t,x) \left(\int_0^x y(t,s)ds\right)^2 dx\, dt$$

$$+\int_0^L y(\tau,x)\phi(\tau,x)dx - \int_0^L y^0(x)\phi(0,x)dx = 0,$$

for every $\tau \in [0,T]$ and for every $\phi \in C^1([0,T] \times [0,L])$ such that

$$\phi(t,L) = 0,\ \forall t \in [0,T].$$

1. *Prove that, for every $u \in L^2(0,T)$ and for every $y^0 \in L^2(0,L)$, the Cauchy problem (4.27)-(4.28)-(4.29) has at most one solution.*
2. *Prove that there exist $\varepsilon_1 > 0$ and $C_4 > 0$ (depending on T and L) such that, for every $u \in L^2(0,T)$ and for every $y^0 \in L^2(0,L)$ such that*

$$\|u\|_{L^2(0,T)} + \|y^0\|_{L^2(0,L)} \leqslant \varepsilon_1,$$

  the Cauchy problem (4.27)-(4.28)-(4.29) has a solution y and this solution satisfies

$$\|y\|_{C^0([0,T];L^2(0,L))} \leqslant C_4(\|u\|_{L^2(0,T)} + \|y^0\|_{L^2(0,L)}).$$

3. *In this question, we assume that $T > L$. Prove that the control system is locally controllable around the equilibrium $(y_e, u_e) := (0,0)$. More precisely, prove that there exist $\varepsilon_2 > 0$ and $C_5 > 0$ such that, for every $y^0 \in L^2(0,L)$ and for every $y^1 \in L^2(0,L)$ such that*

$$\|y^0\|_{L^2(0,L)} + \|y^1\|_{L^2(0,L)} \leqslant \varepsilon_2,$$

  there exists $u \in L^2(0,T)$ satisfying

$$\|u\|_{L^2(0,T)} \leqslant C_5 \left(\|y^0\|_{L^2(0,L)} + \|y^1\|_{L^2(0,L)}\right)$$

  such that the Cauchy problem (4.27)-(4.28)-(4.29) has a unique solution y and this solution satisfies $y(T,\cdot) = y^1$.

Let us end this section by mentioning the following references where one can find other results on the controllability of nonlinear Korteweg-de Vries control systems:

- [136] by Emmanuelle Crépeau,
- [407, 408, 410, 411] by Lionel Rosier,
- [428] by David Russell and Bing-Yu Zhang,
- [513] by Bing-Yu Zhang.

In each of these papers, the authors get local controllability results from the controllability of the linearized control system around the considered equilibrium using similar fixed-point methods.

4.2. The linear test: The case of loss of derivatives

Unfortunately the iterative method presented above needs to be modified for many partial differential control systems. We study in this section two examples:

- The first one on hyperbolic systems (Section 4.2.1).
- The second one on Schrödinger equations (Section 4.2.2).

4.2.1. Loss of derivatives and hyperbolic systems. In order to present the problem of loss of derivatives, we start with a very simple nonlinear transport equation

$$(4.30) \quad y_t + a(y)y_x = 0, \ x \in [0, L], \ t \in [0, T],$$

$$(4.31) \quad y(t, 0) = u(t), \ t \in [0, T],$$

where $a \in C^2(\mathbb{R})$ satisfies

$$(4.32) \quad a(0) > 0.$$

For this control system, at time $t \in [0, T]$, the state is $y(t, \cdot) \in C^1([0, L])$ and the control is $u(t) \in \mathbb{R}$. We could also work with suitable Sobolev spaces (for example $y(t, \cdot) \in H^2(0, L)$ is a suitable space). We are interested in the local controllability of the control system (4.30)-(4.31) at the equilibrium $(\bar{y}, \bar{u}) = (0, 0)$. Hence we first look at the linearized control system at the equilibrium $(\bar{y}, \bar{u}) = (0, 0)$. This linear control system is the following one:

$$(4.33) \quad y_t + a(0)y_x = 0, \ t \in [0, T], \ x \in [0, L],$$

$$(4.34) \quad y(t, 0) = u(t), \ t \in [0, T].$$

For this linear control system, at time $t \in [0, T]$, the state is $y(t, \cdot) \in C^1([0, L])$ and the control is $u(t) \in \mathbb{R}$. Concerning the well-posedness of the Cauchy problem of this linear control system, following the explicit method detailed on page 27, and the uniqueness statement of Theorem 2.4 on page 27, one easily gets the following proposition.

PROPOSITION 4.5. *Let $T > 0$. Let $y^0 \in C^1([0, L])$ and $u \in C^1([0, T])$ be such that the following compatibility conditions hold*

$$(4.35) \quad u(0) = y^0(0),$$

$$(4.36) \quad \dot{u}(0) + a(0)y_x^0(0) = 0.$$

Then the Cauchy problem

(4.37) $$y_t + a(0)y_x = 0, \ t \in [0,T], \ x \in [0,L],$$
(4.38) $$y(t,0) = u(t), \ t \in [0,T],$$
(4.39) $$y(0,x) = y^0(x), \ x \in [0,L],$$

has a unique solution $y \in C^1([0,T] \times [0,L])$.

Of course (4.35) is a consequence of (4.38) and (4.39): it is a necessary condition for the existence of a solution $y \in C^0([0,T] \times [0,L])$ to the Cauchy problem (4.37)-(4.38)-(4.39). Similarly (4.36) is a direct consequence of (4.37), (4.38) and (4.39): it is a necessary condition for the existence of a solution $y \in C^1([0,T] \times [0,L])$ to the Cauchy problem (4.37)-(4.38)-(4.39).

Adapting the explicit strategy detailed in Section 2.1.2.1, one easily gets the following proposition.

PROPOSITION 4.6. *Let* $T > L/a(0)$. *The linear control system (4.33)-(4.34) is controllable in time* T. *In other words, for every* $y^0 \in C^1([0,L])$ *and for every* $y^1 \in C^1([0,L])$, *there exists* $u \in C^1([0,T])$ *such that the solution* y *of the Cauchy problem (4.37)-(4.38)-(4.39) satisfies*

(4.40) $$y(T,x) = y^1(x), \ x \in [0,L].$$

It is interesting to give an example of such a u. Let $u \in C^1([0,T])$ be such that

(4.41) $$u(t) = y^1(a(0)(T-t)), \text{ for every } t \in [T - (L/a(0)), T],$$
(4.42) $$u(0) = y^0(0),$$
(4.43) $$\dot{u}(0) = -a(0)y^0_x(0).$$

Such a u exists since $T > L/a(0)$. Then the solution y of the Cauchy problem (4.37)-(4.38)-(4.39) is given by

(4.44) $$y(t,x) = y^0(x - a(0)t), \ \forall (t,x) \in [0,T] \times [0,L] \text{ such that } a(0)t \leqslant x,$$
(4.45) $$y(t,x) = u(t - (x/a(0))), \ \forall (t,x) \in [0,T] \times [0,L] \text{ such that } a(0)t > x.$$

(The fact that such a y is in $C^1([0,T] \times [0,L])$ follows from the fact that y and u are of class C^1 and from the compatibility conditions (4.42)-(4.43).) From (4.44), one has (4.39). From $T > L/a(0)$, (4.41) and (4.44), one gets (4.40). With this method, one can easily construct a continuous linear map

$$\Gamma : \begin{array}{ccc} C^1([0,L]) & \to & \{u \in C^1([0,T]), \ u(0) = \dot{u}(0) = 0\} \\ y^1 & \mapsto & u \end{array}$$

such that the solution $y \in C^1([0,T] \times [0,L])$ of the Cauchy problem

$$y_t + a(0)y_x = 0, \ (t,x) \in [0,T] \times [0,L],$$
$$y(t,0) = u(t), \ t \in [0,T],$$
$$y(0,x) = 0, \ x \in [0,L],$$

satisfies

$$y(T,x) = y^1(x), \ x \in [0,L].$$

In order to prove the local controllability of the control system (4.30)-(4.31) at the equilibrium $(\bar{y}, \bar{u}) = (0,0)$, let us try to mimic what we have done for the nonlinear KdV equation (4.1)-(4.2) to prove the local controllability around $(\bar{y}, \bar{u}) = (0,0)$

(see Theorem 4.3 on page 161). Let $\mathcal{Y}$ be the linear map which associates to $u : (0,T) \to \mathbb{R}$ the function $y : (0,T) \times (0,L) \to \mathbb{R}$ which is the solution of

$$y_t + a(0)y_x = 0,\ (t,x) \in (0,T) \times (0,L),$$
$$y(t,0) = u(t),\ t \in (0,T),$$
$$y(0,x) = 0,\ x \in (0,L).$$

Let S be the linear map which to $y^0 : (0,L) \to \mathbb{R}$ associates the function y which is the solution of

$$y_t + a(0)y_x = 0,\ (t,x) \in (0,T) \times (0,L),$$
$$y(t,0) = 0,\ t \in (0,T),$$
$$y(0,x) = y^0(x),\ x \in (0,L).$$

Let Ψ be the linear map which associates to $f : (0,T) \times (0,L) \to \mathbb{R}$ the function $y : (0,T) \times (0,L) \to \mathbb{R}$ which is the solution of

(4.46) $$y_t + a(0)y_x = f,\ (t,x) \in (0,T) \times (0,L),$$
(4.47) $$y(t,0) = 0,\ t \in (0,T),$$
(4.48) $$y(0,x) = 0,\ x \in (0,L).$$

Finally, let $\mathfrak{F}$ be the nonlinear map which to $y : (0,T) \times (0,L) \to \mathbb{R}$ associates the function $z : (0,T) \times (0,L) \to \mathbb{R}$ defined by

(4.49) $$z := Sy^0 + \mathcal{Y} \circ \Gamma(y^1 - Sy^0(T,\cdot) + (\Psi((a(y) - a(0))y_x))(T,\cdot))$$
$$- \Psi((a(y) - a(0))y_x).$$

Then, up to regularity problems, a fixed point of $\mathfrak{F}$ is a solution to our controllability problem. But the problem is now to find the good spaces in order to have a well defined continuous map $\mathfrak{F}$. In order to avoid small technicalities coming from compatibility conditions, let us restrict to the case where

(4.50) $$y^0(0) = y_x^0(0) = 0.$$

Concerning $\mathcal{Y}$, the natural choice of spaces is

$$\{u \in C^1([0,T]);\ u(0) = \dot{u}(0) = 0\} \to C^1([0,T] \times [0,L])$$
$$u \mapsto y$$

with the usual topologies. Concerning S, the natural choice of spaces is

$$\{y^0 \in C^1([0,L]);\ y^0(0) = y_x^0(0) = 0\} \to C^1([0,T] \times [0,L])$$
$$y^0 \mapsto y.$$

But a problem appears now with Ψ. Indeed, for $y \in C^1([0,T] \times [0,L])$ with $y(0,0) = y_x(0,0) = 0$, $(a(y) - a(0))y_x$ is expected to be only continuous (and vanishing at $(t,x) = (0,0)$). So we would like to have

$$\Psi :\ \{f \in C^0([0,T] \times [0,L]);\ f(0,0) = 0)\} \to C^1([0,T] \times [0,L])$$
$$f \mapsto y.$$

But this does not hold. Indeed, let y be the solution of the Cauchy problem (in any reasonable sense) (4.46)-(4.47)-(4.48). From (4.46), one gets

$$\frac{d}{dt}y(t, a(0)t + x) = f(t, a(0)t + x),$$

which, together with (4.47) and (4.48), leads to

$$
(4.51) \quad y(t,x) = \int_{t-(x/a(0))}^{t} f(\tau, x - a(0)(t-\tau))d\tau, \text{ if } a(0)t \geqslant x,
$$

$$
(4.52) \quad y(t,x) = \int_{0}^{t} f(\tau, x - a(0)(t-\tau))d\tau, \text{ if } a(0)t < x.
$$

But, there are plenty of $f \in C^0([0,T] \times [0,L])$, even vanishing at $(0,0)$, such that y, defined by (4.51)-(4.52), is not in $C^1([0,T] \times [0,L])$. Note that y is continuous: we have "lost one derivative". This problem of loss of derivatives appears in many situations.

There is a general tool, namely the Nash-Moser method, that we will briefly describe in the next section (Section 4.2.2), which allows us to deal with this problem of loss of derivatives. There are many forms of this method. Let us mention, in particular, the ones given by Mikhael Gromov in [**206**, Section 2.3.2], Lars Hörmander in [**236**], Richard Hamilton in [**216**]; see also the book by Serge Alinhac and Patrick Gérard [**10**]. This approach can also be used in our context, but:

1. It does not give the optimal functional spaces for the state and the control.
2. It is more complicated to apply than the method we want to present here.

The method we want to present here is the one which is used to construct solutions to the Cauchy problem

$$
(4.53) \quad y_t + a(y)y_x = 0, \, t \in [0,T], \, x \in [0,L],
$$

$$
(4.54) \quad y(t,0) = u(t), \, t \in [0,T],
$$

$$
(4.55) \quad y(0,x) = y^0(x), \, x \in [0,L],
$$

and to more general hyperbolic systems (see e.g. [**68**, pages 67–70], [**134**, pages 476–478], [**238**, pages 54–55], [**321**, pages 96–107], [**340**, pages 35–43] or [**443**, pages 106–116]). Concerning the Cauchy problem (4.53)-(4.54)-(4.55), one has the following theorem (see, for a much more general result, [**321**, Chapter 4]).

THEOREM 4.7. *Let $T > 0$. There exist $\varepsilon > 0$ and $C > 0$ such that, for every $y^0 \in C^1([0,L])$ and for every $u \in C^1([0,T])$ satisfying the compatibility conditions (4.35)-(4.36) and such that*

$$
\|y^0\|_{C^1([0,L])} + \|u\|_{C^1([0,T])} \leqslant \varepsilon,
$$

the Cauchy problem (4.53)-(4.54)-(4.55) has one and only one solution in $C^1([0,T] \times [0,L])$. Moreover, this solution y satisfies

$$
(4.56) \quad \|y\|_{C^1([0,T]\times[0,L])} \leqslant C(\|y^0\|_{C^1([0,L])} + \|u\|_{C^1([0,T])}).
$$

The uniqueness part can be proved by using Gronwall's inequality. This is in the existence part that one uses the iterative scheme that we are now going to detail in order to prove the following controllability result.

THEOREM 4.8. *Let us assume that*

$$
(4.57) \quad T > \frac{L}{a(0)}.
$$

Then there exist $\varepsilon > 0$ and $C > 0$ such that, for every $y^0 \in C^1([0,L])$ and for every $y^1 \in C^1([0,L])$ such that

$$
\|y^0\|_{C^1([0,L])} \leqslant \varepsilon \text{ and } \|y^1\|_{C^1([0,L])} \leqslant \varepsilon,
$$

there exists $u \in C^1([0,T])$ *such that*

(4.58) $$\|u\|_{C^1([0,T])} \leqslant C(\|y^0\|_{C^1([0,L])} + \|y^1\|_{C^1([0,L])})$$

and such that the solution of the Cauchy problem (4.53)-(4.54)-(4.55) exists, is of class C^1 *on* $[0,T] \times [0,L]$ *and satisfies*

$$y(T,x) = y^1(x), \, x \in [0,L].$$

One can find similar results for much more general hyperbolic systems in the papers [**95**] by Marco Cirinà, [**322**] by Ta-tsien Li and Bing-Yu Zhang, and [**319**] by Ta-tsien Li and Bo-Peng Rao. In fact, these papers use a method which is different from the one we present now. We sketch below (see page 173) the method used in these papers in the framework of our simple transport equation.

Our method to prove Theorem 4.8 is the following one. For $y \in C^1([0,T] \times [0,L])$, we consider the following *linear* control system

(4.59) $$z_t + a(y)z_x = 0, \, (t,x) \in [0,T] \times [0,L],$$

(4.60) $$z(t,0) = u(t), \, t \in [0,T],$$

where, at time t, the state is $z(t,\cdot) \in C^1([0,L])$ and the control is $u(t) \in \mathbb{R}$. For every function $f : K \to \mathbb{R}$, where K is a compact subset of $\mathbb{R}^n$, and for every $\rho \geqslant 0$, let

$$\omega_\rho(f) := \mathrm{Max}\{|f(\xi_2) - f(\xi_1)|; \, (\xi_1, \xi_2) \in K^2 \text{ and } |\xi_2 - \xi_1| \leqslant \rho\}.$$

For $\varepsilon > 0$, let W_ε be the set of $(y^0, y^1, b) \in C^1([0,L]) \times C^1([0,L]) \times C^1([0,T] \times [0,L])$ such that

$$\|y^0\|_{C^1([0,L])} + \|y^1\|_{C^1([0,L])} \leqslant \varepsilon \text{ and } \|b - a(0)\|_{C^1([0,T] \times [0,L])} \leqslant \varepsilon.$$

The set W_ε is equipped with the usual topology of $C^1([0,L]) \times C^1([0,L]) \times C^1([0,T] \times [0,L])$. Let us admit, for the moment, that one has the following controllability result.

PROPOSITION 4.9. *Let* $T > L/a(0)$. *There exist* $\epsilon_0 > 0$, $C_0 \geqslant 1$ *and a continuous map* Z,

$$\begin{array}{rcl} W_{\varepsilon_0} & \to & C^1([0,T] \times [0,L]) \\ (y^0, y^1, b) & \mapsto & z = Z(y^0, y^1, b), \end{array}$$

such that, for every $(y^0, y^1, b) \in W_{\varepsilon_0}$, z *satisfies*

(4.61) $$z_t + bz_x = 0, \, t \in [0,T], \, x \in [0,L],$$

(4.62) $$z(0,x) = y^0(x), \, z(T,x) = y^1(x), \, x \in [0,L],$$

(4.63) $$\|z\|_{C^1([0,T] \times [0,L])} \leqslant C_0(\|y^0\|_{C^1([0,L])} + \|y^1\|_{C^1([0,L])}),$$

(4.64) $$\omega_\rho(z_t) + \omega_\rho(z_x) \leqslant C_0(\omega_\rho(y^0_x) + \omega_\rho(y^1_x)) + C_0(\rho + \omega_\rho(b_x) + \omega_\rho(b_t))$$
$$\times (\|y^0\|_{C^1([0,L])} + \|y^1\|_{C^1([0,L])}), \, \forall \rho \in [0, +\infty).$$

For $\varepsilon > 0$, let

$$B_\varepsilon := \{y \in C^1([0,T] \times [0,L]); \, \|y\|_{C^1([0,T] \times [0,L])} \leqslant \varepsilon\},$$

which is equipped with the topology defined by the norm $\|\cdot\|_{C^1([0,T] \times [0,L])}$ on $C^1([0,T] \times [0,L])$. Let us now consider the following map $\mathfrak{F}$ defined on B_ε, for $\varepsilon > 0$ and $\|y^0\|_{C^1([0,L])} + \|y^1\|_{C^1([0,L])}$ small enough, by the formula

$$\mathfrak{F}(y) := Z(y^0, y^1, a(y)).$$

By Proposition 4.9, there exists $\varepsilon_1 \in (0, \varepsilon_0]$ such that, for $\varepsilon := \varepsilon_1$, $\mathfrak{F}$ is well defined and continuous if $\|y^0\|_{C^1([0,L])} + \|y^1\|_{C^1([0,L])} \leqslant \varepsilon_1$. Then, by (4.63), if

$$(4.65) \qquad \|y^0\|_{C^1([0,L])} + \|y^1\|_{C^1([0,L])} \leqslant \frac{\varepsilon_1}{C_0} \ (\leqslant \varepsilon_1),$$

we have

$$(4.66) \qquad \mathfrak{F}(B_{\varepsilon_1}) \subset B_{\varepsilon_1}.$$

Since a is assumed to be of class C^2 on $\mathbb{R}$, there exists $C_1 > 0$ such that

$$(4.67) \quad \omega_\rho(a(y)_x) + \omega_\rho(a(y)_t) \leqslant C_1(\omega_\rho(y_x) + \omega_\rho(y_t)), \ \forall y \in B_{\varepsilon_1}, \ \forall \rho \in [0, +\infty).$$

By (4.64) and (4.67), we have, with $z = \mathfrak{F}(y)$ and $y \in B_{\varepsilon_0}$,

$$(4.68) \quad \omega_\rho(z_t) + \omega_\rho(z_x) \leqslant C_0(\omega_\rho(y_x^0) + \omega_\rho(y_x^1))$$
$$+ C_0(\rho + C_1\omega_\rho(y_x) + C_1\omega_\rho(y_t))(\|y^0\|_{C^1([0,L])} + \|y^1\|_{C^1([0,L])}).$$

Let $\Omega : [0, +\infty) \to [0, +\infty)$ be defined by

$$(4.69) \qquad \Omega(\rho) := 2C_0(\omega_\rho(y_x^0) + \omega_\rho(y_x^1)) + 2\rho.$$

Let

$$(4.70) \qquad K := \{y \in B_{\varepsilon_0}; \ \omega_\rho(y_x) + \omega_\rho(y_t) \leqslant \Omega(\rho), \ \forall \rho \in [0, +\infty)\}.$$

By Ascoli's theorem (see, for example, [**419**, A 5 Ascoli's theorem, page 369]), K is a compact subset of B_{ε_0}. Clearly K is convex. Moreover, by (4.66), (4.68), (4.69) and (4.70), if

$$(4.71) \qquad \|y^0\|_{C^1([0,L])} + \|y^1\|_{C^1([0,L])} \leqslant \mathrm{Min}\left\{\varepsilon_1, \frac{2}{C_0(1+2C_1)}, \frac{1}{2C_0C_1}\right\},$$

then

$$\mathfrak{F}(K) \subset K,$$

which, together with the Schauder fixed-point theorem (Theorem B.19 on page 392), implies that $\mathfrak{F}$ has a fixed point. But, if y is a fixed point, then

$$y_t + a(y)y_x = 0, \ t \in [0, T], \ x \in [0, L],$$
$$y(0, x) = y^0(x), \ x \in [0, L],$$
$$y(T, x) = y^1(x), \ x \in [0, L].$$

Hence the control $u(t) := y(t, 0)$ steers the control system (4.30)-(4.31) from the state y^0 to the state y^1 during the time interval $[0, T]$. Inequality (4.58) follows from (4.63). This concludes the proof of Theorem 4.7, assuming Proposition 4.9. ∎

Proof of Proposition 4.9. Let

$$\Phi : (t_1, t_2, x) \in D(\Phi) \subset [0, T] \times [0, T] \times [0, L] \mapsto \Phi(t_1, t_2, x) \in [0, L]$$

be the flow associated to the ordinary differential equation $\dot{\xi} = b(t, \xi)$:

$$\frac{\partial \Phi}{\partial t_1} = b(t_1, \Phi), \ t_1 \in [0, T],$$
$$\Phi(t_2, t_2, x) = x, \ t_2 \in [0, T], \ x \in [0, L].$$

4.2. THE LINEAR TEST: THE CASE OF LOSS OF DERIVATIVES

From now on, we assume that $(y^0, y^1, b) \in W_\varepsilon$. For every $(t_2, x) \in [0, T] \times [0, L]$, $\Phi(\cdot, t_2, x)$ is defined on some closed interval $[\tau_-(t_2, x), \tau_+(t_2, x)] \subset [0, T]$ containing t_2. At least for $\varepsilon > 0$ small enough (which will always be assumed), we have, using (4.57),

$$b(t, x) > 0, \; \forall (t, x) \in [0, T] \times [0, L],$$
$$\Phi(\tau_-(T, L), T, L) = 0, \; \tau_-(T, L) > 0,$$
$$\Phi(\tau_+(0, 0), 0, 0) = L, \; \tau_+(0, 0) < T.$$

Let

$$Q := \{(t, x) \in [0, T] \times [0, L]; \; \Phi(t, T, L) < x < \Phi(t, 0, 0)\}.$$

(See Figure 1.)

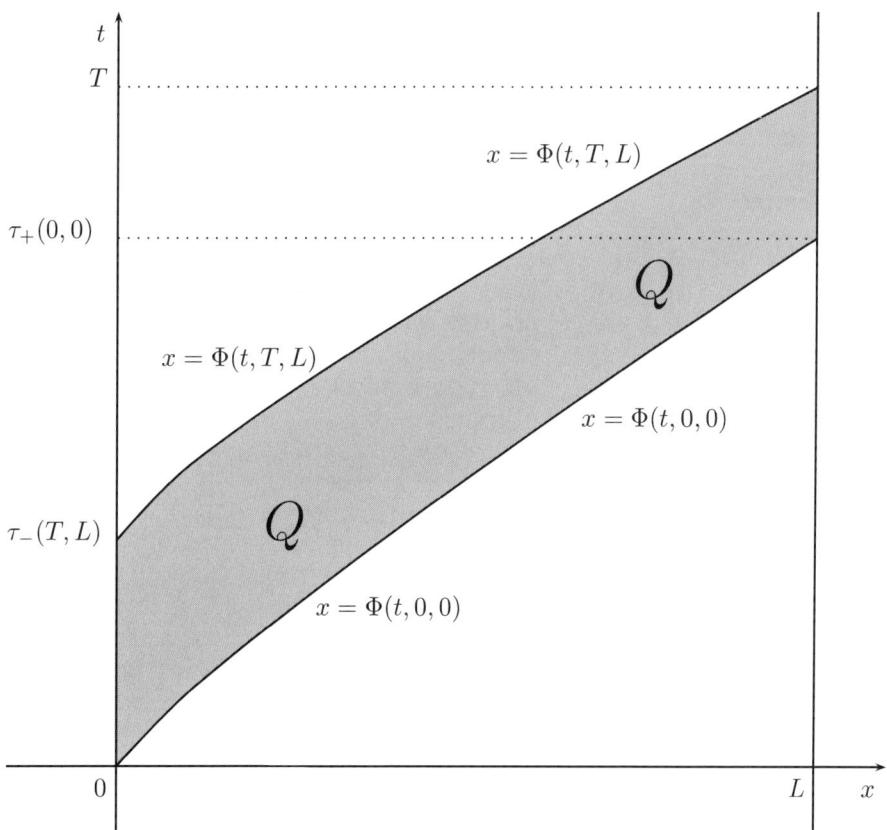

FIGURE 1. Flow Φ, Q, $\tau_-(T, L)$, $\tau_+(0, 0)$

We first define z on the set $([0, T] \times [0, L]) \setminus Q$ by

(4.72) $\quad z(t, x) := y^1(\Phi(T, t, x)), \; \forall (t, x) \in [0, T] \times [0, L]$ such that $x \leqslant \Phi(t, T, L)$,

(4.73) $\quad z(t, x) := y^0(\Phi(0, t, x)), \; \forall (t, x) \in [0, T] \times [0, L]$ such that $x \geqslant \Phi(t, 0, 0)$.

For $i \in \{1,2,3,4\}$, let $\alpha_i \in C^2([0,1])$ be such that
$$\alpha_1(0) = 1,\ \alpha_1'(0) = 0,\ \alpha_1(1) = 0,\ \alpha_1'(1) = 0,$$
$$\alpha_2(0) = 0,\ \alpha_2'(0) = 1,\ \alpha_2(1) = 0,\ \alpha_2'(1) = 0,$$
$$\alpha_3(0) = 0,\ \alpha_3'(0) = 0,\ \alpha_3(1) = 1,\ \alpha_3'(1) = 0,$$
$$\alpha_4(0) = 0,\ \alpha_4'(0) = 0,\ \alpha_4(1) = 0,\ \alpha_4'(1) = 1.$$

Let $u : [0, \tau_-(T,L)] \to \mathbb{R}$ be defined by
$$u(t) := y^0(0)\alpha_1\left(\frac{t}{\tau_-(T,L)}\right) - b(0,0)y_x^0(0)\tau_-(T,L)\alpha_2\left(\frac{t}{\tau_-(T,L)}\right)$$
$$+ z(\tau_-(T,L),0)\alpha_3\left(\frac{t}{\tau_-(T,L)}\right) + z_t(\tau_-(T,L),0)\tau_-(T,L)\alpha_4\left(\frac{t}{\tau_-(T,L)}\right),$$

where $z_t(\tau_-(T,L),0)$ is the left derivative with respect to time at time $\tau_-(T,L)$ of $z(\cdot,0) \in C^1([\tau_-(T,L),T])$; see (4.72). Finally we define

(4.74) $$z(t,x) := u(\tau_-(t,x)),\ \forall (t,x) \in Q.$$

Then, after some rather lengthy but straightforward computations one can check that the function z defined by (4.72), (4.73) and (4.74) satisfies all the required properties if $\varepsilon > 0$ is small enough. This concludes the proof of Proposition 4.9 and also of Theorem 4.8. ∎

REMARK 4.10. As for the Nash-Moser method, the controllability of the linearized control system (4.37)-(4.38)-(4.39) at the equilibrium $(y_e, u_e) := (0,0)$ (see Proposition 4.6 on page 166) is not sufficient for our proof of Theorem 4.8: one needs a controllability result (see Proposition 4.9 on page 169) for linear control systems which are close to the linear control system (4.37)-(4.38)-(4.39). For some other hyperbolic control systems (see for example page 214 and the proof of Proposition 6.29 given in [**116**]) we have not been able to check that *every* linear control system close to the linearized control system at the equilibrium considered are controllable, but we can do it for a large enough family of these linearized control systems. Then one can again get the controllability by a fixed-point method such that the image of the map $\mathfrak{F}$ lies in a suitable family. Note that the control offers some flexibility: we can require, beside (4.61) to (4.63), some extra conditions.

REMARK 4.11. Even if $(y, y(\cdot, 0))$ is a trajectory of the control system (4.53)-(4.54), the linear control system (4.59)-(4.60) is not the linearized control system around this trajectory. Indeed this linearized control system is

(4.75) $$z_t + a(y)z_x + a'(y)zy_x = 0,\ (t,x) \in [0,T] \times [0,L],$$
(4.76) $$z(t,0) = u(t),\ t \in [0,T],$$

where, at time t, the state is $z(t, \cdot) \in C^1([0,L])$ and the state is $u(t) \in \mathbb{R}$. Note that one encounters the problem of loss of derivatives if one uses the linear control system (4.75)-(4.76), together with the usual Newton iterative scheme; see also Remark 6.1 on page 189 as well as Remark 6.30 on page 214.

REMARK 4.12. Our proof of Theorem 4.8 on page 168 does not use the existence part of Theorem 4.7 on page 168 on the solutions to the Cauchy problem (4.53)-(4.54)-(4.55): our method directly proves the existence of a solution. (This is also the case for many other control problems. This is for example the case for the Euler equations of incompressible fluids; see Section 6.2.1.) However, as we mentioned on page 168, the fixed-point method used here is in fact strongly related to the classical method used to prove the existence of a solution to the Cauchy problem (4.53)-(4.54)-(4.55) (or more general hyperbolic systems). Our fixed-point approach is, in particular, very close to the one introduced by Ta-tsien Li and Wen Ci Yu in [**321**, pages 96–107]. As in [**321**, pages 96–107], we get the existence of a fixed point of $\mathfrak{F}$ by establishing estimates on the modulus of continuity (see (4.68)). With some extra estimates, one can use the Banach fixed-point theorem instead of the Schauder fixed-point theorem. Note that the Banach fixed-point theorem is more convenient than the Schauder fixed-point theorem for numerical purposes; see also [**68**, pages 67–70], [**134**, pages 476–478], [**238**, pages 54–55], [**340**, pages 35–43] or [**443**, pages 106–116] for the existence of solutions to the Cauchy problem (4.53)-(4.54)-(4.55).

Finally, let us briefly sketch the alternative method to prove controllability results for hyperbolic systems used by Marco Cirinà in [**95**] and by Ta-tsien Li and Bo-Peng Rao in [**319**]. For simplicity we present it again on our simple nonlinear transport equation (4.30)-(4.31), that is, we show how their method to prove Theorem 4.8 works. For $y \in C^1([0,T] \times [0,L])$, let us denote by Φ_y the flow associated to the differential equation $\dot\xi = a(y(t,\xi))$ (see page 170). For every $(t_2, x) \in [0,T] \times [0,L]$, $\Phi(\cdot, t_2, x)$ is defined on some closed interval $[\tau_{y-}(t_2, x), \tau_{y+}(t_2, x)] \subset [0,T]$ containing t_2. Let also Q_y be the set of $(t, x) \in [0,T] \times [0,L]$ such that

$$x \leqslant \Phi_y(t, T, L) \text{ or } x \geqslant \Phi_y(t, 0, 0).$$

The first step is the proof of the following proposition.

PROPOSITION 4.13. *Let us assume that (4.57) holds. Then there exist $\varepsilon > 0$ and $C > 0$ such that, for every $y^0 \in C^1([0,L])$ and for every $y^1 \in C^1([0,L])$ such that*

$$\|y^0\|_{C^1([0,L])} \leqslant \varepsilon \text{ and } \|y^1\|_{C^1([0,L])} \leqslant \varepsilon,$$

there exists $z \in C^1([0,T] \times [0,L])$ such that

(4.77) $\quad a(z(t,x)) > 0, \ \forall (t,x) \in [0,T] \times [0,L],$

(4.78) $\quad \tau_{z-}(T,L) > 0, \ \Phi_z(0,T,L) = 0,$

(4.79) $\quad \Phi_z(\tau_{z+}(0,0), 0, 0) = L, \ \tau_{z+}(0,0) < T,$

(4.80) $\quad z_t + a(z)z_x = 0 \text{ in } Q_z,$

(4.81) $\quad z(0,x) = y^0(x), \ \forall x \in [0,L],$

(4.82) $\quad z(T,x) = y^1(x), \ \forall x \in [0,L],$

(4.83) $\quad \|z\|_{C^1([0,T]\times[0,L])} \leqslant C(\|y^0\|_{C^1([0,L])} + \|y^1\|_{C^1([0,L])}).$

The proof of this proposition is now classical; see [**321**, Chapter 4] by Ta-tsien Li and Wen Ci Yu for much more general results on hyperbolic systems. As we mention in Remark 4.12 on the preceding page, it can be obtained by the fixed-point strategy we have used above to prove Theorem 4.8 on page 168. We omit it.

Now let $u \in C^1([0,T])$ be such that
$$u(t) = z(t,0), \forall t \in [\tau_{z_-}(T,L), T],$$
$$u(0) = y^0(0),$$
$$\dot{u}(0) = -a(y^0(0))y_x^0(0).$$

We can also impose on u that, for some $C_1 > 0$ independent of y^0 and y^1, provided that $\|y^0\|_{C^1([0,L])} + \|y^1\|_{C^1([0,L])}$ is small enough,

(4.84) $$\|u\|_{C^1([0,T])} \leqslant C_1(\|z(\cdot,0)\|_{C^1([0,T])} + \|y^0\|_{C^1([0,L])}).$$

By (4.83), (4.84) and Theorem 4.7 on page 168, if $\|y^0\|_{C^1([0,L])} + \|y^1\|_{C^1([0,L])}$ is small enough, there exists (a unique) $y \in C^1([0,T] \times [0,L])$ solution to the Cauchy problem

$$y_t + a(y)y_x = 0, t \in [0,T], x \in [0,L],$$
$$y(t,0) = u(t), t \in [0,T],$$
$$y(0,x) = y^0(x), x \in [0,L].$$

Then one checks that $Q_y = Q_z$ and that $y = z$ on Q_y. In particular,
$$y(T,x) = z(T,x) = y^1(x), \forall x \in [0,L].$$

Inequality (4.58) follows from (4.56), (4.83) and (4.84). This concludes the sketch of this alternative proof of Theorem 4.8. ∎

REMARK 4.14. This method is simpler than our first proof of Theorem 4.8 (see pages 168-172). However, in some cases, the alternative method does not seem to apply whereas the first one can be used; see, for example, Proposition 6.29 on page 214.

4.2.2. Loss of derivatives and a Schrödinger equation.

In this section we consider a nonlinear control system where the problem of loss of derivatives appears and where the trick used above in Section 4.2.1 to handle this problem does not seem to be applicable. Let $I = (-1,1)$ and let $T > 0$. We consider the Schrödinger control system

(4.85) $$\psi_t(t,x) = i\psi_{xx}(t,x) + iu(t)x\psi(t,x), (t,x) \in (0,T) \times I,$$
(4.86) $$\psi(t,-1) = \psi(t,1) = 0, t \in (0,T).$$

This is a control system, where, at time $t \in (0,T)$:

- The state is $\psi(t,\cdot) \in L^2(I;\mathbb{C})$ with $\int_I |\psi(t,x)|^2 dx = 1$.
- The control is $u(t) \in \mathbb{R}$.

This system has been introduced by Pierre Rouchon in [**417**]. It models a non-relativistic charged particle in a 1-D moving infinite square potential well. At time t, $\psi(t,\cdot)$ is the wave function of the particle in a frame attached to the potential well. The control $u(t)$ is the acceleration of the potential well at time t. We use the notations introduced in Section 2.6. Note that, with $\psi_{1,\gamma}$ defined in (2.494), we have

$$\psi_{1,\gamma t} = i\psi_{1,\gamma xx} + i\gamma x\psi_{1,\gamma},\ t \in (0,T),\ x \in I,$$
$$\psi_{1,\gamma}(t,-1) = \psi_{1,\gamma}(t,1) = 0,\ t \in (0,T),$$
$$\int_I |\psi_{1,\gamma}(t,x)|^2 dx = 1,\ t \in (0,T).$$

Hence $(\psi_{1,\gamma}, \gamma)$ is a trajectory of the control system (4.85)-(4.86). We are interested in the local controllability around this trajectory. The linearized control system around this trajectory is the control system

(4.87) $$\Psi_t = i\Psi_{xx} + i\gamma x\Psi + iux\psi_{1,\gamma},\ (t,x) \in (0,T) \times I,$$
(4.88) $$\Psi(t,-1) = \Psi(t,1) = 0,\ t \in (0,T).$$

It is a linear control system where, at time $t \in [0,T]$:

- The state is $\Psi(t,\cdot) \in L^2(I;\mathbb{C})$ with $\Psi(t,\cdot) \in T_{\mathbb{S}}(\psi_{1,\gamma}(t,\cdot))$.
- The control is $u(t) \in \mathbb{R}$.

Let us recall that $\mathbb{S}$ is the unit sphere of $L^2(I;\mathbb{C})$ (see (2.490)) and that, for $\phi \in \mathbb{S}$, $T_{\mathbb{S}}(\phi)$ is the tangent space to $\mathbb{S}$ at ϕ (see (2.491)). We have studied the controllability of this linear control system in Section 2.6. Theorem 2.87 tells us that this linear control system is controllable with state in $H^3_{(0)} \cap T_{\mathbb{S}}(\psi_{1,\gamma}(t,\cdot))$ ($H^3_{(0)}$ is defined in (2.499)) and control u in $L^2(0,T)$ if $\gamma > 0$ is small enough (this controllability does not hold for $\gamma = 0$; see page 97). This leads naturally to the following problem, which is still open.

OPEN PROBLEM 4.15. *Let $T > 0$. Does there exist $\gamma_0 > 0$ such that, for every $\gamma \in (0,\gamma_0)$, there exists $\varepsilon > 0$ such that, for every $\psi^0 \in H^3_{(0)}(I;\mathbb{C}) \cap T_{\mathbb{S}}(\psi_{1,\gamma}(0,\cdot))$ and for every $\psi^1 \in H^3_{(0)}(I;\mathbb{C}) \cap T_{\mathbb{S}}(\psi_{1,\gamma}(T,\cdot))$ such that*

$$\|\psi^0 - \psi_{1,\gamma}(0,\cdot)\|_{H^3(I;\mathbb{C})} + \|\psi^1 - \psi_{1,\gamma}(T,\cdot)\|_{H^3(I;\mathbb{C})} \leqslant \varepsilon,$$

there exists $u \in L^2(0,T)$ such that the solution

$$\psi \in C^0([0,T]; H^3(I;\mathbb{C}) \cap H^1_0(I;\mathbb{C})) \cap C^1([0,T]; H^1(I;\mathbb{C}))$$

of the Cauchy problem

(4.89) $$\psi_t(t,x) = i\psi_{xx}(t,x) + iu(t)x\psi(t,x),\ (t,x) \in (0,T) \times I,$$
(4.90) $$\psi(t,-1) = \psi(t,1) = 0,\ t \in (0,T),$$
(4.91) $$\psi(0,x) = \psi^0(x),\ x \in I,$$

satisfies

$$\psi(T,x) = \psi^1(x),\ x \in I?$$

(For the well-posedness of the Cauchy problem (4.89)-(4.90)-(4.91), we refer to [**40**].) However, if one lets

(4.92) $$H^5_{(0)}(I;\mathbb{C}) := \{\varphi \in H^5(I;\mathbb{C});\ \varphi^{(2k)}(-1) = \varphi^{(2k)}(1) = 0,\ \forall k \in \{0,1,2\}\},$$

one has the following theorem due to Karine Beauchard [**40**, Theorem 2].

THEOREM 4.16. *Let $T = 8/\pi$ and let $\eta > 0$. There exists $\varepsilon > 0$ such that, for every $\psi^0 \in H^5_{(0)}(I;\mathbb{C}) \cap H^{5+\eta}(I;\mathbb{C}) \cap T_{\mathbb{S}}(\psi_{1,\gamma}(0,\cdot))$ and for every $\psi^1 \in H^5_{(0)}(I;\mathbb{C}) \cap H^{5+\eta}(I;\mathbb{C}) \cap T_{\mathbb{S}}(\psi_{1,\gamma}(T,\cdot))$ such that*

$$\|\psi^0 - \psi_{1,\gamma}(0,\cdot)\|_{H^{5+\eta}(I;\mathbb{C})} + \|\psi^1 - \psi_{1,\gamma}(T,\cdot)\|_{H^{5+\eta}(I;\mathbb{C})} \leqslant \varepsilon,$$

there exists $u \in H^1_0(0,T)$ such that the solution

$$\psi \in C^0([0,T]; H^3(I;\mathbb{C}) \cap H^1_0(I;\mathbb{C})) \cap C^1([0,T]; H^1(I;\mathbb{C}))$$

of the Cauchy problem (4.89)-(4.90)-(4.91) satisfies

$$\psi(T,x) = \psi^1(x), \ x \in I.$$

REMARK 4.17. In [**40**, Theorem 2], ψ^0 and ψ^1 are assumed to be more regular than in $H^{5+\eta}$. However, Karine Beauchard has proposed in [**41, 42**] a modification which allows us to handle the regularity stated in Theorem 4.16.

The proof of Theorem 4.16 is quite long. Here, our goal is not to give it but only to explain the main difficulty (loss of derivatives) and the method used by Karine Beauchard to solve it. For simplicity, let us study only the case where $\psi^0 = \psi_{1,\gamma}$. Then the local controllability problem is to decide whether the map $\mathcal{G}$ which, to a control $u : [0,T] \to \mathbb{R}$, associates the state $\psi(T,\cdot) \in \mathbb{S}$, where $\psi : (0,T) \times I \to \mathbb{C}$ is the solution of the Cauchy problem (4.89)-(4.90)-(4.91), is locally onto at the point $\bar{u} := \gamma$ or not. The problem is that one does not know a couple of Hilbert spaces $(H, \tilde{H})$ (of infinite dimension) such that

(4.93) $\quad\quad\quad\quad\quad\quad\quad\quad \mathcal{G}$ is of class C^1,

(4.94) $\quad\quad\quad\quad \mathcal{G}'(\bar{u}) : H \to \tilde{H} \cap T_{\mathbb{S}}(\psi_{1,\gamma}(T,\cdot))$ is onto.

Let us recall that, ignoring this problem, we would get a control u steering our control system (4.85)-(4.86) from $\psi^0 = \psi_{1,\gamma}$ to ψ^1 by showing that the sequence

(4.95) $\quad\quad\quad\quad\quad\quad u_{n+1} = u_n - \mathcal{G}'(\bar{u})^{-1} P(\mathcal{G}(u_n) - \psi^1)$

is convergent. In (4.95), $\mathcal{G}'(\bar{u})^{-1}$ denotes a right inverse of $\mathcal{G}'(\bar{u})$ and P is the projection on the tangent space to $\mathbb{S}$ at $\psi_{1,\gamma}$. In Theorem 2.87 on page 96, we have constructed a right inverse $\mathcal{G}'(\bar{u})^{-1}$. But this right inverse sends elements of $H^k(I;\mathbb{C})$ to elements of $H^{k-3}(0,T)$ and, if u is in $H^k(0,T)$, one cannot expect $\mathcal{G}(u)$ to be more regular than $H^{k+2}(I;\mathbb{C})$. So, if $u_n \in H^k(0,T)$, we expect u_{n+1} to be only in $H^{k-1}(0,T)$. (In fact the situation is even worse due to some boundary conditions which have to be handled.) So, at each step, we lose "one derivative". Therefore the sequence $(u_n)_{n\in\mathbb{N}}$ will stop after a finite number of steps, at least if u_1 is not of class C^∞ (and is not expected to be convergent if u_1 is of class C^∞). In order to avoid this problem, one could try to modify (4.95) by putting

(4.96) $\quad\quad\quad\quad\quad\quad u_{n+1} = u_n - S_n \mathcal{G}'(\bar{u})^{-1} P(\mathcal{G}(u_n) - \psi^1),$

where S_n is some "smoothing" operator such that, as $n \to \infty$, $S_n \to \text{Id}$, Id denoting the identity map. But it seems difficult to get the convergence of the u_n's with such an iterative scheme, because, in some sense, the convergence of (4.95), even without the loss of derivatives problem is too slow. One uses instead of (4.95), the Newton method, which ensures a much faster convergence,

(4.97) $\quad\quad\quad\quad\quad\quad u_{n+1} = u_n - \mathcal{G}'(u_n)^{-1} P(\mathcal{G}(u_n) - \psi^1).$

The Newton method modified to take care of the loss of derivatives problem, is

(4.98) $$u_{n+1} = u_n - S_n \mathcal{G}'(u_n)^{-1} P(\mathcal{G}(u_n) - \psi^1).$$

The replacement of (4.97) by (4.98) is the key point of the Nash-Moser method. Now, with very careful (and lengthy) estimates and clever choices of S_n, the sequence $(u_n)_{n \in \mathbb{N}}$ is converging to the desired control.

REMARK 4.18. As in the case of the problem of loss of derivatives we have seen for a simple hyperbolic equation (see Section 4.2.1), it is not sufficient to have the controllability of the linearized control system along the trajectory $(\psi_{1,\gamma}, \gamma)$, but also the controllability of the linearized control systems around the trajectories which are close to the trajectory $(\psi_{1,\gamma}, \gamma)$.

REMARK 4.19. Karine Beauchard has shown in [41] that this Nash-Moser method can also be used to get various controllability results for a nonlinear one-dimensional beam equation.

REMARK 4.20. One can also adapt this Nash-Moser method to get the controllability of the hyperbolic control system (4.30)-(4.31). However, in terms of regularity of the couple state-control, one gets a slightly weaker result than Theorem 4.8. Note that Serge Alinhac and Patrick Gérard have used in [10] the Nash-Moser method to study the Cauchy problem (4.53)-(4.54)-(4.55).

REMARK 4.21. The Nash-Moser method has been introduced by John Nash in [374] to prove that every Riemannian manifold can be isometrically embedded in some $\mathbb{R}^k$ (equipped with the usual metric). It has been developed subsequently by Jürgen Moser in [368, 370, 369], by Mikhael Gromov in [205] and [206, Section 2.3], by Howard Jacobowitz in [249], by Richard Hamilton in [216] and by Lars Hörmander in [236, 237]. Let us point out that Matthias Günther has found in [212, 213] a way to prove the result on isometric embeddings by classical iterative schemes. It would be interesting to know if this is also possible for Theorem 4.16 on the previous page.

4.3. Global controllability for perturbations of linear controllable systems

When the nonlinearity is not too strong, one can get with suitable arguments global controllability results.

In this section, we illustrate in a detailed way, on an example, a method due to Enrique Zuazua [515, 516], which allows us to deal with such cases. For a description of this method in the framework of finite-dimensional control systems, see Section 3.5.1. We explain it on a simple 1-D semilinear wave equation with a boundary control and a nonlinearity which is at most linear (even if one can consider less restrictive growth at infinity as shown in [516]; see also Remark 4.23 on page 179).

The control system we consider is the following one:

(4.99) $$y_{tt} - y_{xx} + f(y) = 0, \ (t,x) \in (0,T) \times (0,L),$$
(4.100) $$y(t,0) = 0, \ y_x(t,L) = u(t), \ t \in (0,T).$$

In this control system, the control at time t is $u(t) \in \mathbb{R}$ and the state at time t is $y(t, \cdot) : (0, L) \to \mathbb{R}$. We assume that the function $f : \mathbb{R} \to \mathbb{R}$ is of class C^1 and satisfies, for some $C > 0$,

(4.101) $$|f(s)| \leqslant C(|s| + 1), \, \forall s \in \mathbb{R}.$$

As above (see Section 2.4), let
$$H^1_{(0)}(0, L) := \{\alpha \in H^1(0, L); \, \alpha(0) = 0\}.$$

Adapting the proof of Theorem 2.53 on page 68 and using classical arguments (see e.g. [84]), it is not difficult to check that, for every $T > 0$, for every $u \in L^2(0, T)$, and for every $(\alpha^0, \beta^0) \in H^1_{(0)} \times L^2(0, L)$, there exists a unique solution $y \in C^0([0, T]; H^1_0(0, L)) \cap C^1([0, T]; L^2(0, L))$ to the Cauchy problem

(4.102) $$y_{tt} - y_{xx} + f(y) = 0, \, (t, x) \in (0, T) \times (0, L),$$

(4.103) $$y(t, 0) = 0, \, y_x(t, L) = u(t), \, t \in (0, T),$$

(4.104) $$y(0, x) = \alpha^0(x) \text{ and } y_t(0, x) = \beta^0(x), \, x \in (0, L).$$

The goal of this section is to prove the following theorem, which is due to Enrique Zuazua [**516**, Theorem 3 and Section 6].

THEOREM 4.22. *Let $T > 2L > 0$. For every $\alpha^0 \in H^1_{(0)}(0, L)$, for every $\beta^0 \in L^2(0, L)$, for every $\alpha^1 \in H^1_{(0)}(0, L)$, and for every $\beta^1 \in L^2(0, L)$, there exists $u \in L^2(0, T)$ such that the solution y of the Cauchy problem (4.102)-(4.103)-(4.104) satisfies*

$$(y(T, \cdot), y_t(T, \cdot)) = (\alpha^1, \beta^1).$$

Proof of Theorem 4.22. Let $T > 2L > 0$. Let $(\alpha^0, \alpha^1) \in H^1_{(0)}(0, L)^2$ and $(\beta^0, \beta^1) \in L^2(0, L)^2$. As in the proof of Theorem 3.40 on page 151, we are going to get u as a fixed point of a map $\mathcal{F}$. This map

(4.105) $$\begin{array}{rccc} \mathcal{F}: & L^\infty((0, T) \times (0, L)) & \to & L^\infty((0, T) \times (0, L)) \\ & y & \mapsto & z = \mathcal{F}(y) \end{array}$$

is defined in the following way. Since $f \in C^1(\mathbb{R})$, there exists $g \in C^0(\mathbb{R})$ such that

(4.106) $$f(y) = f(0) + y g(y), \, \forall y \in \mathbb{R}.$$

Note that, by (4.101),

(4.107) $$g \in L^\infty(\mathbb{R}).$$

Let $\eta \in C^1([0, T]; L^2(0, L)) \cap C^0([0, T]; H^1(0, L))$ be the solution of

(4.108) $$\eta_{tt} - \eta_{xx} + g(y)\eta + f(0) = 0, \, t \in [0, T], \, x \in [0, L],$$

(4.109) $$\eta(t, 0) = \eta_x(t, L) = 0, \, t \in (0, T),$$

(4.110) $$\eta(0, x) = \eta_t(0, x) = 0, \, x \in (0, L).$$

Let $\mathcal{U}$ be the set of $u^* \in L^2(0, T)$ such that the solution z^* of the Cauchy problem (see Definition 2.52 on page 68 and Theorem 2.53 on page 68)

$$z^*_{tt} - z^*_{xx} + g(y) z^* = 0, \, t \in [0, T], \, x \in [0, L],$$
$$z^*(t, 0) = 0, \, z^*_x(t, L) = u^*(t), \, t \in (0, T),$$
$$z^*(0, x) = \alpha^0(x), \, z^*_t(0, x) = \beta^0(x), \, x \in (0, L),$$

satisfies
$$z^*(T,x) = \alpha^1(x) - \eta(T,x),\ z_t^*(T,x) = \beta^1(x) - \eta_t(T,x),\ x \in (0,L).$$

The set $\mathcal{U}$ is a closed affine subspace of $L^2(0,T)$. By Theorem 2.55 on page 72, $\mathcal{U}$ is not empty. Let u be the projection of 0 on $\mathcal{U}$. In other words, u is the least $L^2(0,L)$-norm element of $\mathcal{U}$. Let $\tilde{z} \in C^1([0,T]; L^2(0,L)) \cap C^0([0,T]; H^1_{(0)})$ be the solution of the Cauchy problem
$$\tilde{z}_{tt} - \tilde{z}_{xx} + g(y)\tilde{z} = 0,\ t \in [0,T],\ x \in [0,L],$$
$$\tilde{z}(t,0) = 0,\ \tilde{z}_x(t,L) = u(t),\ t \in (0,T),$$
$$\tilde{z}(0,x) = \alpha^0(x),\ \tilde{z}_t(0,x) = \beta^0(x),\ x \in (0,L).$$

Then we let
$$z := \eta + \tilde{z}.$$

Note that $z \in C^1([0,T]; L^2(0,L)) \cap C^0([0,T]; H^1_{(0)}(0,L))$ is the solution of the Cauchy problem

(4.111) $\quad z_{tt} - z_{xx} + g(y)z + f(0) = 0,\ t \in [0,T],\ x \in [0,L],$

(4.112) $\quad z(t,0) = 0,\ z_x(t,L) = u(t),\ t \in (0,T),$

(4.113) $\quad z(0,x) = \alpha^0(x),\ z_t(0,x) = \beta^0(x),\ x \in (0,L),$

and satisfies
$$z(T,x) = \alpha^1(x),\ z_t(T,x) = \beta^1(x),\ x \in (0,L).$$

Hence it suffices to prove that $\mathcal{F}$ has a fixed point, which gives $y = z$ in (4.111). It is not hard to check that $\mathcal{F}$ is a continuous map from $L^\infty((0,T) \times (0,L))$ into $L^\infty((0,T) \times (0,L))$. Using Theorem 2.53 on page 68, Theorem 2.55 on page 72 and (4.107), one gets the existence of $M > 0$ such that

(4.114) $\quad\quad\quad\quad \mathcal{F}(L^\infty((0,T) \times (0,L))) \subset K,$

where K is the set of $z \in C^1([0,T]; L^2(0,L)) \cap C^0([0,T]; H^1_{(0)}(0,L))$ such that
$$\|z\|_{C^1([0,T]; L^2(0,L))} \leqslant M,\ \|z\|_{C^0([0,T]; H^1_{(0)}(0,L))} \leqslant M.$$

Note that K is a convex subset of $L^\infty((0,T) \times (0,L))$. Moreover, by a theorem due to Jacques Simon [**449**],

(4.115) $\quad\quad\quad K$ is a compact subset of $L^\infty((0,T) \times (0,L)).$

Hence, by the Schauder fixed-point theorem (Theorem B.19 on page 392), $\mathcal{F}$ has a fixed point. This concludes the proof of Theorem 4.22. ∎

REMARK 4.23. It is proved in [**516**] that the growth condition (4.101) can be replaced by the weaker assumption

(4.116) $\quad\quad\quad\quad \limsup_{|s| \to +\infty} \frac{|f(s)|}{|s| \ln^2(|s|)} < \infty.$

Note that (4.116) is nearly optimal: Enrique Zuazua has proved ([**516**, Theorem 2]) that, if
$$\liminf_{s \to +\infty} \frac{-f(s)}{|s| \ln^p(s)} > 0$$

for some $p > 2$, then there are $\alpha^0 \in H^1_{(0)}(0,L)$, $\beta^0 \in L^2(0,L)$, $\alpha^1 \in H^1_{(0)}(0,L)$ and $\beta^1 \in L^2(0,L)$ such that, for every $T > 0$ and every $u \in L^2(0,T)$, the solution y of the Cauchy problem (4.102)-(4.103)-(4.104) does not satisfy $(y(T,\cdot), y_t(T\cdot)) = (\alpha^1, \beta^1)$.

REMARK 4.24. Enrique Zuazua has proposed in [**520**] a numerical method based on this fixed-point method and a two-grid approximation scheme to get a control steering the control system (4.99)-(4.100) from a given state to another given state, as in Theorem 4.22 on page 178. The idea to use a two-grid method in the control framework in order to avoid spurious high frequencies interfering with the mesh is due to Roland Glowinski [**201**].

REMARK 4.25. The above method has been used for many other partial differential equations. In particular, for the controllability of semilinear heat equations, let us mention [**156**] by Caroline Fabre, Jean-Pierre Puel and Enrique Zuazua, [**165**] by Luis Fernández and Enrique Zuazua, [**169**] by Enrique Fernández-Cara and Enrique Zuazua, [**186**, Chapter I, Section 3] by Andrei Fursikov and Oleg Imanuvilov, and [**517, 518**] by Enrique Zuazua.

REMARK 4.26. In order to conclude by using the Schauder fixed-point theorem, one needs some compactness properties ((4.115) here). Irena Lasiecka and Roberto Triggiani have presented in [**300**] a method to conclude by using a global inversion theorem. This allows us to treat other semilinear control systems which cannot be handled by the Schauder fixed-point theorem.

CHAPTER 5

Iterated Lie brackets

In this short chapter we start by explaining why iterated Lie brackets are less powerful in infinite dimension. Then we show on an example of how iterated Lie brackets can sometimes still be useful in infinite dimension.

Let us go back to the simplest control partial differential equation we have been considering, namely the case of the transport equation (see Section 2.1 on page 24). So our control system is

(5.1) $$y_t + y_x = 0, \; x \in [0, L],$$
(5.2) $$y(t, 0) = u(t),$$

where $L > 0$ is fixed and where, at time t, the control is $u(t) \in \mathbb{R}$ and the state is $y(t, \cdot) : [0, L] \to \mathbb{R}$. Let us use the same control $t \mapsto u(t)$ that we have used on page 131 to justify the interest of the Lie bracket $[f_0, f_1]$ for the *finite-dimensional* control system $\dot{x} = f_0(x) + u f_1(x)$. (Let us recall that we have seen in Section 2.3.3.1 how to write the control system (5.1)-(5.2) in the form $\dot{y} = Ay + Bu$.) So, let us consider, for $\varepsilon > 0$, the control defined on $[0, 2\varepsilon]$ by

$$u(t) = -1 \text{ for } t \in (0, \varepsilon),$$
$$u(t) = 1 \text{ for } t \in (\varepsilon, 2\varepsilon).$$

Let $y : (0, 2\varepsilon) \times (0, L) \to \mathbb{R}$ be the solution of the Cauchy problem

$$y_t + y_x = 0, \; t \in (0, 2\varepsilon), \; x \in (0, L),$$
$$y(t, 0) = u(t), \; t \in (0, 2\varepsilon),$$
$$y(0, x) = 0, \; x \in (0, L).$$

Then one readily gets, if $2\varepsilon \leqslant L$,

$$y(2\varepsilon, x) = 1, \; x \in (0, \varepsilon),$$
$$y(2\varepsilon, x) = -1, \; x \in (\varepsilon, 2\varepsilon),$$
$$y(2\varepsilon, x) = 0, \; x \in (2\varepsilon, L).$$

Hence (compare with (3.25))

(5.3) $$\left\| \frac{y(2\varepsilon, x) - y(0, x)}{\varepsilon^2} \right\|_{L^2(0, L)} \to +\infty \text{ as } \varepsilon \to 0^+.$$

Note that, for every $\phi \in H^2(0,L)$,

$$\begin{aligned}
\int_0^L \phi(x)(y(2\varepsilon,x) - y(0,x))dx &= -\left(\int_\varepsilon^{2\varepsilon} \phi(x)dx - \int_0^\varepsilon \phi(x)dx\right) \\
&= -\int_0^\varepsilon (\phi(x+\varepsilon) - \phi(x))dx \\
&= -\int_0^\varepsilon \int_x^{x+\varepsilon} \phi'(s)ds\,dx \\
&= -\varepsilon^2 \phi'(0) + \int_0^\varepsilon \int_x^{x+\varepsilon} (\phi'(0) - \phi'(s))ds\,dx.
\end{aligned}$$

But we have, for every $s \in [0, 2\varepsilon]$,

(5.4) $$|\phi'(s) - \phi'(0)| = \left|\int_0^s \phi''(\tau)d\tau\right| \leqslant \sqrt{2\varepsilon}\|\phi''\|_{L^2(0,L)}.$$

Hence

(5.5) $$\frac{y(2\varepsilon,\cdot) - y(0,\cdot)}{\varepsilon^2} \rightharpoonup \delta_0' \text{ in } (H^2(0,L))' \text{ as } \varepsilon \to \infty,$$

where δ_0 denotes the Dirac mass at 0. So in some sense we could say that, for the control system (5.1)-(5.2), $[f_0, f_1] = \delta_0'$. Unfortunately it is not clear how to use this derivative of a Dirac mass at 0.

However, there are cases where the iterated Lie brackets are (essentially) in good spaces and can indeed be used for studying the controllability of partial differential equations. This is, in particular, the case when there is no boundary. Let us give an example for the control of a quantum oscillator. This example is borrowed from Pierre Rouchon [415] and from Mazyar Mirrahimi and Pierre Rouchon [359]. We consider the following control system

(5.6) $$\psi_t = i\psi_{xx} - ix^2\psi + iux\psi,\ t \in (0,T),\ x \in \mathbb{R}.$$

For this control system, the state at time t is $\psi(t,\cdot) \in L^2(\mathbb{R};\mathbb{C})$ with $\int_\mathbb{R} |\psi(t,x)|^2 dx = 1$ and the control at time t is $u(t) \in \mathbb{R}$. The function $\psi(t,\cdot)$ is the complex amplitude vector. The control $u(t) \in \mathbb{R}$ is a classical control electro-magnetic field. The free Hamiltonian

$$H_0(\psi) := -\psi_{xx} + x^2\psi$$

corresponds to the usual harmonic oscillator. With our previous notations, we define the vector fields

(5.7) $$f_0(\psi) := i\psi_{xx} - ix^2\psi,$$
(5.8) $$f_1(\psi) := ix\psi.$$

There is a little problem with the domain of definition of f_0 and f_1: they are not defined on the whole $\mathbb{S} := \{\psi \in L^2(\mathbb{R};\mathbb{C});\ \int_\mathbb{R} |\psi(t,x)|^2 dx = 1\}$. But let us forget about this problem and remain, for the moment, at a formal level. Using (3.19), (5.7) and (5.8), straightforward computations lead to

$$[f_0, f_1](\psi) = 2\psi_x,$$
$$[f_0, [f_0, f_1]](\psi) = -4ix\psi = -4f_1(\psi),$$
$$[f_1, [f_0, f_1]](\psi) = 2i\psi.$$

Hence the Lie algebra generated by f_0 and f_1 is of dimension 4: it is the $\mathbb{R}$-linear space generated by f_0, f_1, $[f_0, f_1]$ and iId. Hence by Corollary 3.26 on page 141 one would expect that the dimension of the space which can be reached from a given space should be of dimension at most 4. Since we are in infinite dimension and, more importantly, due to the fact that f_0, f_1, are not smooth and not even defined on all of $\mathbb{S}$, one cannot apply directly Corollary 3.26. But let us check, by direct computations which are written-up in [**415**] and [**359**] (see, also, [**78**, pages 51–52] and [**39**]) that the intuition given by the above arguments is indeed correct.

For simplicity, we do not specify the regularity of $u : [0, T] \to \mathbb{R}$, the regularity of $\psi : [0, T] \times \mathbb{R} \to \mathbb{C}$ as well as its decay to 0 as $|x| \to +\infty$: they are assumed to be good enough to perform the computations below. Let $q : [0, T] \to \mathbb{R}$ and $p : [0, T] \to \mathbb{R}$ be defined by

$$q(t) := \int_{\mathbb{R}} x |\psi(t, x)|^2 dx,$$

$$p(t) := -2 \int_{\mathbb{R}} \Im(\bar{\psi}_x \psi) dx,$$

where $\Im(z)$ denotes the imaginary part of $z \in \mathbb{C}$. Note that $q(t)$ is the average position at time t and $p(t)$ is the average momentum of the quantum system. The dynamics of the couple (p, q) is given by the Ehrenfest theorem (see, for example, [**353**, (I) page 182, vol. I])

(5.9) $$\dot{q} = p, \ \dot{p} = -4q + 2u.$$

The linear system (5.9) can be considered as a control system, where, at time t, the state is $(p(t), q(t))^{tr}$, with $p(t) \in \mathbb{R}$ and $q(t) \in \mathbb{R}$, and the control is $u(t) \in \mathbb{R}$. Note that, by the Kalman rank condition (Theorem 1.16 on page 9), the linear control system (5.9) is controllable. Following Anatoliy Butkovskiy and Yu. Samoĭlenko in [**78**, page 51], let us define $\phi \in C^0([0, T]; \mathbb{S})$ by

(5.10) $$\phi(t, x) := \psi(t, x + q) \exp(-ix(p/2) + ir),$$

with

(5.11) $$r(t) := \int_0^t (q^2(\tau) - \frac{3}{4} p^2(\tau) - u(\tau) q(\tau)) d\tau.$$

Then straightforward computations lead to

(5.12) $$\phi_t = i \phi_{xx} - ix^2 \phi, \ t \in (0, T), \ x \in \mathbb{R}.$$

Note that the evolution of ϕ does not depend on the control. With this precise description of ψ, it is not hard to check that, given $\psi^0 \in \mathbb{S}$, the set

$$\{\psi(T, \cdot); \ T > 0, \ u : (0, T) \to \mathbb{R}, \ \psi \text{ is a solution of (5.6) such that } \psi(0, \cdot) = \psi^0\}$$

is contained in a submanifold of $\mathbb{S}$ of dimension 4; see also the following exercise.

EXERCISE 5.1. *Let us consider the control system*

(5.13) $$\dot{q} = p, \ \dot{p} = -4q + 2u, \ \dot{r} = q^2 - \frac{3}{4} p^2 - uq,$$

where the state is $x := (q, p, r)^{tr} \in \mathbb{R}^3$ and the control is $u \in \mathbb{R}$. Check that this control system satisfies the Lie algebra rank condition at the equilibrium $(x_e, u_e) := (0, 0) \in \mathbb{R}^3 \times \mathbb{R}$ (see Definition 3.16 on page 134). Prove that this control system is not locally controllable at (x_e, u_e). (**Hint.** Use Theorem 3.36 on page 145 with $k = 1$.)

Let $T > \pi$. Prove that, for every $\varepsilon > 0$, there exists $\eta > 0$ such that for every $(q^0, p^0, r^0) \in \mathbb{R}^3$ and every $(q^1, p^1, r^1) \in \mathbb{R}^3$ with

$$|q^0| + |p^0| + |r^0| + |q^1| + |p^1| + |r^1| \leqslant \eta,$$

there exists $u \in L^\infty((0,T); \mathbb{R})$ with

$$|u(t)| \leqslant \varepsilon,\ t \in (0, T),$$

such that the solution of the Cauchy problem (5.13) with the initial condition

(5.14) $\qquad q(0) = q^0,\ p(0) = p^0,\ r(0) = r^0,$

satisfies

(5.15) $\qquad q(T) = q^1,\ p(T) = p^1,\ r(T) = r^1.$

(**Hint.** See Example 6.4 on page 190.)

Prove that the control system (5.13) is globally controllable in large time in the following sense: for every $(q^0, p^0, r^0) \in \mathbb{R}^3$ and every $(q^1, p^1, r^1) \in \mathbb{R}^3$, there exist $T > 0$ and $u \in L^\infty((0, T); \mathbb{R})$ such that the solution of the Cauchy problem (5.13)-(5.14) satisfies (5.15).

REMARK 5.2. Let us consider the following modal approximation of the control system (5.6):

(5.16) $\qquad \dot{z} = -iH_0 z - iuH_1 z,$

with

$$H_0 = \begin{pmatrix} \frac{1}{2} & 0 & & \cdots & & 0 \\ 0 & \frac{3}{2} & 0 & \cdots & & 0 \\ 0 & 0 & \frac{5}{2} & 0 & & \vdots \\ \vdots & & & \ddots & & \vdots \\ 0 & 0 & \cdots & & 0 & \frac{2n+1}{2} \end{pmatrix},$$

$$H_1 = \begin{pmatrix} 0 & 1 & 0 & \cdots & & 0 \\ 1 & 0 & \sqrt{2} & \ddots & & \vdots \\ 0 & \sqrt{2} & 0 & \sqrt{3} & & \\ \vdots & & \sqrt{3} & \ddots & \ddots & \\ \vdots & & & \ddots & \ddots & \sqrt{n+1} \\ 0 & \cdots & & 0 & \sqrt{n+1} & 0 \end{pmatrix}.$$

In (5.16), $z \in \mathbb{S}_{2n+1} := \{z \in \mathbb{C}^{n+1};\ |z| = 1\}$. Hongchen Fu, Sonia Schirmer and Allan Solomon have proved in [**180**] that, for every $n \in \mathbb{N}$, this modal approximation is globally controllable in large time: for every $z^0 \in \mathbb{S}_{2n+1}$ and for every $z^1 \in \mathbb{S}_{2n+1}$, there exist $T > 0$ and $u \in L^\infty(0, T)$ such that the solution of the Cauchy problem

$$\dot{z} = -iH_0 z - iuH_1 z,\ z(0) = z^0$$

satisfies

$$z(T) = z^1.$$

This result shows that one has to be careful with finite-dimensional approximations in order to get controllability for infinite-dimensional control systems.

REMARK 5.3. Our application of iterated Lie brackets presented here (for the control system (5.6)) was for a *non-controllability* result for partial differential equations. There are applications of iterated Lie brackets to get (global) controllability results for a partial differential equation. Let us, for example, mention the control by means of low modes forcing for the Navier-Stokes equations and the Euler equations of incompressible fluids obtained by Andrei Agrachev and Andrei Sarychev in [**9**] and by Armen Shirikyan in [**446**].

CHAPTER 6

Return method

6.1. Description of the method

The return method has been introduced in [**103**] for a stabilization problem, that we detail in Section 11.2.1. It has been used for the first time in [**104, 112**] for the controllability of a partial differential equation (see Section 6.2.1).

In order to explain this method, let us first consider the problem of local controllability of a control system in finite dimension. Thus we consider the control system
$$\dot{x} = f(x, u),$$
where $x \in \mathbb{R}^n$ is the state and $u \in \mathbb{R}^m$ is the control; we assume that f is of class C^∞ and satisfies
$$f(0,0) = 0.$$
Note that all the controllability conditions given in the previous chapter rely on a Lie bracket formulation. Unfortunately, this geometric tool does not seem to give good results for distributed control systems; in this case x belongs to an infinite-dimensional space. The main reason is that, for many interesting distributed control systems, (3.22) and (3.25) do not hold. More precisely, the left hand sides of (3.22) and (3.25) divided by ε^2 have no limit as $\varepsilon \to 0$ in many interesting cases. See Chapter 5 for a very simple example. On the other hand, as we have already seen in Chapter 2, for *linear* distributed control systems, there are powerful methods to prove controllability. The return method consists of reducing the local controllability of a nonlinear control system to the existence of (suitable) periodic (or "almost periodic"; see below the cases of the Navier-Stokes control system, the shallow water equations and the Schrödinger equation) trajectories and to the controllability of *linear* systems (see also Remark 6.1 on page 189). The idea is the following one: Assume that, for every positive real number T and every positive real number ε, there exists a measurable bounded function $\bar{u} : [0,T] \to \mathbb{R}^m$ with $\|\bar{u}\|_{L^\infty(0,T)} \leqslant \varepsilon$ such that, if we denote by $\bar{x}$ the (maximal) solution of $\dot{\bar{x}} = f(\bar{x}, \bar{u}(t))$, $\bar{x}(0) = 0$, then

(6.1) $$\bar{x}(T) = 0,$$

(6.2) the linearized control system around $(\bar{x}, \bar{u})$ is controllable on $[0,T]$.

Then, from Theorem 3.6 on page 127, one gets the existence of $\eta > 0$ such that, for every $x^0 \in \mathbb{R}^n$ and for every $x^1 \in \mathbb{R}^n$ such that
$$|x^0| < \eta, \ |x^1| < \eta,$$
there exists $u \in L^\infty((0,T); \mathbb{R}^m)$ such that
$$|u(t) - \bar{u}(t)| \leqslant \varepsilon, \ t \in [0,T]$$

and such that, if $x : [0, T] \to \mathbb{R}^n$ is the solution of the Cauchy problem
$$\dot{x} = f(x, u(t)), \; x(0) = x^0,$$
then
$$x(T) = x^1;$$
see Figure 1. Since $T > 0$ and $\varepsilon > 0$ are arbitrary, one gets that $\dot{x} = f(x, u)$ is small-time locally controllable at the equilibrium $(0, 0) \in \mathbb{R}^n \times \mathbb{R}^m$. (For the definition of small-time local controllability, see Definition 3.2 on page 125.)

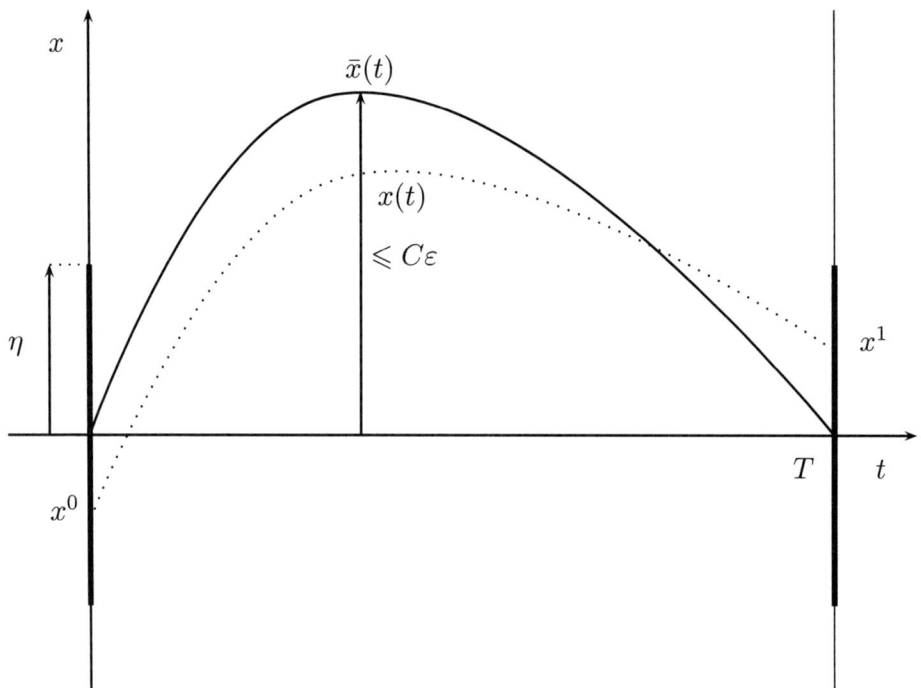

FIGURE 1. Return method

Let us recall (see Definition 3.5 on page 127) that the linearized control system around $(\bar{x}, \bar{u})$ is the time-varying control system
(6.3) $$\dot{y} = A(t)y + B(t)v,$$
where the state is $y \in \mathbb{R}^n$, the control is $v \in \mathbb{R}^m$ and
$$A(t) := \frac{\partial f}{\partial x}(\bar{x}(t), \bar{u}(t)), \; B(t) := \frac{\partial f}{\partial u}(\bar{x}(t), \bar{u}(t)).$$

Let us recall (Definition 1.2 on page 4), that, for the linear control system (6.3), controllability on $[0, T]$ means that, for every y_0 and y_1 in $\mathbb{R}^n$, there exists a bounded measurable function $v : [0, T] \to \mathbb{R}^m$ such that if $\dot{y} = A(t)y + B(t)v$ and $y(0) = y_0$, then $y(T) = y_1$.

Let us also recall (Theorem 1.18 on page 11) that if A and B are of class C^∞, and if, for some $\bar{t} \in [0, T]$,
(6.4) $$\text{Span } \{B_i(\bar{t})u; \; u \in \mathbb{R}^m, \; i \in \mathbb{N}\} = \mathbb{R}^n,$$

6.1. DESCRIPTION OF THE METHOD

with $B_i \in C^\infty([T_0, T_1]; \mathcal{L}(\mathbb{R}^m; \mathbb{R}^n))$ defined by induction on i by
$$B_0(t) := B(t),\ B_i(t) := \dot{B}_{i-1}(t) - A(t)B_{i-1}(t),$$
then the linearized control system (6.3) is controllable on $[0, T]$.

Note that if one takes $\bar{u} = 0$, then the above method just gives the well-known fact that, if the time-invariant linear control system
$$\dot{y} = \frac{\partial f}{\partial x}(0,0)y + \frac{\partial f}{\partial u}(0,0)v,$$
is controllable, then the nonlinear control system $\dot{x} = f(x, u)$ is small-time locally controllable (Theorem 3.8 on page 128). However, it may happen that (6.2) does not hold for $\bar{u} = 0$, but holds for other choices of $\bar{u}$.

REMARK 6.1. In fact, for many nonlinear partial differential equations, one does not use the linearized control system along the trajectory $(\bar{x}, \bar{u})$ to get the local controllability result along $(\bar{x}, \bar{u})$. One uses other adapted methods. These methods can rely on the controllability of some linear control systems which are *not* the linearized control system along the trajectory $(\bar{x}, \bar{u})$ (see Remark 4.11 on page 172 as well as Remark 6.30 on page 214). It can also rely on some specific methods which do not use the controllability of any linear control system. This last case appears in the papers [**239**] by Thierry Horsin and [**200**] by Olivier Glass.

Let us show on simple examples in finite dimension how the return method can be used.

EXAMPLE 6.2. *We take* $n = 2$, $m = 1$ *and again consider (see Example 3.34 on page 144 above) the control system*
$$(6.5) \qquad \dot{x}_1 = x_2^3,\ \dot{x}_2 = u.$$
For $\bar{u} = 0$ *and* $\bar{x} = 0$, *the linearized control system around* $(\bar{x}, \bar{u})$ *is*
$$\dot{y}_1 = 0,\ \dot{y}_2 = v,$$
which is clearly not controllable (one cannot change y_1). Let us now take $T > 0$ and $\bar{u} \in C^\infty([0, T]; \mathbb{R})$ such that
$$\int_0^{T/2} \bar{u}(t)dt = 0,$$
$$\bar{u}(T - t) = \bar{u}(t),\ \forall t \in [0, T].$$
Then one easily checks that
$$\bar{x}_2(T/2) = 0,$$
$$\bar{x}_2(T - t) = -\bar{x}_2(t),\ \forall t \in [0, T],$$
$$\bar{x}_1(T - t) = \bar{x}_1(t),\ \forall t \in [0, T].$$
In particular, we have
$$\bar{x}_1(T) = 0,\ \bar{x}_2(T) = 0.$$
The linearized control system around $(\bar{x}, \bar{u})$ is
$$\dot{y}_1 = 3\bar{x}_2^2(t)y_2,\ \dot{y}_2 = v.$$
Hence
$$A(t) = \begin{pmatrix} 0 & 3\bar{x}_2(t)^2 \\ 0 & 0 \end{pmatrix},\ B(t) = \begin{pmatrix} 0 \\ 1 \end{pmatrix},$$

and one easily sees that (6.4) holds if and only if

(6.6) $$\exists i \in \mathbb{N} \text{ such that } \frac{d^i \bar{x}_2}{dt^i}(\bar{t}) \neq 0.$$

Note that (6.6) holds for at least a $\bar{t}$ in $[0,T]$ if (and only if) $\bar{u} \neq 0$. So (6.2) holds if (and only if) $\bar{u} \neq 0$. We recover that the control system (6.5) is small-time locally controllable at the equilibrium $(0,0) \in \mathbb{R}^2 \times \mathbb{R}$.

EXAMPLE 6.3. *We return to the nonholonomic integrator (3.41) considered in Example 3.20 on page 135; we take $n = 3$, $m = 2$ and the control system is*

(6.7) $$\dot{x}_1 = u_1, \ \dot{x}_2 = u_2, \ \dot{x}_3 = x_1 u_2 - x_2 u_1,$$

where the state is $x = (x_1, x_2, x_3)^{tr} \in \mathbb{R}^3$ and the control is $u = (u_1, u_2)^{tr} \in \mathbb{R}^2$. Again one can check that the linearized control system around the trajectory $(\bar{x}, \bar{u})$ is controllable on $[0,T]$ if and only if $\bar{u} \neq 0$.

Note that, for the control system (6.7), it is easy to achieve the "return condition" (6.1). Indeed, let us impose that

(6.8) $$\bar{u}(T-t) = -\bar{u}(t), \ \forall t \in [0,T].$$

Then

(6.9) $$\bar{x}(T-t) = \bar{x}(t), \ \forall t \in [0,T].$$

Indeed, let $y : [0,T] \to \mathbb{R}^3$ be defined by

$$y(t) = \bar{x}(T-t), \ \forall t \in [0,T].$$

Then y, as $\bar{x}$, satisfies the ordinary differential equation

$$\dot{x}_1 = \bar{u}_1, \ \dot{x}_2 = \bar{u}_2, \ \dot{x}_3 = x_1 \bar{u}_2 - x_2 \bar{u}_1.$$

Hence, since y and $\bar{x}$ are equal at the time $T/2$, $y = \bar{x}$, which gives (6.9). From (6.9), we have

$$\bar{x}(T) = \bar{x}(0) = 0.$$

So the control system (6.7) is again small-time locally controllable at the equilibrium $(0,0) \in \mathbb{R}^3 \times \mathbb{R}^2$ (a result which follows also directly from the Rashevski-Chow Theorem 3.19 on page 135).

EXAMPLE 6.4. *We go back to the control system considered in Example 3.38 on page 148: the control system we consider is given by*

(6.10) $$\dot{x}_1 = x_2, \ \dot{x}_2 = -x_1 + u, \ \dot{x}_3 = x_4, \ \dot{x}_4 = -x_3 + 2x_1 x_2,$$

where the state is $(x_1, x_2, x_3, x_4)^{tr} \in \mathbb{R}^4$ and the control is $u \in \mathbb{R}$. One easily checks that this system satisfies the Lie algebra rank condition at the equilibrium $(0,0) \in \mathbb{R}^4 \times \mathbb{R}$ (Definition 3.16 on page 134). In Example 3.38 on page 148, we saw that the control system (6.10) is not small-time locally controllable at the equilibrium $(x_e, u_e) = (0,0) \in \mathbb{R}^4 \times \mathbb{R}$. Here we want to see that the return method can be used to get large-time local controllability at the equilibrium $(x_e, u_e) = (0,0) \in \mathbb{R}^4 \times \mathbb{R}$.

Let us first give a new proof that this control system is not small-time locally controllable at the equilibrium $(x_e, u_e) = (0,0) \in \mathbb{R}^4 \times \mathbb{R}$. Indeed, if $(x,u) : [0,T] \to$

$\mathbb{R}^4 \times \mathbb{R}$ is a trajectory of the control system (6.10) such that $x(0) = 0$, then

(6.11) $$x_3(T) = \int_0^T x_1^2(t) \cos(T-t) dt,$$

(6.12) $$x_4(T) = x_1^2(T) - \int_0^T x_1^2(t) \sin(T-t) dt.$$

In particular, if $x_1(T) = 0$ and $T \leqslant \pi$, then $x_4(T) \leqslant 0$ with equality if and only if $x = 0$. So, if for $T > 0$ we denote by $\mathcal{P}(T)$ the following controllability property

$\mathcal{P}(T)$ There exists $\varepsilon > 0$ such that, for every $x^0 \in \mathbb{R}^4$ and for every $x^1 \in \mathbb{R}^4$ both of norm less than ε, there exists a bounded measurable function $u : [0,T] \to \mathbb{R}$ such that, if x is the (maximal) solution of (6.10) which satisfies $x(0) = x^0$, then $x(T) = x^1$,

then, for every $T \in (0, \pi]$, $\mathcal{P}(T)$ is false.

Let us show how the return method can be used to prove that

(6.13) $$\mathcal{P}(T) \text{ holds for every } T \in (\pi, +\infty).$$

Let $T > \pi$. Let

$$\eta = \frac{1}{2} \text{Min} \{T - \pi, \pi\}.$$

Let $\bar{x}_1 : [0,T] \to \mathbb{R}$ be a function of class C^∞ such that

(6.14) $$\bar{x}_1(t) = 0, \forall t \in [\eta, \pi] \cup [\pi + \eta, T],$$

(6.15) $$\bar{x}_1(t + \pi) = \bar{x}_1(t), \forall t \in [0, \eta].$$

Let $\bar{x}_2 : [0,T] \to \mathbb{R}$ and $\bar{u} : [0,T] \to \mathbb{R}$ be such that

$$\bar{x}_2 := \dot{\bar{x}}_1, \bar{u} := \dot{\bar{x}}_2 + \bar{x}_1.$$

In particular,

(6.16) $$\bar{x}_2(t) = 0, \forall t \in [\eta, \pi] \cup [\pi + \eta, T],$$

(6.17) $$\bar{x}_2(t + \pi) = x_2(t), \forall t \in [0, \eta].$$

Let $\bar{x}_3 : [0,T] \to \mathbb{R}$ and $\bar{x}_4 : [0,T] \to \mathbb{R}$ be defined by

(6.18) $$\dot{\bar{x}}_3 = \bar{x}_4, \dot{\bar{x}}_4 = -\bar{x}_3 + 2\bar{x}_1 \bar{x}_2,$$

(6.19) $$\bar{x}_3(0) = 0, \bar{x}_4(0) = 0.$$

So $(\bar{x}, \bar{u})$ is a trajectory of the control system (6.10). Then, using (6.11), (6.12), (6.14), (6.16), (6.15), (6.17), one sees that

$$\bar{x}(T) = 0.$$

If $\bar{x}_1 = 0$, $(\bar{x}, \bar{u}) = 0$ and then the linearized control system around $(\bar{x}, \bar{u})$ is not controllable. However, as one easily checks by using the Kalman-type sufficient condition for the controllability of linear time-varying control system given in Theorem 1.18 on page 11, if $\bar{x}_1 \neq 0$, then the linearized control system around $(\bar{x}, \bar{u})$ is controllable. This establishes (6.13).

One may wonder if the local controllability of $\dot{x} = f(x, u)$ implies the existence of u in $C^\infty([0,T]; \mathbb{R}^m)$ such that (6.1) and (6.2) hold. This is indeed the case, as proved by Eduardo Sontag in [**452**]. Let us also remark that the above examples suggest that, for many choices of $\bar{u}$, (6.2) holds. This in fact holds in general. More precisely, let us assume that

$$(6.20) \qquad \left\{ h(0); h \in \text{Lie} \left\{ \frac{\partial f}{\partial u^\alpha}(\cdot, 0), \, \alpha \in \mathbb{N}^m \right\} \right\} = \mathbb{R}^n,$$

where Lie $\mathcal{F}$ denotes the Lie algebra generated by the vector fields in $\mathcal{F}$ (see Definition 3.13 on page 133). Then, for generic u in $C^\infty([0,T]; \mathbb{R}^m)$, (6.2) holds. This is proved in [**106**], and in [**456**] if f is analytic. Let us recall that by Theorem 3.17 on page 134, (6.20) is a necessary condition for small-time local controllability at the equilibrium $(0,0) \in \mathbb{R}^n \times \mathbb{R}^m$ if f is analytic.

Note that the controllability results obtained in Example 6.2 and Example 6.3 by means of the return method can also be obtained by already known results:

- The small-time local controllability in Example 6.2 on page 189 follows from the Hermes condition (see page 143 and Theorem 3.29 on page 143) [**228, 469**].
- The small-time local controllability in Example 6.3 follows from Rashevski-Chow's Theorem 3.19 on page 135.

It is not clear how to deduce the large-time local controllability of Example 6.4 (more precisely (6.13)) from already known results. More interestingly, the return method gives some new results in the following cases:

- for the stabilization of driftless control affine systems (see Section 11.2.1),
- for the controllability of distributed control systems (see Section 6.2 and Section 6.3).

6.2. Controllability of the Euler and Navier-Stokes equations

In this section, we show how the return method can be used to give controllability results for the Euler and Navier-Stokes equations. Let us introduce some notation. Let $l \in \{2, 3\}$ and let Ω be a bounded nonempty connected open subset of $\mathbb{R}^l$ of class C^∞. Let Γ_0 be an open subset of $\Gamma := \partial \Omega$ and let Ω_0 be an open subset of Ω. We assume that

$$(6.21) \qquad \Gamma_0 \cup \Omega_0 \neq \emptyset.$$

The set Γ_0, resp. Ω_0, is the part of the boundary Γ, resp. of the domain Ω, on which the control acts. The fluid that we consider is incompressible, so that the velocity field y satisfies

$$\text{div } y = 0.$$

On the part of the boundary $\Gamma \backslash \Gamma_0$ where there is no control, the fluid does not cross the boundary: it satisfies

$$(6.22) \qquad y \cdot n = 0 \text{ on } \Gamma \backslash \Gamma_0,$$

where n denotes the outward unit normal vector field on Γ. When the fluid is viscous, it satisfies on $\Gamma \backslash \Gamma_0$, besides (6.22), some extra conditions which will be specified later on. For the moment, let us just denote by BC all the boundary conditions (*including* (6.22)) satisfied by the fluid on $\Gamma \backslash \Gamma_0$.

Let us introduce the following definition.

DEFINITION 6.5. A *trajectory of the Navier-Stokes control system (resp. Euler control system) on the time interval* $[0,T]$ is a map $y : [0,T] \times \overline{\Omega} \to \mathbb{R}^l$ such that, for some function $p : [0,T] \times \overline{\Omega} \to \mathbb{R}$,

$$\frac{\partial y}{\partial t} - \nu \Delta y + (y \cdot \nabla)y + \nabla p = 0 \text{ in } [0,T] \times (\overline{\Omega} \setminus \Omega_0), \tag{6.23}$$

$$(\text{resp. } \frac{\partial y}{\partial t} + (y \cdot \nabla)y + \nabla p = 0 \text{ in } [0,T] \times (\overline{\Omega} \setminus \Omega_0)) \tag{6.24}$$

$$\operatorname{div} y = 0 \text{ in } [0,T] \times \overline{\Omega}, \tag{6.25}$$

(6.26) $y(t,\cdot)$ satisfies the boundary conditions BC on $\Gamma \setminus \Gamma_0$, $\forall t \in [0,T]$.

The real number $\nu > 0$ appearing in (6.23) is the viscosity. In Definition 6.5 and throughout the whole book, for $A : \Omega \to \mathbb{R}^l$ and $B : \Omega \to \mathbb{R}^l$, $(A \cdot \nabla)B : \Omega \to \mathbb{R}^l$ is defined by

$$((A \cdot \nabla)B)^k := \sum_{j=1}^{l} A^j \frac{\partial B^k}{\partial x_j}, \; \forall k \in \{1, \ldots, l\}.$$

Moreover, in Definition 6.5 and throughout this section, we do not specify in general the regularities of the functions considered. For the precise regularities, see the references mentioned in this section.

Jacques-Louis Lions's problem of approximate controllability is the following one.

OPEN PROBLEM 6.6. *Let* $T > 0$, *let* $y^0 : \overline{\Omega} \to \mathbb{R}^l$ *and* $y^1 : \overline{\Omega} \to \mathbb{R}^l$ *be two functions such that*

$$\operatorname{div} y^0 = 0 \text{ in } \overline{\Omega}, \tag{6.27}$$

$$\operatorname{div} y^1 = 0 \text{ in } \overline{\Omega}, \tag{6.28}$$

(6.29) $\quad\quad\quad y^0$ *satisfies the boundary conditions BC on* $\Gamma \setminus \Gamma_0$,

(6.30) $\quad\quad\quad y^1$ *satisfies the boundary conditions BC on* $\Gamma \setminus \Gamma_0$.

Does there exist a trajectory y *of the Navier-Stokes or the Euler control system such that*

$$y(0,\cdot) = y^0 \text{ in } \overline{\Omega}, \tag{6.31}$$

and, for an appropriate topology (see [**327**, **328**]*),*

$$y(T,\cdot) \text{ is "close" to } y^1 \text{ in } \overline{\Omega}? \tag{6.32}$$

In other words, starting with the initial data y^0 *for the velocity field, we ask whether there are trajectories of the considered control system (Navier-Stokes if* $\nu > 0$, *Euler if* $\nu = 0$*) which, at a fixed time* T, *are arbitrarily close to the given velocity field* y^1.

If this problem always has a solution, one says that the considered control system is approximately controllable.

Note that (6.23), (6.25), (6.26) and (6.31) have many solutions. In order to have uniqueness one needs to add extra conditions. These extra conditions are the controls.

In the case of the Euler control system, one can even require instead of (6.32) the stronger condition

(6.33) $$y(T, \cdot) = y^1 \text{ in } \overline{\Omega}.$$

If y still exists with this stronger condition, one says that the Euler control system is globally controllable. Of course, due to the smoothing effects of the Navier-Stokes equations, one cannot expect to have (6.33) instead of (6.32) for general y^1. We have already encountered this problem for the control of the heat equation in Section 2.5. Proceeding as in Section 2.5, we replace, in Section 6.2.2, (6.33) by another condition in order to recover a natural definition of controllability of the Navier-Stokes equation.

This section is organized as follows:

1. In Section 6.2.1, we consider the case of the Euler control system.
2. In Section 6.2.2, we consider the case of the Navier-Stokes control system.

6.2.1. Controllability of the Euler equations. In this section the boundary conditions BC in (6.26), (6.29), and (6.30) are, respectively,

(6.34) $$y(t,x) \cdot n(x) = 0, \; \forall (t,x) \in [0,T] \times (\Gamma \backslash \Gamma_0),$$

(6.35) $$y^0(x) \cdot n(x) = 0, \; \forall x \in \Gamma \backslash \Gamma_0,$$

(6.36) $$y^1(x) \cdot n(x) = 0, \; \forall x \in \Gamma \backslash \Gamma_0.$$

For simplicity we assume that

$$\Omega_0 = \emptyset,$$

i.e., we study the case of boundary control (see [**112**] when $\Omega_0 \neq \emptyset$ and $l = 2$). In this case the control is

1. $y \cdot n$ on Γ_0 with $\int_{\Gamma_0} y \cdot n = 0$,
2. curl y if $l = 2$ and the tangent vector (curl y) $\times n$ if $l = 3$ at the points of $[0,T] \times \Gamma_0$ where $y \cdot n < 0$.

These boundary conditions, (6.34), and the initial condition (6.31) imply the uniqueness of the solution to the Euler equations (6.24)—up to an arbitrary function of t which may be added to p and for suitable regularities; see also the papers [**257**] by Victor Judovič and [**271**] by Alexandre Kazhikov, for results on uniqueness and existence of solutions.

Let us first point out that, in order to have controllability, one needs that

(6.37) Γ_0 intersects every connected component of Γ.

Indeed, let $\mathcal{C}$ be a connected component of Γ which does not intersect Γ_0 and assume that, for some smooth oriented Jordan curve γ_0 on $\mathcal{C}$ (if $l = 2$, one takes $\gamma_0 = \mathcal{C}$),

(6.38) $$\mathbb{R} \ni \int_{\gamma_0} y^0 \cdot \vec{ds} \neq 0,$$

but that

(6.39) $$y^1(x) = 0, \forall x \in \mathcal{C}.$$

Then there is no solution to our problem, that is, there is no $y : [0,T] \times \overline{\Omega} \to \mathbb{R}^2$ and $p : [0,T] \times \overline{\Omega} \to \mathbb{R}$ such that (6.24), (6.25), (6.31), (6.33), and (6.34) hold.

Indeed, if such a solution (y, p) exists, then, by the Kelvin circulation theorem (see, for example, [**38**, Equality (5.3.1), page 273]),

(6.40) $$\int_{\gamma(t)} y(t, \cdot) \cdot \vec{ds} = \int_{\gamma_0} y^0 \cdot \vec{ds}, \, \forall t \in [0, T],$$

where $\gamma(t)$ is the Jordan curve obtained, at time $t \in [0, T]$, from the points of the fluids which, at time 0, were on γ_0; in other words, $\gamma(t)$ is the image of γ_0 by the flow map associated to the time-varying vector field y. But, if γ_0 does not intersect Γ_0, (6.34) shows that $\gamma(t) \subset \mathcal{C}$, for every $t \in [0, T]$. Thus (6.33), (6.38), (6.39) and (6.40) are in contradiction.

Conversely, if (6.37) holds, then the Euler control system is globally controllable:

THEOREM 6.7. *Assume that Γ_0 intersects every connected component of Γ. Then the Euler control system is globally controllable.*

Theorem 6.7 has been proved in

1. [**104**] when Ω is simply-connected and $l = 2$,
2. [**112**] when Ω is multi-connected and $l = 2$,
3. [**194**], by Olivier Glass, when Ω is contractible and $l = 3$,
4. [**195**], by Olivier Glass, when Ω is not contractible and $l = 3$.

The strategy of the proof of Theorem 6.7 relies on the "return method". Applied to the controllability of the Euler control system, the return method consists of looking for $(\bar{y}, \bar{p})$ such that (6.24), (6.25), (6.31), (6.33) hold, with $y = \bar{y}, p = \bar{p}, y^0 = y^1 = 0$ and such that the linearized control system along the trajectory $\bar{y}$ is controllable under the assumptions of Theorem 6.7. With such a $(\bar{y}, \bar{p})$ one may hope that there exists (y, p)—close to $(\bar{y}, \bar{p})$—satisfying the required conditions, at least if y^0 and y^1 are "small". Finally, by using some scaling arguments, one can deduce from the existence of (y, p) when y^0 and y^1 are "small", the existence of (y, p) even though y^0 and y^1 are not "small".

Let us emphasize that one cannot take $(\bar{y}, \bar{p}) = (0, 0)$. Indeed, with such a choice of $(\bar{y}, \bar{p})$, (6.24), (6.25), (6.31), (6.33) hold, with $y = \bar{y}, p = \bar{p}, y^0 = y^1 = 0$, but the linearized control system around $\bar{y} = 0$ is not at all controllable. Indeed, the linearized control system around $\bar{y} = 0$ is

(6.41) $$\text{div } z = 0 \text{ in } [0, T] \times \overline{\Omega},$$

(6.42) $$\frac{\partial z}{\partial t} + \nabla \pi = 0 \text{ in } [0, T] \times \overline{\Omega},$$

$$z(t, x) \cdot n(x) = 0, \, \forall (t, x) \in [0, T] \times (\Gamma \backslash \Gamma_0).$$

Taking the curl of (6.42), one gets

$$\frac{\partial \text{curl } z}{\partial t} = 0,$$

which clearly shows that the linearized control system is not controllable (one cannot modify the curl using the control).

Thus, one needs to consider other $(\bar{y}, \bar{p})$'s. Let us briefly explain how one constructs "good" $(\bar{y}, \bar{p})$'s when $l = 2$ and Ω is simply connected. In such a case,

one easily checks the existence of a harmonic function θ in $C^\infty(\overline{\Omega})$ such that
$$\nabla\theta(x) \neq 0, \; \forall x \in \overline{\Omega},$$
$$\frac{\partial \theta}{\partial n} = 0 \text{ on } \Gamma\backslash\Gamma_0.$$
Let $\alpha \in C^\infty([0,T])$ vanishing at 0 and T. Let

(6.43) $\quad (\bar{y}, \bar{p})(t,x) = (\alpha(t)\nabla\theta(x), -\alpha'(t)\theta(x) - \frac{1}{2}\alpha^2(t)|\nabla\theta(x)|^2).$

Then (6.24), (6.25), (6.31) and (6.33) hold, with $y = \bar{y}, p = \bar{p}, y^0 = y^1 = 0$. Moreover, using arguments relying on an extension method similar to the one described in Section 2.1.2.2 (see also [**426**, Proof of Theorem 5.3, pages 688–690] by David Russell and [**332**] by Walter Littman), one can see that the linearized control system around $\bar{y}$ is controllable.

When Γ_0 does not intersect all the connected components of Γ, one can get, if $l = 2$, approximate controllability and even controllability outside every arbitrarily small neighborhood of the union $\Gamma^\star$ of the connected components of Γ which do not intersect Γ_0. More precisely, one has the following theorem, which is proved in [**112**].

THEOREM 6.8. *Assume that $l = 2$. There exists a constant c_0 depending only on Ω such that, for every Γ_0 as above, every $T > 0$, every $\varepsilon > 0$, and every y^0, y^1 in $C^\infty(\overline{\Omega}; \mathbb{R}^2)$ satisfying (6.27), (6.28), (6.35) and (6.36), there exists a trajectory y of the Euler control system on $[0,T]$ satisfying (6.31), such that*

(6.44) $\quad y(T,x) = y^1(x), \; \forall x \in \overline{\Omega} \text{ such that } dist(x, \Gamma^\star) \geqslant \varepsilon,$

(6.45) $\quad \|y(T,\cdot)\|_{L^\infty} \leqslant c_0(\|y^0\|_{L^2} + \|y^1\|_{L^2} + \|\operatorname{curl} y^0\|_{L^\infty} + \|\operatorname{curl} y^1\|_{L^\infty}).$

In (6.44), $\operatorname{dist}(x, \Gamma^\star)$ denotes the distance of x to $\Gamma^\star$, i.e.,

(6.46) $\quad \operatorname{dist}(x, \Gamma^\star) = \operatorname{Min}\{|x - x^\star|; x^\star \in \Gamma^\star\}.$

We use the convention $\operatorname{dist}(x, \emptyset) = +\infty$ and so Theorem 6.8 implies Theorem 6.7. In (6.45), $\|\cdot\|_{L^r}$, for $r \in [1, +\infty]$, denotes the L^r-norm on Ω.

Let us point out that, y^0, y^1 and T as in Theorem 6.8 being given, it follows from (6.44) and (6.45) that, for every r in $[1, +\infty)$,

(6.47) $\quad \lim_{\varepsilon \to 0^+} \|y(T,\cdot) - y^1\|_{L^r} = 0.$

Therefore, Theorem 6.8 implies approximate controllability in the space L^r, for every r in $[1, +\infty)$.

Let us also notice that, if $\Gamma^\star \neq \emptyset$, then, again by Kelvin's circulation theorem, approximate controllability for the L^∞-norm does not hold. More precisely, let us consider the case $l = 2$ and let us denote by $\Gamma^\star_1, \ldots, \Gamma^\star_k$ the connected components of Γ which do not meet Γ_0. Let y^0, y^1 in $C^\infty(\overline{\Omega}; \mathbb{R}^2)$ satisfy (6.27), (6.28), (6.35) and (6.36). Assume that, for some $i \in \{1, \ldots, k\}$,

$$\int_{\Gamma^\star_i} y^0 \cdot \vec{ds} \neq \int_{\Gamma^\star_i} y^1 \cdot \vec{ds}.$$

Then, for $\varepsilon > 0$ small enough, there is no trajectory y of the Euler control system on $[0,T]$ satisfying (6.31) such that

(6.48) $\quad \|y(T,\cdot) - y^1\|_{L^\infty} \leqslant \varepsilon.$

One may wonder what happens if, on the contrary, one assumes that

(6.49) $$\int_{\Gamma_i^\star} y^0 \cdot \vec{ds} = \int_{\Gamma_i^\star} y^1 \cdot \vec{ds}, \; \forall i \in \{1, \ldots, k\}.$$

Then Olivier Glass proved that one has approximate controllability in L^∞ and even in the Sobolev spaces $W^{1,p}$ for every $p \in [1, +\infty)$. Indeed he proved in [**196**] the following theorem.

THEOREM 6.9. *Assume that $l = 2$. For every $T > 0$, and every y^0, y^1 in $C^\infty(\overline{\Omega}; \mathbb{R}^2)$ satisfying (6.27), (6.28), (6.35), (6.36) and (6.49), there exists a sequence $(y^k)_{k \in \mathbb{N}}$ of trajectories of the Euler control system on $[0, T]$ satisfying (6.31) such that*

(6.50) $\quad y^k(T, x) = y^1(x), \; \forall x \in \overline{\Omega}$ *such that* $\text{dist}(x, \Gamma^\star) \geqslant 1/k, \; \forall k \in \mathbb{N},$

(6.51) $\quad y^k(T, \cdot) \to y^1$ *in* $W^{1,p}(\Omega)$ *as* $k \to +\infty, \; \forall p \in [1, +\infty).$

Again the convergence in (6.51) is optimal. Indeed, since the vorticity curl y is conserved along the trajectories of the vector field y, one cannot have the convergence in $W^{1,\infty}$. In order to have convergence in $W^{1,\infty}$, one needs to add a relation between curl y^0 and curl y^1 on the Γ_i for $i \in \{1, \ldots, l\}$. In this direction, Olivier Glass proved in [**196**]:

THEOREM 6.10. *Assume that $l = 2$. Let $T > 0$, and let y^0, y^1 in $C^\infty(\overline{\Omega}; \mathbb{R}^2)$ be such that (6.27), (6.28), (6.35), (6.36) and (6.49) hold. Assume that, for every $i \in \{1, \ldots, l\}$, there exists a diffeomorphism D_i of $\Gamma_i^\star$ preserving the orientation such that*

$$\text{curl } y^1 = (\text{curl } y^0) \circ D_i.$$

Then there exists a sequence $(y^k)_{k \in \mathbb{N}}$ of trajectories of the Euler control system on $[0, T]$ satisfying (6.31), (6.50) and

(6.52) $\quad y^k(T, \cdot) \to y^1$ *in* $W^{2,p}(\Omega)$ *as* $k \to +\infty, \; \forall p \in [1, +\infty).$

Again, one cannot expect a convergence in $W^{2,\infty}$ without an extra assumption on y^0 and y^1; see [**196**].

Let us end this section by mentioning that, for the Vlasov-Poisson system and for the 1-D isentropic Euler equation, Olivier Glass, using the return method, obtained controllability results in [**198, 200**].

6.2.2. Controllability of the Navier-Stokes equations.

In this section, $\nu > 0$. We now need to specify the boundary conditions BC. Three types of conditions are considered:

- Stokes boundary condition,
- Navier boundary condition,
- Curl condition.

The Stokes boundary condition is the well-known no-slip boundary condition

(6.53) $$y = 0 \text{ on } \Gamma \backslash \Gamma_0,$$

which of course implies (6.22).

The Navier boundary condition, introduced in [**375**], imposes condition (6.22), which is always assumed, and

$$
(6.54) \qquad \overline{\sigma} y \cdot \tau + (1-\overline{\sigma}) \sum_{i=1,j=1}^{i=l,j=l} n^i \left(\frac{\partial y^i}{\partial x^j} + \frac{\partial y^j}{\partial x^i} \right) \tau^j = 0 \text{ on } \Gamma \backslash \Gamma_0, \forall \tau \in T\Gamma,
$$

where $\overline{\sigma}$ is a constant in $[0,1)$. In (6.54), $n = (n^1, \ldots, n^l)$, $\tau = (\tau^1, \ldots, \tau^l)$ and $T\Gamma$ is the set of tangent vector fields on the boundary Γ. Note that the Stokes boundary condition (6.53) corresponds to the case $\overline{\sigma} = 1$, which we will not include in the Navier boundary condition considered here. The boundary condition (6.54) with $\overline{\sigma} = 0$ corresponds to the case where the fluid slips on the wall without friction. It is the appropriate physical model for some flow problems; see, for example, the paper [**192**] by Giuseppe Geymonat and Enrique Sánchez-Palencia. The case $\overline{\sigma} \in (0,1)$ corresponds to a case where the fluid slips on the wall with friction; it is also used in models of turbulence with rough walls; see, e.g., [**304**]. Note that in [**101**] François Coron derived rigorously the Navier boundary condition (6.54) from the boundary condition at the kinetic level (Boltzmann equation) for compressible fluids. Let us also recall that Claude Bardos, François Golse, and David Levermore have derived in [**32, 33**] the incompressible Navier-Stokes equations from a Boltzmann equation.

Let us point out that, using (6.22), one sees that, if $l = 2$ and if τ is the unit tangent vector field on Γ such that (n, τ) is a direct basis of $\mathbb{R}^2$, (6.54) is equivalent to

$$\sigma y \cdot \tau + \operatorname{curl} y = 0 \text{ on } \Gamma \backslash \Gamma_0,$$

with $\sigma \in C^\infty(\Gamma; \mathbb{R})$ defined by

$$
(6.55) \qquad \sigma(x) := \frac{-2(1-\overline{\sigma})\kappa(x) + \overline{\sigma}}{1-\overline{\sigma}}, \forall x \in \Gamma,
$$

where κ is the curvature of Γ defined through the relation

$$\frac{\partial n}{\partial \tau} = \kappa \tau.$$

In fact, we will not use this particular form of (6.55) in our considerations: Theorem 6.14 below holds for every $\sigma \in C^\infty(\Gamma; \mathbb{R})$.

Finally, the curl condition is considered in dimension $l = 2$. This condition is condition (6.22), which is always assumed, and

$$(6.56) \qquad \operatorname{curl} y = 0 \text{ on } \Gamma \backslash \Gamma_0.$$

It corresponds to the case $\sigma = 0$ in (6.55).

As mentioned at the beginning of Section 6.2, due to the smoothing property of the Navier-Stokes equation, one cannot expect to get (6.33), at least for general y^1. For these equations, the good notion for controllability is not passing from a given state y^0 to another given state y^1. As already mentioned (see Section 2.5.2), the good definition for controllability, which is due to Andrei Fursikov and Oleg Imanuvilov [**185, 187**], is passing from a given state y^0 to a given *trajectory* $\hat{y}^1$. This leads to the following, still open, problem of controllability of the Navier-Stokes equation with the Stokes or Navier condition.

OPEN PROBLEM 6.11. *Let $T > 0$. Let $\hat{y}^1$ be a trajectory of the Navier-Stokes control system on $[0,T]$. Let $y^0 \in C^\infty(\overline{\Omega}; \mathbb{R}^l)$ satisfy (6.27) and (6.29). Does there* *exist a trajectory y of the Navier-Stokes control system on $[0, T]$ such that*

(6.57) $$y(0, x) = y^0(x), \, \forall x \in \overline{\Omega},$$

(6.58) $$y(T, x) = \hat{y}^1(T, x), \, \forall x \in \overline{\Omega}?$$

Let us point out that the (global) approximate controllability of the Navier-Stokes control system is also an open problem. Related to Open Problem 6.11, one knows two types of results:
- local results,
- global results,

which we briefly describe in Section 6.2.2.1 and Section 6.2.2.2, respectively.

6.2.2.1. *Local results*. These results do not rely on the return method, but on observability inequalities related to the one obtained for the heat equation in Section 2.5 (see (2.398)). Compared to the case of the heat equation considered in Section 2.5, the main new difficulty is to estimate the pressure. Let us introduce the following definition.

DEFINITION 6.12. *The Navier-Stokes control system is locally (for the Sobolev H^1-norm) controllable along the trajectory $\hat{y}^1$ on $[0, T]$ of the Navier-Stokes control system if there exists $\varepsilon > 0$ such that, for every $y^0 \in C^\infty(\overline{\Omega}; \mathbb{R}^l)$ satisfying (6.27), (6.29) and*

$$\|y^0 - \hat{y}^1(0, \cdot)\|_{H^1(\Omega)} < \varepsilon,$$

there exists a trajectory y of the Navier-Stokes control system on $[0, T]$ satisfying (6.57) and (6.58).

Then one has the following result.

THEOREM 6.13. *The Navier-Stokes control system is locally controllable (for the Navier or Stokes condition) along every trajectory $\hat{y}^1$ of the Navier-Stokes control system.*

Many mathematicians have contributed to this result. Let us mention the main papers:
- Andrei Fursikov and Oleg Imanuvilov in [**184, 187**] treated the case $l = 2$ and the Navier boundary condition.
- Andrei Fursikov in [**181**] treated the case where $\Gamma_0 = \Gamma$.
- Oleg Imanuvilov in [**244, 245**] treated in full generality the case of the Stokes condition. In [**244, 245**], Enrique Fernández-Cara, Sergio Guerrero, Oleg Imanuvilov and Jean-Pierre Puel in [**167**] weakened some regularity assumptions.
- Sergio Guerrero in [**207**] treated in full generality the case of the Navier boundary condition.

6.2.2.2. *Global results*. Let $d \in C^0(\overline{\Omega}; \mathbb{R})$ be defined by

$$d(x) = \text{dist}\,(x, \Gamma) = \text{Min}\,\{|x - x'|; x' \in \Gamma\}.$$

In [**111**] the following theorem is proved.

THEOREM 6.14. *Let $T > 0$. Let y^0 and y^1 in $H^1(\Omega, \mathbb{R}^2)$ be such that (6.27) and (6.28) hold and*

$$\text{curl}\, y^0 \in L^\infty(\Omega) \text{ and } \text{curl}\, y^1 \in L^\infty(\Omega),$$

$$y^0 \cdot n = 0, \, y^1 \cdot n = 0 \text{ on } \Gamma \setminus \Gamma_0.$$

Let us also assume that (6.54) holds for $y := y^0$ and for $y := y^1$. Then there exists a sequence $(y_k, p_k)_{k \in \mathbb{N}}$ of solutions of the Navier-Stokes control system on $[0, T]$ with the Navier boundary condition (6.54) such that

(6.59) $$y_k(0, \cdot) = y^0,$$

(6.60) $\qquad\qquad y_k$ and p_k are of class C^∞ in $(0, T] \times \overline{\Omega}$,

(6.61) $$y_k \in L^2((0, T), H^2(\Omega)^2),$$

(6.62) $$y_{kt} \in L^2((0, T), L^2(\Omega)^2),$$

(6.63) $$p_k \in L^2((0, T), H^1(\Omega)),$$

and such that, as $k \to +\infty$,

(6.64) $$\int_\Omega d(x)^\mu |y_k(T, x) - y^1(x)| dx \to 0, \ \forall \mu > 0,$$

(6.65) $$\|y_k(T, \cdot) - y^1\|_{W^{-1,\infty}(\Omega)} \to 0,$$

and, for every compact K contained in $\Omega \cup \Gamma_0$,

(6.66) $$\|y_k(T, \cdot) - y^1\|_{L^\infty(K)} + \|\mathrm{curl}\ y_k(T, \cdot) - \mathrm{curl}\ y^1\|_{L^\infty(K)} \to 0.$$

In this theorem, $W^{-1,\infty}(\Omega)$ denotes the usual Sobolev space of first derivatives of functions in $L^\infty(\Omega)$ and $\|\cdot\|_{W^{-1,\infty}(\Omega)}$ one of its usual norms, for example the norm given in [**3**, Section 3.12, page 64].

REMARK 6.15. In [**111**] we said that, if y^0 is, moreover, of class C^∞ in $\overline{\Omega}$ and satisfies (6.29), then the (y_k, p_k) could be chosen to be of class C^∞ in $[0, T] \times \overline{\Omega}$ (compared with (6.60)). This is wrong. In order to have regularity near $t = 0$, one needs to require (as in the case where there is no control) some extra nonlocal compatibility conditions on y^0. One can easily adapt the paper [**479**] by Roger Temam (which deals with the case of the Stokes conditions) to find these compatibility conditions. When there is a control, the compatibility conditions are weaker than in the case without control but they still do exist if $\Gamma_0 \neq \Gamma$.

For the proof of Theorem 6.14, as in the proof of the controllability of the 2-D Euler equations of incompressible inviscid fluids (see Section 6.2.1), one uses the return method. Let us recall that it consists of looking for a trajectory of the Navier-Stokes control system $\bar{y}$ such that

(6.67) $$\bar{y}(0, \cdot) = \bar{y}(T, \cdot) = 0 \text{ in } \overline{\Omega},$$

and such that the linearized control system around the trajectory $\bar{y}$ has a controllability in a "good" sense. With such a $\bar{y}$ one may hope that there exists y—close to $\bar{y}$—satisfying the required conditions, at least if y^0 and y^1 are "small". Note that the linearized control system around $\bar{y}$ is

(6.68) $$\frac{\partial z}{\partial t} - \nu \Delta z + (\bar{y} \cdot \nabla) z + (z \cdot \nabla) \bar{y} + \nabla \pi = 0 \text{ in } [0, T] \times (\overline{\Omega} \setminus \Omega_0),$$

(6.69) $$\mathrm{div}\ z = 0 \text{ in } [0, T] \times \overline{\Omega},$$

(6.70) $$z \cdot n = 0 \text{ on } [0, T] \times (\Gamma \setminus \Gamma_0),$$

(6.71) $$\sigma z \cdot \tau + \mathrm{curl}\ z = 0 \text{ on } [0, T] \times (\Gamma \setminus \Gamma_0).$$

In [**184, 187**] Andrei Fursikov and Oleg Imanuvilov proved that this linear control system is controllable (see also [**323**] for the approximate controllability). Of course it is tempting to consider the case $\bar{y} = 0$. Unfortunately, it is not clear how to deduce from the controllability of the linear system (6.68) with $\bar{y} = 0$, the existence of a trajectory y of the Navier-Stokes control system (with the Navier boundary condition) satisfying (6.31) and (6.32) if y^0 and y^1 are *not small*. (The nonlinearity sounds too big to apply the method of Section 4.3.) For this reason, one does not use $\bar{y} = 0$, but $\bar{y}$ similar to the one defined by (6.43) to prove the controllability of the 2-D Euler equations of incompressible inviscid fluids. These $\bar{y}$'s are chosen to be "large" so that, in some sense, "Δ" is small compared to "$(\bar{y} \cdot \nabla) + (\cdot \nabla)\bar{y}$".

REMARK 6.16. In fact, with the $\bar{y}$ we use, one does not have (6.67): one only has the weaker property

(6.72) $\quad\quad\quad\quad \bar{y}(0, \cdot) = 0,\ \bar{y}(T, \cdot)$ is "close" to 0 in $\overline{\Omega}$.

However, the controllability of the linearized control system around $\bar{y}$ is strong enough to take care of the fact that $\bar{y}(\cdot, T)$ is not equal to 0 but only close to 0. A similar observation can be found, in a different setting, in [**471**]. With the notation introduced by Héctor Sussmann in [**471**], $\bar{y}$ plays the role of ξ_* in the proof of [**471**, Theorem 12] and the Euler control system plays for the Navier-Stokes control system the role played by $\mathcal{G}$ for $\mathcal{F}$.

Note that (6.64), (6.65), and (6.66) are not strong enough to imply

(6.73) $\quad\quad\quad\quad\quad\quad \|y_k(T, \cdot) - y^1\|_{L^2(\Omega)} \to 0,$

i.e., to get the approximate controllability in L^2 of the Navier-Stokes control system. But, in the special case where $\Gamma_0 = \Gamma$, (6.64), (6.65), and (6.66) are strong enough to imply (6.73). Moreover, combining together a proof of Theorem 6.13 (see [**184, 187**]) and the proof of Theorem 6.14, one gets the following theorem.

THEOREM 6.17 ([**123**]). *The Open Problem 6.11 has a positive answer when $\Gamma_0 = \Gamma$ and $l = 2$.*

This result has been recently generalized to the case $l = 3$ by Andrei Fursikov and Oleg Imanuvilov in [**188**]. Let us also mention that, in [**155**], Caroline Fabre obtained, in every dimension, an approximate controllability result of two natural "cut off" Navier-Stokes equations. Her proof relies on a general method introduced by Enrique Zuazua in [**516**] to prove approximate controllability of semilinear wave equations. This general method is based on the HUM (Hilbert Uniqueness Method, due to Jacques-Louis Lions [**326**]; see Section 1.4) and on a fixed-point technique introduced by Enrique Zuazua in [**515, 516**] and that we described in Section 3.5.1 and in Section 4.3.

REMARK 6.18. Roughly speaking, we have deduced global controllability results for the Navier-Stokes equations from the controllability of the Euler equations. This is possible because the Euler equations are quadratic and the Navier-Stokes equations are the Euler equations plus a linear "perturbation". See Section 3.5.2, where we have studied such phenomena for control systems in finite dimension.

REMARK 6.19. As already mentioned in Remark 5.3 on page 185, Andrei Agrachev and Andrei Sarychev in [**9**], and Armen Shirikyan in [**446**] obtained global approximate controllability results for the Navier-Stokes equations and the Euler equations of incompressible fluids when the controls are on some low modes and Ω is a torus. They also obtained the controllability for the Navier-Stokes equations from the controllability for the Euler equations by the same scaling arguments. But their proof of the controllability for the Euler equations is completely different. It relies on iterated Lie brackets and a regularity property of the Navier-Stokes and Euler equations for controls in $W^{-1,\infty}(0,T)$.

REMARK 6.20. It is usually accepted that the viscous Burgers equation provides a realistic simplification of the Navier-Stokes system in Fluid Mechanics. But Ildefonso Diaz proved in [**145**] that the viscous Burgers equation is not approximately controllable for boundary controls; see also [**185**], by Andrei Fursikov and Oleg Imanuvilov, and [**208**] by Sergio Guerrero and Oleg Imanuvilov. Enrique Fernández-Cara and Sergio Guerrero proved in [**166**] that the viscous Burgers equation is not small-time null controllable for large data (and boundary control). But we proved in [**118**] that there exists a time T such that, whatever the initial datum is, there exists a control which can steer the Burgers control system from this initial datum to rest in time T.

For the nonviscous Burgers equation, results have been obtained by Fabio Ancona and Andrea Marson in [**16**] and, using the return method, by Thierry Horsin in [**239**] (see also [**69**] by Alberto Bressan and Giuseppe Coclite, and [**15**] by Fabio Ancona and Giuseppe Coclite for hyperbolic systems and BV-solutions). Note that, with boundary control, the nonviscous Burgers equation is not controllable (even if the control is on the whole boundary of Ω, which is now an open bounded nonempty interval of $\mathbb{R}$; see [**239**]). By Remark 6.18, this might be the reason why the viscous Burgers equation is not approximately controllable for boundary control (even if the control is on the whole boundary of Ω; see [**208**]) and could explain the difference, for the controllability, between the Navier-Stokes equations and the viscous Burgers equation. Recently, Marianne Chapouly, using the return method, obtained in [**87**] a global controllability result in arbitrary time for the nonviscous Burgers equation when, besides control on the whole boundary, there is a constant control inside. It seems that this constant control plays the role of the pressure in the Euler equations.

REMARK 6.21. It would be quite interesting to understand the controllability of incompressible fluids in Lagrangian coordinates. Recently Thierry Horsin obtained results in this direction for the viscous Burgers equation in [**241**] and for the heat equation in [**240**].

REMARK 6.22. Recently, there has been a lot of research on the interaction fluid with other materials. Let us mention, in particular,
  - the paper [**330**] by Jacques Louis Lions and Enrique Zuazua, the papers [**377, 378**] by Axel Osses and Jean-Pierre Puel, on the controllability of an incompressible fluid interacting with an elastic structure,
  - the papers [**147, 147**] by Anna Doubova and Enrique Fernández-Cara on the controllability of one-dimensional nonlinear system which models the interaction of a fluid and a particle,
  - the paper [**514**] by Xu Zhang and Enrique Zuazua on the controllability of a linearized and simplified 1-D model for fluidstructure interaction. This model consists of a wave and a heat equation and two bounded intervals, coupled by transmission conditions at the point of interface.

6.3. Local controllability of a 1-D tank containing a fluid modeled by the Saint-Venant equations

In this section, we consider a 1-D tank containing an inviscid incompressible irrotational fluid. The tank is subject to one-dimensional horizontal moves. We assume that the horizontal acceleration of the tank is small compared to the gravity constant and that the height of the fluid is small compared to the length of the tank. These physical considerations motivate the use of the Saint-Venant equations [430] (also called shallow water equations) to describe the motion of the fluid; see e.g. [144, Sec. 4.2]. Hence the considered dynamics equations are (see the paper [150] by François Dubois, Nicolas Petit and Pierre Rouchon)

(6.74) $$H_t(t,x) + (Hv)_x(t,x) = 0,\, t \in [0,T],\, x \in [0,L],$$

(6.75) $$v_t(t,x) + \left(gH + \frac{v^2}{2}\right)_x (t,x) = -u(t),\, t \in [0,T],\, x \in [0,L],$$

(6.76) $$v(t,0) = v(t,L) = 0,\, t \in [0,T],$$

(6.77) $$\frac{ds}{dt}(t) = u(t),\, t \in [0,T],$$

(6.78) $$\frac{dD}{dt}(t) = s(t),\, t \in [0,T],$$

where (see Figure 2),
- L is the length of the 1-D tank,
- $H(t,x)$ is the height of the fluid at time t and at the position $x \in [0,L]$,
- $v(t,x)$ is the horizontal water velocity of the fluid *in a referential attached to the tank* at time t and at the position $x \in [0,L]$ (in the shallow water model, all the points on the same vertical have the same horizontal velocity),
- $u(t)$ is the horizontal acceleration of the tank in the absolute referential,
- g is the gravity constant,
- s is the horizontal velocity of the tank,
- D is the horizontal displacement of the tank.

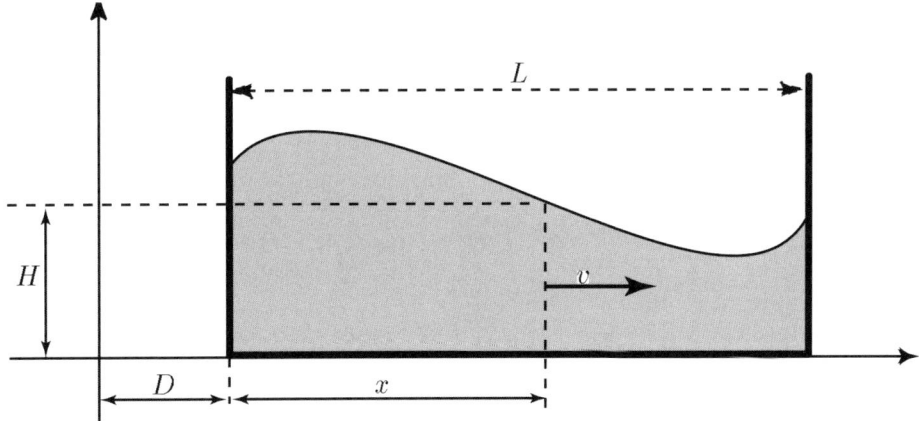

FIGURE 2. Fluid in the 1-D tank

This is a control system, denoted Σ, where, at time $t \in [0,T]$,

- the state is $Y(t) = (H(t,\cdot), v(t,\cdot), s(t), D(t))$,
- the control is $u(t) \in \mathbb{R}$.

Our goal is to study the local controllability of the control system Σ around the equilibrium point
$$(Y_e, u_e) := ((H_e, 0, 0, 0), 0).$$

This problem has been raised by François Dubois, Nicolas Petit and Pierre Rouchon in [**150**].

Of course, the total mass of the fluid is conserved so that, for every solution of (6.74) to (6.76),

$$\text{(6.79)} \qquad \frac{\mathrm{d}}{\mathrm{d}t} \int_0^L H(t, x)\, \mathrm{d}x = 0.$$

(One gets (6.79) by integrating (6.74) on $[0, L]$ and by using (6.76) together with an integration by parts.) Moreover, if H and v are of class C^1, it follows from (6.75) and (6.76) that

$$\text{(6.80)} \qquad H_x(t, 0) = H_x(t, L),$$

which is also $-u(t)/g$. Therefore we introduce the vector space E of functions $Y = (H, v, s, D) \in C^1([0, L]) \times C^1([0, L]) \times \mathbb{R} \times \mathbb{R}$ such that

$$\text{(6.81)} \qquad H_x(0) = H_x(L),$$
$$\text{(6.82)} \qquad v(0) = v(L) = 0,$$

and we consider the affine subspace $\mathcal{Y} \subset E$ consisting of elements $Y = (H, v, s, D) \in E$ satisfying

$$\text{(6.83)} \qquad \int_0^L H(x)\mathrm{d}x = LH_e.$$

The vector space E is equipped with the natural norm

$$|Y| := \|H\|_{C^1([0,L])} + \|v\|_{C^1([0,L])} + |s| + |D|.$$

With these notations, we can define a trajectory of the control system Σ.

DEFINITION 6.23. Let T_1 and T_2 be two real numbers satisfying $T_1 \leqslant T_2$. A function $(Y, u) = ((H, v, s, d), u) : [T_1, T_2] \to \mathcal{Y} \times \mathbb{R}$ is a *trajectory of the control system* Σ if

(i) the functions H and v are of class C^1 on $[T_1, T_2] \times [0, L]$,
(ii) the functions s and D are of class C^1 on $[T_1, T_2]$ and the function u is continuous on $[T_1, T_2]$,
(iii) the equations (6.74) to (6.78) hold for every $(t, x) \in [T_1, T_2] \times [0, L]$.

Our main result states that the control system Σ is locally controllable around the equilibrium point (Y_e, u_e). More precisely, one has the following theorem.

THEOREM 6.24 ([**116**]). *There exist $T > 0$, $C_0 > 0$ and $\eta > 0$ such that, for every $Y^0 = (H^0, v^0, s^0, D^0) \in \mathcal{Y}$, and for every $Y^1 = (H^1, v^1, s^1, D^1) \in \mathcal{Y}$ such that*

$$\|H^0 - H_e\|_{C^1([0,L])} + \|v^0\|_{C^1([0,L])} < \eta, \|H^1 - H_e\|_{C^1([0,L])} + \|v^1\|_{C^1([0,L])} < \eta,$$
$$|s^1 - s^0| + |D^1 - s^0 T - D^0| < \eta,$$

there exists a trajectory $(Y, u) : t \in [0, T] \mapsto ((H(t), v(t), s(t), D(t)), u(t)) \in \mathcal{Y} \times \mathbb{R}$ of the control system Σ, such that

(6.84) $$Y(0) = Y^0 \text{ and } Y(T) = Y^1,$$

and, for every $t \in [0, T]$,

(6.85) $\|H(t) - H_e\|_{C^1([0,L])} + \|v(t)\|_{C^1([0,L])} + |u(t)| \leqslant$
$$C_0 \left(\sqrt{\|H^0 - H_e\|_{C^1([0,L])} + \|v^0\|_{C^1([0,L])} + \|H^1 - H_e\|_{C^1([0,L])} + \|v^1\|_{C^1([0,L])}} \right)$$
$$+ C_0 \left(|s^1 - s^0| + |D^1 - s^0 T - D^0| \right).$$

As a corollary of this theorem, any steady state $Y^1 = (H_e, 0, 0, D^1)$ can be reached from any other steady state $Y^0 = (H_e, 0, 0, D^0)$. More precisely, one has the following corollary.

COROLLARY 6.25 ([**116**]). *Let T, C_0 and η be as in Theorem 6.24. Let D^0 and D^1 be two real numbers and let $\eta_1 \in (0, \eta]$. Then there exists a trajectory*

$$(Y, u) : [0, T(|D^1 - D^0| + \eta_1)/\eta_1] \to \mathcal{Y} \times \mathbb{R}$$
$$t \mapsto ((H(t), v(t), s(t), D(t)), u(t))$$

of the control system Σ, such that

(6.86) $$Y(0) = (H_e, 0, 0, D^0) \text{ and } Y(T(|D^1 - D^0| + \eta_1)/\eta_1) = (H_e, 0, 0, D^1),$$

and, for every $t \in [0, T(|D^1 - D^0| + \eta_1)/\eta_1]$,

(6.87) $$\|H(t) - H_e\|_{C^1([0,L])} + \|v(t)\|_{C^1([0,L])} + |u(t)| \leqslant C_0 \eta_1.$$

We will first explain, on a toy model in finite dimension, the main ideas for proving Theorem 6.24. The rest of the section will then be devoted to a more detailed description of the proof of Theorem 6.24.

In order to motivate our toy model, recalling that H^0 and H^1 have to be close to H_e, let us write

$$H^0 = H_e + \varepsilon h_1^0 + \varepsilon^2 h_2^0, \, v^0 = \varepsilon v_1^0 + \varepsilon^2 v_2^0, \, s^0 = \varepsilon s_1^0, \, D^0 = \varepsilon d_1^0,$$
$$H^1 = H_e + \varepsilon h_1^1 + \varepsilon^2 h_2^1, \, v^1 = \varepsilon v_1^1 + \varepsilon^2 v_2^1, \, s^1 = \varepsilon s_1^1, \, D^1 = \varepsilon d_1^1, \, u = \varepsilon u^1,$$

where $\varepsilon > 0$ is small, and let us study the problem of second order in ε. (In fact, we should also write the terms in ε^2 for u, s, D but they turn out to be useless.) Let us expand the solution (H, v, s, D) of (6.74) to (6.78) together with

$$H(0, x) = H^0(x), v(0, x) = v^0(x), s(0) = s^0, D(0) = D^0,$$

as power series in ε:

$$H = H_e + \varepsilon h_1 + \varepsilon^2 h_2 + \ldots, \, v = \varepsilon v_1 + \varepsilon^2 v_2 + \ldots, \, s = \varepsilon s_1, \, D = \varepsilon D_1, u = \varepsilon u_1.$$

Identifying the power series expansions up to order 2 in ε, one gets

(6.88) $$\begin{cases} h_{1t} + H_e v_{1x} = 0, \, t \in [0,T], \, x \in [0,L], \\ v_{1t} + gh_{1x} = -u_1(t), \, t \in [0,T], \, x \in [0,L], \\ v_1(t,0) = v_1(t,L) = 0, \, t \in [0,T], \\ \dfrac{ds_1}{dt}(t) = u_1(t), \, t \in [0,T], \\ \dfrac{dD_1}{dt}(t) = s_1(t), \, t \in [0,T], \end{cases}$$

(6.89) $$\begin{cases} h_{2t} + H_e v_{2x} = -(h_1 v_1)_x, \, t \in [0,T], \, x \in [0,L], \\ v_{2t} + gh_{2x} = -v_1 v_{1x}, \, t \in [0,T], \, x \in [0,L], \\ v_1(t,0) = v_1(t,L) = 0, \, t \in [0,T], \end{cases}$$

together with the initial conditions
$$h_1(0,x) = h_1^0(x), \, v_1(0,x) = v_1^0(x), \, x \in [0,L],$$
$$s_1(0) = s^1, \, D_1(0) = D^1, \, h_2(0,x) = h_2^0(x), \, v_2(0,x) = v_2^0(x), \, x \in [0,L].$$

We consider (6.88) as a control system where, at time $t \in [0,T]$, the control is $u_1(t)$ and the state is $(h_1(t,\cdot), v_1(t,\cdot), s_1(t), D_1(t)) \in C^1([0,L]) \times C^1([0,L]) \times \mathbb{R} \times \mathbb{R}$ and satisfies

$$v_1(t,0) = v_1(t,L) = 0, \, h_1(t, L-x) = -h_1(t,x), \, v_1(t, L-x) = v_1(t,x), \, x \in [0,L].$$

Note that, by [**150**], this linear control system is controllable in any time $T > L/\sqrt{H_e g}$. This can be proved by means of explicit solutions as for a transport equation in Section 2.1.2.1. (For controllability in Sobolev spaces, one can also get this controllability by proving an observability inequality by means of the multiplier method as in Section 2.1.2.3 for a transport equation.) Similarly, we consider (6.89) as a control system where, at time $t \in [0,T]$, the state is $(h_2(t,\cdot), v_2(t,\cdot)) : [0,L] \to \mathbb{R}^2$ and satisfies

(6.90) $\quad h_2(t, L-x) = h_2(t,x), \, h_{2x}(t,0) = h_{2x}(t,L) = 0, \, \int_0^L h_2(t,x)dx = 0,$

(6.91) $\quad v_2(t, L-x) = -v_2(t,x), \, x \in [0,L], \, v_2(t,0) = v_2(t,L) = 0.$

The control in this system is $(h_1 v_1)_x(t,\cdot)$ and $v_1 v_{1x}(t,\cdot)$. It depends only on the state of the control system (6.88) and is quadratic with respect to this state. The first two equations of (6.88) and (6.89) are the usual wave equations. A natural analogue of the wave equation in finite dimension is the oscillator equation. Hence a natural analogue of our control system (6.88) and (6.89) is

(6.92) $\qquad\qquad\qquad \dot{x}_1 = x_2, \, \dot{x}_2 = -x_1 + u,$

(6.93) $\qquad\qquad\qquad \dot{x}_3 = x_4, \, \dot{x}_4 = -x_3 + 2x_1 x_2,$

(6.94) $\qquad\qquad\qquad \dot{s} = u, \, \dot{D} = s,$

where the state is $(x_1, x_2, x_3, x_4, s, D) \in \mathbb{R}^6$ and the control $u \in \mathbb{R}$. This control system is our toy model. We call it $\mathfrak{T}$.

The linearized control system of $\mathfrak{T}$ at the equilibrium $(0,0) \in \mathbb{R}^6 \times \mathbb{R}$ is

(6.95) $\qquad \dot{x}_1 = x_2, \, \dot{x}_2 = -x_1 + u, \, \dot{x}_3 = x_4, \, \dot{x}_4 = -x_3, \, \dot{s} = u, \, \dot{D} = s.$

The linear control system (6.95) is not controllable. However, as one easily checks, (6.95) is steady-state controllable for arbitrary time T, that is, for every $(D^0, D^1) \in \mathbb{R}^2$ and for every $T > 0$, there exists a trajectory $((x_1, x_2, x_3, x_4, s, D), u) : [0, T] \to \mathbb{R}^6 \times \mathbb{R}$ of the linear control system (6.95), such that

$$x_1(0) = x_2(0) = x_3(0) = x_4(0) = s(0) = 0, \ D(0) = D^0,$$
$$x_1(T) = x_2(T) = x_3(T) = x_4(T) = s(T) = 0, \ D(T) = D^1.$$

But the same does not hold for the nonlinear control system $\mathfrak{T}$. Indeed, it follows from Example 6.4 on page 190 that if $((x_1, x_2, x_3, x_4, s, D), u) : [0, T] \to \mathbb{R}^6 \times \mathbb{R}$ is a trajectory of the control system $\mathfrak{T}$, such that

$$x_1(0) = x_2(0) = x_3(0) = x_4(0) = s(0) = 0,$$
$$x_1(T) = x_2(T) = x_3(T) = x_4(T) = s(T) = 0, \ D(T) \neq D(0),$$

then $T > \pi$. This minimal time for steady state controllability is optimal. This can be seen by using the return method as in Example 6.4 on page 190. Indeed, let $T > \pi$ and let $(\bar{x}_1, \bar{x}_2, \bar{x}_3, \bar{x}_4, \bar{u})$ be as in Example 6.4, except that one also requires

$$(6.96) \qquad \int_0^\eta \bar{x}_1(t) dt = 0, \ \int_0^\eta \int_0^t \bar{x}_1(s) ds dt = 0.$$

Coming back to the controllability of $\mathfrak{T}$, let us define $(\bar{s}, \bar{D}) : [0, T] \to \mathbb{R}^2$ by

$$\dot{\bar{s}} = \bar{u}, \ \bar{s}(0) = 0, \ \dot{\bar{D}} = \bar{s}, \ \bar{D}(0) = 0.$$

Then, using in particular (6.96), one easily sees that

$$\bar{s}(T) = 0, \ \bar{D}(T) = 0.$$

Moreover, straightforward computations show that the linearized control system along the trajectory $((\bar{x}_1, \bar{x}_2, \bar{x}_3, \bar{x}_4, \bar{s}, \bar{D}), \bar{u})$ is controllable if $\bar{x}_1$ is not identically equal to 0. Hence the return method gives the local controllability of $\mathfrak{T}$ in time $T > \pi$.

Unfortunately, this proof for checking the local controllability heavily relies on explicit computations, which we have not been able to perform in the case of the control system Σ defined on page 204. Let us now present a more flexible use of the return method to prove the following large-time local controllability of $\mathfrak{T}$.

PROPOSITION 6.26. *There exist $T > 0$ and $\delta > 0$ such that, for every $a \in \mathbb{R}^6$ and every $b \in \mathbb{R}^6$ with $|a| < \delta$ and $|b| < \delta$, there exists $u \in L^\infty(0, T)$ such that, if $x = (x_1, x_2, x_3, x_4, s, D) : [0, T] \to \mathbb{R}^6$ is the solution of the Cauchy problem*

$$\dot{x}_1 = x_2, \ \dot{x}_2 = -x_1 + u,$$
$$\dot{x}_3 = x_4, \ \dot{x}_4 = -x_3 + 2x_1 x_2,$$
$$\dot{s} = u, \ \dot{D} = s,$$
$$x(0) = a,$$

then $x(T) = b$.

Proof of Proposition 6.26. We prove this proposition by using the return method. In order to use this method, one needs, at least, to know trajectories of the control system $\mathfrak{T}$, such that the linearized control systems around these trajectories are controllable. The simplest trajectories one can consider are the trajectories

$$(6.97) \qquad ((x_1^\gamma, x_2^\gamma, x_3^\gamma, x_4^\gamma, s^\gamma, D^\gamma), u^\gamma) = ((\gamma, 0, 0, 0, \gamma t, \gamma t^2/2), \gamma),$$

where γ is any real number different from 0 and $t \in [0, \tau_1]$, with $\tau_1 > 0$ fixed. The linearized control system around the trajectory

$$(x^\gamma, u^\gamma) := ((x_1^\gamma, x_2^\gamma, x_3^\gamma, x_4^\gamma, s^\gamma, D^\gamma), u^\gamma)$$

is the linear control system

(6.98)
$$\begin{cases} \dot{x}_1 = x_2, \ \dot{x}_2 = -x_1 + u, \\ \dot{x}_3 = x_4, \ \dot{x}_4 = -x_3 + 2\gamma x_2, \\ \dot{s} = u, \ \dot{D} = s. \end{cases}$$

Using the Kalman rank condition (Theorem 1.16 on page 9), one easily checks that this linear control system (6.98) is controllable if and only if $\gamma \neq 0$. Let us now choose $\gamma \neq 0$. Then, since the linearized control system around (x^γ, u^γ) is controllable, there exists $\delta_1 > 0$ such that, for every

$$a \in B(x^\gamma(0), \delta_1) := \{x \in \mathbb{R}^6; |x - x^\gamma(0)| < \delta_1\}$$

and for every

$$b \in B(x^\gamma(\tau_1), \delta_1) := \{x \in \mathbb{R}^6; |x - x^\gamma(\tau_1)| < \delta_1\},$$

there exists $u \in L^\infty((0, \tau_1); \mathbb{R})$ such that

$$\left(\begin{cases} \dot{x}_1 = x_2, \ \dot{x}_2 = -x_1 + u, \\ \dot{x}_3 = x_4, \ \dot{x}_4 = -x_3 + 2x_1x_2, \dot{s} = u, \ \dot{D} = S, \ x(0) = a \end{cases}\right) \Rightarrow (x(\tau_1) = b).$$

Hence, in order to prove Proposition 6.26, it suffices to check that:

(i) *There exist $\tau_2 > 0$ and a trajectory $(\tilde{x}, \tilde{u}) : [0, \tau_2] \to \mathbb{R}^6 \times \mathbb{R}$ of the control system $\mathfrak{T}$, such that $\tilde{x}(0) = 0$ and $|\tilde{x}(\tau_2) - x^\gamma(0)| < \delta_1$.*

(ii) *There exist $\tau_3 > 0$ and a trajectory $(\hat{x}, \hat{u}) : [0, \tau_3] \to \mathbb{R}^6 \times \mathbb{R}$ of the control system $\mathfrak{T}$, such that $\hat{x}(\tau_3) = 0$ and $|\hat{x}(0) - x^\gamma(\tau_1)| < \delta_1$.*

Indeed, by the continuity of the solutions of the Cauchy problem with respect to the initial condition, there exists $\delta > 0$ such that

$$\left(\begin{cases} \dot{x}_1 = x_2, \ \dot{x}_2 = -x_1 + \tilde{u}, \\ \dot{x}_3 = x_4, \ \dot{x}_4 = -x_3 + 2x_1x_2, \ \dot{s} = \tilde{u}, \ \dot{D} = S, \\ |x(0)| \leqslant \delta \end{cases}\right) \Rightarrow (|x(\tau_2) - x^\gamma(0)| < \delta_1),$$

and

$$\left(\begin{cases} \dot{x}_1 = x_2, \ \dot{x}_2 = -x_1 + \tilde{u}, \\ \dot{x}_3 = x_4, \ \dot{x}_4 = -x_3 + 2x_1x_2, \\ \dot{s} = \hat{u}(t - \tau_2 - \tau_1), \ \dot{D} = S, \\ |x(\tau_2 + \tau_1 + \tau_3)| \leqslant \delta \end{cases}\right) \Rightarrow (|x(\tau_2 + \tau_1) - x^\gamma(\tau_1)| < \delta_1).$$

Then it suffices to take $T = \tau_2 + \tau_1 + \tau_3$; see Figure 3.

Proof of (i). In order to prove (i), we consider quasi-static deformations (see Chapter 7 for more details on this method). Let $g \in \mathcal{C}^2([0, 1]; \mathbb{R})$ be such that

(6.99)
$$g(0) = 0, \ g(1) = 1.$$

Let $\tilde{u} : [0, 1/\varepsilon] \to \mathbb{R}$ be defined by

(6.100)
$$\tilde{u}(t) = \gamma g(\varepsilon t), \ t \in [0, 1/\varepsilon].$$

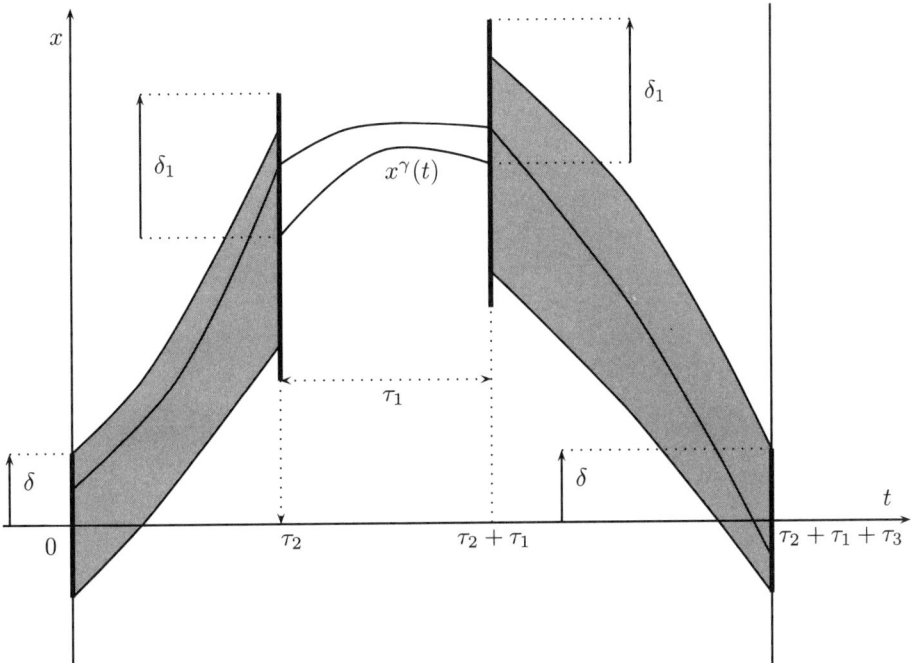

FIGURE 3. Return method and quasi-static deformations

Let $\tilde{x} := (\tilde{x}_1, \tilde{x}_2, \tilde{x}_3, \tilde{x}_4, \tilde{s}, \tilde{D}) : [0, 1/\varepsilon] \to \mathbb{R}^6$ be defined by
$$\dot{\tilde{x}}_1 = \tilde{x}_2, \ \dot{\tilde{x}}_2 = -\tilde{x}_1 + \tilde{u},$$
$$\dot{\tilde{x}}_3 = \tilde{x}_4, \ \dot{\tilde{x}}_4 = -\tilde{x}_3 + 2\tilde{x}_1\tilde{x}_2,$$
$$\dot{\tilde{s}} = \tilde{u}, \ \dot{\tilde{D}} = \tilde{s}, \ \tilde{x}(0) = 0.$$

One has

(6.101) $$\tilde{x}_1 + i\tilde{x}_2 = i\gamma \int_0^t e^{-i(t-s)} g(\varepsilon s) ds,$$

(6.102) $$\tilde{x}_3 + i\tilde{x}_4 = 2i \int_0^t e^{-i(t-s)} \tilde{x}_1(s) \tilde{x}_2(s) ds.$$

From (6.101) and after two integrations by parts, one gets

(6.103) $$\tilde{x}_1 + i\tilde{x}_2 = \gamma g(\varepsilon t) - \gamma g(0) e^{-it} + i\varepsilon \gamma g'(\varepsilon t)$$
$$- i\varepsilon \gamma g'(0) e^{-it} - i\varepsilon^2 \gamma \int_0^t e^{i(s-t)} g''(\varepsilon s) ds,$$

which, together with (6.99), gives

(6.104) $$(\tilde{x}_1(1/\varepsilon), \tilde{x}_2(1/\varepsilon)) \to (\gamma, 0) \text{ as } \varepsilon \to 0.$$

Similarly, using (6.99), (6.102), (6.103) and integration by parts, one gets

(6.105) $$(\tilde{x}_3(1/\varepsilon), \tilde{x}_4(1/\varepsilon)) \to 0 \text{ as } \varepsilon \to 0.$$

Finally,

$$\tilde{s}(1/\varepsilon) = \frac{\gamma}{\varepsilon} \int_0^1 g(z)dz, \tag{6.106}$$

$$\tilde{D}(1/\varepsilon) = \frac{\gamma}{\varepsilon^2} \int_0^1 \left(\int_0^{z_1} g(z_2)dz_2 \right) dz_1. \tag{6.107}$$

Hence, if

$$\int_0^1 g(z)dz = 0, \int_0^1 \left(\int_0^{z_1} g(z_2)dz_2 \right) dz_1 = 0, \tag{6.108}$$

one gets, using (6.104), (6.105), (6.106) and (6.107), that

$$\tilde{x}(1/\varepsilon) \to (\gamma, 0, 0, 0, 0, 0) \text{ as } \varepsilon \to 0,$$

which concludes the proof of (i). ∎

Proof of (ii). One needs to modify, a little bit, the above construction. Now let $g \in C^2([0,1]; \mathbb{R})$ be such that $g(0) = 1$ and $g(1) = 0$. Let $g_1 \in C^1([0,1]; \mathbb{R})$ be such that

$$\int_0^1 g_1(z)dz = \tau_1, \int_0^1 \left(\int_{z_1}^1 g_1(z_2)dz_2 \right) dz_1 = 0. \tag{6.109}$$

Also let $g_2 \in C^0([0,1]; \mathbb{R})$ be such that

$$\int_0^1 \left(\int_{z_1}^1 g_2(z_2)dz_2 \right) dz_1 = \tau_1^2/2. \tag{6.110}$$

Finally, let $\hat{u} : [0, 1/\varepsilon] \to \mathbb{R}$ be defined by

$$\hat{u}(t) = \gamma(g(\varepsilon t) - \varepsilon g_1(\varepsilon t) + \varepsilon^2 g_2(\varepsilon t)), \ t \in [0, 1/\varepsilon]. \tag{6.111}$$

Then similar computations as for $\tilde{x}$ show that

$$|\hat{x}(0) - (\gamma, 0, 0, 0, \gamma\tau_1, \gamma\tau_1^2/2)| \to 0 \text{ as } \varepsilon \to 0,$$

where $\hat{x} := (\hat{x}_1, \hat{x}_2, \hat{x}_3, \hat{x}_4, \hat{s}, \hat{D}) : [0, 1/\varepsilon] \to \mathbb{R}^6$ is defined by

$$\dot{\hat{x}}_1 = \hat{x}_2, \ \dot{\hat{x}}_2 = -\hat{x}_1 + \hat{u},$$
$$\dot{\hat{x}}_3 = \hat{x}_4, \ \dot{\hat{x}}_4 = -\hat{x}_3 + 2\hat{x}_1\hat{x}_2,$$
$$\dot{\hat{s}} = \hat{u}, \dot{\hat{D}} = \hat{s},$$
$$\hat{x}(1/\varepsilon) = 0.$$

This concludes the proof of (ii) and of Proposition 6.26. ∎

We now come back to our description of the proof of Theorem 6.24 on page 205. Note that the analogue of Theorem 6.24 for the control system $\mathfrak{T}$ is the following proposition.

PROPOSITION 6.27. *There exists $T > 0$ such that, for every $\varepsilon > 0$, there exists $\eta > 0$, such that, for every $a = (a_1, a_2, a_3, a_4, s^0, D^0) \in \mathbb{R}^6$ and for every $b = (b_1, b_2, b_3, b_4, s^1, D^1) \in \mathbb{R}^6$ satisfying*

$$|a_1| + |a_2| + |a_3| + |a_4| < \eta,$$
$$|b_1| + |b_2| + |b_3| + |b_4| < \eta,$$
$$|s^0| + |s^1| + |D^0| + |D^1| < \eta,$$

there exists a trajectory

$$(x, u) : [0, T] \to \mathbb{R}^6 \times \mathbb{R},$$
$$t \mapsto (x_1(t), x_2(t), x_3(t), x_4(t), s(t), D(t), u(t))$$

of the control system $\mathfrak{T}$, such that

$$x(0) = a \text{ and } x(T) = b,$$

and, for every $t \in [0, T]$,

(6.112) $\quad |x_1(t)| + |x_2(t)| + |x_3(t)| + |x_4(t)| + |S(t)| + |D(t)| + |u(t)| \leqslant \varepsilon.$

This proposition is stronger than Proposition 6.26, because we added condition (6.112). The proof of Proposition 6.27 requires some new estimates. In particular, one needs an estimate on δ_1 compared to γ. See [**116**] for the estimates for the control system Σ.

Let us now sketch how one adapts, for our control system Σ defined on page 204, what we have done for the toy control system $\mathfrak{T}$ defined on page 206. Let us first point out that by scaling arguments one can assume without loss of generality that

(6.113) $\qquad\qquad\qquad L = g = H_e = 1.$

Indeed, if we let

$$H^*(t, x) := \frac{1}{H_e} H\left(\frac{Lt}{\sqrt{H_e g}}, Lx\right), \quad v^*(t, x) := \frac{1}{\sqrt{H_e g}} v\left(\frac{Lt}{\sqrt{H_e g}}, Lx\right),$$

$$u^*(t) := \frac{L}{H_e g} u\left(\frac{Lt}{\sqrt{H_e g}}\right), \quad s^*(t) := \frac{1}{\sqrt{H_e g}} s\left(\frac{Lt}{\sqrt{H_e g}}\right), \quad D^*(t) := \frac{1}{L} D\left(\frac{Lt}{\sqrt{H_e g}}\right),$$

with $x \in [0, 1]$, then equations (6.74) to (6.78) are equivalent to

$$H_t^*(t, x) + (H^* v^*)_x(t, x) = 0,$$
$$v_t^*(t, x) + \left(H^* + \frac{v^{*2}}{2}\right)_x (t, x) = -u^*(t),$$
$$v^*(t, 0) = v^*(t, 1) = 0,$$
$$\frac{ds^*}{dt}(t) = u^*(t),$$
$$\frac{dD^*}{dt}(t) = s^*(t).$$

From now on, we always assume that we have (6.113). Since $(Y, u) = ((H, v, s, D), u)$ is a trajectory of the control system Σ if and only if $((H, v, s - a, D - at - b), u)$ is a trajectory of the control system Σ, we may also assume without loss of generality that $s^0 = D^0 = 0$.

The proof of Theorem 6.24 relies again on the return method. So one looks for a trajectory $(\bar{Y}, \bar{u}) : [0, T] \to \mathcal{Y} \times \mathbb{R}$ of the control system Σ satisfying

(6.114) $$\bar{Y}(0) = \bar{Y}(T) = Y_e,$$

(6.115) the linearized control system around $(\bar{Y}, \bar{u})$ is controllable.

Let us point out that, as already noticed by François Dubois, Nicolas Petit and Pierre Rouchon in [**150**] (see also [**385**] by Nicolas Petit and Pierre Rouchon), property (6.115) does not hold for the natural trajectory $(\bar{Y}, \bar{u}) = (Y_e, u_e)$. Indeed the linearized control system around (Y_e, u_e) is

(6.116) $$\Sigma_0 \begin{cases} h_t + v_x = 0, \\ v_t + h_x = -u(t), \\ v(t, 0) = v(t, 1) = 0, \\ \dfrac{ds}{dt}(t) = u(t), \\ \dfrac{dD}{dt}(t) = s(t), \end{cases}$$

where the state is $(h, v, s, D) \in \mathcal{Y}_0$, with

$$\mathcal{Y}_0 := \left\{ (h, v, s, D) \in E; \int_0^L h\, dx = 0 \right\},$$

and the control is $u \in \mathbb{R}$. But (6.116) implies that, if

$$h(0, 1-x) = -h(0, x) \text{ and } v(0, 1-x) = v(0, x), \forall x \in [0, 1],$$

then

$$h(t, 1-x) = -h(t, x) \text{ and } v(t, 1-x) = v(t, x), \forall x \in [0, 1], \forall t.$$

REMARK 6.28. Even if the control system (6.116) is not controllable, one can move, as it is proved in [**150**], from every steady state $(h^0, v^0, s^0, D^0) := (0, 0, 0, D^0)$ to every steady state $(h^1, v^1, s^1, D^1) := (0, 0, 0, D^1)$ for this control system (see also [**385**] when the tank has a non-straight bottom). This does not a priori imply that the related property (move from $(1, 0, 0, D^0)$ to $(1, 0, 0, D^1)$) also holds for the nonlinear control system Σ, but it follows in fact from Corollary 6.25 that this property indeed also holds for the nonlinear control system Σ. Moreover, the fact that, for the control system (6.116), it is possible (see [**150**]) to move from every steady state $(h^0, v^0, s^0, D^0) := (0, 0, s^0, D^0)$ to every steady state $(h^1, v^1, s^1, D^1) := (0, 0, s^1, D^1)$ explains why in the right hand side of (6.85) one has $|s^1 - s^0| + |D^1 - s^0 T - D^0|$ and not $\left(|s^1 - s^0| + |D^1 - s^0 T - D^0| \right)^{1/2}$.

As in [**103, 104, 112, 123, 188, 194, 195, 457**], one has to look for more complicated trajectories $(\bar{Y}, \bar{u})$ in order to get (6.115). In fact, as in [**111**] and for our toy model $\mathfrak{T}$, one can require, instead of (6.114), the weaker property

(6.117) $$\bar{Y}(0) = Y_e \text{ and } \bar{Y}(T) \text{ is close to } Y_e,$$

and hope that, as it happens for the Navier-Stokes control system (see Remark 6.16 on page 201 and [**111**]), the controllability around $(\bar{Y}, \bar{u})$ will be strong enough to tackle the problem that $\bar{Y}(T)$ is not Y_e but only close to Y_e. Moreover, since, as it is proved in [**150**], one can move, for the linear control system Σ_0, from $Y_e = (0, 0, 0, 0)$

to $(0,0,s^1,D^1)$, it is natural to try not to "return" to Y_e, but to require instead of (6.117) the property

(6.118) $\qquad \bar{Y}(0) = Y_e$ and $\bar{Y}(T)$ is close to $(1,0,s^1,D^1)$.

In order to use this method, one first needs to have trajectories of the control system Σ, such that the linearized control systems around these trajectories are controllable. Let us give an example of a family of such trajectories. Let us fix a positive real number T^* in $(2,+\infty)$. For $\gamma \in (0,1]$ and $(a,b) \in \mathbb{R}^2$, let us define $(Y^{\gamma,a,b}, u^\gamma) : [0,T^*] \to \mathcal{Y} \times \mathbb{R}$ by

(6.119) $\quad Y^{\gamma,a,b}(t) := \left(x \mapsto 1 + \gamma \left(\frac{1}{2} - x \right), 0, \gamma t + a, \gamma \frac{t^2}{2} + at + b \right), \forall t \in [0, T^*],$

(6.120) $\qquad\qquad\qquad u^\gamma(t) := \gamma, \forall t \in [0, T^*].$

Then $(Y^{\gamma,a,b}, u^\gamma)$ is a trajectory of the control system Σ. The linearized control system around this trajectory is the control system

(6.121) $\quad \Sigma_\gamma \begin{cases} h_t + \left(\left(1 + \gamma \left(\frac{1}{2} - x \right) \right) v \right)_x = 0, \\ v_t + h_x = -u(t), \\ v(t,0) = v(t,1) = 0, \\ \dfrac{ds}{dt}(t) = u(t), \\ \dfrac{dD}{dt}(t) = s(t), \end{cases}$

where, at time t, the state is $(h(t,\cdot), v(t,\cdot), s(t), D(t)) \in \mathcal{Y}_0$ and the control is $u(t) \in \mathbb{R}$. This linear control system Σ_γ is controllable if $\gamma > 0$ is small enough (see [**116**] for a proof). Unfortunately, the controllability of Σ_γ does not seem to imply directly the local controllability of the control system Σ along the trajectory $(Y^{\gamma,a,b}, u^\gamma)$. Indeed, the map from $\mathcal{Y} \times C^0([0,T])$ into $\mathcal{Y}$ which, to any initial data $Y^0 = (H^0, v^0, s^0, D^0) \in \mathcal{Y}$ and to any $u \in C^0([0,T])$ such that

$$H^0_x(0) = H^0_x(1) = -u(0),$$

associates the state $Y(T) \in \mathcal{Y}$, where $Y = (H, v, s, D) : [0,T] \to \mathcal{Y}$ satisfies (6.74) to (6.78) and $Y(0) = Y^0$, is well-defined and continuous on a small open neighborhood of $(Y_e, 0)$ (see e.g. [**321**]) but is not of class C^1 on this neighborhood. See also Section 4.2.1 for a transport equation and Section 4.2.2 for a one-dimensional Schrödinger equation. So one cannot use the classical inverse function theorem to get the desired local controllability. To handle this problem, one proceeds as for the proof of Theorem 4.8 on page 168: one adapts the usual iterative scheme used to prove the existence of solutions to hyperbolic systems (see, e.g., [**68**, pages 67–70], [**134**, pages 476–478], [**238**, pages 54–55], [**321**, pages 96–107], [**340**, pages 35–43] or [**443**, pages 106–116]; see also [**104**, **112**, **123**, **188**, **194**, **195**] for the Euler and the Navier control systems for incompressible fluids): one uses the inductive procedure

$$(h^n, v^n, s^n, D^n, u^n) \mapsto (h^{n+1}, v^{n+1}, s^{n+1}, D^{n+1}, u^{n+1})$$

such that

(6.122) $$h_t^{n+1} + v^n h_x^{n+1} + \left(1 + \gamma\left(\frac{1}{2} - x\right) + h^n\right) v_x^{n+1} - \gamma v^{n+1} = 0,$$

(6.123) $$v_t^{n+1} + h_x^{n+1} + v^n v_x^{n+1} = -u^{n+1}(t),$$

(6.124) $$v^{n+1}(t,0) = v^{n+1}(t,1) = 0,$$

(6.125) $$\frac{\mathrm{d}s^{n+1}}{\mathrm{d}t}(t) = u^{n+1}(t),$$

(6.126) $$\frac{\mathrm{d}D^{n+1}}{\mathrm{d}t}(t) = s^{n+1}(t),$$

and $\left(h^{n+1}, v^{n+1}, s^{n+1}, D^{n+1}, u^{n+1}\right)$ has the required value for $t = 0$ and for $t = T^*$. Unfortunately, we have only been able to prove that the control system (6.122)-(6.126), where the state is $\left(h^{n+1}, v^{n+1}, s^{n+1}, D^{n+1}\right)$ and the control is u^{n+1}, is controllable under a special assumption on (h^n, v^n); see [**116**]. Hence one has to ensure that, at each iterative step, (h^n, v^n) satisfies this condition, which turns out to be possible. So one gets the following proposition, which is proved in [**116**].

PROPOSITION 6.29. *There exist $C_1 > 0$, $\mu > 0$ and $\gamma_0 \in (0, 1]$ such that, for every $\gamma \in [0, \gamma_0]$, for every $(a, b) \in \mathbb{R}^2$, and for every $(Y^0, Y^1) \in \mathcal{Y}^2$ satisfying*

$$|Y^0 - Y^{\gamma,a,b}(0)| \leqslant \mu\gamma^2 \text{ and } |Y^1 - Y^{\gamma,a,b}(T^*)| \leqslant \mu\gamma^2,$$

there exists a trajectory $(Y, u) : [0, T^*] \to \mathcal{Y} \times \mathbb{R}$ of the control system Σ, such that

$$Y(0) = Y^0 \text{ and } Y(T^*) = Y^1,$$

(6.127) $$\left|Y(t) - Y^{\gamma,a,b}(t)\right| + |u(t)| \leqslant C_1\gamma, \ \forall t \in [0, T^*].$$

REMARK 6.30. Let

$$H^n := \left(1 + \gamma\left(\frac{1}{2} - x\right) + h^n\right).$$

Let us assume that

$$t \in [0, T^*] \mapsto ((H^n(t, \cdot), v^n(t, \cdot), s^n(t), D^n(t)), u^n(t))$$

is a trajectory of the control system Σ. The linearized control system (for Σ) around this trajectory is the linear control system

(6.128) $$h_t + v^n h_x + v_x^n h + vH_x^n + v_x H^n = 0,$$

(6.129) $$v_t + h_x + v^n v_x + vv_x^n = -u(t),$$

(6.130) $$v(t,0) = v(t,1) = 0,$$

(6.131) $$\frac{\mathrm{d}s}{\mathrm{d}t}(t) = u(t),$$

(6.132) $$\frac{\mathrm{d}D}{\mathrm{d}t}(t) = s(t).$$

Comparing (6.122) with (6.128) and (6.123) with (6.129), one sees that a solution $t \in [0, T^*] \mapsto ((H^{n+1}(t, \cdot), v^{n+1}(t, \cdot), s^{n+1}(t), D^{n+1}(t)), u^{n+1}(t))$ of (6.122)-(6.123)-(6.124)-(6.125)-(6.126) is *not*, in general, a trajectory of the linearized control system (6.128)-(6.129)-(6.130)-(6.131)-(6.132). Hence one does not get the local controllability stated in Proposition 6.29 from the controllability of linearized control systems, but from the controllability of other linear control systems. This is in fact also the case for many other applications of the return method in infinite dimension, for example, to the Euler equations of incompressible fluids. (For clarity of the exposition, in Section 6.2.1, we have skipped this technical point that is detailed in the papers mentioned in this section; see, in particular, [**112, 195**].) See also Remark 4.11 on page 172 as well as Remark 6.1 on page 189.

One now needs to construct, for every given small enough $\gamma > 0$, trajectories $(Y, u) : [0, T^0] \to \mathcal{Y} \times \mathbb{R}$ of the control system Σ satisfying

(6.133) $\qquad Y(0) = (1, 0, 0, 0)$ and $|Y(T^0) - Y^{\gamma,a,b}(0)| \leqslant \mu\gamma^2$,

and trajectories $(Y, u) : [T^0 + T^*, T^0 + T^* + T^1] \to \mathcal{Y} \times \mathbb{R}$ of the control system Σ, such that

(6.134)
$$Y\left(T^0 + T^1 + T^*\right) = \left(1, 0, s^1, D^1\right) \text{ and } \left|Y\left(T^0 + T^*\right) - Y^{\gamma,a,b}\left(T^*\right)\right| \leqslant \mu\gamma^2,$$

for a suitable choice of $(a, b) \in \mathbb{R}^2$, $T^0 > 0$, $T^1 > 0$.

Let us first point out that it follows from [**150**] that one knows explicit trajectories $(Y^l, u^l) : [0, T^0] \to \mathcal{Y} \times \mathbb{R}$ of *the linearized control system Σ_0 around* $(0, 0)$, satisfying $Y^l(0) = 0$ and $Y^l(T^0) = Y^{\gamma,a,b}(0)$. (In fact, François Dubois, Nicolas Petit and Pierre Rouchon have proved in [**150**] that the linear control system Σ_0 is flat—a notion introduced by Michel Fliess, Jean Lévine, Philippe Martin and Pierre Rouchon in [**175**]; see also Section 2.5.3 for an heat equation. They have given a complete explicit parametrization of the trajectories of Σ_0 by means of an arbitrary function and a 1-periodic function.) Then the idea is that, if, as in the proof of (i) on page 208, one moves "slowly", the same control u^l gives a trajectory $(Y, u) : [0, T^0] \to \mathcal{Y} \times \mathbb{R}$ of the control system Σ such that (6.133) holds.

More precisely, let $f_0 \in C^4([0, 4])$ be such that

(6.135) $\qquad f_0 = 0$ in $[0, 1/2] \cup [3, 4]$,

(6.136) $\qquad f_0(t) = t/2, \ \forall t \in [1, 3/2]$,

(6.137) $$\int_0^4 f_0(t_1) dt_1 = 0.$$

Let $f_1 \in C^4([0, 4])$ and $f_2 \in C^4([0, 4])$ be such that

(6.138) $\qquad f_1 = 0$ in $[0, 1/2] \cup [1, 3/2]$ and $f_1 = 1/2$ in $[3, 4]$,

(6.139) $$\int_0^3 f_1(t_1) dt_1 = 0,$$

(6.140) $\qquad f_2 = 0$ in $[0, 1/2] \cup [1, 3/2] \cup [3, 4]$,

(6.141) $$\int_0^4 f_2(t_1) dt_1 = 1/2.$$

Let
$$\mathbb{D} := \left\{ (\bar{s}, \bar{D}) \in \mathbb{R}^2; \, |\bar{s}| \leqslant 1, \, |\bar{D}| \leqslant 1 \right\}.$$

For $(\bar{s}, \bar{D}) \in \mathbb{D}$, let $f_{\bar{s}, \bar{D}} \in C^4([0, 4])$ be defined by

(6.142) $\qquad f_{\bar{s}, \bar{D}} := f_0 + \bar{s} f_1 + \bar{D} f_2.$

For $\varepsilon \in (0, 1/2]$ and for $\gamma \in \mathbb{R}$, let $u_{\bar{s}, \bar{D}}^{\varepsilon, \gamma} : [0, 3/\varepsilon] \to \mathbb{R}$ be defined by

(6.143) $\qquad u_{\bar{s}, \bar{D}}^{\varepsilon, \gamma}(t) := \gamma f'_{\bar{s}, \bar{D}}(\varepsilon t) + \gamma f'_{\bar{s}, \bar{D}}(\varepsilon(t + 1)).$

Let $\left(h_{\bar{s},\bar{D}}^{\varepsilon,\gamma}, v_{\bar{s},\bar{D}}^{\varepsilon,\gamma}, s_{\bar{s},\bar{D}}^{\varepsilon,\gamma}, D_{\bar{s},\bar{D}}^{\varepsilon,\gamma}\right) : [0, 3/\varepsilon] \to C^1([0,1]) \times C^1([0,1]) \times \mathbb{R} \times \mathbb{R}$ be such that (6.116) holds for $(h, v, s, D) = \left(h_{\bar{s},\bar{D}}^{\varepsilon,\gamma}, v_{\bar{s},\bar{D}}^{\varepsilon,\gamma}, s_{\bar{s},\bar{D}}^{\varepsilon,\gamma}, D_{\bar{s},\bar{D}}^{\varepsilon,\gamma}\right)$, $u = u_{\bar{s},\bar{D}}^{\varepsilon,\gamma}$ and
$$\left(h_{\bar{s},\bar{D}}^{\varepsilon,\gamma}(0,\cdot), v_{\bar{s},\bar{D}}^{\varepsilon,\gamma}(0,\cdot), s_{\bar{s},\bar{D}}^{\varepsilon,\gamma}(0), D_{\bar{s},\bar{D}}^{\varepsilon,\gamma}(0)\right) = (0,0,0,0).$$

From [**150**] one gets that

(6.144) $\quad h_{\bar{s},\bar{D}}^{\varepsilon,\gamma}(t,x) = -\dfrac{\gamma}{\varepsilon} f_{\bar{s},\bar{D}}(\varepsilon(t+x)) + \dfrac{\gamma}{\varepsilon} f_{\bar{s},\bar{D}}(\varepsilon(t+1-x)),$

(6.145) $\quad v_{\bar{s},\bar{D}}^{\varepsilon,\gamma}(t,x) = \dfrac{\gamma}{\varepsilon} f_{\bar{s},\bar{D}}(\varepsilon(t+x)) + \dfrac{\gamma}{\varepsilon} f_{\bar{s},\bar{D}}(\varepsilon(t+1-x))$
$\qquad\qquad\qquad - \dfrac{\gamma}{\varepsilon} f_{\bar{s},\bar{D}}(\varepsilon t) - \dfrac{\gamma}{\varepsilon} f_{\bar{s},\bar{D}}(\varepsilon(t+1)),$

(6.146) $\quad s_{\bar{s},\bar{D}}^{\varepsilon,\gamma}(t) = \dfrac{\gamma}{\varepsilon} f_{\bar{s},\bar{D}}(\varepsilon t) + \dfrac{\gamma}{\varepsilon} f_{\bar{s},\bar{D}}(\varepsilon(t+1)),$

(6.147) $\quad D_{\bar{s},\bar{D}}^{\varepsilon,\gamma}(t) = \dfrac{\gamma}{\varepsilon^2} F_{\bar{s},\bar{D}}(\varepsilon t) + \dfrac{\gamma}{\varepsilon^2} F_{\bar{s},\bar{D}}(\varepsilon(t+1)),$

with
$$F_{\bar{s},\bar{D}}(t) := \int_0^t f_{\bar{s},\bar{D}}(t_1)\mathrm{d}t_1.$$

In particular, using also (6.135) to (6.141), one gets

(6.148) $\quad h_{\bar{s},\bar{D}}^{\varepsilon,\gamma}\left(\dfrac{1}{\varepsilon}+t, x\right) = \gamma\left(\dfrac{1}{2}-x\right), \forall t \in \left[0, \dfrac{1-2\varepsilon}{2\varepsilon}\right], \forall x \in [0,1],$

(6.149) $\quad v_{\bar{s},\bar{D}}^{\varepsilon,\gamma}\left(\dfrac{1}{\varepsilon}+t, x\right) = 0, \forall t \in \left[0, \dfrac{1-2\varepsilon}{2\varepsilon}\right], \forall x \in [0,1],$

(6.150) $\quad s_{\bar{s},\bar{D}}^{\varepsilon,\gamma}\left(\dfrac{1}{\varepsilon}+t\right) = \dfrac{\gamma}{\varepsilon} + \dfrac{\gamma}{2} + \gamma t, \forall t \in \left[0, \dfrac{1-2\varepsilon}{2\varepsilon}\right],$

(6.151) $\quad D_{\bar{s},\bar{D}}^{\varepsilon,\gamma}\left(\dfrac{1}{\varepsilon}+t\right) = D_{\bar{s},\bar{D}}^{\varepsilon,\gamma}\left(\dfrac{1}{\varepsilon}\right) + \left(\dfrac{\gamma}{\varepsilon}+\dfrac{\gamma}{2}\right)t + \dfrac{\gamma}{2}t^2, \forall t \in \left[0, \dfrac{1-2\varepsilon}{2\varepsilon}\right],$

(6.152) $\quad h_{\bar{s},\bar{D}}^{\varepsilon,\gamma}\left(\dfrac{3}{\varepsilon}, x\right) = 0 \text{ and } v_{\bar{s},\bar{D}}^{\varepsilon,\gamma}\left(\dfrac{3}{\varepsilon}, x\right) = 0, \forall x \in [0,1],$

(6.153) $\quad s_{\bar{s},\bar{D}}^{\varepsilon,\gamma}\left(\dfrac{3}{\varepsilon}\right) = \dfrac{\gamma}{\varepsilon}\bar{s} \text{ and } D_{\bar{s},\bar{D}}^{\varepsilon,\gamma}\left(\dfrac{3}{\varepsilon}\right) = \dfrac{\gamma}{2\varepsilon}\bar{s} + \dfrac{\gamma}{\varepsilon^2}\bar{D}.$

Let $H_{\bar{s},\bar{D}}^{\varepsilon,\gamma} = 1 + h_{\bar{s},\bar{D}}^{\varepsilon,\gamma}$ and $Y_{\bar{s},\bar{D}}^{\varepsilon,\gamma} = \left(H_{\bar{s},\bar{D}}^{\varepsilon,\gamma}, v_{\bar{s},\bar{D}}^{\varepsilon,\gamma}, s_{\bar{s},\bar{D}}^{\varepsilon,\gamma}, D_{\bar{s},\bar{D}}^{\varepsilon,\gamma}\right)$. Let us introduce
$$a_{\varepsilon,\gamma} := \dfrac{\gamma}{\varepsilon} f_{\bar{s},\bar{D}}(1) + \dfrac{\gamma}{\varepsilon} f_{\bar{s},\bar{D}}(1+\varepsilon) = \dfrac{\gamma}{\varepsilon} + \dfrac{\gamma}{2},$$
$$b_{\varepsilon,\gamma}^{\bar{s},\bar{D}} := \dfrac{\gamma}{\varepsilon^2} F_{\bar{s},\bar{D}}(1) + \dfrac{\gamma}{\varepsilon^2} F_{\bar{s},\bar{D}}(1+\varepsilon) = D_{\bar{s},\bar{D}}^{\varepsilon,\gamma}\left(\dfrac{1}{\varepsilon}\right).$$

Using (6.119), (6.148), (6.149), (6.150) and (6.151), one gets

(6.154) $\quad Y_{\bar{s},\bar{D}}^{\varepsilon,\gamma}\left(\dfrac{1}{\varepsilon}\right) = Y^{\gamma, a_{\varepsilon,\gamma}, b_{\varepsilon,\gamma}^{\bar{s},\bar{D}}}(0),$

and, if $\varepsilon \in (0, 1/(2(T^*+1))]$,

(6.155) $\quad Y_{\bar{s},\bar{D}}^{\varepsilon,\gamma}\left(\dfrac{1}{\varepsilon}+T^*\right) = Y^{\gamma, a_{\varepsilon,\gamma}, b_{\varepsilon,\gamma}^{\bar{s},\bar{D}}}(T^*).$

The next proposition, which is proved in [**116**], shows that one can achieve (6.133) with $u = u_{\bar{s},\bar{D}}^{\varepsilon,\gamma}$ for suitable choices of T^0, ε and γ.

PROPOSITION 6.31. *There exists a constant $C_2 > 2$ such that, for every $\varepsilon \in (0, 1/C_2]$, for every $(\bar{s}, \bar{D}) \in \mathbb{D}$ and for every $\gamma \in [0, \varepsilon/C_2]$, there exists one and only one map $\tilde{Y}_{\bar{s},\bar{D}}^{\varepsilon,\gamma} : [0, 1/\varepsilon] \to \mathcal{Y}$ satisfying the two following conditions:*

$$(\tilde{Y}_{\bar{s},\bar{D}}^{\varepsilon,\gamma}, u_{\bar{s},\bar{D}}^{\varepsilon,\gamma}) \text{ is a trajectory of the control system } \Sigma \text{ on } \left[0, \frac{1}{\varepsilon}\right],$$

$$\tilde{Y}_{\bar{s},\bar{D}}^{\varepsilon,\gamma}(0) = (1, 0, 0, 0),$$

and furthermore this unique map $\tilde{Y}_{\bar{s},\bar{D}}^{\varepsilon,\gamma}$ verifies

(6.156) $$\left|\tilde{Y}_{\bar{s},\bar{D}}^{\varepsilon,\gamma}(t) - Y_{\bar{s},\bar{D}}^{\varepsilon,\gamma}(t)\right| \leqslant C_2 \varepsilon \gamma^2, \; \forall t \in \left[0, \frac{1}{\varepsilon}\right].$$

In particular, by (6.148) and (6.149),

(6.157) $$\left\|\tilde{v}_{\bar{s},\bar{D}}^{\varepsilon,\gamma}\left(\frac{1}{\varepsilon}\right)\right\|_{C^1([0,L])} + \left\|\tilde{h}_{\bar{s},\bar{D}}^{\varepsilon,\gamma}\left(\frac{1}{\varepsilon}\right) - \gamma\left(\frac{1}{2} - x\right)\right\|_{C^1([0,L])} \leqslant C_2 \varepsilon \gamma^2.$$

Similarly, one has the following proposition, which shows that (6.134) is achieved with $u = u_{\bar{s},\bar{D}}^{\varepsilon,\gamma}$ for suitable choices of T^1, ε and γ.

PROPOSITION 6.32. *There exists a constant $C_3 > 2(T^* + 1)$ such that, for every $\varepsilon \in (0, 1/C_3]$, for every $(\bar{s}, \bar{D}) \in \mathbb{D}$, and for every $\gamma \in [0, \varepsilon/C_3]$, there exists one and only one map $\hat{Y}_{\bar{s},\bar{D}}^{\varepsilon,\gamma} : [(1/\varepsilon) + T^*, 3/\varepsilon] \to \mathcal{Y}$ satisfying the two following conditions:*

$$\left(\hat{Y}_{\bar{s},\bar{D}}^{\varepsilon,\gamma}, u_{\bar{s},\bar{D}}^{\varepsilon,\gamma}\right) \text{ is a trajectory of the control system } \Sigma \text{ on } \left[\frac{1}{\varepsilon} + T^*, \frac{3}{\varepsilon}\right],$$

$$\hat{Y}_{\bar{s},\bar{D}}^{\varepsilon,\gamma}\left(\frac{3}{\varepsilon}\right) = \left(1, 0, \frac{\gamma}{\varepsilon}\bar{s}, \frac{\gamma}{2\varepsilon}\bar{s} + \frac{\gamma}{\varepsilon^2}\bar{D}\right) = Y_{\bar{s},\bar{D}}^{\varepsilon,\gamma}\left(\frac{3}{\varepsilon}\right),$$

and furthermore this unique map $\hat{Y}_{\bar{s},\bar{D}}^{\varepsilon,\gamma}$ verifies

(6.158) $$\left|\hat{Y}_{\bar{s},\bar{D}}^{\varepsilon,\gamma}(t) - Y_{\bar{s},\bar{D}}^{\varepsilon,\gamma}(t)\right| \leqslant C_3 \varepsilon \gamma^2, \; \forall t \in \left[\frac{1}{\varepsilon} + T^*, \frac{3}{\varepsilon}\right].$$

In particular, by (6.148) and (6.149),

(6.159) $$\left\|\hat{v}_{\bar{s},\bar{D}}^{\varepsilon,\gamma}\left(\frac{1}{\varepsilon}\right)\right\|_{C^1([0,1])} + \left\|\hat{h}_{\bar{s},\bar{D}}^{\varepsilon,\gamma}\left(\frac{1}{\varepsilon}\right) - \gamma\left(\frac{1}{2} - x\right)\right\|_{C^1([0,1])} \leqslant C_2 \varepsilon \gamma^2.$$

Let us choose

(6.160) $$\varepsilon := \text{Min}\left(\frac{1}{C_2}, \frac{1}{C_3}, \frac{\mu}{2C_2}, \frac{\mu}{2C_3}\right) \leqslant \frac{1}{2}.$$

Let us point out that there exists $C_4 > 0$ such that, for every $(\bar{s}, \bar{D}) \in \mathbb{D}$ and for every $\gamma \in [-\varepsilon, \varepsilon]$,

(6.161) $$\left\|H_{\bar{s},\bar{D}}^{\varepsilon,\gamma}\right\|_{C^2([0,3/\varepsilon]\times[0,1])} + \left\|v_{\bar{s},\bar{D}}^{\varepsilon,\gamma}\right\|_{C^2([0,3/\varepsilon]\times[0,1])} \leqslant C_4,$$

which, together with straightforward estimates, leads to the next proposition, whose proof is omitted.

PROPOSITION 6.33. *There exists $C_5 > 0$ such that, for every $(\bar{s}, \bar{D}) \in \mathbb{D}$, for every $Y^0 = (H^0, v^0, s^0, D^0) \in \mathcal{Y}$ with*

$$|Y^0 - Y_e| \leqslant \frac{1}{C_5}, \; s^0 = 0, \; D^0 = 0,$$

and for every $\gamma \in [0, \varepsilon/C_2]$, there exists one and only one $Y : [0, 1/\varepsilon] \to \mathcal{Y}$ such that

$$(Y, u^{\varepsilon,\gamma}_{\bar{s},\bar{D}} - H^0_x(0)) \text{ is a trajectory of the control system } \Sigma,$$
$$Y(0) = Y^0,$$

and this unique map Y satisfies

$$\left| Y(t) - \tilde{Y}^{\varepsilon,\gamma}_{\bar{s},\bar{D}}(t) \right| \leqslant C_5 |Y^0 - Y_e|, \; \forall t \in \left[0, \frac{1}{\varepsilon}\right].$$

Similarly, (6.161) leads to the following proposition.

PROPOSITION 6.34. *There exists $C_6 > 0$ such that, for every $(\bar{s}, \bar{D}) \in \mathbb{D}$, for every $\gamma \in [0, \varepsilon/C_3]$, and for every $Y^1 = (H^1, v^1, s^1, D^1) \in \mathcal{Y}$ such that*

$$\left| Y^1 - \left(1, 0, \frac{\gamma}{\varepsilon}\bar{s}, \frac{\gamma}{2\varepsilon}\bar{s} + \frac{\gamma}{\varepsilon^2}\bar{D}\right) \right| \leqslant \frac{1}{C_6}, \; s^1 = \frac{\gamma}{\varepsilon}\bar{s}, \; D^1 = \frac{\gamma}{2\varepsilon}\bar{s} + \frac{\gamma}{\varepsilon^2}\bar{D},$$

there exists one and only one $Y : [(1/\varepsilon) + T^*, 3/\varepsilon] \to \mathcal{Y}$ such that

$$(Y, u^{\varepsilon,\gamma}_{\bar{s},\bar{D}} - H^1_x(0)) \text{ is a trajectory of the control system } \Sigma,$$
$$Y(3/\varepsilon) = Y^1,$$

and this unique map Y satisfies

$$\left| Y(t) - \hat{Y}^{\varepsilon,\gamma}_{\bar{s},\bar{D}}(t) \right| \leqslant C_6 \left| Y^1 - Y^{\varepsilon,\gamma}_{\bar{s},\bar{D}}\left(\frac{3}{\varepsilon}\right) \right|, \; \forall t \in \left[\frac{1}{\varepsilon} + T^*, \frac{3}{\varepsilon}\right].$$

Finally, let us define

(6.162) $$T := \frac{3}{\varepsilon},$$

(6.163)
$$\eta := \text{Min}\left(\frac{\varepsilon^2 \mu}{2C_5(C_3^2 + C_2^2)}, \frac{\varepsilon^2 \mu}{2C_6(C_3^2 + C_2^2)}, \frac{\varepsilon}{C_2}, \frac{\varepsilon}{C_3}, \frac{1}{C_5}, \frac{1}{C_6}, \frac{\gamma_0^2 \mu}{2C_5}, \frac{\gamma_0^2 \mu}{2C_6}, \gamma_0\right).$$

We want to check that Theorem 6.24 holds with these constants for a large enough C_0. Let $Y^0 = (H^0, v^0, 0, 0) \in \mathcal{Y}$ and $Y^1 = (H^1, v^1, s^1, D^1) \in \mathcal{Y}$ be such that

(6.164) $$\|H^0 - 1\|_{C^1([0,1])} + \|v^0\|_{C^1([0,1])} \leqslant \eta,$$
(6.165) $$\|H^1 - 1\|_{C^1([0,1])} + \|v^1\|_{C^1([0,1])} \leqslant \eta,$$
(6.166) $$|s^1| + |D^1| \leqslant \eta.$$

Let

(6.167) $$\kappa_0 := \sqrt{\frac{2C_5}{\mu}} \sqrt{\|H^0 - 1\|_{C^1([0,L])} + \|v^0\|_{C^1([0,L])}},$$

(6.168) $$\kappa_1 := \sqrt{\frac{2C_6}{\mu}} \sqrt{\|H^1 - 1\|_{C^1([0,L])} + \|v^1\|_{C^1([0,L])}},$$

(6.169) $$\gamma := \text{Max}\left(\kappa_0, \kappa_1, |s^1| + |D^1|\right),$$

(6.170) $$\bar{s} := \frac{\varepsilon}{\gamma} s^1, \quad \bar{D} := \frac{\varepsilon^2}{\gamma}\left(D^1 - \frac{s^1}{2}\right),$$

so that, thanks to (6.153),

(6.171) $$s_{\bar{s},\bar{D}}^{\varepsilon,\gamma}\left(\frac{3}{\varepsilon}\right) = s^1, \quad D_{\bar{s},\bar{D}}^{\varepsilon,\gamma}\left(\frac{3}{\varepsilon}\right) = D^1.$$

Note that, by (6.160), (6.166), (6.169) and (6.170),

(6.172) $$(\bar{s}, \bar{D}) \in \mathbb{D}.$$

By (6.163), (6.164), (6.165), (6.166), (6.167), (6.168) and (6.169), we obtain that

(6.173) $$\gamma \in \left[0, \text{Min}\left(\frac{\varepsilon}{C_2}, \frac{\varepsilon}{C_3}\right)\right].$$

Then, by Proposition 6.33, (6.163), (6.164) and (6.173), there exists a function $Y_0 = (H_0, v_0, s_0, D_0) : [0, 1/\varepsilon] \to \mathcal{Y}$ such that

(6.174) $(Y_0, u_{\bar{s},\bar{D}}^{\varepsilon,\gamma} - H_x^0(0))$ is a trajectory of the control system Σ on $\left[0, \frac{1}{\varepsilon}\right]$,

(6.175) $$Y_0(0) = Y^0,$$

(6.176) $$\left|Y_0(t) - \tilde{Y}_{\bar{s},\bar{D}}^{\varepsilon,\gamma}(t)\right| \leqslant C_5|Y^0 - Y_e|, \quad \forall t \in \left[0, \frac{1}{\varepsilon}\right].$$

By (6.167), (6.169), and (6.176),

(6.177) $$\left|Y_0\left(\frac{1}{\varepsilon}\right) - \tilde{Y}_{\bar{s},\bar{D}}^{\varepsilon,\gamma}\left(\frac{1}{\varepsilon}\right)\right| \leqslant \frac{\mu\gamma^2}{2}.$$

By Proposition 6.31, (6.160) and (6.173),

$$\left|\tilde{Y}_{\bar{s},\bar{D}}^{\varepsilon,\gamma}\left(\frac{1}{\varepsilon}\right) - Y^{\gamma,a_\varepsilon,\gamma,b_{\varepsilon,\gamma}^{\bar{s},\bar{D}}}(0)\right| \leqslant C_2\varepsilon\gamma^2 \leqslant \frac{\mu\gamma^2}{2},$$

which, together with (6.177), gives

(6.178) $$\left|Y_0\left(\frac{1}{\varepsilon}\right) - Y^{\gamma,a_\varepsilon,\gamma,b_{\varepsilon,\gamma}^{\bar{s},\bar{D}}}\right| \leqslant \mu\gamma^2.$$

Similarly, by Propositions 6.32 and 6.34, (6.160), (6.162), (6.163), (6.165), (6.168), (6.169), (6.171) and (6.173), there exists $Y_1 = (H_1, v_1, s_1, D_1) : [(1/\varepsilon) + T^*, T] \to \mathcal{Y}$

such that
(6.179)

$(Y_1, u_{\bar{s},\bar{D}}^{\varepsilon,\gamma} - H_x^1(0))$ is a trajectory of the control system Σ on $[(1/\varepsilon) + T^*, T]$,

(6.180) $$Y_1(T) = Y^1,$$

(6.181) $$\left|Y_1(t) - \tilde{Y}_{\bar{s},\bar{D}}^{\varepsilon,\gamma}(t)\right| \leqslant C_6 \left|Y_1 - (1, 0, s^1, D^1)\right|, \forall t \in \left[\frac{1}{\varepsilon} + T^*, T\right],$$

(6.182) $$\left|Y_1\left(\frac{1}{\varepsilon} + T^*\right) - Y^{\gamma, a_\varepsilon, \gamma, b_{\varepsilon,\gamma}^{\bar{s},\bar{D}}}(T^*)\right| \leqslant \mu\gamma^2.$$

By (6.163), (6.164), (6.165), (6.166), (6.167), (6.168) and (6.169),

(6.183) $$\gamma \leqslant \gamma_0.$$

From Proposition 6.29, (6.178), (6.182) and (6.183), there exists a trajectory
$$(Y_*, u_*) : [0, T^*] \to \mathcal{Y}$$
of the control system Σ satisfying

(6.184) $$Y_*(0) = Y^0\left(\frac{1}{\varepsilon}\right),$$

(6.185) $$\left|Y_*(t) - Y^{\gamma, a_\varepsilon, \gamma, b_{\varepsilon,\gamma}^{\bar{s},\bar{D}}}(t)\right| \leqslant C_1 \mu\gamma, \forall t \in [0, T^*],$$

(6.186) $$Y_*(T^*) = Y^1\left(\frac{1}{\varepsilon} + T^*\right).$$

The map $(Y, u) : [0, T] \to \mathcal{Y}$ defined by

$$(Y(t), u(t)) = (Y_0(t), u_{\bar{s},\bar{D}}^{\varepsilon,\gamma}(t) - H_x^0(0)), \forall t \in \left[0, \frac{1}{\varepsilon}\right],$$

$$(Y(t), u(t)) = \left(Y_*\left(t - \frac{1}{\varepsilon}\right), u_*\left(t - \frac{1}{\varepsilon}\right)\right), \forall t \in \left[\frac{1}{\varepsilon}, \frac{1}{\varepsilon} + T^*\right],$$

$$(Y(t), u(t)) = (Y_1(t), u_{\bar{s},\bar{D}}^{\varepsilon,\gamma}(t) - H_x^1(0)), \forall t \in \left[\frac{1}{\varepsilon} + T^*, T\right],$$

is a trajectory of the control system Σ which, by (6.175) and (6.180), satisfies (6.84). Finally, the existence of $C_0 > 0$ such that (6.85) holds follows from the construction of (Y, u), (6.127), (6.144) to (6.147), (6.156), (6.158), (6.164), (6.165), (6.166), (6.167), (6.168), (6.169), (6.176), (6.181) and (6.185). This concludes our sketch of the proof of Theorem 6.24. ∎

Let us notice that, since $T^* > 2$, our method requires, in Theorem 6.24, the inequality $T > 2L/\sqrt{gH_e}$. It is natural to raise the following open problem.

OPEN PROBLEM 6.35. *Does Theorem 6.24 holds for every $T > 2L/\sqrt{gH_e}$? Let $T \leqslant 2L/\sqrt{gH_e}$. Is it true that Theorem 6.24 does not hold? Is it true that the steady-state motion problem (move from $(0, 0, 0, D^0)$ to $(0, 0, 0, D^1)$) is not possible for the time interval $[0, T]$ if $\|u\|_{C^0([0,T])}$ is small and $D^1 \neq D^0$? More precisely, does there exist $\varepsilon > 0$ such that, whatever $(D^0, D^1) \in \mathbb{R}^2$ with $D^0 \neq D^1$ is, there is no control $u \in C^0([0, T])$ such that $\|u\|_{C^0([0,T])} \leqslant \varepsilon$ steering the control system Σ from $(0, 0, 0, D^0)$ to $(0, 0, 0, D^1)$?*

Concerning this open problem, let us recall that, for the linearized control system Σ_0, $T > L/\sqrt{gH_e}$ is enough for the steady-state controllability (see [**150**]). See also Theorem 9.8 on page 251 and Section 9.3 for a related phenomenon.

CHAPTER 7

Quasi-static deformations

The quasi-static deformation method has in fact already been used for the controllability of the 1-D tank (see Section 6.3). Note that, in connection with our toy model $\mathfrak{T}$ of the 1-D tank, the linearized system of

$$\dot{x}_1 = x_2,\ \dot{x}_2 = -x_1 + \gamma,\ \dot{x}_3 = x_4,\ \dot{x}_4 = -x_3 + 2x_1 x_2$$

at the equilibrium $(\gamma, 0, 0, 0)$ is

$$\dot{y} = A_\gamma y,$$

with

(7.1) $$A_\gamma = \begin{pmatrix} 0 & 1 & 0 & 0 \\ -1 & 0 & 0 & 0 \\ 0 & 0 & 0 & 1 \\ 0 & 2\gamma & -1 & 0 \end{pmatrix}.$$

The eigenvalues of the matrix A_γ are i and $-i$, which have nonpositive real parts. This is why the quasi-static deformation is so easy to perform (even though 0 is in fact not stable for $\dot{y} = A_\gamma y$ if γ is not 0). If A_γ had eigenvalues of strictly positive real part, it would still be possible to perform in some cases the quasi-static deformation by stabilizing the equilibria by suitable feedbacks, as we are going to see in this chapter.

7.1. Description of the method

For the sake of simplicity, we explain the method in finite dimension. Let us consider our usual control system

(7.2) $$\dot{x} = f(x, u),$$

where the state is $x \in \mathbb{R}^n$, the control is $u \in \mathbb{R}^m$ and $f : \mathbb{R}^n \times \mathbb{R}^m \to \mathbb{R}^n$ is of class C^1. Let $(x^0, u^0) \in \mathbb{R}^n \times \mathbb{R}^m$ and $(x^1, u^1) \in \mathbb{R}^n \times \mathbb{R}^m$ be two equilibria of the control system (7.2), that is (see Definition 3.1 on page 125)

$$f(x^0, u^0) = 0,\ f(x^1, u^1) = 0.$$

We assume that (x^0, u^0) and (x^1, u^1) belong to the same C^1-path connected component of the set of zeros of f in $\mathbb{R}^n \times \mathbb{R}^m$. Our aim is to steer the control system (7.2) from x^0 to x^1 in some (possibly large) time $T > 0$. The method is divided into three steps:

First step: construction of an almost trajectory. Choose a C^1-map $\tau \in [0,1] \mapsto (\bar{x}(\tau), \bar{u}(\tau)) \in \mathbb{R}^n \times \mathbb{R}^m$ such that

$$(\bar{x}(0), \bar{u}(0)) = (x^0, u^0), \ (\bar{x}(1), \bar{u}(1)) = (x^1, u^1),$$
$$f(\bar{x}(\tau), \bar{u}(\tau)) = 0, \ \forall \tau \in [0,1].$$

Of course this map is not in general a trajectory of the control system (7.2), but, if $\varepsilon > 0$ is small enough, then the C^1-map

$$(x_\varepsilon, u_\varepsilon): \ [0, 1/\varepsilon] \ \to \ \mathbb{R}^n \times \mathbb{R}^m$$
$$t \ \mapsto \ (x_\varepsilon(t), u_\varepsilon(t)) := (\bar{x}(\varepsilon t), \bar{u}(\varepsilon t))$$

is "almost" a trajectory of the control system (7.2). Indeed, as $\varepsilon \to 0$,

$$\|\dot{x}_\varepsilon - f(x_\varepsilon, u_\varepsilon)\|_{C^0([0,1/\varepsilon];\mathbb{R}^n)} = O(\varepsilon).$$

Second step: stabilization procedure. This quasi-static almost trajectory is not in general stable, and thus has to be stabilized. To this aim, introduce the following change of variables:

$$z(t) = x(t) - x_\varepsilon(t),$$
$$v(t) = u(t) - u_\varepsilon(t),$$

where $t \in [0, 1/\varepsilon]$. In the new variables z, v, one gets the existence of $C > 0$ such that, if (7.2) holds and $|z(t)| + |v(t)| + \varepsilon$ is small enough, then

(7.3) $\quad |\dot{z}(t) - (A(\varepsilon t)z(t) + B(\varepsilon t)v(t))| \leqslant C(|z(t)|^2 + |v(t)|^2 + \varepsilon),$

where $t \in [0, 1/\varepsilon]$, and where

$$A(\tau) := \frac{\partial f}{\partial x}(\bar{x}(\tau), \bar{u}(\tau)) \quad \text{and} \quad B(\tau) := \frac{\partial f}{\partial u}(\bar{x}(\tau), \bar{u}(\tau)),$$

with $\tau = \varepsilon t \in [0, 1]$. Therefore we have to "stabilize" the origin for the *slowly-varying in time* linear control system

(7.4) $\quad \dot{z}(t) = A(\varepsilon t)z(t) + B(\varepsilon t)v(t).$

We refer to [**273**, Section 5.3] by Hassan Khalil for this classical problem. In particular, it is sufficient to asymptotically stabilize, for every $\tau \in [0,1]$, the origin for the control system $\dot{\tilde{z}} = A(\tau)\tilde{z} + B(\tau)v$ by some feedback $\tilde{z} \mapsto K_\tau \tilde{z}$, with $K_\tau \in \mathcal{M}_{m,n}(\mathbb{R})$ depending smoothly on τ, and then define $v = K_{\varepsilon t}z$ to "stabilize" the origin for the control system (7.4). (Here "stabilize" is not in the usual sense since $t \in [0, 1/\varepsilon]$ instead of $t \in [0, \infty)$.)

Third step: local controllability at the end. If the control system (7.2) is *locally controllable* (in small time see Definition 3.2 on page 125) or in a fixed time) at the equilibrium (x^1, u^1), we conclude that it is possible, if $\varepsilon > 0$ is small enough, to steer the control system (7.2) in finite time from the point $x(1/\varepsilon)$ to the desired target x^1.

REMARK 7.1. Of course the method of quasi-static deformations can be used in more general situations than controllability between equilibria. It can be used for the controllability between periodic trajectories (see, for example, [**40**, Theorem 11, page 926] for a Schrödinger equation) or even more complicated trajectories (see, for example, Proposition 6.31 on page 217 for Saint-Venant equations and page 260 or [**43**, Proposition 12] for a Schrödinger equation).

7.2. Application to a semilinear heat equation

Let us now explain how to apply this method on a specific control system modeled by a semilinear heat equation.

Let $L > 0$ be fixed and let $f : \mathbb{R} \to \mathbb{R}$ be a function of class C^2. Let us consider the nonlinear control system

(7.5)
$$\begin{cases} y_t = y_{xx} + f(y), \, t \in (0,T), \, x \in (0,L), \\ y(t,0) = 0, \, y(t,L) = u(t), \, t \in (0,T), \end{cases}$$

where at time $t \in [0,T]$, the state is $y(t,\cdot) : [0,L] \to \mathbb{R}$ and the control is $u(t) \in \mathbb{R}$.

Let us first define the notion of steady-state.

DEFINITION 7.2. *A function $y \in C^2([0,L])$ is a steady-state of the control system (7.5) if*
$$\frac{d^2 y}{dx^2} + f(y) = 0, \, y(0) = 0.$$

We denote by $\mathcal{S}$ the set of steady-states, endowed with the C^2 topology.

In other words, with Definition 3.1 on page 125, $y \in C^2([0,L])$ is a steady-state of the control system (7.5) if there exists a control $u_e \in \mathbb{R}$ such that (y, u_e) is an equilibrium of our control system (7.5).

Our goal is to show that one can move from any given steady-state to any other one belonging to the same connected component of the set of steady-states. To be more precise, let us define the notion of steady-state.

Let us also introduce the Banach space

(7.6)
$$Y_T = \Big\{ y : (0,T) \times (0,L) \to \mathbb{R}; y \in L^2((0,T); H^2(0,L)) \\ \text{and } y_t \in L^2((0,T) \times (0,L)) \Big\}$$

endowed with the norm
$$\|y\|_{Y_T} = \|y\|_{L^2((0,T); H^2(0,L))} + \|y_t\|_{L^2((0,T) \times (0,L))}.$$

Notice that Y_T is continuously imbedded in $L^\infty((0,T) \times (0,L))$.

The main result of this section is the following.

THEOREM 7.3 (**[132]**). *Let y^0 and y^1 be two steady-states belonging to the same connected component of $\mathcal{S}$. There exist a time $T > 0$ and a control function $u \in H^1(0,T)$ such that the solution $y(t,x)$ in Y_T of*

(7.7)
$$\begin{cases} y_t - y_{xx} = f(y), \, t \in (0,T), \, x \in (0,L), \\ y(t,0) = 0, \, y(t,L) = u(t), \, t \in (0,T), \\ y(0,x) = y^0(x), \, x \in (0,L), \end{cases}$$

satisfies $y(T,\cdot) = y^1(\cdot)$.

REMARK 7.4. In **[132]** we prove that, if y^0 and y^1 belong to distinct connected components of $\mathcal{S}$, then it is actually impossible to move either from y^0 to y^1 or from y^1 to y^0, whatever the time and the control are.

REMARK 7.5. For every $T > 0$ and every $u \in H^1(0,T)$, there is at most one solution of (7.7) in the Banach space Y_T.

REMARK 7.6. Concerning the global controllability problem, one of the main results in [**156**], by Caroline Fabre, Jean-Pierre Puel and Enrique Zuazua, asserts that, if f is globally Lipschitzian, then this control system is approximately globally controllable; see also [**243**] by Oleg Imanuvilov for (exact) controllability. When f is superlinear the situation is still widely open, in particular, because of possible blowing up. Indeed, it is well known that, if $yf(y) > 0$ as $y \neq 0$, then blow-up phenomena may occur for the Cauchy problem

(7.8) $$\begin{cases} y_t - y_{xx} = f(y), \, t \in (0,T), \, x \in (0,L), \\ y(t,0) = 0, \, y(t,L) = 0, \, t \in (0,T), \\ y(0,x) = y^0(x), \, x \in (0,L). \end{cases}$$

For example, if $f(y) = y^3$, then for numerous initial data there exists $T > 0$ such that the unique solution to the previous Cauchy problem is well defined on $[0,T) \times [0,L]$ and satisfies

$$\lim_{t \to T} \|y(t,\cdot)\|_{L^\infty(0,L)} = +\infty;$$

see for instance [**45, 84, 352, 507**] and references therein.

One may ask if, acting on the boundary of $[0,L]$, one could avoid the blow-up phenomenon. Actually, the answer to this question is negative in general, see the paper [**169**] by Enrique Fernández-Cara and Enrique Zuazua where it is proved that, for every $p > 2$, there exists some nonlinear function f satisfying

$$|f(y)| \sim |y| \log^p(1+|y|) \quad \text{as } |y| \to +\infty$$

such that, for every time $T > 0$, there exist initial data which lead to blow-up before time T, whatever the control function u is. Notice, however, that if

$$|f(y)| = o\left(|y| \log^{3/2}(1+|y|)\right) \quad \text{as } |y| \to +\infty,$$

then the blow-up (which could occur in the absence of control) can be avoided by means of boundary controls; see again [**169**]. The situation with $p \in [3/2, 2]$ is open as well as the case where $f(y) = -y^3$ (see Section 7.2.4 for this last case).

REMARK 7.7. This is a (partial) global exact controllability result. The time needed in our proof is large, but on the other hand, there are indeed cases where the time T of controllability cannot be taken arbitrarily small. For instance, in the case where $f(y) = -y^3$, any solution of (7.7) starting from 0 satisfies the inequality

$$\int_0^L (L-x)^4 y(T,x)^2 dx \leqslant 8LT,$$

and hence, if $y^0 = 0$, a minimal time is needed to reach a given $y^1 \neq 0$. This result is due to Alain Bamberger (see the thesis [**220**] by Jacques Henry); see also [**185**, Lemma 2.1].

In order to prove Theorem 7.3, we follow the steps described previously in Section 7.1.

7.2.1. Construction of an almost trajectory. The following lemma is obvious.

LEMMA 7.8. *Let $\phi_0, \phi_1 \in \mathcal{S}$. Then ϕ_0 and ϕ_1 belong to the same connected component of $\mathcal{S}$ if and only if, for any real number α between $\phi_0'(0)$ and $\phi_1'(0)$, the maximal solution of*

$$\frac{d^2y}{dx^2} + f(y) = 0, \ y(0) = 0, y'(0) = \alpha,$$

denoted by $y_\alpha(\cdot)$, is defined on $[0, L]$.

Now let y^0 and y^1 be in the same connected component of $\mathcal{S}$. Let us construct in $\mathcal{S}$ a C^1 path $(\bar{y}(\tau, \cdot), \bar{u}(\tau))$, $0 \leqslant \tau \leqslant 1$, joining y^0 to y^1. For each $i = 0, 1$, set

$$\alpha_i = y^{i\prime}(0).$$

Then with our previous notations: $y^i(\cdot) = y_{\alpha_i}(\cdot), i = 0, 1$. Now set

$$\bar{y}(\tau, x) = y_{(1-\tau)\alpha_0 + \tau\alpha_1}(x) \text{ and } \bar{u}(\tau) = \bar{y}(\tau, L),$$

where $\tau \in [0, 1]$ and $x \in [0, L]$. By construction we have

$$\bar{y}(0, \cdot) = y^0(\cdot), \ \bar{y}(1, \cdot) = y^1(\cdot) \text{ and } \bar{u}(0) = \bar{u}(1) = 0,$$

and thus $(\bar{y}(\tau, \cdot), \bar{u}(\tau))$ is a C^1 path in $\mathcal{S}$ connecting y^0 to y^1.

7.2.2. Stabilization procedure. This is the most technical part of the proof. Let $\varepsilon > 0$. We set $x \in [0, L]$,

(7.9) $$\begin{cases} z(t, x) := y(t, x) - \bar{y}(\varepsilon t, x), \ t \in [0, 1/\varepsilon], \ x \in [0, L], \\ v(t) := u(t) - \bar{u}(\varepsilon t), \ t \in [0, 1/\varepsilon]. \end{cases}$$

Then, from the definition of $(\bar{y}, \bar{u})$, we infer that z satisfies the initial-boundary problem

(7.10) $$\begin{cases} z_t = z_{xx} + f'(\bar{y})z + z^2 \int_0^1 (1-s) f''(\bar{y} + sz) ds - \varepsilon \bar{y}_\tau, \\ z(t, 0) = 0, \ z(t, L) = v(t), \\ z(0, x) = 0. \end{cases}$$

Now, in order to deal with a Dirichlet-type problem, we set

(7.11) $$w(t, x) := z(t, x) - \frac{x}{L} v(t), \ t \in [0, 1/\varepsilon], \ x \in [0, L],$$

and we suppose that the control v is differentiable. This leads to the equations

(7.12) $$\begin{cases} w_t = w_{xx} + f'(\bar{y})w + \frac{x}{L} f'(\bar{y})v - \frac{x}{L} v' + r(\varepsilon, t, x), \\ w(t, 0) = w(t, L) = 0, \\ w(0, x) = -\frac{x}{L} v(0), \end{cases}$$

where

(7.13) $$r(\varepsilon, t, x) = -\varepsilon \bar{y}_\tau + \left(w + \frac{x}{L} v\right)^2 \int_0^1 (1-s) f''\left(\bar{y} + s(w + \frac{x}{L} v)\right) ds,$$

and the next step is to prove that there exist ε small enough and a pair (v, w) solution of (7.12) such that $w(1/\varepsilon, \cdot)$ belongs to some arbitrary neighborhood of 0 in H_0^1-topology. To achieve this, one constructs an appropriate control function and a Lyapunov functional which stabilizes system (7.12) to 0. In fact, as we shall see, the control v will be chosen in $H^1(0, 1/\varepsilon)$ such that $v(0) = 0$.

In order to motivate what follows, let us first notice that if the residual term r and the control v were equal to zero, then (7.12) would reduce to
$$w_t = w_{xx} + f'(\bar{y})w,$$
$$w(t,0) = w(t,L) = 0.$$
This suggests introducing the *one-parameter family of linear operators*
(7.14) $$A(\tau) = \Delta + f'(\bar{y}(\tau,\cdot))\text{Id}, \quad \tau \in [0,1],$$
defined on $H^2(0,L) \cap H^1_0(0,L)$, Id denoting the identity map on $L^2(0,L)$. For every $\tau \in [0,1]$, $A(\tau)$ has compact resolvent (see page 95) and $A(\tau)^* = A(\tau)$. Hence the eigenvalues of A_τ are real and the Hilbert space $L^2(0,L)$ has a complete orthonormal system of eigenfunctions for the operator $A(\tau)$ (see, for example, [**71**, Théorème VI.11, page 97] or [**267**, page 277]).

Let $(e_j(\tau,\cdot))_{j \geqslant 1}$ be a Hilbertian basis of $L^2(0,L)$ consisting of eigenfunctions of $A(\tau)$ and let $(\lambda_j(\tau))_{j \geqslant 1}$ denote the corresponding eigenvalues. Of course, for $j \geqslant 1$ and each $\tau \in [0,1]$,
$$e_j(\tau,\cdot) \in H^1_0(0,L) \cap C^2([0,L]).$$
Note that, since the variable x is of dimension 1, the eigenvalues are simple (this is no longer true in higher dimension). We order these eigenvalues: for each $\tau \in [0,1]$,
$$-\infty < \ldots < \lambda_n(\tau) < \ldots < \lambda_1(\tau) \quad \text{and} \quad \lambda_n(\tau) \underset{n \to +\infty}{\to} -\infty.$$
A standard application of the min-max principle (see for instance [**401**, Theorem XIII.1, pages 76–77]) shows that these eigenfunctions and eigenvalues are C^1 functions of τ. We can define an integer n as the maximal number of eigenvalues taking at least a nonnegative value as $\tau \in [0,1]$, i.e., there exists $\eta > 0$ such that
(7.15) $$\forall t \in [0, 1/\varepsilon], \forall k > n, \lambda_k(\varepsilon t) < -\eta < 0.$$

REMARK 7.9. Note that the integer n can be arbitrarily large. For example, if $f(y) = y^3$, then
$$\left(\frac{dy^1}{dx}(0) \to +\infty\right) \Rightarrow (n \to +\infty).$$

We also set, for any $\tau \in [0,1]$ and $x \in [0,L]$,
$$a(\tau, x) = \frac{x}{L} f'(\bar{y}(\tau, x)) \quad \text{and} \quad b(x) = -\frac{x}{L}.$$
In these notations, system (7.12) leads to
(7.16) $$w_t(t,\cdot) = A(\varepsilon t)w(t,\cdot) + a(\varepsilon t, \cdot)v(t) + b(\cdot)v'(t) + r(\varepsilon, t, \cdot).$$
Any solution $w(t,\cdot) \in H^2(0,L) \cap H^1_0(0,L)$ of (7.16) can be expanded as a series in the eigenfunctions $e_j(\varepsilon t, \cdot)$, convergent in $H^1_0(0,L)$,
$$w(t,\cdot) = \sum_{j=1}^{\infty} w_j(t) e_j(\varepsilon t, \cdot).$$
In fact, the w_j's depend on ε and should be called, for example, w_j^ε. For simplicity we omit the index ε, and we shall also omit the index ε for other functions.

In what follows we are going to move, by means of an appropriate *feedback control*, the n first eigenvalues of the operator $A(\varepsilon t)$, without moving the others, in order to make all eigenvalues negative. This pole-shifting process is the first part

7.2. APPLICATION TO A SEMILINEAR HEAT EQUATION

of the stabilization procedure. It is related to the prior works [**483**] by Roberto Triggiani and [**433**] by Yoshiyuki Sakawa; see also [**426**, page 711] by David Russell.

For any $\tau \in [0,1]$, let $\pi_1(\tau)$ denote the orthogonal projection onto the subspace of $L^2(0,L)$ spanned by $e_1(\tau,\cdot), \ldots, e_n(\tau,\cdot)$, and let

$$(7.17) \qquad w^1(t) = \pi_1(\varepsilon t) w(t,\cdot) = \sum_{j=1}^n w_j(t) e_j(\varepsilon t, \cdot).$$

It is clear that, for every $\tau \in [0,1]$, the operators $\pi_1(\tau)$ and $A(\tau)$ commute, and moreover, for every $y \in L^2(0,L)$, we have

$$\pi_1'(\tau) y = \sum_{j=1}^n (y, e_j(\tau, \cdot))_{L^2(0,L)} \frac{\partial e_j}{\partial \tau}(\tau, \cdot) + \sum_{j=1}^n \left(y, \frac{\partial e_j}{\partial \tau}(\tau, \cdot) \right)_{L^2(0,L)} e_j(\tau, \cdot).$$

Hence, derivating (7.17) with respect to t, we get

$$\sum_{j=1}^n w_j'(t) e_j(\varepsilon t, \cdot) = \pi_1(\varepsilon t) w_t(t, \cdot) + \varepsilon \sum_{j=1}^n \left(w(t, \cdot), \frac{\partial e_j}{\partial \tau}(\varepsilon t, \cdot) \right)_{L^2(0,L)} e_j(\varepsilon t, \cdot).$$

On the other hand,

$$A(\varepsilon t) w^1(t) = \sum_{j=1}^n \lambda_j(\varepsilon t) w_j(t) e_j(\varepsilon t, \cdot),$$

and thus (7.16) yields

$$(7.18) \qquad \sum_{j=1}^n w_j'(t) e_j(\varepsilon t, \cdot) = \sum_{j=1}^n \lambda_j(\varepsilon t) w_j(t) e_j(\varepsilon t, \cdot) + \pi_1(\varepsilon t) a(\varepsilon t, \cdot) v(t) \\ + \pi_1(\varepsilon t) b(\cdot) v'(t) + r^1(\varepsilon, t, \cdot),$$

where

$$(7.19) \qquad r^1(\varepsilon, t, \cdot) = \pi_1(\varepsilon t) r(\varepsilon, t, \cdot) + \varepsilon \sum_{j=1}^n \left(w, \frac{\partial e_j}{\partial \tau}(\varepsilon t, \cdot) \right)_{L^2(0,L)} e_j(\varepsilon t, \cdot).$$

Let us give an upper bound for the residual term r^1. First, it is not difficult to check that there exists a constant C such that, if $|v(t)|$ and $\|w(t,\cdot)\|_{L^\infty(0,L)}$ are less than 1, then the inequality

$$\|r(\varepsilon, t, \cdot)\|_{L^\infty(0,L)} \leqslant C(\varepsilon + v(t)^2 + \|w(t,\cdot)\|_{L^\infty(0,L)}^2)$$

holds, where r is defined by (7.13). Therefore we easily get

$$\|r^1(\varepsilon, t, \cdot)\|_{L^\infty(0,L)} \leqslant C_1(\varepsilon + v(t)^2 + \|w(t,\cdot)\|_{L^\infty(0,L)}^2).$$

Moreover, since $H_0^1(0,L)$ is continuously imbedded in $C^0([0,L])$, we can assert that there exists a constant C_2 such that, if $|v(t)|$ and $\|w(t,\cdot)\|_{L^\infty(0,L)}$ are less than 1, then

$$(7.20) \qquad \|r^1(\varepsilon, t, \cdot)\|_{L^\infty(0,L)} \leqslant C_2(\varepsilon + v(t)^2 + \|w(t,\cdot)\|_{H_0^1(0,L)}^2).$$

Now projecting (7.18) on each e_i, $i = 1, \ldots, n$, one comes to

$$(7.21) \qquad w_i'(t) = \lambda_i(\varepsilon t) w_i(t) + a_i(\varepsilon t) v(t) + b_i(\varepsilon t) v'(t) + r_i^1(\varepsilon, t), \quad i = 1, \ldots, n,$$

where

$$r_i^1(\varepsilon, t) = (r^1(\varepsilon, t, \cdot), e_i(\varepsilon t, \cdot))_{L^2(0,L)},$$

(7.22) $$a_i(\varepsilon t) = (a(\varepsilon t, \cdot), e_i(\varepsilon t, \cdot))_{L^2(0,L)} = \frac{1}{L}\int_0^L x f'(\bar{y}(\varepsilon t, x)) e_i(\varepsilon t, x) dx,$$

$$b_i(\varepsilon t) = (b(\cdot), e_i(\varepsilon t, \cdot))_{L^2(0,L)} = -\frac{1}{L}\int_0^L x e_i(\varepsilon t, x) dx.$$

The n equations (7.21) form a differential system controlled by v, v'. Set

(7.23) $$\alpha(t) = v'(t),$$

and consider now $v(t)$ as a state and $\alpha(t)$ as a control. Then the former finite-dimensional system may be rewritten as

(7.24) $$\begin{cases} v' = \alpha, \\ w_1' = \lambda_1 w_1 + a_1 v + b_1 \alpha + r_1^1, \\ \vdots \\ w_n' = \lambda_n w_n + a_n v + b_n \alpha + r_n^1. \end{cases}$$

If we introduce the matrix notations

$$X_1(t) = \begin{pmatrix} v(t) \\ w_1(t) \\ \vdots \\ w_n(t) \end{pmatrix}, \quad R_1(\varepsilon, t) = \begin{pmatrix} 0 \\ r_1^1(\varepsilon, t) \\ \vdots \\ r_n^1(\varepsilon, t) \end{pmatrix},$$

$$A_1(\tau) = \begin{pmatrix} 0 & 0 & \cdots & 0 \\ a_1(\tau) & \lambda_1(\tau) & \cdots & 0 \\ \vdots & \vdots & \ddots & \vdots \\ a_n(\tau) & 0 & \cdots & \lambda_n(\tau) \end{pmatrix}, \quad B_1(\tau) = \begin{pmatrix} 1 \\ b_1(\tau) \\ \vdots \\ b_n(\tau) \end{pmatrix},$$

then equations (7.24) yield the *finite-dimensional linear control system*

(7.25) $$X_1'(t) = A_1(\varepsilon t) X_1(t) + B_1(\varepsilon t) \alpha(t) + R_1(\varepsilon, t).$$

Let us now prove the following lemma.

LEMMA 7.10. *For each $\tau \in [0,1]$, the pair $(A_1(\tau), B_1(\tau))$ satisfies the Kalman rank condition (Theorem 1.16 on page 9), i.e.,*

(7.26) $$\operatorname{rank}\bigl(B_1(\tau), A_1(\tau) B_1(\tau), \ldots, A_1(\tau)^{n-1} B_1(\tau)\bigr) = n.$$

Proof of Lemma 7.10. Let $\tau \in [0,1]$ be fixed. We compute directly

(7.27) $$\det\bigl(B_1, A_1 B_1, \ldots, A_1^{n-1} B_1\bigr) = \prod_{j=1}^n (a_j + \lambda_j b_j) \operatorname{VdM}(\lambda_1, \ldots, \lambda_n),$$

where $\operatorname{VdM}(\lambda_1, \ldots, \lambda_n)$ is a Vandermonde determinant, and thus is never equal to zero since the $\lambda_i(\tau), i = 1, \ldots, n$, are distinct, for any $\tau \in [0,1]$. On the other hand, using the fact that each $e_j(\tau,\cdot)$ is an eigenfunction of $A(\tau)$ and belongs to $H_0^1(0,L)$, we compute

$$\begin{aligned} a_j(\tau) + \lambda_j(\tau)b_j(\tau) &= \frac{1}{L}\int_0^L x\left(f'(\bar{y}(\tau,x))e_j(\tau,x) - \lambda_j(\tau)e_j(\tau,x)\right)dx \\ &= -\frac{1}{L}\int_0^L x\frac{\partial^2 e_j}{\partial x^2}(\tau,x)dx \\ &= -\frac{\partial e_j}{\partial x}(\tau,L). \end{aligned}$$

But this quantity is never equal to zero since $e_j(\tau,L) = 0$ and $e_j(\tau,\cdot)$ is a nontrivial solution of a linear second-order scalar differential equation. Therefore the determinant (7.27) is never equal to zero and we are done. ∎

As we shall recall below (Theorem 10.1 on page 275), the Kalman rank condition (7.26) implies a *pole-shifting* property. To be more precise, from (7.26), Theorem 10.1 on page 275 and Remark 10.5 on page 279, we get the following corollary.

COROLLARY 7.11. *For each $\tau \in [0,1]$, there exist scalars $k_0(\tau),\ldots,k_n(\tau)$ such that, if we let*

$$K_1(\tau) := (k_0(\tau),\ldots,k_n(\tau)),$$

then the matrix $A_1(\tau) + B_1(\tau)K_1(\tau)$ admits -1 as an eigenvalue with order $n+1$.

Moreover, the functions k_i, $i \in \{0,\ldots,n\}$, are of class C^1 on $[0,T]$ and therefore there exists a C^1 application $\tau \mapsto P(\tau)$ on $[0,1]$, where $P(\tau)$ is a $(n+1)\times(n+1)$ symmetric positive definite matrix, such that the identity

(7.28) $\quad P(\tau)\left(A_1(\tau) + B_1(\tau)K_1(\tau)\right) + \left(A_1(\tau) + B_1(\tau)K_1(\tau)\right)^{tr}P(\tau) = -\mathrm{Id}_{n+1}$

holds for any $\tau \in [0,1]$.

In fact, one has (see for example [**309**, page 198]),

$$P(\tau) = \int_0^\infty e^{t(A_1(\tau)+B_1(\tau)K_1(\tau))^{tr}}e^{t(A_1(\tau)+B_1(\tau)K_1(\tau))}dt.$$

We are now able to construct a control Lyapunov functional in order to stabilize system (7.16). Let $c > 0$ be chosen later on. For any $t \in [0, 1/\varepsilon]$, $v \in \mathbb{R}$ and $w \in H^2(0,L) \cap H_0^1(0,L)$, we set

(7.29) $\quad V(t,v,w) = c\,X_1(t)P(\varepsilon t)X_1(t) - \frac{1}{2}(w, A(\varepsilon t)w)_{L^2(0,L)},$

where $X_1(t)$ denotes the following column vector in $\mathbb{R}^{n+1}$:

$$X_1(t) := \begin{pmatrix} v \\ w_1(t) \\ \vdots \\ w_n(t) \end{pmatrix}$$

and

$$w_i(t) := (w, e_i(\varepsilon t, \cdot))_{L^2(0,L)}.$$

In particular, we have

(7.30) $$V(t,v,w) = c\, X_1(t) P(\varepsilon t) X_1(t) - \frac{1}{2}\sum_{j=1}^{\infty} \lambda_j(\varepsilon t) w_j(t)^2.$$

Let $V_1(t) := V(t, v(t), w(t))$. After some straightforward estimates (see [**132**]), one gets the existence of $\sigma > 0$ and of $\rho \in (0, \sigma]$ such that, for every $\varepsilon \in (0,1]$ and for every $t \in [0, 1/\varepsilon]$ such that $V_1(t) \leqslant \rho$,

$$V_1'(t) \leqslant \sigma \varepsilon^2.$$

Hence, since $V_1(0) = 0$, we get, if $\varepsilon \in (0, \rho/\sigma]$, that

$$V_1(t) \leqslant \sigma\varepsilon, \quad \forall t \in [0, 1/\varepsilon],$$

and in particular,

$$V_1\left(\frac{1}{\varepsilon}\right) \leqslant \sigma\varepsilon.$$

Coming back to definitions (7.9) and (7.11), we have

(7.31) $$\left\| y\left(\frac{1}{\varepsilon}, \cdot\right) - y^1(\cdot) \right\|_{H^1(0,L)} \leqslant \gamma\varepsilon,$$

where $y^1(\cdot) = \bar{y}(1,\cdot)$ is the final target and γ is a positive constant which does not depend on $\varepsilon \in (0, \rho/\sigma]$. This concludes the second step.

7.2.3. Local controllability at the end-point. The last step in the proof of Theorem 7.3 consists of solving a local exact controllability result: from (7.31), $y(1/\varepsilon, \cdot)$ belongs to an arbitrarily small neighborhood of y^1 in the H^1-topology if ε is small enough, and our aim is now to construct a trajectory $t \in [0, T] \mapsto (q(t,\cdot), u(t))$ of the control system (7.5) for some time $T > 0$ (for instance $T = 1$) such that

$$q(0, x) = y\left(\frac{1}{\varepsilon}, x\right),\ q(T, x) = y^1(x),\ x \in (0, L).$$

Existence of such a (q, u), with $u \in H^1(0,t)$ satisfying $u(0) = v(1/\varepsilon)$, is given by [**243**, Theorem 3.3], with $\tilde{y}(t,x) = y^1(x)$; see also [**242**]. Actually, in [**243**], the function f is assumed to be globally Lipschitzian, but the local result one needs here readily follows from the proofs and the estimates contained in [**243**]. The key step for this controllability result is an observability inequality similar to (2.398). This completes the proof of Theorem 7.3 on page 225. ∎

REMARK 7.12. One could also deduce Theorem 7.3 from [**243**, Theorem 3.3] and a straightforward compactness argument. But the method presented here leads to a control which can be easily implemented. See [**132**] for an example of numerical simulation.

REMARK 7.13. The previous method has also been applied in [**133**] for a semilinear wave equation and in [**437**] by Michael Schmidt and Emmanuel Trélat for Couette flows.

7.2.4. Open problems.
In this section we give open problems related to the previous results. We take
$$f(y) := y^3.$$
As mentioned above, it is one of the cases where, for numerous initial Cauchy data $y^0 \in H_0^1(0, L)$, there are T such that there is no solution $y \in Y_T$ of
$$(7.32) \qquad y_t - y_{xx} = y^3, \ (t, x) \in (0, T) \times (0, L),$$
satisfying the initial condition
$$(7.33) \qquad y(0, x) = y^0(x), \ x \in (0, L)$$
and the Dirichlet boundary condition
$$(7.34) \qquad y(t, 0) = y(t, L) = 0, \ t \in (0, +\infty).$$
One may ask, if *for this special nonlinearity*, one can avoid the blow-up phenomenon with a control at $x = 1$, that is:

OPEN PROBLEM 7.14. *Let $y^0 \in H^1(0, L)$ be such that $y^0(0) = 0$ and let $T > 0$. Does there exist $y \in Y_T$ satisfying the heat equation (7.32), the initial condition (7.33) and the boundary condition*
$$(7.35) \qquad y(t, 0) = 0, \ t \in (0, T)?$$

Note that this seems to be a case where one cannot use the proof of [**169**, Theorem 1.1] to get a negative answer to Open Problem 7.14. One could be even more "ambitious" and ask if the nonlinear heat equation (7.32) with a control at $x = L$ is exactly controllable in an arbitrary time, i.e., ask the following question.

OPEN PROBLEM 7.15. *Let $T > 0$. Let $y^0 \in H_0^1(0, L)$ satisfy $y^0(0) = 0$ and let $\hat{y} \in Y_T$ be a solution of the nonlinear heat equation (7.32) satisfying*
$$\hat{y}(t, 0) = 0, \ t \in (0, T).$$
Does there exist a solution $y \in Y_T$ of the nonlinear heat equation (7.32) on $(0, T) \times (0, L)$ such that
$$y(0, x) = y^0(x), \ x \in (0, L),$$
$$y(t, 0) = 0, \ t \in (0, T),$$
$$y(T, x) = \hat{y}(T, x), \ x \in (0, L)?$$

Let us recall (see Remark 7.7) that, if in (7.32) one replaces y^3 by $-y^3$, the answer to this question is negative.

One can also ask questions about the connectedness of $\mathcal{S}$ in higher dimension. Indeed, the proof of Theorem 7.3 given in [**132**] can be easily adapted to higher dimensions. Let us recall (see the proof of Proposition 3.1 in [**132**]) that, for every nonempty bounded open interval $\Omega \subset \mathbb{R}$,
$$\mathfrak{S}(\Omega) := \left\{ y \in C^2(\overline{\Omega}); \ -y'' = y^3 \right\} \subset C^2(\overline{\Omega})$$
is a connected subset of $C^2(\overline{\Omega})$. One may ask if the same holds in higher dimension. In particular:

OPEN PROBLEM 7.16. *Let Ω be a smooth bounded open set of $\mathbb{R}^2$. Let*
$$\mathfrak{S}(\Omega) := \left\{ y \in C^2(\overline{\Omega}); \ -\Delta y = y^3 \right\} \subset C^2(\overline{\Omega}).$$
Is $\mathfrak{S}(\Omega)$ a connected subset of $C^2(\overline{\Omega})$?

CHAPTER 8

Power series expansion

8.1. Description of the method

We again explain first the method on the control system of finite dimension

(8.1) $$\dot{x} = f(x, u),$$

where the state is $x \in \mathbb{R}^n$ and the control is $u \in \mathbb{R}^m$. Here f is a function of class C^∞ on a neighborhood of $(0,0) \in \mathbb{R}^n \times \mathbb{R}^m$ and we assume that $(0,0) \in \mathbb{R}^n \times \mathbb{R}^m$ is an equilibrium of the control system (8.1), i.e $f(0,0) = 0$. Let

$$H := \mathrm{Span}\,\{A^i B u;\, u \in \mathbb{R}^m,\, i \in \{0, \ldots, n-1\}\}$$

with

$$A := \frac{\partial f}{\partial x}(0,0),\ B := \frac{\partial f}{\partial u}(0,0).$$

If $H = \mathbb{R}^n$, the linearized control system around $(0,0)$ is controllable (Theorem 1.16 on page 9) and therefore the nonlinear control system (8.1) is small-time locally controllable at $(0,0) \in \mathbb{R}^n \times \mathbb{R}^m$ (Theorem 3.8 on page 128). Let us look at the case where the dimension of H is $n-1$. Let us make a (formal) power series expansion of the control system (8.1) in (x, u) around the constant trajectory $t \mapsto (0,0) \in \mathbb{R}^n \times \mathbb{R}^m$. We write

$$x = y^1 + y^2 + \ldots,\ u = v^1 + v^2 + \ldots.$$

The order 1 is given by (y^1, v^1); the order 2 is given by (y^2, v^2) and so on. The dynamics of these different orders are given by

(8.2) $$\dot{y}^1 = \frac{\partial f}{\partial x}(0,0) y^1 + \frac{\partial f}{\partial u}(0,0) v^1,$$

(8.3) $$\dot{y}^2 = \frac{\partial f}{\partial x}(0,0) y^2 + \frac{\partial f}{\partial u}(0,0) v^2 + \frac{1}{2} \frac{\partial^2 f}{\partial x^2}(0,0)(y^1, y^1)$$
$$+ \frac{\partial^2 f}{\partial x \partial u}(0,0)(y^1, v^1) + \frac{1}{2} \frac{\partial^2 f}{\partial u^2}(0,0)(v^1, v^1),$$

and so on. Let $e_1 \in H^\perp$. Let $T > 0$. Let us assume that there are controls $v^1_\pm$ and $v^2_\pm$, both in $L^\infty((0,T); \mathbb{R}^m)$, such that, if $y^1_\pm$ and $y^2_\pm$ are solutions of

$$\dot{y}^1_\pm = \frac{\partial f}{\partial x}(0,0) y^1_\pm + \frac{\partial f}{\partial u}(0,0) v^1_\pm,$$
$$y^1_\pm(0) = 0,$$

$$\dot{y}_\pm^2 = \frac{\partial f}{\partial x}(0,0)y_\pm^2 + \frac{\partial f}{\partial u}(0,0)v_\pm^2 + \frac{1}{2}\frac{\partial^2 f}{\partial x^2}(0,0)(y_\pm^1, y_\pm^1)$$
$$+ \frac{\partial^2 f}{\partial x \partial u}(0,0)(y_\pm^1, u_\pm^1) + \frac{1}{2}\frac{\partial^2 f}{\partial u^2}(0,0)(u_\pm^1, u_\pm^1),$$
$$y_\pm^2(0) = 0,$$

then
$$y_\pm^1(T) = 0,$$
$$y_\pm^2(T) = \pm e_1.$$

Let $(e_i)_{i \in \{2,\ldots n\}}$ be a basis of H. By the definition of H and a classical result about the controllable part of a linear system (see e.g. [**458**, Section 3.3]), there are $(u_i)_{i=2,\ldots,n}$, all in $L^\infty(0,T)^m$, such that, if $(x_i)_{i=2,\ldots,n}$ are the solutions of

$$\dot{x}_i = \frac{\partial f}{\partial x}(0,0)x_i + \frac{\partial f}{\partial u}(0,0)u_i,$$
$$x_i(0) = 0,$$

then, for every $i \in \{2,\ldots,n\}$,
$$x_i(T) = e_i.$$

Now let
$$b = \sum_{i=1}^n b_i e_i$$

be a point in $\mathbb{R}^n$. Let $v^1 \in L^\infty((0,T); \mathbb{R}^m)$ be defined by the following
- If $b_1 \geqslant 0$, then $v^1 := v_+^1$ and $v^2 := v_+^2$.
- If $b_1 < 0$, then $v^1 := v_-^1$ and $v^2 := v_-^2$.

Then let $u : (0,T) \to \mathbb{R}^m$ be defined by
$$u(t) := |b_1|^{1/2} v^1(t) + |b_1| v^2(t) + \sum_{i=2}^n b_i u_i(t).$$

Let $x : [0,T] \to \mathbb{R}^n$ be the solution of
$$\dot{x} = f(x, u(t)), \; x(0) = 0.$$

Then one has, as $b \to 0$,
(8.4) $$x(T) = b + o(b).$$

Hence, using the Brouwer fixed-point theorem (see Theorem B.15 on page 390) and standard estimates on ordinary differential equations, one gets the local controllability of $\dot{x} = f(x,u)$ (around $(0,0) \in \mathbb{R}^n \times \mathbb{R}^m$) in time T, that is, for every $\varepsilon > 0$, there exists $\eta > 0$ such that, for every $(a,b) \in \mathbb{R}^n \times \mathbb{R}^n$ with $|a| < \eta$ and $|b| < \eta$, there exists a trajectory $(x,u) : [0,T] \to \mathbb{R}^n \times \mathbb{R}^m$ of the control system (8.1) such that
$$x(0) = a, \; x(T) = b,$$
$$|u(t)| \leqslant \varepsilon, \; t \in (0,T).$$

(One can also get this local controllability result from (8.4) by using degree theory as in the proof of Theorem 11.26 on page 301.)

This method is classical to study the local controllability of control systems in finite dimension. In fact, more subtle tools have been introduced: for example different scalings on the components of the control and of the state as well as scaling on time. Furthermore, the case where the codimension of H is strictly larger than 1 has been considered; see e.g. [**5, 55, 57, 56, 226, 228, 270, 468, 469, 482**] and the references therein.

In the next section we apply this method to a partial differential equation, namely the Korteweg-de Vries equation. In fact, for this equation, an expansion to order 2 is not sufficient: we obtain the local controllability by means of an expansion to order 3.

8.2. Application to a Korteweg-de Vries equation

Let $L > 0$. Let us consider, as in Section 4.1, the following Korteweg-de Vries (KdV) control system

(8.5) $$y_t + y_x + y_{xxx} + yy_x = 0, \ t \in (0,T), \ x \in (0,L),$$

(8.6) $$y(t,0) = y(t,L) = 0, \ y_x(t,L) = u(t), \ t \in (0,T),$$

where, at time $t \in [0,T]$, the control is $u(t) \in \mathbb{R}$ and the state is $y(t,\cdot) : (0,L) \to \mathbb{R}$.

Let us recall that the well-posedness of the Cauchy problem

$$y_t + y_x + y_{xxx} + yy_x = 0, \ t \in (0,T), \ x \in (0,L),$$
$$y(t,0) = y(t,L) = 0, y_x(t,L) = u(t), \ t \in (0,T),$$
$$y(0,\cdot) = y^0(x),$$

for $u \in L^2(0,T)$, has been established in Section 4.1.1 (see, in particular, Definition 4.1 on page 160 and Theorem 4.2 on page 161).

We are interested in the local controllability of the control system (8.5)-(8.6) around the equilibrium $(y_e, u_e) := (0,0)$. Let us recall (see Theorem 4.3 on page 161) that this local controllability holds if

(8.7) $$L \notin \mathcal{N} := \left\{ 2\pi\sqrt{\frac{j^2 + l^2 + jl}{3}}; \ j \in \mathbb{N}\setminus\{0\}, \ l \in \mathbb{N}\setminus\{0\} \right\}.$$

The aim of this section is to study the local exact controllability around the equilibrium $(y_e, u_e) := (0,0) \in L^2(0,L) \times \mathbb{R}$ of the nonlinear KdV equation when $L = 2k\pi \in \mathcal{N}$ (take $j = l = k$ in (4.12)). The main theorem is the following one.

THEOREM 8.1 ([**122**, Theorem 2]). *Let $T > 0$ and $k \in \mathbb{N}\setminus\{0\}$. Let us assume that*

(8.8) $$\left(j^2 + l^2 + jl = 3k^2 \text{ and } (j,l) \in (\mathbb{N}\setminus\{0\})^2\right) \Rightarrow (j = l = k).$$

Let $L = 2k\pi$. (Thus, in particular, $L \in \mathcal{N}$.) Then there exist $C > 0$ and $r_1 > 0$ such that for any $y^0, y^1 \in L^2(0,L)$, with $\|y^0\|_{L^2(0,L)} < r_1$ and $\|y^1\|_{L^2(0,L)} < r_1$, there exist

$$y \in C^0([0,T]; L^2(0,L)) \cap L^2((0,T); H^1(0,L))$$

and $u \in L^2(0,T)$ satisfying (8.5)-(8.6), such that

(8.9) $$y(0,\cdot) = y^0,$$
(8.10) $$y(T,\cdot) = y^1,$$
(8.11) $$\|u\|_{L^2(0,T)} \leqslant C(\|y^0\|_{L^2(0,L)} + \|y^1\|_{L^2(0,L)})^{1/3}.$$

REMARK 8.2. Assumption (8.8) holds for $k \leqslant 6$ and for an infinite number of positive integers (see Proposition 8.3 below). However, there are also infinitely many positive integers k such that (8.8) does not hold: note that
$$x^2 + xy + y^2 = 3z^2$$
if
$$x := a^2 + 4ab + b^2,\ y := a^2 - 2ab - 2b^2,\ z := a^2 + b^2 + ab,$$
and that $x > 0$, $y > 0$ and $z > 0$ if $a > 0$ and
$$(-2 + \sqrt{3})a < b < \frac{-1 + \sqrt{3}}{2}a.$$
Unfortunately, we have forgotten assumption (8.8) in [**122**, Theorem 2]. This is a mistake: the proof of [**122**, Theorem 2] requires (8.8).

PROPOSITION 8.3. *There are infinitely many positive integers k satisfying (8.8).*

Proof of Proposition 8.3. Let $k > 2$ be a prime integer. Let us assume that there are positive integers x and y such that

(8.12) $\qquad (x, y) \neq (k, k),$

(8.13) $\qquad x^2 + xy + y^2 = 3k^2.$

Let us prove that

(8.14) $\qquad -3$ is a square in $\mathbb{Z}/k\mathbb{Z}$,

that is, there exists $p \in \mathbb{Z}$ such that $3 + p^2$ is divisible by k. Indeed, from (8.13), one gets

(8.15) $\qquad -3xy \equiv (x - y)^2 \,(\mathrm{mod}\ k),$

(8.16) $\qquad xy \equiv (x + y)^2 \,(\mathrm{mod}\ k).$

Note that

(8.17) $\qquad x \not\equiv 0 \,(\mathrm{mod}\ k).$

Indeed, let us assume that

(8.18) $\qquad x \equiv 0 \,(\mathrm{mod}\ k).$

From (8.16) and (8.18), one gets that
$$x + y \equiv 0 \,(\mathrm{mod}\ k),$$
which, together with (8.18), implies that

(8.19) $\qquad x \equiv 0 \,(\mathrm{mod}\ k),\ y \equiv 0 \,(\mathrm{mod}\ k).$

However, (8.12) and (8.19) implies that
$$x^2 + xy + y^2 > 3k^2,$$
in contradiction with (8.13). Hence (8.17) holds. Similarly,

(8.20) $\qquad y \not\equiv 0 \,(\mathrm{mod}\ k).$

From (8.16), (8.17) and (8.20), one gets that

(8.21) $\qquad x + y \not\equiv 0 \,(\mathrm{mod}\ k).$

Finally, from (8.15), (8.16) and (8.21), we have
$$-3 = \left((x+y)^{-1}(x-y)\right)^2 \text{ in } \mathbb{Z}/k\mathbb{Z},$$
which shows (8.14).

Let us prove, following [444, 445], that the set of prime integers k not satisfying condition (8.14) is infinite, and more precisely, its density inside the set of prime integers is equal to $1/2$.

First of all, let us introduce the Legendre symbol $\left(\frac{x}{k}\right)$, where k is a prime and $x \in \mathbb{Z}$ is an integer not divisible by k. We put
$$\begin{cases} \left(\frac{x}{k}\right) := 1, & \text{if } x \text{ is a square modulo } k, \\ \left(\frac{x}{k}\right) := -1, & \text{if } x \text{ is not a square modulo } k. \end{cases}$$

Note that

(8.22) $$\left(\frac{xy}{k}\right) = \left(\frac{x}{k}\right)\left(\frac{y}{k}\right),$$

for $x, y \in \mathbb{Z}$ coprime to k. Indeed, if one of x or y is a square modulo k, this is obvious. Otherwise, note that the subgroup $G \subset (\mathbb{Z}/k\mathbb{Z})^*$ consisting of squares has the property that the quotient $(\mathbb{Z}/k\mathbb{Z})^*/G$ has order 2, because G is the image of the group morphism $x \mapsto x^2$ from $(\mathbb{Z}/k\mathbb{Z})^*$ into itself, and this group morphism has for kernel the order 2 group $\{1, -1\}$. Thus if neither x nor y is a square modulo k, the classes of x and y modulo G are equal, and their product modulo k belongs to G.

If $p > 2$ and $k > 2$ are two distinct prime integers, the quadratic reciprocity law due to Gauss (see, e.g., [444, Theorem 6, Chapter I, page 7] or [445, Théorème 6, Chapitre I, page 16]) says the following:

(8.23) $$\left(\frac{p}{k}\right) = \left(\frac{k}{p}\right)(-1)^{\varepsilon(p)\varepsilon(k)},$$

where
$$\begin{cases} \varepsilon(p) = 0, & \text{if } p = 1 \,(\text{mod } 4), \\ \varepsilon(p) = 1, & \text{if } p = -1 \,(\text{mod } 4). \end{cases}$$

Let us now prove the following lemma.

LEMMA 8.4. *Let $k > 2$ be a prime integer. One has*

(8.24) $$(-1)^{\varepsilon(k)} = \left(\frac{-1}{k}\right).$$

Proof of Lemma 8.4. Let us consider the multiplicative group $(\mathbb{Z}/k\mathbb{Z})^*$. The nonzero squares in $\mathbb{Z}/k\mathbb{Z}$ are exactly the elements of the subgroup $G \subset (\mathbb{Z}/k\mathbb{Z})^*$ defined as the image of the group morphism $x \mapsto x^2$. This group G has cardinality $(k-1)/2$. Indeed, the kernel of the morphism above consists of the two elements $1, -1$, which are the only solutions of the polynomial equation $x^2 = 1$ in $\mathbb{Z}/k\mathbb{Z}$.

Note that -1 is a square modulo k if and only if $-1 \in G$, and that if this is the case, multiplication by -1 induces a bijection of G. Now we introduce
$$\alpha := \prod_{g \in G} g.$$

Then, if $-1 \in G$, we have

$$(-1)^{\frac{k-1}{2}}\alpha = \prod_{g \in G}(-g) = \alpha$$

and thus $(-1)^{\frac{k-1}{2}} = 1$, i.e., $\varepsilon(k) = 0$. Hence (8.24) holds in this case.

If, on the other hand, -1 is not a square in $\mathbb{Z}/k\mathbb{Z}$, we find that multiplication by -1 induces a bijection between G and its complement, denoted ^cG. Thus we get in this case

$$(-1)^{\frac{k-1}{2}}\alpha = \prod_{g \in G}(-g) = \prod_{g \in {^cG}} g.$$

Observe to conclude that in $(\mathbb{Z}/k\mathbb{Z})^*$ we have the relation

(8.25) $$\prod_{g \in (\mathbb{Z}/k\mathbb{Z})^*} g = -1,$$

which implies that

$$\prod_{g \in (\mathbb{Z}/k\mathbb{Z})^* \setminus G} g = -\prod_{g \in G} g.$$

Thus, in the second case, we get $(-1)^{(k-1)/2}\alpha = -\alpha$ and hence $(-1)^{(k-1)/2} = -1$. Therefore $\varepsilon(k) = 1$ and (8.24) holds in this case also.

Relation (8.25) is proved by noticing that in this product, we find the terms 1, -1, and that all the other terms $g \notin \{1, -1\}$ appear in pairs $\{g, g^{-1}\}$, with $g \neq g^{-1}$. This concludes the proof of Lemma 8.4. ∎

We return to the proof of Proposition 8.3. From (8.24), (8.23) and (8.22), we have

(8.26) $$\left(\frac{-p}{k}\right) = \left(\frac{k}{p}\right)(-1)^{(\varepsilon(p)+1)\varepsilon(k)}.$$

Let us apply this to $p = 3$ and $k > 3$ (we recall that k is a prime integer). We have $\varepsilon(3) = 1$. By definition

$$\left(\frac{-3}{k}\right) = 1$$

if and only if -3 is a square modulo k. By (8.26), this is equivalent to

$$\left(\frac{k}{3}\right) = 1,$$

that is, $k \equiv 1 \pmod 3$.

In conclusion, if $k > 3$ is a prime integer such that $k \equiv -1 \pmod 3$, then k satisfies (8.8). As there are two possible nonzero congruences modulo 3, the Dirichlet density theorem (see, e.g., [**444**, Theorem 2, Chapter VI, page 73] or [**445**, Théorème 2, Chapitre VI, page 122]) now says that the set of prime integers k such that $k \equiv -1 \pmod 3$ has density equal to $1/2$ in the set of prime integers. In particular, there are infinitely many prime integers k such that $k \equiv -1 \pmod 3$. This concludes the proof of Proposition 8.3. ∎

8.2. APPLICATION TO A KORTEWEG-DE VRIES EQUATION

Proof of Theorem 8.1. We assume that $L = 2k\pi$. The linearized control system of the nonlinear control system (8.5)-(8.6) around $(y_e, u_e) := (0,0) \in L^2(0, 2k\pi) \times \mathbb{R}$ is

$$(\text{KdVL}) \begin{cases} y_t + y_x + y_{xxx} = 0, \ t \in (0,T), \ x \in (0, 2k\pi), \\ y(t,0) = y(t, 2k\pi) = 0, \ y_x(t, 2k\pi) = 0, \ t \in (0,T). \end{cases}$$

As we have seen previously (see Theorem 2.25 on page 42), this linear control system is not controllable. Indeed, for every solution

$$(y, u) \in C^0([0,T]; L^2(0, 2k\pi)) \times L^2(0,T)$$

of (KdVL), one has, using simple integrations by parts and density arguments (see also (2.181)), that

$$\frac{\mathrm{d}}{\mathrm{d}t} \int_0^{2k\pi} (1 - \cos(x)) y \, dx = 0.$$

To prove that the nonlinear term yy_x gives the local controllability, a first idea could be to use the exact controllability of the linearized equation around nontrivial stationary solutions proved by Emmanuelle Crépeau in [**136**] and to apply the method we have used to get the local controllability of the 1-D tank (that is, use the return method together with quasi-static deformations; see Section 6.3). However, with this method, we could only obtain the local exact controllability in *large* time. In order to prove Theorem 8.1, we use a different strategy that we have briefly described in the above section (Section 8.1).

We first point out that, in this theorem, we may assume that $y^0 = 0$: this follows easily from the invariance of the control system (8.5)-(8.6) by the change of variables $\tau = T - t$, $\xi = 2k\pi - x$ (and a suitable change for the control). Then we use the following result for the linear control system, due to Lionel Rosier.

THEOREM 8.5 ([**407**]). *Let $T > 0$ and*

$$(8.27) \qquad H := \left\{ y \in L^2(0, 2k\pi), \int_0^{2k\pi} y(1 - \cos(x)) dx = 0 \right\}.$$

Let us assume that $k \in \mathbb{N} \setminus \{0\}$ is such that (8.8) holds. Then, for every $(y^0, y^1) \in H \times H$, there exist

$$y \in C^0([0,T]; L^2(0, 2k\pi)) \cap L^2((0,T); H^1(0, 2k\pi)), \ u \in L^2(0,T),$$

satisfying $(KdVL)$, such that $y(0, \cdot) = y^0$ and $y(T, \cdot) = y^1$.

Next, as we are going to see, the nonlinear term yy_x allows us to "move" in the two (oriented) directions $\pm(1 - \cos(x))$ which are missed by the linearized control system (KdVL). Finally, we briefly sketch how to derive Theorem 8.1 from these motions by means of a fixed-point theorem. Let us now explain how the nonlinear term yy_x allows us to "move" in the two (oriented) directions $\pm(1 - \cos(x))$. Let $L > 0$. We first recall some properties about the following linear KdV Cauchy problem

$$(8.28) \qquad\qquad y_t + y_x + y_{xxx} = f,$$
$$(8.29) \qquad\qquad y(t,0) = y(t,L) = 0,$$
$$(8.30) \qquad\qquad y_x(t,L) = h(t),$$
$$(8.31) \qquad\qquad y(T_0, x) = y^0(x).$$

It follows from what we have seen in Section 4.1.2 that, for every $y^0 \in L^2(0, L)$, for every $u \in L^2(T_0, T_1)$ and for every $f \in L^1((T_0, T_1); L^2(0, L))$, there exists a unique solution $y \in C^0([T_0, T_1]; L^2(0, L))$ of (8.28) to (8.31) (see page 162 for the definition of a solution of (8.28) to (8.31)). Let us also recall, as we have seen in the same section, that

$$y \in B_{T_0,T_1} := C^0([T_0, T_1]; L^2(0, L)) \cap L^2((T_0, T_1); H^1(0, L));$$

see the paper [**407**] by Lionel Rosier for complete details.

In [**122**] we proved the following proposition.

PROPOSITION 8.6. *Let $T > 0$ and let $k \in \mathbb{N} \setminus \{0\}$, satisfying (8.8). Then there exists $(u_\pm, v_\pm, w_\pm)$ in $L^2(0, T)^3$ such that, if $\alpha_\pm, \beta_\pm, \gamma_\pm$ are respectively the solutions of*

(8.32) $$\alpha_{\pm t} + \alpha_{\pm x} + \alpha_{\pm xxx} = 0,$$
(8.33) $$\alpha_\pm(t, 0) = \alpha_\pm(t, 2k\pi) = 0,$$
(8.34) $$\alpha_{\pm x}(t, 2k\pi) = u_\pm(t),$$
(8.35) $$\alpha_\pm(0, x) = 0,$$

(8.36) $$\beta_{\pm t} + \beta_{\pm x} + \beta_{\pm xxx} = -\alpha_\pm \alpha_{\pm x},$$
(8.37) $$\beta_\pm(t, 0) = \beta_\pm(t, 2k\pi) = 0,$$
(8.38) $$\beta_{\pm x}(t, 2k\pi) = v_\pm(t),$$
(8.39) $$\beta_\pm(0, x) = 0,$$

(8.40) $$\gamma_{\pm t} + \gamma_{\pm x} + \gamma_{\pm xxx} = -(\alpha_\pm \beta_\pm)_x,$$
(8.41) $$\gamma_\pm(t, 0) = \gamma_\pm(t, 2k\pi) = 0,$$
(8.42) $$\gamma_{\pm x}(t, 2k\pi) = w_\pm(t),$$
(8.43) $$\gamma_\pm(0, x) = 0,$$

then

(8.44) $$\alpha_\pm(T, x) = 0, \ \beta_\pm(T, x) = 0 \ \text{and} \ \gamma_\pm(T, x) = \pm(1 - \cos(x)).$$

REMARK 8.7. It would have been quite natural to look for the existence of $(u_\pm, v_\pm)$ in $L^2(0, T)^2$ such that, if $\alpha_\pm, \beta_\pm$ are respectively the solutions of

$$\alpha_{\pm t} + \alpha_{\pm x} + \alpha_{\pm xxx} = 0,$$
$$\alpha_\pm(t, 0) = \alpha_\pm(t, 2k\pi) = 0,$$
$$\alpha_{\pm x}(t, 2k\pi) = u_\pm(t),$$
$$\alpha_\pm(0, x) = 0,$$

$$\beta_{\pm t} + \beta_{\pm x} + \beta_{\pm xxx} = -\alpha_\pm \alpha_{\pm x},$$
$$\beta_\pm(t, 0) = \beta_\pm(t, 2k\pi) = 0,$$
$$\beta_{\pm x}(t, 2k\pi) = v_\pm(t),$$
$$\beta_\pm(0, x) = 0,$$

then
$$\alpha_{\pm}(T,x) = 0 \text{ and } \beta_{\pm}(T,x) = \pm(1 - \cos(x)).$$

The existence of such $(u_\pm, v_\pm)$ would also have implied Theorem 8.1. Unfortunately, as it is proved in [**122**], such $(u_\pm, v_\pm)$ does not exist. At this point we see that, as mentioned earlier on page 237, an expansion to order 2 is not sufficient: this is the expansion to order 3 which gives the local controllability.

Let us now explain how to deduce Theorem 8.1 from Proposition 8.6. As pointed out above, by the invariance of the control system (8.5)-(8.6) by the change of variables $\tau = T - t$, $\xi = 2k\pi - x$ (together with a suitable change for the control), we can content ourselves to prove that, for every $T > 0$, there exists $r'_1 > 0$ such that, for every $y^1 \in L^2(0, 2k\pi)$ with $\|y^1\|_{L^2(0,2k\pi)} \leqslant r'_1$, there exists $u \in L^2(0,T)$ such that the solution y of

(8.45) $$y_t + y_x + y_{xxx} + yy_x = 0,$$
(8.46) $$y(t,0) = y(t, 2k\pi) = 0,$$
(8.47) $$y_x(t, 2k\pi) = u(t),$$
(8.48) $$y(0, x) = 0,$$

satisfies $y(T, \cdot) = y^1$. Of course, by "y is a solution of (8.45), (8.46), (8.47) and (8.48)", we mean that y is in $B_{0,T}$ and is the solution of

$$y_t + y_x + y_{xxx} = f,$$
$$y(t,0) = y(t, 2k\pi) = 0,$$
$$y_x(t, 2k\pi) = u(t),$$
$$y(0, x) = 0,$$

with $f := -yy_x$ (note that, if y is in $B_{0,T}$, then $yy_x \in L^1((0,T); L^2(0, 2k\pi))$). It is proved in [**122**] that, for a given $u \in L^2(0,T)$, there exists at most one solution of (8.45), (8.46), (8.47) and (8.48), and that such a solution exists if $\|u\|_{L^2(0,T)}$ is small enough (the smallness depending on T and k).

By (the proof of) Theorem 8.5 on page 241 and Remark 2.33 on page 48 (see also [**407**, Remark 3.10]), there exists a continuous linear map Γ,

(8.49) $$\Gamma : h \in H \subset L^2(0, 2k\pi) \mapsto \Gamma(h) \in L^2(0,T),$$

such that the solution of the linear system

$$y_t + y_x + y_{xxx} = 0,$$
$$y(t,0) = y(t, 2k\pi) = 0,$$
$$y_x(t, 2k\pi) = \Gamma(h)(t),$$
$$y(0, x) = 0,$$

satisfies $y(T, x) = h(x)$. (One can take for Γ the control obtained by means of HUM; see [**407**, Remark 3.10].)

Let $y^1 \in L^2(0, 2k\pi)$ be such that $\|y^1\|_{L^2(0,2k\pi)} \leqslant r$, $r > 0$ small enough so that the maps below are well defined in a neighborhood of 0. Let $\mathcal{T}_{y^1}$ denote the map,

$$\mathcal{T}_{y^1} : \begin{array}{ccc} L^2(0, 2k\pi) & \to & L^2(0, 2k\pi) \\ z & \mapsto & z + y^1 - F(G(z)), \end{array}$$

with
$$F: L^2(0,T) \to L^2(0, 2k\pi)$$
$$u \mapsto y(T, \cdot),$$
where y is the solution of (8.45), (8.46), (8.47) and (8.48), and where
$$G: L^2(0, 2k\pi) \to L^2(0,T)$$
is defined as follows. We write
$$z = \mathcal{P}_H(z) + \rho(z)(1 - \cos(x))$$
with
$$\int_0^{2k\pi} \mathcal{P}_H(z)(1 - \cos(x))dx = 0.$$
(In other words, $\mathcal{P}_H(z)$ is the orthogonal projection of z on H for the $L^2(0, 2k\pi)$-scalar product.) Then:

1. If $\rho(z) \geqslant 0$, $G(z) := \Gamma(\mathcal{P}_H(z)) + \rho^{1/3}(z)u_+ + \rho^{2/3}(z)v_+ + \rho(z)w_+$.
2. If $\rho(z) \leqslant 0$, $G(z) := \Gamma(\mathcal{P}_H(z)) + |\rho(z)|^{1/3}u_- + |\rho(z)|^{2/3}v_- + |\rho(z)|w_-$.

(The functions $u_\pm$, $v_\pm$ and $w_\pm$ have already been defined in Proposition 8.6.)

Clearly, each fixed point z^* of $\mathcal{T}_{y^1}$ satisfies $F(G(z^*)) = y^1$, and the control $u = G(z^*)$ is a solution to our problem. In order to prove the existence of a fixed point to $\mathcal{T}_{y^1}$, at least if $\|y^1\|_{L^2(0,2k\pi)}$ is small enough, one first proves the following estimate (see [**122**] for the proof): There exist $K > 0$ and $R > 0$ such that

$$\|\mathcal{T}_0 z\|_{L^2(0,2k\pi)} \leqslant K\|z\|_{L^2(0,2k\pi)}^{4/3} \text{ for every } z \in L^2(0, 2k\pi) \text{ with } \|z\|_{L^2(0,2k\pi)} \leqslant R.$$

Then, using the continuity of $\mathcal{T}_{y^1}(z)$ with respect to y^1 and z, we could conclude to the existence of a fixed point to $\mathcal{T}_{y^1}$ at least if $\|y^1\|_{L^2(0,2k\pi)}$ were small enough and if we were in finite dimension or had, more generally, a suitable compactness property. Unfortunately, we do not have this compactness property. So we proceed in a different manner. We use the Banach fixed-point theorem for the part in H and the intermediate value theorem in $H^\perp$. (If $H^\perp$ had dimension in $\mathbb{N} \setminus \{0, 1\}$, we would have used the Brouwer fixed-point theorem instead of the intermediate value theorem: In dimension 1 the Brouwer fixed-point theorem reduces to the intermediate value theorem.) See [**122**] for the detailed proofs. ∎

8.2.1. Open problems. For the other critical lengths, the situation is more complicated: there are now more noncontrollable (oriented) directions of the linearized control system around $(0, 0)$. It is natural to ask if Theorem 8.1 also holds for all the other critical lengths, that is, to ask if the answer to the following open problem is positive.

OPEN PROBLEM 8.8. *Let $L \in \mathcal{N}$ and $T > 0$. Do there exist $C > 0$ and $r > 0$ such that, for every y^0, $y^1 \in L^2(0, L)$ with $\|y^0\|_{L^2(0,L)} < r$ and $\|y^1\|_{L^2(0,L)} < r$, there exist $y \in C^0([0, T]; L^2(0, L)) \cap L^2((0, T); H^1(0, L))$ and $u \in L^2(0, T)$ satisfying (8.5)-(8.6) such that (8.9), (8.10) and (8.11) hold?*

REMARK 8.9. One could also ask the same question by requiring a weaker smallness condition than (8.11) or even simply by omitting this condition. In fact, our intuition is that for $L \notin 2\pi\mathbb{N}$, a power series expansion to the order 2 could be sufficient and therefore $1/3$ could be replaced by $1/2$ for these lengths. It is also possible that the local controllability holds only for large enough time T (that is, one should replace "Let $L > 0$ and $T > 0$. Do there exist $C > 0$ and $r > 0$ ..." by "Let $L > 0$. Do there exist $T > 0$, $C > 0$ and $r > 0$ ..." in Open Problem 8.8). Note that this is indeed the case for the nonlinear Schrödinger equation (see Theorem 9.8 on page 251 and Section 9.3). Eduardo Cerpa obtained in [**85**] a positive answer to Open problem 8.8 if there are exactly 4 uncontrollable (oriented) directions for the linearized control system around the equilibrium $(y_e, u_e) := (0,0)$ (instead of 2 for $L \in 2\pi\mathbb{N}$ satisfying (8.8), namely $1 - \cos(x)$ and $\cos(x) - 1$) if $T > 0$ is large enough. His proof also relies on the power series expansion method. In this case a power series expansion to order 2 is indeed sufficient. Note that, using Remark 8.2 on page 238, one can prove that, for every $n \in \mathbb{N}$, there are infinitely many L's such that there are at least n uncontrollable (oriented) directions for the linearized control system around the equilibrium $(y_e, u_e) := (0,0) \in L^2(0, L) \times \mathbb{R}$.

All the previous results are local controllability results. Concerning global controllability results, one has the following one due to Lionel Rosier [**409**].

THEOREM 8.10. *Let $L > 0$. For every $y^0 \in L^2(0, L)$, for every $y^1 \in L^2(0, L)$, there exist $T > 0$ and $y \in C^0([0, T]; L^2(0, L)) \cap L^2((0, T); H^1(0, L))$ satisfying*

$$y_t + y_x + y_{xxx} + yy_x = 0 \text{ in } \mathcal{D}'((0, T) \times (0, L))$$

such that $y(0, \cdot) = y^0$ and $y(T, \cdot) = y^1$.

Note that in this theorem
1. One does not require $y(t, 0) = y(t, L) = 0$,
2. A priori the time T depends on y^0 and y^1.

It is natural to wonder if one can remove these restrictions. For example, one may ask:

OPEN PROBLEM 8.11. *Let $L > 0$, $T > 0$, $y^0 \in L^2(0, L)$ and $y^1 \in L^2(0, L)$. Do there exist $y \in C^0([0, T]; L^2(0, L)) \cap L^2((0, T); H^1(0, L))$ and $u \in L^2(0, T)$ satisfying (8.5)-(8.6) such that $y(0, \cdot) = y^0$ and $y(T, \cdot) = y^1$?*

If one does not care about T (i.e., we allow T to be as large as we want, depending on y^0 and y^1), a classical way to attack this open problem is the following one (this is the method which is already used by Lionel Rosier to prove Theorem 8.10).

Step 1. Use the reversibility with respect to time of the equation to show that one may assume that $y^1 = 0$. (In fact, this part holds even if one deals with the case where one wants $T > 0$ to be small.)

Step 2. Use a suitable stabilizing feedback to go from y^0 into a given neighborhood of 0.

Step 3. Conclude by using a suitable local controllability around $(\bar{y}, \bar{u}) := 0$.

Step 1 indeed holds (perform the change of variables $(\tilde{t}, \tilde{x}) := (T - t, L - x)$). Let us assume that $L \notin \mathcal{N}$ or $L = 2\pi k$ with $k \in \mathbb{N} \setminus \{0\}$ satisfying (8.8). Then, applying Theorem 4.3 or Theorem 8.1, one can perform Step 3. Concerning Step 2, a natural choice is to consider the simpler feedback, namely $y_x(t, L) := 0$, which leads to the following Cauchy problem:

(8.50) $\qquad y_t + y_x + y_{xxx} + yy_x = 0, \ t \in (0, +\infty), \ x \in (0, L),$

(8.51) $\qquad y(t, 0) = y(t, L) = y_x(t, L) = 0, \ t \in (0, +\infty),$

(8.52) $\qquad y(0, x) = y^0(x), \ x \in (0, L).$

where y^0 is given in $L^2(0, L)$ and one requires

(8.53) $$y \in C([0, +\infty); L^2(0, L)) \cap L^2_{\text{loc}}([0, +\infty); H^1(0, L)).$$

Straightforward computations shows that, at least if y is smooth enough,

$$\frac{d}{dt} \int_0^L |y(t, x)|^2 dx = -|y_x(t, 0)|^2.$$

Hence one can expect that the Cauchy problem (8.50)-(8.51)-(8.52) has a solution. This turns out to be true: It has been proved by Gustavo Perla Menzala, Carlos Vasconcellos and Enrique Zuazua in [**382**] that the the Cauchy problem (8.50)-(8.51)-(8.52) has one and only one solution. (See also [**122**, Appendix A].) However, it is still not known (even for $L \notin \mathcal{N}$) if the following open problem has a positive answer.

OPEN PROBLEM 8.12. *Does one have, for every $y^0 \in L^2(0, L)$,*

(8.54) $$\lim_{t \to +\infty} \int_0^L |y(t, x)|^2 dt = 0?$$

There are available results showing that (8.54) holds if there is some interior damping, more precisely with (8.50) replaced by

$$y_t + y_x + y_{xxx} + yy_x + a(x)y = 0, \; t \in (0, +\infty), \; x \in (0, L),$$

where $a \in L^\infty(0, L)$ is such that

$$a(x) \geqslant 0, \; x \in (0, L),$$

the support of a has a nonempty interior.

These results are in [**382**] by Gustavo Perla Menzala, Carlos Vasconcellos and Enrique Zuazua, in [**380**] by Pazoto, and in [**413**] by Lionel Rosier and Bing-Yu Zhang.

CHAPTER 9

Previous methods applied to a Schrödinger equation

9.1. Controllability and uncontrollability results

In this chapter we consider a control system which is related to the one-dimensional Schrödinger control system (4.85)-(4.86) considered in Section 4.2.2. Let $I = (-1, 1)$. We consider the Schrödinger control system

(9.1) $\quad\quad\quad\quad \psi_t = i\psi_{xx} + iu(t)x\psi,\ (t, x) \in (0, T) \times I,$

(9.2) $\quad\quad\quad\quad \psi(t, -1) = \psi(t, 1) = 0,\ t \in (0, T),$

(9.3) $\quad\quad\quad\quad \dot{S}(t) = u(t),\ \dot{D}(t) = S(t),\ t \in (0, T).$

This is a control system, where, at time $t \in [0, T]$,
- the state is $(\psi(t, \cdot), S(t), D(t)) \in L^2(I; \mathbb{C}) \times \mathbb{R} \times \mathbb{R}$ with $\int_I |\psi(t, x)|^2 dx = 1$,
- the control is $u(t) \in \mathbb{R}$.

Compared to the one-dimensional Schrödinger control system (4.85)-(4.86) considered in Section 4.2.2, we have only added the variables S and D. This system has been introduced by Pierre Rouchon in [**417**]. It models a nonrelativistic charged particle in a 1-D moving infinite square potential well. At time t, $\psi(t, \cdot)$ is the wave function of the particle in a frame attached to the potential well, $S(t)$ is the speed of the potential well and $D(t)$ is the displacement of the potential well. The control $u(t)$ is the acceleration of the potential well at time t. (For other related control models in quantum chemistry, let us mention the paper [**306**] by Claude Le Bris and the references therein.) We want to control at the same time the wave function ψ, the speed S and the position D of the potential well. Let us first recall some important trajectories of the above control system when the control is 0. Let

(9.4) $\quad\quad\quad\quad \psi_n(t, x) := \varphi_n(x) e^{-i\lambda_n t}, n \in \mathbb{N} \setminus \{0\}.$

Here

(9.5) $\quad\quad\quad\quad \lambda_n := (n\pi)^2/4$

are the eigenvalues of the operator A defined on

$$D(A) := H^2 \cap H_0^1(I; \mathbb{C})$$

by

$$A\varphi := -\varphi'',$$

and the functions φ_n are the associated eigenvectors:

(9.6) $\quad\quad\quad\quad \varphi_n(x) := \sin(n\pi x/2),\ n \in \mathbb{N} \setminus \{0\},\ \text{if}\ n\ \text{is even},$

(9.7) $\quad\quad\quad\quad \varphi_n(x) := \cos(n\pi x/2),\ n \in \mathbb{N} \setminus \{0\},\ \text{if}\ n\ \text{is odd}.$

Let us recall that
$$\mathbb{S} := \{\varphi \in L^2(I;\mathbb{C}); \|\varphi\|_{L^2(I;\mathbb{C})} = 1\},$$
and let us also introduce the following notation:
$$H^7_{(0)}(I;\mathbb{C}) := \{\varphi \in H^7(I;\mathbb{C}); \varphi^{(2k)}(-1) = \varphi^{(2k)}(1) = 0 \text{ for } k = 0,1,2,3\}.$$
With these notations, one has the following result, proved in [**43**].

THEOREM 9.1. *For every $n \in \mathbb{N} \setminus \{0\}$, there exists $\eta_n > 0$ such that, for every $n_0, n_f \in \mathbb{N} \setminus \{0\}$ and for every $(\psi^0, S^0, D^0), (\psi^1, S^1, D^1) \in (\mathbb{S} \cap H^7_{(0)}(I;\mathbb{C})) \times \mathbb{R} \times \mathbb{R}$ with*

(9.8) $$\|\psi^0 - \varphi_{n_0}\|_{H^7(I;\mathbb{C})} + |S^0| + |D^0| < \eta_{n_0},$$

(9.9) $$\|\psi^1 - \varphi_{n_f}\|_{H^7(I;\mathbb{C})} + |S^1| + |D^1| < \eta_{n_f},$$

there exist a time $T > 0$ and (ψ, S, D, u) such that

(9.10) $$\psi \in C^0([0,T]; H^2 \cap H^1_0(I;\mathbb{C})) \cap C^1([0,T]; L^2(I;\mathbb{C})),$$

(9.11) $$u \in H^1_0(0,T),$$

(9.12) $$S \in C^1([0,T]), D \in C^2([0,T]),$$

(9.13) $$(9.1), (9.2) \text{ and } (9.3) \text{ hold},$$

(9.14) $$(\psi(0), S(0), D(0)) = (\psi^0, S^0, D^0),$$

(9.15) $$(\psi(T), S(T), D(T)) = (\psi^1, S^1, D^1).$$

REMARK 9.2. Using standard methods, one can check that ψ is in fact more regular than stated by (9.10): One has

(9.16) $$\psi \in C^0([0,T]; H^3(I;\mathbb{C}) \cap H^1_0(I;\mathbb{C})) \cap C^1([0,T]; H^1(I;\mathbb{C})).$$

Let us emphasize that, if one does not care about S and D and if $(n_0, n_f) = (1,1)$, then Theorem 9.1 is due to Karine Beauchard [**40**].

From Theorem 9.1, we have the following corollary.

COROLLARY 9.3. *For every $n_0, n_f \in \mathbb{N} \setminus \{0\}$, there exist $T > 0$ and (ψ, S, D, u) satisfying (9.10)-(9.11)-(9.12)-(9.13) such that*
$$(\psi(0), S(0), D(0)) = (\varphi_{n_0}, 0, 0),$$
$$(\psi(T), S(T), D(T)) = (\varphi_{n_f}, 0, 0).$$

Note that, as pointed out by Gabriel Turinici in [**488**], it follows from a general uncontrollability theorem [**28**, Theorem 3.6] due to John Ball, Jerrold Marsden and Marshall Slemrod that Theorem 9.1 does not hold if one replaces $H^7_{(0)}(I;\mathbb{C})$ by $H^2(I;\mathbb{C}) \cap H^1_0(I;\mathbb{C})$ and $H^1_0(0,T)$ by $L^2(0,T)$; see also Remark 2.14 on page 34. In the framework of Hilbert spaces, the general uncontrollability theorem [**28**, Theorem 3.6] is the following one.

THEOREM 9.4. *Let H be a Hilbert space and let A be the infinitesimal generator of a strongly continuous semigroup $S(t)$, $t \in [0, +\infty)$, of continuous linear operators on H. Let $B : H \to H$ be a linear bounded operator. Let $x^0 \in H$ and let $\mathcal{R}(x^0)$ be the set of $x^1 \in H$ such that there exist $T, r > 0$ and $u \in L^r(0,T)$ such that the solution $x \in C^0([0,T]; H)$ of*

(9.17) $$\dot{x} = Ax + u(t)Bx, \ x(0) = x^0,$$

satisfies $x(T) = x^1$. Then $\mathcal{R}(x^0)$ is contained in a countable union of compact subsets of H. In particular, if H is of infinite dimension, $\mathcal{R}(x^0)$ has an empty interior.

The definition of "strongly continuous semigroup $S(t)$, $t \in [0, +\infty)$, of continuous linear operators" is given in Definition A.5 and Definition A.6 on page 374. For the meaning of "A is the infinitesimal generator of a strongly continuous semigroup $S(t)$, $t \in [0, +\infty)$ of continuous linear operators", see Definition A.9 on page 375.

By "$x : [0, T] \to H$ is a solution of (9.17)", one means that $x \in C^0([0, T]; H)$ and satisfies

$$(9.18) \qquad x(t) = S(t)x^0 + \int_0^t u(s)S(t-s)Bx(s)ds, \ \forall t \in [0, T].$$

For the existence and uniqueness of solutions to the Cauchy problem (9.17), see, for example, [**28**, Theorem 2.5].

Proof of Theorem 9.4. Let $n \in \mathbb{N}$. Let K_n be the set of $b \in H$ such that, for some $u \in L^{1+(1/n)}(0, n)$ satisfying

$$(9.19) \qquad \|u\|_{L^{1+(1/n)}(0,n)} \leqslant n,$$

and for some $\tau \in [0, n]$, one has $x(\tau) = b$, where x is the solution of

$$(9.20) \qquad \dot{x} = Ax + u(t)Bx(t), \ x(0) = x^0.$$

In order to prove Theorem 9.4, it suffices to check that

$$(9.21) \qquad K_n \text{ is a compact subset of } H.$$

Let us prove (9.21). Let $(b_k)_{k \in \mathbb{N}}$ be a sequence of elements in K_n. For $k \in \mathbb{N}$, let $u_k \in L^{1+(1/n)}(0, n)$ such that

$$(9.22) \qquad \|u_k\|_{L^{1+(1/n)}(0,n)} \leqslant n,$$

and $\tau_k \in [0, n]$ such that $x_k(\tau_k) = b_k$, where x_k is the solution of

$$\dot{x}_k = Ax_k + u_k(t)Bx_k, \ x_k(0) = x^0.$$

In other words,

$$(9.23) \qquad x_k(t) = S(t)x^0 + \int_0^t u_k(s)S(t-s)Bx_k(s)ds, \ \forall t \in [0, n].$$

Without loss of generality, we may assume that, for some $\tau \in [0, n]$ and some $u \in L^{1+(1/n)}(0, n)$ satisfying (9.19),

$$(9.24) \qquad \tau_k \to \tau \text{ as } k \to +\infty,$$

$$(9.25) \qquad u_k \rightharpoonup u \text{ weakly in } L^{1+(1/n)}(0, n) \text{ as } k \to +\infty.$$

Let $x \in C^0([0, n]; H)$ be the solution of (9.20) associated to this control u. In other words, one has (9.18) for $T := n$. It suffices to check that

$$(9.26) \qquad x_k(\tau_k) \to x(\tau) \text{ as } k \to +\infty.$$

Let $y_k \in C^0([0, n]; H)$ be defined by

$$(9.27) \qquad y_k(t) := x_k(t) - x(t), \ \forall t \in [0, n].$$

Since $\tau_k \to \tau$ as $k \to +\infty$, in order to prove (9.26), it suffices to check that

$$(9.28) \qquad \|y_k\|_{C^0([0,n]; H)} \to 0 \text{ as } k \to \infty.$$

From (9.18), (9.23) and (9.27), we get
(9.29)
$$y_k(t) = \int_0^t (u_k(s) - u(s))S(t-s)Bx(s)ds + \int_0^t u_k(s)S(t-s)By_k(s)ds, \forall t \in [0, n].$$
Let
(9.30) $$\rho_k := \text{Max}\left\{\left\|\int_0^t (u_k(s) - u(s))S(t-s)Bx(s)ds\right\|_H ; t \in [0, n]\right\}.$$
Let us check that
(9.31) $$\lim_{k \to +\infty} \rho_k = 0.$$
Let $t_k \in [0, n]$ be such that
(9.32) $$\rho_k = \left\|\int_0^{t_k} (u_k(s) - u(s))S(t_k - s)Bx(s)ds\right\|_H.$$
Without loss of generality, we may assume that there exists $t \in [0, n]$ such that
(9.33) $$t_k \to t \text{ as } k \to +\infty.$$
Let us extend $S : [0, +\infty) \to \mathcal{L}(H; H)$ to a map $S : \mathbb{R} \to \mathcal{L}(H; H)$ by requiring
$$S(s)a = a, \forall s \in (-\infty, 0), \forall a \in H.$$
Since $S(s)$, $s \in [0, +\infty)$, is a strongly continuous semigroup of continuous linear operators, there exists $C \in (0, +\infty)$ such that
(9.34) $$\|S(s)\|_{\mathcal{L}(H,H)} \leqslant C, \forall s \in [0, n];$$
see Theorem A.8 on page 375. From (9.34) and Definition A.6 on page 374, one sees that the map $(s_1, s_2) \in \mathbb{R} \times [0, n] \mapsto S(s_1)Bx(s_2) \in H$ is continuous. Hence, also using (9.19) and (9.22) (for u_k and for u), one gets
(9.35) $$\left\|\int_0^{t_k} (u_k(s) - u(s))(S(t_k - s)Bx(s) - f(s))ds\right\|_H \to 0 \text{ as } k \to +\infty,$$
with $f(s) := S(t-s)Bx(s)$, $s \in [0, n]$. Let $\varepsilon > 0$. Since $f \in C^0([0, n]; H)$, there exist $l \in \mathbb{N}$, $0 = s_1 < s_2 < \ldots < s_l = n$, and l elements $a_1, a_2, \ldots, a_l$ in H such that
(9.36) $$\|f - g\|_{L^\infty((0,n);H)} \leqslant \varepsilon,$$
where $g \in L^\infty((0, n); H)$ is defined by
(9.37) $$g(s) = a_i, \forall s \in (s_i, s_{i+1}), \forall i \in \{1, 2, \ldots, l-1\}.$$
Using (9.25), (9.33) and (9.37), one gets
(9.38) $$\left\|\int_0^{t_k} (u_k(s) - u(s))g(s)ds\right\|_H \to 0 \text{ as } k \to +\infty.$$
From (9.19), (9.22), (9.35), (9.36) and (9.38), one gets
$$\limsup_{k \to +\infty} \rho_k \leqslant 2n^{(n+2)/(n+1)}\varepsilon,$$
which, since $\varepsilon > 0$ is arbitrary, gives (9.31).

From (9.29), (9.30), (9.34) and the Gronwall inequality, one gets
$$\|y_k(t)\|_H \leqslant \rho_k \exp\left(C\|B\|_{\mathcal{L}(H,H)} \int_0^t |u_k(s)|ds\right),$$

which, together with (9.22) and (9.31), proves (9.28). This concludes the proof of (9.21) and also of Theorem 9.4. ∎

Coming back to Theorem 9.1 on page 248, there are at least three points that one might want to improve in this result:
- Remove the assumptions (9.8) and (9.9) on the initial and final data.
- Weaken the regularity assumptions on the initial and final data.
- Get information about the time T of controllability.

Concerning the first two points, one can propose, for example, the following open problems.

OPEN PROBLEM 9.5. *Let $(\psi^0, S^0, D^0), (\psi^1, S^1, D^1) \in [\mathbb{S} \cap H^7_{(0)}(I; \mathbb{C})] \times \mathbb{R} \times \mathbb{R}$. Do there exist $T > 0$ and (ψ, S, D, u) such that (9.10) to (9.15) hold?*

OPEN PROBLEM 9.6. *Let*
$$H^3_{(0)}(I; \mathbb{C}) := \{\varphi \in H^3(I; \mathbb{C}); \varphi^{(2k)}(-1) = \varphi^{(2k)}(1) = 0 \text{ for } k = 0, 1\}.$$
Let $(\psi^0, S^0, D^0), (\psi^1, S^1, D^1) \in (\mathbb{S} \cap H^3_{(0)}(I; \mathbb{C})) \times \mathbb{R} \times \mathbb{R}$. Do there exist $T > 0$ and (ψ, S, D, u) with $u \in L^2(0, T)$ such that (9.10), (9.12), (9.13), (9.14) and (9.15) hold?

REMARK 9.7. The conjectured regularity in Open Problem 9.6 comes from the regularities required for the controllability of linearized control systems (see Theorem 2.87 on page 96, Proposition 9.11 on page 256 and Proposition 9.15 on page 262).

Concerning the time of controllability, one may wonder if, for every $\varepsilon > 0$, there exists $\eta > 0$ such that, for every $(\psi^0, S^0, D^0), (\psi^1, S^1, D^1) \in (\mathbb{S} \cap H^7_{(0)}(I; \mathbb{C})) \times \mathbb{R} \times \mathbb{R}$ satisfying

(9.39) $$\|\psi^0 - \psi_1(0, \cdot)\|_{H^7(I; \mathbb{C})} + |S^0| + |D^0| < \eta,$$

(9.40) $$\|\psi^1 - \psi_1(\varepsilon, \cdot)\|_{H^7(I; \mathbb{C})} + |S^1| + |D^1| < \eta,$$

there exists (ψ, S, D, u) such that (9.10), (9.11), (9.12), (9.13), (9.14) and (9.15) hold with $T = \varepsilon$ and $\|u\|_{H^1(0, \varepsilon)} \leqslant \varepsilon$. (Let us recall that ψ_1 is defined by (9.5), (9.4) and (9.7).) This question has a negative answer. More precisely, one has the following theorem (see [**117**, Theorem 1.2] for a prior weaker result).

THEOREM 9.8. *Let $T > 0$ be such that*

(9.41) $$T < \frac{2}{\pi^2}.$$

Then there exists $\varepsilon > 0$ such that for every $\bar{D} \neq 0$, there is no $u \in L^2((0, T); (-\varepsilon, \varepsilon))$ such that the solution $(\psi, S, D) \in C^0([0, T]; H^1_0(I; \mathbb{C})) \times C^0([0, T]; \mathbb{R}) \times C^1([0, T]; \mathbb{R})$ of the Cauchy problem

(9.42) $$\psi_t = i\psi_{xx} + iu(t)x\psi, \ (t, x) \in (0, T) \times I,$$

(9.43) $$\psi(t, -1) = \psi(t, 1) = 0, \ t \in (0, T),$$

(9.44) $$\dot{S}(t) = u(t), \ \dot{D}(t) = S(t), \ t \in (0, T),$$

(9.45) $$\psi(0, x) = \psi_1(0, x), \ x \in I,$$

(9.46) $$S(0) = 0, \ D(0) = 0,$$

satisfies

(9.47) $$\psi(T,x) = \psi_1(T,x),\ x \in I,$$
(9.48) $$S(T) = 0,\ D(T) = \bar{D},$$

The remainder of this chapter is organized in the following way:
- In Section 9.2 we sketch the main ingredients of the proof of Theorem 9.1,
- In Section 9.3 we first give, in Section 9.3.1, some comments and explain the "raison d'être" of Theorem 9.8. Then, in Section 9.3.2, we give the proof of Theorem 9.8.

9.2. Sketch of the proof of the controllability in large time

In this section we sketch the proof of Theorem 9.1. Using the reversibility with respect to time of the control system (9.1)-(9.2)-(9.3), it is sufficient to prove Theorem 9.1 for $n^1 = n_0 + 1$. We prove it with $n_0 = 1$ and $n^1 = 2$ to simplify the notations (the proof of the general case is similar).

The key step is then the study of the large time local controllability of the control system (9.1)-(9.2)-(9.3) around the trajectory $(Y^\theta, u = 0)$ for every $\theta \in [0,1]$, where

(9.49) $$Y^\theta(t) := (f_\theta(t), 0, 0) \in L^2(I; \mathbb{C}) \times \mathbb{R} \times \mathbb{R},\ \forall \theta \in [0,1],\ \forall t \in \mathbb{R},$$

together with

(9.50) $$f_\theta(t) := \sqrt{1-\theta}\psi_1(t) + \sqrt{\theta}\psi_2(t),\ \forall \theta \in [0,1],\ \forall t \in \mathbb{R}.$$

Here and in the following, we write $\psi_1(t)$ for $\psi_1(t,\cdot)$, $\psi_2(t)$ for $\psi_2(t,\cdot)$, etc. In order to state such a controllability result, for $\theta \in [0,1]$ and $\nu > 0$, let

$$D(\theta,\nu) := \{(\varphi, S, D) \in (H^7_{(0)}(I;\mathbb{C}) \cap \mathbb{S}) \times \mathbb{R} \times \mathbb{R};\ \|\varphi - f_\theta(0)\|_{H^7(I;\mathbb{C})} + |S| + |D| \leqslant \nu\}.$$

Let us also point out that

$$f_\theta(t + \tau_0) = f_\theta(t),\ \forall t \in \mathbb{R},\ \forall \theta \in [0,1],$$

with

$$\tau_0 := \frac{8}{\pi}.$$

The large time controllability of the control system (9.1)-(9.2)-(9.3) around the trajectory $(Y^\theta, u=0)$ we use is the following.

THEOREM 9.9. *For every $\theta \in [0,1]$, there exist $n_\theta \in \mathbb{N} \setminus \{0\}$ and $\nu_\theta > 0$ such that, for every $(\psi^0, S^0, D^0) \in D(\theta, \nu_\theta)$ and for every $(\psi^1, S^1, D^1) \in D(\theta, \nu_\theta)$, there exists (ψ, S, D, u) such that*

(9.51) $$\psi \in C^0([0, n_\theta \tau_0]; H^3_{(0)}(I;\mathbb{C})) \cap C^1([0, n_\theta \tau_0]; H^1_0(I;\mathbb{C})),$$
(9.52) $$u \in H^1_0(0, n_\theta \tau_0),$$
(9.53) $$S \in C^1([0, n_\theta \tau_0]),\ D \in C^2([0, n_\theta \tau_0]),$$
(9.54) $$(9.1),\ (9.2)\ and\ (9.3)\ hold,$$
(9.55) $$(\psi(0), S(0), D(0)) = (\psi^0, S^0, D^0),$$
(9.56) $$(\psi(n_\theta \tau_0), S(n_\theta \tau_0), D(n_\theta \tau_0)) = (\psi^1, S^1, D^1).$$

Let us assume, for the moment, that this theorem holds and we conclude the proof of Theorem 9.1. Let us first notice that

(9.57) $$f_\theta(0) \in H^7_{(0)}(I;\mathbb{C}) \cap \mathbb{S}, \forall \theta \in [0,1],$$

(9.58) the map $\theta \in [0,1] \mapsto Y^\theta(0) \in H^7(I;\mathbb{C}) \times \mathbb{R} \times \mathbb{R}$ is continuous.

For $\theta \in [0,1]$, let N_θ be the minimum of the set of the $n_\theta \in \mathbb{N} \setminus \{0\}$ such that there exists $\nu_\theta > 0$ such that the conclusion of Theorem 9.9 holds. From (9.57) and (9.58), it is not hard to see that the map

$$\theta \in [0,1] \mapsto N_\theta \in \mathbb{N} \setminus \{0\}$$

is upper semicontinuous. In particular, since $[0,1]$ is compact, this map is bounded. Let us choose $N \in \mathbb{N} \setminus \{0\}$ such that

$$N_\theta \leqslant N, \forall \theta \in [0,1].$$

Straightforward continuity arguments show that, for every $\theta \in [0,1]$, there exists $\nu_\theta > 0$ such that, for every $(\psi^0, S^0, D^0) \in D(\theta, \nu_\theta)$ and for every $(\psi^1, S^1, D^1) \in D(\theta, \nu_\theta)$, there exists (ψ, S, D, u) satisfying (9.51) to (9.56) with $n_\theta := N$. Let $\varepsilon_\theta \in (0, +\infty]$ be the supremum of the set of such ν_θ. Again it is not hard to check that the map

$$\theta \in [0,1] \mapsto \varepsilon_\theta \in (0,+\infty]$$

is lower semicontinuous. Hence, since $[0,1]$ is compact, there exists $\varepsilon > 0$ such that

$$\varepsilon < \varepsilon_\theta, \forall \theta \in [0,1].$$

By construction, for every $\theta \in [0,1]$, for every $(\psi^0, S^0, D^0) \in D(\theta, \varepsilon)$, and for every $(\psi^1, S^1, D^1) \in D(\theta, \varepsilon)$, there exists (ψ, S, D, u) satisfying

$$\psi \in C^0([0, N\tau_0]; H^3_{(0)}(I;\mathbb{C})) \cap C^1([0, N\tau_0]; H^1_0(I;\mathbb{C})),$$
$$u \in H^1_0(0, N\tau_0),$$
$$S \in C^1([0, N\tau_0]), D \in C^2([0, N\tau_0]),$$
$$(9.1), (9.2) \text{ and } (9.3) \text{ hold},$$
$$(\psi(0), S(0), D(0)) = (\psi^0, S^0, D^0),$$
$$(\psi(N\tau_0), S(N\tau_0), D(N\tau_0)) = (\psi^1, S^1, D^1).$$

Let $k \in \mathbb{N} \setminus \{0\}$ be large enough so that

(9.59) $$D(j/k, \varepsilon) \cap D((j+1)/k, \varepsilon) \neq \emptyset, \forall j \in \{0, \ldots, k-1\}.$$

Let $(\psi^0, S^0, D^0) \in D(0, \varepsilon)$, $(\psi^1, S^1, D^1) \in D(1, \varepsilon)$ and

(9.60) $$T := (k+1)N\tau_0.$$

Let us check that there exists (ψ, S, D, u) such that (9.10) to (9.15) hold. By (9.59), there exists $(\tilde\psi^j, \tilde S^j, \tilde D^j)$ such that

$$(\tilde\psi^j, \tilde S^j, \tilde D^j) \in D(j/k, \varepsilon) \cap D((j+1)/k, \varepsilon), \forall j \in \{0, \ldots, k-1\}$$

which implies, for every $j \in \{0, \ldots, k-2\}$, the existence of $(\tilde{\psi}_j, \tilde{S}_j, \tilde{D}_j, \tilde{u}_j)$ satisfying

$$\tilde{\psi}_j \in C^0([0, N\tau_0]; H^3_{(0)}(I; \mathbb{C})) \cap C^1([0, N\tau_0]; H^1_0(I; \mathbb{C})),$$
$$\tilde{u}_j \in H^1_0(0, N\tau_0),$$
$$\tilde{S}_j \in C^1([0, N\tau_0]), \tilde{D}_j \in C^2([0, N\tau_0]),$$
(9.1), (9.2) and (9.3) hold for $(\psi, S, D, u, T) := (\tilde{\psi}_j, \tilde{S}_j, \tilde{D}_j, \tilde{u}_j, N\tau_0)$,
$$(\tilde{\psi}_j(0), \tilde{S}_j(0), \tilde{D}_j(0)) = (\psi^j, S^j, D^j),$$
$$(\tilde{\psi}_j(N\tau_0), \tilde{S}_j(N\tau_0), \tilde{D}_j(N\tau_0)) = (\tilde{\psi}^{j+1}, \tilde{S}^{j+1}, \tilde{D}^{j+1}).$$

Since
$$(\psi^0, S^0, D^0) \in D(0, \varepsilon) \text{ and } (\tilde{\psi}^0, \tilde{S}^0, \tilde{D}^0) \in D(0, \varepsilon),$$
there exists $(\tilde{\psi}_{-1}, \tilde{S}_{-1}, \tilde{D}_{-1}, \tilde{u}_{-1})$ satisfying

$$\tilde{\psi}_{-1} \in C^0([0, N\tau_0]; H^3_{(0)}(I; \mathbb{C})) \cap C^1([0, N\tau_0]; H^1_0(I; \mathbb{C})),$$
$$\tilde{u}_{-1} \in H^1_0(0, N\tau_0),$$
$$\tilde{S}_{-1} \in C^1([0, N\tau_0]), \tilde{D}_{-1} \in C^2([0, N\tau_0]),$$
(9.1), (9.2) and (9.3) hold for $(\psi, S, D, u, T) := (\tilde{\psi}_{-1}, \tilde{S}_{-1}, \tilde{D}_{-1}, \tilde{u}_{-1}, N\tau_0)$,
$$(\tilde{\psi}_{-1}(0), \tilde{S}_{-1}(0), \tilde{D}_{-1}(0)) = (\psi^0, S^0, D^0),$$
$$(\tilde{\psi}_{-1}(N\tau_0), \tilde{S}_{-1}(N\tau_0), \tilde{D}_{-1}(N\tau_0)) = (\tilde{\psi}^0, \tilde{S}^0, \tilde{D}^0).$$

Similarly, since
$$(\tilde{\psi}^{k-1}, \tilde{S}^{k-1}, \tilde{D}^{k-1}) \in D(1, \varepsilon) \text{ and } (\psi^1, S^1, D^1) \in D(1, \varepsilon),$$
there exists $(\tilde{\psi}_{k-1}, \tilde{S}_{k-1}, \tilde{D}_{k-1}, \tilde{u}_{k-1})$ satisfying

$$\tilde{\psi}_{k-1} \in C^0([0, N\tau_0]; H^3_{(0)}(I; \mathbb{C})) \cap C^1([0, N\tau_0]; H^1_0(I; \mathbb{C})),$$
$$\tilde{u}_{k-1} \in H^1_0(0, N\tau_0),$$
$$\tilde{S}_{k-1} \in C^1([0, N\tau_0]), \tilde{D}_{k-1} \in C^2([0, N\tau_0]),$$
(9.1), (9.2) and (9.3) hold for $(\psi, S, D, u, T) := (\tilde{\psi}_{k-1}, \tilde{S}_{k-1}, \tilde{D}_{k-1}, \tilde{u}_{k-1}, N\tau_0)$,
$$(\tilde{\psi}_{k-1}(0), \tilde{S}_{k-1}(0), \tilde{D}_{k-1}(0)) = (\tilde{\psi}^{k-1}, \tilde{S}^{k-1}, \tilde{D}^{k-1}),$$
$$(\tilde{\psi}_{k-1}(N\tau_0), \tilde{S}_{k-1}(N\tau_0), \tilde{D}_{k-1}(N\tau_0)) = (\psi^1, S^1, D^1).$$

Then, it suffices to define $(\psi, S, D, u) : [0, T] \to H^3_{(0)}(I; \mathbb{C}) \times \mathbb{R} \times \mathbb{R} \times \mathbb{R}$ by requiring, besides (9.60), for every $j \in \{-1, 0, \ldots, k-1\}$ and for every $t \in [(j+1)N\tau_0, (j+2)N\tau_0]$,

$$\psi(t) := (\tilde{\psi}_j(t - (j+1)N\tau_0),$$
$$S(t) := \tilde{S}_j(t - (j+1)N\tau_0),$$
$$D(t) := \tilde{D}_j(t - (j+1)N\tau_0),$$
$$u(t) := \tilde{u}_j(t - (j+1)N\tau_0)).$$

This concludes the proof of Theorem 9.1 assuming Theorem 9.9. ∎

It remains to prove Theorem 9.9 on page 252. The strategy for $\theta \in (0,1)$ is different from the one for $\theta \in \{0,1\}$. In the next sections, we detail both strategies. We start (Section 9.2.1) with the simplest case, namely $\theta \in (0,1)$. Then, in Section 9.2.2, we treat the more complicated case, namely $\theta \in \{0,1\}$.

9.2.1. Local controllability of the control system (9.1)-(9.2)-(9.3) around $(Y^\theta, u = 0)$ for $\theta \in (0,1)$. As we have seen earlier, for example in Section 3.1 and in Section 4.1, a classical approach to prove the local controllability around a trajectory consists of proving the controllability of the linearized system along the studied trajectory and concluding with an inverse mapping theorem. This strategy does not work here because the linearized system around $(Y^\theta(t), u = 0)$ is not controllable. Let us check this noncontrollability. For $\psi \in \mathbb{S}$, the tangent space $T_\mathbb{S}(\psi)$ to the $L^2(I; \mathbb{C})$-sphere at the point ψ is

$$(9.61) \qquad T_\mathbb{S}\psi := \left\{ \varphi \in L^2(I; \mathbb{C}); \Re\langle \varphi, \psi \rangle = 0 \right\},$$

where

$$\langle \varphi, \psi \rangle := \int_I \varphi(x)\overline{\psi(x)}dx, \ \forall (\varphi, \psi) \in L^2(I; \mathbb{C})^2,$$

and, for $z \in \mathbb{C}$,

$$\Re z \text{ is the real part of } z.$$

Let us also recall that $\Im z$ denotes the imaginary part of $z \in \mathbb{C}$.

The linearized control system around the trajectory $(Y^\theta, u = 0)$ is

$$(\Sigma_\theta^l) \begin{cases} \Psi_t = i\Psi_{xx} + iwx f_\theta, \ (t, x) \in (0, T) \times I, \\ \Psi(t, -1) = \Psi(t, 1) = 0, \ t \in (0, T), \\ \dot{s} = w, \\ \dot{d} = s. \end{cases}$$

It is a control system where, at time t,
- The state is $(\Psi(t, \cdot), s(t), d(t)) \in L^2(I; \mathbb{C}) \times \mathbb{R} \times \mathbb{R}$ with $\Psi(t, \cdot) \in T_\mathbb{S}(f_\theta(t, \cdot))$,
- The control is $w(t) \in \mathbb{R}$.

PROPOSITION 9.10. *Let $T > 0$ and $((\Psi, s, d), w)$ be a trajectory of (Σ_θ^l) on $[0, T]$. Then the function*

$$t \mapsto \Im(\langle \Psi(t), \sqrt{1-\theta}\psi_1(t) - \sqrt{\theta}\psi_2(t)\rangle),$$

is constant on $[0, T]$. In particular, the control system (Σ_θ^l) is not controllable.

(In this proposition and, more generally, in this section as well as in Section 9.2.2, we do not specify the regularities of trajectories and solutions of Cauchy problems: these regularities are the classical ones.)

Proof of Proposition 9.10. Let us consider the function

$$(9.62) \qquad \xi_\theta(t, \cdot) := \sqrt{1-\theta}\psi_1(t, \cdot) - \sqrt{\theta}\psi_2(t, \cdot).$$

We have

$$\xi_{\theta t} = i\xi_{\theta xx}, \ (t, x) \in (0, T) \times I,$$

$$\frac{d}{dt}\left(\Im\langle \Psi(t), \xi_\theta(t)\rangle\right) = \Im(iw\langle x f_\theta(t), \xi_\theta(t)\rangle).$$

The explicit expressions of f_θ (see (9.50)) and ξ_θ (see (9.62)) show that, for every $t \in [0, T]$,
$$\langle x f_\theta(t), \xi_\theta(t) \rangle \in i\mathbb{R},$$
which gives the conclusion.

Let $T > 0$, and $\Psi^0 \in T_\mathbb{S}(f_\theta(0))$, $\Psi^1 \in T_\mathbb{S}(f_\theta(T))$. A necessary condition for the existence of a trajectory of (Σ_θ^l) satisfying $\Psi(0) = \Psi^0$ and $\Psi(T) = \Psi^1$ is
$$\Im(\langle \Psi^1, \sqrt{1-\theta}\psi_1(T) - \sqrt{\theta}\psi_2(T)\rangle) = \Im(\langle \Psi^0, \sqrt{1-\theta}\varphi_1 - \sqrt{\theta}\varphi_2\rangle).$$
This equality does not hold for an arbitrary choice of Ψ^0 and Ψ^1. Thus (Σ_θ^l) is not controllable. ∎

Let us now show that the linearized control system (Σ_θ^l) misses exactly two (oriented) directions on the tangent space to the sphere $\mathbb{S}$, which are $(\Psi, S, D) := (\pm i\xi_\theta, 0, 0)$. (So the situation is the same as for the nonlinear Korteweg-de Vries control system (8.5)-(8.6) with the critical length $L = 2\pi k$ with $k \in \mathbb{N}\setminus\{0\}$ satisfying (8.8); see Section 8.2.)

PROPOSITION 9.11. *Let* $T > 0$, $(\Psi^0, s^0, d^0), (\Psi^1, s^1, d^1) \in H^3_{(0)}(I, \mathbb{R}) \times \mathbb{R} \times \mathbb{R}$ *be such that*

(9.63) $$\Re\langle \Psi^0, f_\theta(0)\rangle = \Re\langle \Psi^1, f_\theta(T)\rangle = 0,$$

(9.64) $$\Im\langle \Psi^1, \sqrt{1-\theta}\varphi_1 e^{-i\lambda_1 T} - \sqrt{\theta}\varphi_2 e^{-i\lambda_2 T}\rangle = \Im\langle \Psi^0, \sqrt{1-\theta}\varphi_1 - \sqrt{\theta}\varphi_2\rangle.$$

There exists $w \in L^2(0, T)$ *such that the solution*
$$(\Psi, s, d) : [0, T] \to L^2(I; \mathbb{C}) \times \mathbb{R} \times \mathbb{R}$$
of (Σ_θ^l) *with control w and such that*
$$(\Psi(0), s(0), d(0)) = (\Psi^0, s^0, d^0)$$
satisfies
$$(\Psi(T), s(T), d(T)) = (\Psi^1, s^1, d^1).$$

Before giving the proof of this proposition, let us point out that, by (9.61), the condition (9.63) just means that
$$\Psi^0 \in T_\mathbb{S}(f_\theta(0)) \text{ and } \Psi^1 \in T_\mathbb{S}(f_\theta(T)).$$

Proof of Proposition 9.11. The proof is similar to the proof of Theorem 2.87 on page 96. It also relies on moment theory. Let $(\Psi^0, s^0, d^0) \in L^2(I, \mathbb{R}) \times \mathbb{R} \times \mathbb{R}$, with $\Psi^0 \in T_\mathbb{S}(f_\theta(0))$ and $T > 0$. Let (Ψ, s, d) be a solution of (Σ_θ^l) with $(\Psi(0), s(0), d(0)) = (\Psi^0, s^0, d^0)$ and a control $w \in L^2(0, T)$. We have the following equality in $L^2(I; \mathbb{C})$:
$$\Psi(t) = \sum_{k=1}^\infty x_k(t)\varphi_k \text{ where } x_k(t) := \langle \Psi(t), \varphi_k\rangle, \forall k \in \mathbb{N} \setminus \{0\}.$$
Using the equation satisfied by Ψ, we get

(9.65) $$x_{2k}(t) = \left(\langle \Psi^0, \varphi_{2k}\rangle + i\sqrt{1-\theta}b_{2k}\int_0^t w(\tau)e^{i(\lambda_{2k}-\lambda_1)\tau}d\tau\right)e^{-i\lambda_{2k}t},$$

(9.66) $$x_{2k-1}(t) = \left(\langle \Psi^0, \varphi_{2k-1}\rangle + i\sqrt{\theta}c_{2k-1}\int_0^t w(\tau)e^{i(\lambda_{2k-1}-\lambda_2)\tau}d\tau\right)e^{-i\lambda_{2k-1}t},$$

9.2. SKETCH OF THE PROOF OF THE CONTROLLABILITY IN LARGE TIME   257

where, for every $k \in \mathbb{N} \setminus \{0\}$, $b_k := \langle x\varphi_k, \varphi_1 \rangle$ and $c_k := \langle x\varphi_k, \varphi_2 \rangle$. Using the explicit expression of the functions φ_k (see (9.6)-(9.7)), we get

$$
(9.67) \qquad b_k = \begin{cases} 0 \text{ if } k \text{ is odd}, \\ \dfrac{-16(-1)^{k/2}k}{\pi^2(1+k)^2(1-k)^2} \text{ if } k \text{ is even}, \end{cases}
$$

$$
(9.68) \qquad c_k = \begin{cases} \dfrac{32(-1)^{(k-1)/2}k}{\pi^2(k+2)^2(k-2)^2} \text{ if } k \text{ is odd}, \\ 0 \text{ if } k \text{ is even}. \end{cases}
$$

Let $(\Psi^1, s^1, d^1) \in L^2(I, \mathbb{R}) \times \mathbb{R} \times \mathbb{R}$ with $\Psi^1 \in T_\mathbb{S}(f_\theta(T))$. The equality $(\Psi(T), s(T), d(T)) = (\Psi^1, s^1, d^1)$ is equivalent to the following moment problem on w,

$$
(9.69) \quad \int_0^T w(t) e^{i(\lambda_{2k} - \lambda_1)t} dt
$$
$$
= \frac{-i}{\sqrt{1-\theta} b_{2k}} \left(\langle \Psi^1, \varphi_{2k} \rangle e^{i\lambda_{2k}T} - \langle \Psi^0, \varphi_{2k} \rangle \right), \forall k \in \mathbb{N} \setminus \{0\},
$$

$$
(9.70) \quad \int_0^T w(t) e^{i(\lambda_{2k-1} - \lambda_2)t} dt
$$
$$
= \frac{-i}{\sqrt{\theta} c_{2k-1}} \left(\langle \Psi^1, \varphi_{2k-1} \rangle e^{i\lambda_{2k-1}T} - \langle \Psi^0, \varphi_{2k-1} \rangle \right), \forall k \in \mathbb{N} \setminus \{0\},
$$

$$
(9.71) \qquad \int_0^T w(t) dt = s^1 - s^0,
$$

$$
(9.72) \qquad \int_0^T (T-t) w(t) dt = d^1 - d^0 - s^0 T.
$$

In (9.69)-(9.70) with $k = 1$, the left hand sides are complex conjugate numbers because w is real valued. Note also that, by (9.67) and (9.68), $b_2 = c_1$. Hence a necessary condition on Ψ^0 and Ψ^1 for the existence of $w \in L^2(0, T)$ solution of (9.69)-(9.70) is

$$
(9.73) \quad \frac{1}{\sqrt{1-\theta}} \left(\overline{\langle \Psi^1, \varphi_2 \rangle} e^{-i\lambda_2 T} - \overline{\langle \Psi^0, \varphi_2 \rangle} \right)
$$
$$
= \frac{-1}{\sqrt{\theta}} \left(\langle \Psi^1, \varphi_1 \rangle e^{i\lambda_1 T} - \langle \Psi^0, \varphi_1 \rangle \right).
$$

The equality of the real parts of the two sides in (9.73) is guaranteed by (9.63). The equality of the imaginary parts of the two sides in (9.73) is equivalent to (9.64). Under the assumption $\Psi^0, \Psi^1 \in H^3_{(0)}(I; \mathbb{C})$, the right hand sides of (9.69)-(9.70) define a sequence in $l^2(\mathbb{C})$. Then the existence, for every $T > 0$, of $w \in L^2(0, T)$ solution of (9.69) to (9.72) is a classical result on trigonometric moment problems; see, for example, [**422**, Section 2], or [**282**, Section 1.2.2]. In fact, if one does not care about condition (9.72), the existence of $w \in L^2(0, T)$ follows from Theorem 2.88 on page 98. ∎

Coming back to our controllability statement, let us try to follow the approach (power series expansion) we used for the nonlinear control system (8.5)-(8.6) (see page 237) with the critical lengths $L = 2\pi k$ with $k \in \mathbb{N} \setminus \{0\}$ satisfying (8.8) (see Section 8.2 and [**122**]). We first try to move in the (oriented) directions which are missed by the linearized control system (Σ_θ^l). This can be done and this is simpler than for the nonlinear KdV control system (8.5)-(8.6), since the order 2 is sufficient to do this motion. Indeed, one has the following proposition.

PROPOSITION 9.12 ([**43**], Theorem 6]). *Let* $T := 8/\pi$. *There exist* $w_\pm \in H^4 \cap H_0^3(0,T)$, $\nu_\pm \in H_0^3(0,T)$ *such that the solutions of*

(9.74) $\begin{cases} \Psi_{\pm t} = i\Psi_{\pm xx} + iw_\pm x f_\theta, \ (t,x) \in (0,T) \times I, \\ \Psi_\pm(0,x) = 0, \ x \in I, \\ \Psi_\pm(t,-1) = \Psi_\pm(t,1) = 0, \ t \in (0,T), \\ \dot{s}_\pm = w_\pm, s_\pm(0) = 0, \\ \dot{d}_\pm = s_\pm, d_\pm(0) = 0, \end{cases}$

(9.75) $\begin{cases} \xi_{\pm t} = i\xi_{\pm xx} + iw_\pm x\Psi_\pm + i\nu_\pm x f_\theta, \\ \xi_\pm(0,x) = 0, \ x \in I, \\ \xi_\pm(t,-1) = \xi_\pm(t,1) = 0, \ t \in (0,T), \\ \dot{\sigma}_\pm = \nu_\pm, \sigma_\pm(0) = 0, \\ \dot{\delta}_\pm = \sigma_\pm, \delta_\pm(0) = 0, \end{cases}$

satisfy $\Psi_\pm(T,\cdot) = 0$, $s_\pm(T) = 0$, $d_\pm(T) = 0$, $\xi_\pm(T) = \pm i\varphi_1$, $\sigma_\pm = 0$, $\delta_\pm = 0$.

The time $T = 8/\pi$ is not optimal, but our guess is that T cannot be arbitrarily small (see also Theorem 9.8 on page 251 and Open Problem 9.18 on page 268).

Unfortunately, the fixed-point argument we used for the nonlinear KdV control system (8.5)-(8.6) with the critical lengths $L = 2\pi k$, with $k \in \mathbb{N} \setminus \{0\}$ satisfying (8.8) (see Section 8.2) no longer works. The reason is the following one. Let us introduce the following closed subspace of $L^2(I;\mathbb{C})$,

$$\Pi := \overline{\mathrm{Span}\{\varphi_k; k \geqslant 2\}}$$

and the orthogonal projection $\mathcal{P} : L^2(I;\mathbb{C}) \to \Pi$. Let Φ be the map defined by

$$\Phi : (\psi^0, S^0, D^0, u) \mapsto (\psi^0, S^0, D^0, \mathcal{P}\psi(T), S(T), D(T)),$$

where ψ solves

$$\begin{cases} \dot\psi = i\psi'' + iux\psi, \\ \dot S = w, \\ \dot D = S, \end{cases}$$

with $(\psi(0), S(0), D(0)) = (\psi^0, S^0, D^0)$. Let $H_{(0)}^2(I;\mathbb{C})$ be defined by

$$H_{(0)}^2(I;\mathbb{C}) := \{\varphi \in H^2(I;\mathbb{C}); \varphi \in H_0^1(I;\mathbb{C})\}.$$

The map Φ is of class C^1 between the spaces

$$\Phi : [\mathbb{S} \cap H_{(0)}^2(I;\mathbb{C})] \times \mathbb{R} \times \mathbb{R} \times L^2(0,T)$$
$$\to [\mathbb{S} \cap H_{(0)}^2(I;\mathbb{C})] \times \mathbb{R} \times \mathbb{R} \times [\Pi \cap H_{(0)}^2(I;\mathbb{C})] \times \mathbb{R} \times \mathbb{R},$$

$$\Phi : [\mathbb{S} \cap H_{(0)}^3(I;\mathbb{C})] \times \mathbb{R} \times \mathbb{R} \times H_0^1(0,T)$$
$$\to [\mathbb{S} \cap H_{(0)}^3(I;\mathbb{C})] \times \mathbb{R} \times \mathbb{R} \times [\Pi \cap H_{(0)}^3(I;\mathbb{C})] \times \mathbb{R} \times \mathbb{R}.$$

So we would like to have a right inverse to the map $d\Phi(f_\theta(0), 0, 0, 0)$ which is a map between the spaces

$$[T_{\mathbb{S}}(f_\theta(0)) \cap H^2_{(0)}(I; \mathbb{C})] \times \mathbb{R} \times \mathbb{R} \times [\Pi \cap H^2_{(0)}(I; \mathbb{C})] \times \mathbb{R} \times \mathbb{R}$$
$$\to [T_{\mathbb{S}}(f_\theta(0)) \cap H^2_{(0)}(I; \mathbb{C})] \times \mathbb{R} \times \mathbb{R} \times L^2(0, T),$$

or

$$[T_{\mathbb{S}}(f_\theta(0)) \cap H^3_{(0)}(I; \mathbb{C})] \times \mathbb{R} \times \mathbb{R} \times [\Pi \cap H^3_{(0)}(I; \mathbb{C})] \times \mathbb{R} \times \mathbb{R}$$
$$\to [T_{\mathbb{S}}(f_\theta(0)) \cap H^3_{(0)}(I; \mathbb{C})] \times \mathbb{R} \times \mathbb{R} \times H^1_0(0, T).$$

The controllability up to codimension one given in Proposition 9.11 for the linearized system around $(Y^\theta, u = 0)$ only provides a right inverse for $d\Phi(f_\theta(0), 0, 0, 0)$ which is a map between the spaces

$$[T_{\mathbb{S}}(f_\theta(0)) \cap H^3_{(0)}(I; \mathbb{C})] \times \mathbb{R} \times \mathbb{R} \times [\Pi \cap H^3_{(0)}(I; \mathbb{C})] \times \mathbb{R} \times \mathbb{R}$$
$$\to [T_{\mathbb{S}}(f_\theta(0)) \cap H^3_{(0)}(I; \mathbb{C})] \times \mathbb{R} \times \mathbb{R} \times L^2(0, T).$$

In order to deal with this loss of regularity in the controllability of the linearized system around $(Y^\theta, u = 0)$, we use a Nash-Moser fixed-point method directly adapted from the one given by Lars Hörmander in [**238**]. See also Section 4.2.2 and the paper [**40**] by Karine Beauchard, where the Nash-Moser method has been used for the first time to solve a controllability problem. This requires some lengthy computations which are detailed in [**43**].

9.2.2. Local controllability of the control system (9.1)-(9.2)-(9.3) around $(Y^k, u = 0)$ for $k \in \{0, 1\}$. Again, the classical approach does not work because the linearized system around $(Y^k, u = 0)$ is not controllable for $k \in \{0, 1\}$. This result was proved by Pierre Rouchon in [**417**]. (He also proved that this linearized system is steady-state controllable, but this result does not imply the same property for the nonlinear system.) The situation is even worse than the previous one because the linearized system misses an infinite number of (oriented) directions (half of the projections). Indeed, the linearized system around $(Y^0, u = 0)$ is

$$(\Sigma_0^l) \begin{cases} \Psi_t = i\Psi_{xx} + iwx\psi_1, \ (t, x) \in (0, T) \times I, \\ \Psi(t, -1) = \Psi(t, 1) = 0, \ t \in (0, T), \\ \dot{s} = w, \\ \dot{d} = s. \end{cases}$$

It is a control system where, at time $t \in [0, T]$,
- the state is $(\Psi(t, \cdot), s(t), d(t)) \in L^2(I; \mathbb{C}) \times \mathbb{R} \times \mathbb{R}$ with $\Psi(t) \in T_{\mathbb{S}}(\psi_1(t))$ for every t,
- the control is $w(t) \in \mathbb{R}$.

Let $(\Psi^0, s^0, d^0) \in T_{\mathbb{S}}(\psi_1(0)) \times \mathbb{R} \times \mathbb{R}$ and (Ψ, s, d) be the solution of (Σ_0^l) such that $(\Psi(0), s(0), d(0)) = (\Psi^0, s^0, d^0)$, with some control $w \in L^2(0, T)$. We have the following equality in $L^2(I; \mathbb{C})$:

$$\Psi(t) = \sum_{k=1}^{\infty} x_k(t)\varphi_k, \text{ where } x_k(t) := \langle \Psi(t), \varphi_k \rangle, \forall k \in \mathbb{N} \setminus \{0\}.$$

Using the evenness of the functions φ_{2k+1} and the equation satisfied by Ψ, we get

$$\dot{x}_{2k+1} = -i\lambda_{2k+1} x_{2k+1}, \forall k \in \mathbb{N}.$$

Half of the components have a dynamic independent of the control w. Thus the control system (Σ_0^l) is not controllable and the dimension of the uncontrollable part is infinite.

The proof of the local controllability of the control system (9.1)-(9.2)-(9.3) around Y^k for $k \in \{0, 1\}$ relies on:

1. the return method (see Chapter 6),
2. quasi-static deformations (see Chapter 7),
3. power series expansion (see Chapter 8).

Let us explain the main steps in the case of Y^0; everything works similarly with Y^1 instead of Y^0. First, for $\gamma \neq 0$ and $(\alpha, \beta) \in \mathbb{R}^2$ we construct a trajectory $(Y^{\gamma,\alpha,\beta}, u = \gamma)$ such that the control system (9.1)-(9.2)-(9.3) is locally controllable around $(Y^{\gamma,\alpha,\beta}, \gamma)$ on $[0, T^*]$ for some T^*. Then one deduces the local controllability (for suitable norms) around Y^0 by using quasi-static deformations, in the same way as in Section 6.3 (see also [**40**, Theorem 11, page 916]). Let Y_0 be given close to $Y^0(0)$ and Y_1 be given close to $Y^0(0)$. We use quasi-static deformations in order to steer the control system:

- from Y_0 to a point $\tilde{Y}_0$ which is close to $Y^{\gamma,\alpha,\beta}(0)$, for some real constants α, β, γ,
- from a point $\tilde{Y}_1$, which is close to $Y^{\gamma,\alpha,\beta}(T^*)$, to Y_1.

By "quasi-static deformations", we mean that we use controls $t \mapsto u(t)$ which change slowly as in Section 6.3 (see in particular Proposition 6.31 on page 217 and Proposition 6.32 on page 217) and as in Chapter 7. Using the local controllability around $Y^{\gamma,\alpha,\beta}$, we can then steer the control system from $\tilde{Y}_0$ to $\tilde{Y}_1$ in time T^*, and this gives the conclusion: We can steer the control system from Y_0 to $\tilde{Y}_1$, by steering it successively to $\tilde{Y}_0$, to $\tilde{Y}_1$ and, finally, to Y_1.

Let us give the construction of $Y^{\gamma,\alpha,\beta}$. Let $\gamma \in \mathbb{R} \setminus \{0\}$. The ground state for $u = \gamma$ is the function
$$\psi_{1,\gamma}(t,x) := \varphi_{1,\gamma}(x) e^{-i\lambda_{1,\gamma} t},$$
where $\lambda_{1,\gamma}$ is the first eigenvalue and $\varphi_{1,\gamma}$ the associated normalized eigenvector of the operator A_γ defined on
$$D(A_\gamma) := H^2 \cap H_0^1(I; \mathbb{C})$$
by
$$A_\gamma \varphi := -\varphi'' - \gamma x \varphi.$$
When $\alpha, \beta \in \mathbb{R}$, the function
$$Y^{\gamma,\alpha,\beta}(t) := (\psi_{1,\gamma}(t, \cdot), \alpha + \gamma t, \beta + \alpha t + \gamma t^2/2)$$
solves (9.1)-(9.2)-(9.3) with $u = \gamma$. We define $T := 8/\pi$, $T^* := 2T$ and the space
$$H_{(\gamma)}^7(I; \mathbb{C}) := \{\varphi \in H^7(I; \mathbb{C}); A_\gamma^n \varphi \text{ exists and is in } H_0^1(I; \mathbb{C}) \text{ for } n = 0, 1, 2, 3\}.$$
Then one has the following theorem.

THEOREM 9.13 ([**43**, Theorem 8]). *There exists $\gamma_0 > 0$ such that, for every $\gamma \in (0, \gamma_0)$, there exists $\delta = \delta(\gamma) > 0$, such that, for every $(\psi^0, S^0, D^0), (\psi^1, S^1, D^1) \in [\mathbb{S} \cap H_{(\gamma)}^7(I; \mathbb{C})] \times \mathbb{R} \times \mathbb{R}$ and for every $(\alpha, \beta) \in \mathbb{R}^2)$ with*
$$\|\psi^0 - \psi_{1,\gamma}(0)\|_{H^7(I;\mathbb{C})} + |S^0 - \alpha| + |D^0 - \beta| < \delta,$$
$$\|\psi^1 - \psi_{1,\gamma}(T^*)\|_{H^7(I;\mathbb{C})} + |S^1 - \alpha - \gamma T^*| + |D^1 - \beta - \alpha T^* - \gamma T^{*2}/2| < \delta,$$

9.2. SKETCH OF THE PROOF OF THE CONTROLLABILITY IN LARGE TIME

for some real constants α, β, there exists $v \in H_0^1((0,T^*),\mathbb{R})$ such that the unique solution
$$(\Psi, S, D) \in (C^0([0,T^*];H^3(I;\mathbb{C})) \cap C^1([0,T^*];H_0^1(I;\mathbb{C}))) \times C^1([0,T^*]) \times C^0([0,T^*])$$
of (9.1)-(9.2)-(9.3) on $[0,T^*]$, with control $u := \gamma + v$, such that
$$(\psi(0), S(0), D(0)) = (\psi^0, S^0, D^0),$$
satisfies
$$(\psi(T^*), S(T^*), D(T^*)) = (\psi^1, S^1, D^1).$$

Let us explain the main ingredients (and main difficulties) of the proof of this theorem. The linearized control system around $(Y^{\gamma,\alpha,\beta}, u = \gamma)$ is

$$(\Sigma_\gamma^l) \begin{cases} \Psi_t = i\Psi_{xx} + i\gamma x \Psi + iwx\psi_{1,\gamma}, \ (t,x) \in (0,T) \times I, \\ \Psi(t,-1) = \Psi(t,1) = 0, \ t \in (0,T), \\ \dot{s} = w, \\ \dot{d} = s. \end{cases}$$

It is a control system where, at time t,
- The state is $(\Psi(t,\cdot), s(t), d(t)) \in L^2(I;\mathbb{C})$ with $\Psi(t,\cdot) \in T_\mathbb{S}(\psi_{1,\gamma}(t,\cdot))$,
- The control is $w(t) \in \mathbb{R}$.

Let us study the controllability of the linear control system (Σ_γ^l). Let us recall (see page 95) that the space $L^2(I;\mathbb{C})$ has a complete orthonormal system $(\varphi_{k,\gamma})_{k \in \mathbb{N} \setminus \{0\}}$ of eigenfunctions for the operator A_γ. One has
$$A_\gamma \varphi_{k,\gamma} = \lambda_{k,\gamma} \varphi_{k,\gamma},$$
where $(\lambda_{k,\gamma})_{k \in \mathbb{N} \setminus \{0\}}$ is an increasing sequence of positive real numbers. Let
$$b_{k,\gamma} := \langle \varphi_{k,\gamma}, x\varphi_{1,\gamma} \rangle.$$

By Proposition 2.90 on page 98 (see also [40, Proposition 1, Section 3.1] or [40, Proposition 41, page 937-938]), for γ small enough and different from zero, $b_{k,\gamma}$ is different from zero for every $k \in \mathbb{N} \setminus \{0\}$ and, roughly speaking, behaves like $1/k^3$ when $k \to +\infty$. Let us first give a negative result showing that the linear control system (Σ_γ^l) is not controllable (in contrast to the case where one does not care about s and d; see Theorem 2.87 on page 96).

PROPOSITION 9.14. *Let $T > 0$ and (Ψ, s, d) be a trajectory of (Σ_γ^l) on $[0,T]$. Then, for every $t \in [0,T]$, we have*

(9.76) $$s(t) = s(0) + \frac{1}{ib_{1,\gamma}} \left(\langle \Psi(t), \varphi_{1,\gamma} \rangle e^{i\lambda_{1,\gamma} t} - \langle \Psi(0), \varphi_{1,\gamma} \rangle \right).$$

Thus, the control system (Σ_γ^l) is not controllable.

Proof of Proposition 9.14. Let $x_1(t) := \langle \Psi(t), \varphi_{1,\gamma} \rangle$. We have
$$\dot{x}_1(t) = \langle \frac{\partial \Psi}{\partial t}(t), \varphi_{1,\gamma} \rangle = \langle -iA_\gamma \Psi(t) + iw(t)x\psi_{1,\gamma}(t,\cdot), \varphi_{1,\gamma} \rangle,$$
which leads to
$$x_1(t) = \left(x_1(0) + ib_{1,\gamma} \int_0^t w(\tau)d\tau \right) e^{-i\lambda_{1,\gamma} t}.$$
We get (9.76) by using
$$s(t) = s(0) + \int_0^t w(\tau)d\tau.$$

Let $T > 0$, $\Psi^0 \in T_\mathbb{S}(\psi_{1,\gamma}(0))$, $\Psi^1 \in T_\mathbb{S}(\psi_{1,\gamma}(T))$, $s^0, s^1 \in \mathbb{R}$. A necessary condition for the existence of a trajectory of (Σ_γ^l) such that $\Psi(0) = \Psi^0$, $s(0) = s^0$, $\Psi(T) = \Psi^1$, $s(T) = s^1$ is

$$(9.77) \qquad s^1 - s^0 = \frac{1}{ib_{1,\gamma}} \left(\langle \Psi^1, \varphi_{1,\gamma} \rangle e^{i\lambda_{1,\gamma} T} - \langle \Psi^0, \varphi_{1,\gamma} \rangle \right).$$

This equality does not hold for an arbitrary choice of Ψ^0, Ψ^1, s^0, s^1. Thus (Σ_γ^l) is not controllable. This concludes the proof of Proposition 9.14. ∎

Our next proposition shows that (9.77) is in fact the only obstruction to controllability for the linear control system (Σ_γ^l).

PROPOSITION 9.15. *Let $T > 0$, $(\Psi^0, s^0, d^0), (\Psi^1, s^1, d^1) \in H^3_{(0)}(I; \mathbb{C}) \times \mathbb{R} \times \mathbb{R}$ be such that*

$$(9.78) \qquad \Re\langle \Psi^0, \psi_{1,\gamma}(0) \rangle = \Re\langle \Psi^1, \psi_{1,\gamma}(T) \rangle = 0,$$

$$(9.79) \qquad s^1 - s^0 = \frac{i}{b_{1,\gamma}} \left(\langle \Psi^0, \varphi_{1,\gamma} \rangle - \langle \Psi^1, \varphi_{1,\gamma} \rangle e^{i\lambda_{1,\gamma} T} \right).$$

Then there exists $w \in L^2(0,T)$ such that the solution of (Σ_γ^l) with control w and such that $(\Psi(0), s(0), d(0)) = (\Psi^0, s^0, d^0)$ satisfies $(\Psi(T), s(T), d(T)) = (\Psi^1, s^1, d^1)$.

REMARK 9.16. We can control Ψ and d but we cannot control s. We miss only two (oriented) directions which are $(\Psi, s, d) = (0, \pm 1, 0)$.

Proof of Proposition 9.15. Let $(\Psi^0, s^0, d^0) \in T_\mathbb{S}(\psi_{1,\gamma}(0)) \times \mathbb{R} \times \mathbb{R}$ and $T > 0$. Let (Ψ, s, d) be a solution of (Σ_γ^l) with $(\Psi(0), s(0), d(0)) = (\Psi^0, s^0, d^0)$ and a control $w \in L^2(0,T)$. Let $(\Psi^1, s^1, d^1) \in T_\mathbb{S}(\psi_{1,\gamma}(T)) \times \mathbb{R} \times \mathbb{R}$. The equality $(\Psi(T), s(T), d(T)) = (\Psi^1, s^1, d^1)$ is equivalent to the following moment problem on w:

$$(9.80) \qquad \int_0^T w(t) e^{i(\lambda_{k,\gamma} - \lambda_{1,\gamma})t} dt$$
$$= \frac{i}{b_{k,\gamma}} \left(\langle \Psi^0, \varphi_{k,\gamma} \rangle - \langle \Psi^1, \varphi_{k,\gamma} \rangle e^{i\lambda_{k,\gamma} T} \right), \forall k \in \mathbb{N} \setminus \{0\},$$

$$(9.81) \qquad \int_0^T w(t) dt = s^1 - s^0,$$

$$(9.82) \qquad \int_0^T (T - t) w(t) dt = d^1 - d^0 - s^0 T.$$

The left hand sides of (9.80) and (9.81) with $k = 1$ are equal, the equality of the corresponding right hand sides is guaranteed by (9.79). Under the assumption $\Psi^0, \Psi^1 \in H^3_{(0)}(I; \mathbb{C})$, the right hand side of (9.81) defines a sequence in $l^2(\mathbb{C})$. Thus, under the assumptions (9.79) and $\Psi^0, \Psi^1 \in H^3_{(0)}(I; \mathbb{C})$, the existence of a solution $w \in L^2(0,T)$ of the moment problem (9.80)-(9.81)-(9.81) can be proved using the same theorem on moment problems as in the proof of Proposition 9.11 on page 256 (see also [**40**, Theorem 5]). This concludes the proof of Proposition 9.15. ∎

Two difficulties remain:

1. First one has to take care of the fact that the linearized control system misses two (oriented) directions. In order to take care of this problem, one uses again power series expansion as in Section 9.2.1.
2. As in Section 4.2.2, for the Schrödinger control system (4.85)-(4.86) (i.e., our control system (9.1)-(9.2)-(9.3) but without taking care about S and D) there is a problem of loss of derivatives. One handles this problem with a Nash-Moser iterative scheme briefly sketched in Section 4.2.2 and due to Karine Beauchard [**40**].

We refer to [**43**] for the full details. This concludes the sketch of the proof of Theorem 9.1 on page 248. ∎

9.3. Proof of the nonlocal controllability in small time

In this section we give the proof of Theorem 9.8 on page 251. First, we make some comments about this theorem.

9.3.1. Comments on and "raison d'être" of Theorem 9.8 on page 251.

Since we are looking for a local statement, it seems natural to first look at the case where one replaces the control system (9.1)-(9.2)-(9.3) with its linear approximation along the trajectory $((\psi_1, 0, 0), 0)$. This linear approximation is the control system

$$\psi_t = i\psi_{xx} + iux\psi_1, \ (t, x) \in (0, T) \times I, \tag{9.83}$$

$$\psi(t, -1) = \psi(t, 1) = 0, \ t \in (0, T), \tag{9.84}$$

$$\dot{S}(t) = u(t), \ \dot{D}(t) = S(t), \ t \in (0, T). \tag{9.85}$$

For this control system, the control at time t is $u(t) \in \mathbb{R}$ and the state is $\psi(t, \cdot) \in L^2(I; \mathbb{C})$, with now

$$\int_I (\psi(t,x)\bar{\psi}_1(t,x) + \bar{\psi}(t,x)\psi_1(t,x))dx = 2. \tag{9.86}$$

However, it has been proved by Pierre Rouchon in [**417**] that Theorem 9.8 does not hold for the linear control system (9.83)-(9.84)-(9.85). More precisely, Pierre Rouchon proves in [**417**] the following theorem.

THEOREM 9.17. *Let $T > 0$. Then there exists $C > 0$ such that, for every $\bar{D} \in \mathbb{R}$, there exists $u \in L^\infty(0, T)$ such that the solution*

$$(\psi, S, D) \in C^0([0, T]; H_0^1(I; \mathbb{C})) \times C^0([0, T]; \mathbb{R}) \times C^1([0, T]; \mathbb{R})$$

of the Cauchy problem

$$\psi_t = i\psi_{xx} + iux\psi_1, \ (t, x) \in (0, T) \times I, \tag{9.87}$$

$$\psi(t, -1) = \psi(t, 1) = 0, \ t \in (0, T), \tag{9.88}$$

$$\dot{S}(t) = u(t), \ \dot{D}(t) = S(t), \ t \in (0, T), \tag{9.89}$$

$$\psi(0, x) = \psi_1(0, x), \ x \in I, \tag{9.90}$$

$$S(0) = 0, \ D(0) = 0, \tag{9.91}$$

satisfies

$$\psi(T, x) = \psi_1(T, x), \ x \in I, \tag{9.92}$$

$$S(T) = 0, \ D(T) = \bar{D}. \tag{9.93}$$

Furthermore, u satisfies

$$\|u\|_{L^\infty(0,T)} \leqslant C|\bar{D}|.$$

Let us point out that it is also proved in [**417**] that the linear control system (9.83)-(9.84)-(9.85) is not controllable on $[0,T]$, whatever $T > 0$ is. But, by Theorem 9.1 on page 248, the nonlinear control system (9.1)-(9.2)-(9.3) is large-time locally controllable along the trajectory $((\psi_1, 0, 0), 0)$. More precisely, by (the proof of) Theorem 9.1, for every $\varepsilon > 0$, there exist $\eta > 0$ and $T > 0$ such that, for every $(\psi^0, S^0, D^0), (\psi^1, S^1, D^1) \in (\mathbb{S} \cap H^7_{(0)}(I; \mathbb{C})) \times \mathbb{R} \times \mathbb{R}$ with

$$\|\psi^0 - \psi_1(0)\|_{H^7(I;\mathbb{C})} + |S^0| + |D^0| < \eta,$$

$$\|\psi^1 - \psi_1(T)\|_{H^7(I;\mathbb{C})} + |S^1| + |D^1| < \eta,$$

there exists (ψ, S, D, u) such that

$$\psi \in C^0([0,T]; H^3_{(0)}(I;\mathbb{C})) \cap C^1([0,T]; H^1_0(I;\mathbb{C})),$$

$$u \in H^1_0(0,T),$$

$$S \in C^1([0,T]), \ D \in C^2([0,T]),$$

$$(9.1), (9.2) \text{ and } (9.3) \text{ hold},$$

$$(\psi(0), S(0), D(0)) = (\psi^0, S^0, D^0),$$

$$(\psi(T), S(T), D(T)) = (\psi^1, S^1, D^1),$$

$$\|\psi(t) - \psi_1(t)\|_{H^3(I;\mathbb{C})} + |S(t)| + |D(t)| + |u(t)| \leqslant \varepsilon, \ \forall t \in [0,T].$$

So, in some sense, the nonlinearity helps to recover local controllability but prevents us from doing some specific natural motions if the time T is too small; motions which are possible for the linear control system (9.83)-(9.84)-(9.85) even if $T > 0$ is small.

Let us now explain why Theorem 9.8 is in fact rather natural if one looks at the obstruction to small-time local controllability given by Theorem 3.36 on page 145 for finite-dimensional control systems. In order to deal with an equilibrium (as in Theorem 3.36 on page 145), let $\theta(t,x) = e^{i\lambda_1 t}\psi(t,x)$. Then the control system for θ is the following one:

$$\theta_t = i\theta_{xx} + i\lambda_1\theta + ixu\theta, \ (t,x) \in (0,T) \times I,$$

$$\theta(t,-1) = \theta(t,1) = 0, \ t \in (0,T),$$

$$\dot{S}(t) = u(t), \ \dot{D}(t) = S(t), \ t \in (0,T).$$

At a formal level, it is an affine system similar to the control system (3.84), with

$$f_0((\theta, S, D)^{\text{tr}}) = (i\theta_{xx} + i\lambda_1\theta, 0, S)^{\text{tr}}, \ f_1((\theta, S, D)^{\text{tr}}) = (ix\theta, 1, 0)^{\text{tr}}.$$

Note that

$$f_0((\varphi_1, 0, 0)^{\text{tr}}) = (0, 0, 0)^{\text{tr}}.$$

Hence (3.83) holds (at $(\varphi_1, 0, 0)^{\text{tr}}$ instead of 0). Formally again (see in particular (3.19)), we get

(9.94) $\qquad [f_1, f_0]((\theta, S, D)^{\text{tr}}) = (-2\theta_x, 0, 1)^{\text{tr}},$

(9.95) $\qquad \text{ad}^2_{f_1} f_0((\theta, S, D)^{\text{tr}}) = (-2i\theta, 0, 0)^{\text{tr}}.$

In particular,

(9.96) $\qquad \text{ad}^2_{f_1} f_0((\varphi_1, 0, 0)^{\text{tr}}) = (-2i\varphi_1, 0, 0)^{\text{tr}}.$

(For the definition of $\mathrm{ad}_{f_1}^p f_0$, $p \in \mathbb{N}$, see Definition 3.11 on page 130.) If one wants to check the assumption of Theorem 3.36 on page 145, one needs to compute $\mathrm{ad}_{f_0}^k f_1$. Unfortunately, these iterated Lie brackets are not well defined; even at a formal level, Dirac masses appear on the boundary of I if $k \geqslant 2$ (see also (5.5)). However, at least in finite dimension, $\mathrm{Span}\{\mathrm{ad}_{f_0}^k f_1(x_0); k \in \mathbb{N}\}$ is the linear space of states which can be reached from the state 0 for the linearized control system around the equilibrium x_0. Let us point out that the linearized control system around the equilibrium $((\varphi_1, 0, 0)^{\mathrm{tr}}, 0)$ is the control system

(9.97) $$\Theta_t = i\Theta_{xx} + i\lambda_1 \Theta + ixu\varphi_1,\ (t,x) \in (0,T) \times I,$$

(9.98) $$\Theta(t,-1) = \Theta(t,1) = 0,\ t \in (0,T),$$

(9.99) $$\dot{S}(t) = u(t),\ \dot{D}(t) = S(t),\ t \in (0,T),$$

where, at time $t \in [0,T]$, the state is $(\Theta(t,\cdot), S, D) \in H_0^1(I; \mathbb{C}) \times \mathbb{R} \times \mathbb{R}$ with $\int_I (\Theta(t,x) + \bar{\Theta}(t,x))\varphi_1(x)dx = 0$ and the control is $u(t) \in \mathbb{R}$. For this linear control system, by symmetry, for every $u \in L^2(0,T)$, the solution $\Theta \in C^0([0,T]; H_0^1(I; \mathbb{C}))$ of (9.97)-(9.98)-(9.99) such that $\Theta(0,\cdot) = 0$ satisfies

$$\Theta(t,-x) = -\Theta(t,x),\ (t,x) \in (0,T) \times I.$$

(In fact, using moments method as in the proof of Proposition 9.11 on page 256 and as in the proof of Proposition 9.15 on page 262, it is not hard to check that, at least if one does not take care about regularity problems, the set of $(\theta, S, D)^{\mathrm{tr}}$ such that the linear control system (9.97)-(9.98)-(9.99) can be steered from $(0,0,0)^{\mathrm{tr}}$ to $(\theta, S, D)^{\mathrm{tr}}$ is the set of all the $(\theta, S, D)^{\mathrm{tr}}$'s such that θ is odd.) Hence, still at a formal level,

$$\mathrm{Span}\{\mathrm{ad}_{f_0}^k f_1((\varphi_1, 0, 0)^{\mathrm{tr}}); k \in \mathbb{N}\} \subset \{(\varphi, S, D)^{\mathrm{tr}};\ \varphi\ \text{is odd},\ S \in \mathbb{R},\ D \in \mathbb{R}\},$$

which, together with (9.95), implies that

$$\mathrm{ad}_{f_1}^2 f_0((\varphi_1, 0, 0)^{\mathrm{tr}}) \notin \mathrm{Span}\{\mathrm{ad}_{f_0}^k f_1((\varphi_1, 0, 0)^{\mathrm{tr}}); k \in \mathbb{N}\},$$

which, by Theorem 3.36 on page 145 for $k = 1$, motivates the uncontrollability result stated in Theorem 9.8.

However, the previous motivation is purely formal. Let us try to follow the proof of Theorem 3.36 sketched above (see in particular page 145). One first needs to choose a suitable $\phi : (H_0^1(I; \mathbb{C}) \cap \mathbb{S}) \times \mathbb{R} \times \mathbb{R} \to \mathbb{R}$. This ϕ must be such that

(9.100) $$\phi(\varphi_1, 0, 0) = 0,$$

(9.101) $\phi'(\varphi_1, 0, 0)(\theta, 0, 0) = 0$, $\forall \theta \in H_0^1(I; \mathbb{C})$ such that $\theta(-x) = -\theta(x)$, $\forall x \in I$,

(9.102) $$\phi'(\varphi_1, 0, 0)(2i\varphi_1, 0, 0) = 1,$$

(9.103) $$\phi'(\theta, S, D)(ix\theta, 1, 0) = 0, \forall (\theta, S, D) \in (H_0^1(I; \mathbb{C}) \cap \mathbb{S}) \times \mathbb{R} \times \mathbb{R}$$
$$\text{close enough to } (\varphi_1, 0, 0).$$

There are many ϕ's which satisfy (9.100), (9.101), (9.102) and (9.103). The simplest one seems to be

(9.104) $$\phi(\theta, S, D) := -\frac{1}{4i}(\theta(0) - \bar{\theta}(0)).$$

But if one takes such a ϕ, it seems that some problems appear in order to get (3.108). These problems come from the fact that the norms $H^k(I; \mathbb{C})$, $k \in \mathbb{N}$ are not equivalent. We will choose another ϕ (ϕ is one half of the real part of V, where V is defined by (9.116)). This ϕ satisfies (9.100), (9.101) and (9.102), but does not satisfy (9.103).

9.3.2. Proof of Theorem 9.8. Let $T > 0$ be such that (9.41) holds. Let $\varepsilon \in (0,1]$. Let $u \in L^2((0,T);\mathbb{R})$ be such that

(9.105) $$|u(t)| < \varepsilon,\, t \in (0,T).$$

Let $(\psi, S, D) \in C^0([0,T]; H_0^1(I;\mathbb{C})) \times C^0([0,T]) \times C^1([0,T])$ be the solution of the Cauchy problem

(9.106) $$\psi_t = i\psi_{xx} + iu(t)x\psi,\, (t,x) \in (0,T) \times I,$$
(9.107) $$\psi(t,-1) = \psi(t,1) = 0,\, t \in (0,T),$$
(9.108) $$\dot{S}(t) = u(t),\, \dot{D}(t) = S(t),\, t \in (0,T),$$
(9.109) $$\psi(0,x) = \psi_1(0,x),\, x \in I,$$
(9.110) $$S(0) = 0,\, D(0) = 0.$$

(Let us recall that ψ_1 is defined in (9.4) and (9.7).) We assume that

(9.111) $$S(T) = 0.$$

Let $\theta \in C^0([0,T]; H_0^1(I;\mathbb{C}))$ be defined by

(9.112) $$\theta(t,x) := e^{i\lambda_1 t}\psi(t,x),\, (t,x) \in (0,T) \times I.$$

From (9.4), (9.7), (9.106), (9.107), (9.109) and (9.112), we have

(9.113) $$\theta_t = i\theta_{xx} + i\lambda_1\theta + iux\theta,\, (t,x) \in (0,T) \times I,$$
(9.114) $$\theta(t,-1) = \theta(t,1) = 0,\, t \in (0,T),$$
(9.115) $$\theta(0,x) = \varphi_1(x),\, x \in I.$$

Let $V \in C^0([0,T];\mathbb{C})$ be defined by

(9.116) $$V(t) := -i + i\int_I \theta(t,x)\varphi_1(x)dx,\, t \in [0,T].$$

From (9.7), (9.115) and (9.116), we have

(9.117) $$V(0) = 0.$$

From (9.7), (9.113), (9.114) and (9.116), we get, using integrations by parts,

(9.118) $$\begin{aligned}\dot{V} &= -\int_I (\theta_{xx} + \lambda_1\theta + ux\theta)\varphi_1 dx \\ &= -u\int_I \theta x\varphi_1 dx.\end{aligned}$$

Let $V_1 \in C^0([0,T];\mathbb{C})$ be defined by

(9.119) $$V_1(t) := -\int_I \theta(t,x)x\varphi_1(x)dx,\, t \in [0,T].$$

9.3. PROOF OF THE NONLOCAL CONTROLLABILITY IN SMALL TIME

From (9.7), (9.113), (9.114) and (9.116), we get, using integrations by parts,

$$
\begin{aligned}
\dot V_1 &= -i\int_I (\theta_{xx} + \lambda_1 \theta + ux\theta)x\varphi_1 dx \\
&= -i\int_I \theta((x\varphi_1)_{xx} + \lambda_1 x\varphi_1)dx - iu\int_I x^2\theta\varphi_1 dx \\
&= -2i\int_I \theta\varphi_{1x}dx - iu\int_I \theta x^2 \varphi_1 dx.
\end{aligned}
\tag{9.120}
$$

From (9.108), (9.110), (9.111), (9.117), (9.118), (9.119), (9.120) and using integrations by parts, one gets

$$
V(T) = \int_0^T S(t)V_{20}(t)dt + \int_0^T S(t)^2 V_{21}(t)dt,
\tag{9.121}
$$

where $V_{20} \in C^0([0,T];\mathbb{C})$ and $V_{21} \in L^\infty((0,T);\mathbb{C})$ are defined by

$$
V_{20}(t) := 2i\int_I \theta(t,x)\varphi_{1x}(x)dx,\ t \in [0,T],
\tag{9.122}
$$

$$
V_{21}(t) := -\frac{i}{2}\int_I \theta_t(t,x)x^2\varphi_1(x)dx,\ t \in [0,T].
\tag{9.123}
$$

Let us first estimate $V_{20}(t)$. Let $f \in C^0([0,T]; H_0^1(I;\mathbb{C}))$ be defined by

$$
f(t,x) := \varphi_1(x)e^{ixS(t)},\ (t,x) \in [0,T] \times I.
\tag{9.124}
$$

Let $g \in C^0([0,T]; L^2(I;\mathbb{C}))$ be the solution of the Cauchy problem

$$
g_t = ig_{xx} + i\lambda_1 g - 2S(t)\varphi_{1x},\ (t,x) \in (0,T) \times I,
\tag{9.125}
$$

$$
g(t,-1) = g(t,1) = 0,\ t \in (0,T),
\tag{9.126}
$$

$$
g(0,x) = 0,\ x \in I.
\tag{9.127}
$$

Finally, let $r \in C^0([0,T]; L^2(I;\mathbb{C}))$ be defined by

$$
r(t,x) := \theta(t,x) - f(t,x) - g(t,x),\ (t,x) \in [0,T] \times I.
\tag{9.128}
$$

From (9.7), (9.108), (9.110), (9.113), (9.114), (9.115), (9.124), (9.125) (9.126), (9.127) and (9.128), we get that

$$
\begin{aligned}
r_t =&\, ir_{xx} + i\lambda_1 r + iuxr - iS^2\varphi_1 e^{ixS} \\
&+ 2S\varphi_{1x}(1 - e^{ixS}) + iuxg,\ (t,x) \in (0,T) \times I,
\end{aligned}
\tag{9.129}
$$

$$
r(t,-1) = r(t,1) = 0,\ t \in (0,T),
\tag{9.130}
$$

$$
r(0,x) = 0,\ x \in I.
\tag{9.131}
$$

From (9.7), (9.125), (9.126) and (9.127), we have

$$
\|g(t,\cdot)\|_{L^2(I;\mathbb{C})} \leqslant 2\|\varphi_{1x}\|_{L^2(I)}\int_0^t |S(\tau)|d\tau = \pi\int_0^t |S(\tau)|d\tau.
\tag{9.132}
$$

From (9.105), (9.108), (9.110), (9.129), (9.130), (9.131) and (9.132), we get

$$
\begin{aligned}
\|r(t,\cdot)\|_{L^2(I;\mathbb{C})} &\leqslant \int_0^t S^2(\tau) + 2S^2(\tau)\|\varphi_{1x}\|_{L^2(I)} + |u(\tau)|\|g(\tau,\cdot)\|_{L^2(I;\mathbb{C})}d\tau \\
&\leqslant C\varepsilon\|S\|_{L^2(0,T)},\ t \in [0,T].
\end{aligned}
\tag{9.133}
$$

In (9.133) and in the following, C denotes various positive constants which may vary from line to line but are independent of $T > 0$ satisfying (9.41), of $t \in [0, T]$ and of $u \in L^2(0, T)$ satisfying (9.105).

From (9.105), (9.108) and (9.124), we get

$$(9.134) \qquad \left| \int_I f(t, x) \varphi_{1x}(x) dx + (iS(t)/2) \right| \leqslant C\varepsilon |S(t)|, \; t \in [0, T].$$

Therefore, using (9.122), (9.132), (9.133) and (9.134), we have

$$(9.135) \qquad |V_{20}(t) - S(t)| \leqslant 2\pi \|\varphi_{1x}\|_{L^2(I)} \int_0^t |S(\tau)| d\tau$$
$$+ C\varepsilon (\|S\|_{L^2(0,T)} + |S(t)|), \; t \in [0, T].$$

Let us now estimate $V_{21}(t)$. From (9.7), (9.113), (9.123) and using integrations by parts, we get

$$(9.136) \qquad 2V_{21} = \int_I (\theta(2\varphi_1 + 4x\varphi_{1x}) + u\theta x^3 \varphi_1) dx.$$

From (9.105), (9.124), (9.128), (9.132) and (9.133), we get

$$(9.137) \qquad \|\theta(t, \cdot) - \varphi_1\|_{L^2(I;\mathbb{C})} \leqslant C\varepsilon, \; t \in [0, T].$$

Using (9.7), one easily checks that

$$(9.138) \qquad \int_I \varphi_1 (2\varphi_1 + 4x\varphi_{1x}) dx = 0,$$

which, together with (9.105), (9.136) and (9.137), leads to

$$(9.139) \qquad |V_{21}(t)| \leqslant C\varepsilon, \; t \in [0, T].$$

From (9.121), (9.135) and (9.139), one gets

$$(9.140) \qquad |V(T) - \|S\|_{L^2(0,T)}^2| \leqslant \pi \|\varphi_{1x}\|_{L^2(I)} \left(\int_0^T |S(t)| dt \right)^2 + C\varepsilon \|S\|_{L^2(0,T)}^2$$
$$\leqslant \left(\frac{\pi^2}{2} T + C\varepsilon \right) \|S\|_{L^2(0,T)}^2.$$

(Note that (9.140) is similar to (3.108).) Since

$$(V(T) \neq 0) \Rightarrow (\theta(T, \cdot) \neq \varphi_1) \Rightarrow (\psi(T, \cdot) \neq \psi_1(T, \cdot)),$$

this concludes the proof of Theorem 9.8. ∎

Let us end this section with two open problems and a remark.

Looking at our result (Theorem 6.24 on page 205) on the the 1-D tank control system (see Section 6.3), one may state the following open problem.

OPEN PROBLEM 9.18. *Is it possible to remove assumption (9.41) in Theorem 9.8?*

Note that Theorem 6.24 on page 205 tells us that Open Problem 9.18 adapted to the 1-D tank control system has a negative answer. Note, however, that, for a natural adaptation to the Euler equations of incompressible fluids, the answer to Open problem 9.18 is in general positive, at least for strong enough topologies. More precisely, with the notations of Section 6.2.1, even with $\Gamma_0 = \partial \Omega$ but with

$\Omega \not\subset \overline{\Omega}_0$, given $T > 0$ and $k \in \mathbb{N} \setminus \{0\}$, there exists $\varepsilon > 0$ such that, for every $\eta > 0$, there exists $y^0 \in C^k(\overline{\Omega})$ vanishing in a neighborhood of $\partial\Omega$ satisfying the divergence free condition (6.27) as well as $\|y^0\|_{C^k(\overline{\Omega})} \leqslant \eta$ such that there is no trajectory $(y, p) \in C^1([0,T] \times \overline{\Omega})$ of the Euler control system (6.24) satisfying

(9.141) $$y(0, \cdot) = y^0, \ y(T, \cdot) = 0,$$
(9.142) $$\|y(t, \cdot)\|_{C^1(\overline{\Omega})} \leqslant \varepsilon, \ \forall t \in [0, T].$$

This can be easily checked by looking at the evolution of curl $y(t, \cdot)$ during the interval of time $[0, T]$.

Looking at Theorem 9.1 on page 248 and Theorem 9.8 on page 251, one may also ask if one has small-time local controllability with *large* controls, i.e.,

OPEN PROBLEM 9.19. *Let $T > 0$. Does there exist $\eta > 0$ such that, for every* $(\psi^0, S^0, D^0), (\psi^1, S^1, D^1) \in [\mathbb{S} \cap H^7_{(0)}(I; \mathbb{C})] \times \mathbb{R} \times \mathbb{R}$ *with*

$$\|\psi^0 - \psi_1(0)\|_{H^7} + |S^0| + |D^0| < \eta,$$
$$\|\psi^1 - \psi_1(T)\|_{H^7} + |S^1| + |D^1| < \eta,$$

there exists (ψ, S, D, u) such that

$$\psi \in C^0([0,T]; H^2 \cap H^1_0(I; \mathbb{C})) \cap C^1([0,T]; L^2(I; \mathbb{C})),$$
$$u \in H^1_0(0, T),$$
$$S \in C^1([0,T]), \ D \in C^2([0,T]),$$
$$(9.1), (9.2) \text{ and } (9.3) \text{ hold},$$
$$(\psi(0), S(0), D(0)) = (\psi^0, S^0, D^0),$$
$$(\psi(T), S(T), D(T)) = (\psi^1, S^1, D^1)?$$

REMARK 9.20. Let us consider the following subsystem of our control system (9.1)-(9.2)-(9.3):

(9.143) $$\psi_t(t, x) = i\psi_{xx}(t, x) + iu(t)x\psi(t, x), \ (t, x) \in (0, T) \times I,$$
(9.144) $$\psi(t, -1) = \psi(t, 1) = 0, \ t \in (0, T).$$

For this new control system, at time t, the state is $\psi(t, \cdot)$ with $\int_I |\psi(t, x)|^2 dx = 1$ and the control is $u(t) \in \mathbb{R}$. Then $(\psi, u) := (\psi_1, 0)$ is a trajectory of this control system. Our proof of Theorem 9.8 can be adapted to prove that the control system (9.143) is not small-time locally controllable along this trajectory. Indeed, if $S(T) \neq 0$, then (9.121) just has to be replaced by

(9.145) $$V(T) = \frac{S(T)}{2}\left(-2\int_I \theta(T, x)x\varphi_1(x)dx + iS(T)\int_I \theta(T, x)x^2\varphi_1(x)dx\right)$$
$$+ \int_0^T S(t)V_{20}(t)dt + \int_0^T S(t)^2 V_{21}(t)dt.$$

Note that

$$\int_I \varphi_1(x)x\varphi_1(x)dx = 0,$$
$$\Re\left(i\int_I \varphi_1(x)x^2\varphi_1(x)dx\right) = 0.$$

Hence, in every neighborhood of φ_1 for the H^7-topology (for example) there exists θ_1 in this neighborhood such that

(9.146) $$\int_I \theta_1(x) x \varphi_1(x) dx = 0,$$

(9.147) $$\Re\left(i \int_I \theta_1(x) x^2 \varphi_1(x) dx\right) = 0,$$

(9.148) $$\theta_1 \in \mathbb{S} \cap H^7_{(0)},$$

(9.149) $$\Re\left(-i + i \int_I \theta_1(x) \varphi_1(x) dx\right) < 0.$$

Moreover, it follows from our proof of Theorem 9.8 that, if (9.41) holds, then there exists $\varepsilon > 0$ such that, for every $\theta_1 \in L^2(I; \mathbb{C})$ satisfying (9.146), (9.147) and (9.149), there is no $u \in L^2(0, T)$, satisfying (9.105), such that the solution $\psi \in C^0([0, T]; L^2(I; \mathbb{C}))$ of (9.106), (9.107) and (9.109) satisfies $\psi(T, \cdot) = e^{-i\lambda_1 T} \theta_1$.

Part 3

Stabilization

The two previous parts dealt with the controllability problem, which asks if one can move from a first given state to a second given state. The control that one gets is an open-loop control: it depends on time and on the two given states, it *does not* depend on the state during the evolution of the control system. In many practical situations one prefers closed-loop controls, i.e., controls which do not depend on the initial state but depend, at time t, on the state x at this time which (asymptotically) stabilizes the point one wants to reach. Usually such closed-loop controls (or feedback laws) have the advantage of being more robust to disturbances. The main issue of this part is the question of whether or not a controllable system can be (asymptotically) stabilized.

This part is divided into four chapters which we will briefly describe.

Chapter 10. This chapter is mainly devoted to the stabilization of *linear* control systems in *finite dimension*. We first start by recalling the classical pole-shifting theorem (Theorem 10.1 on page 275). A consequence of this theorem is that every controllable linear system can be stabilized by means of linear feedback laws. This implies that, if the linearized controllable system at an equilibrium of a nonlinear control system is controllable, then this equilibrium can be stabilized by smooth feedback laws. We give an application to the stabilization of the attitude of a rigid spacecraft (Example 10.15 on page 282).

Chapter 11. The subject of this chapter is the stabilization of *finite-dimensional nonlinear* control systems, mainly in the case where the nonlinearity plays a key role. In particular, it deals with the case where the linearized control system around the equilibrium that one wants to stabilize is no longer controllable. Then there are obstructions to stabilizability by smooth feedback laws even for controllable systems. We recall some of these obstructions (Theorem 11.1 on page 289 and Theorem 11.6 on page 292). We give an application to the stabilization of the attitude of an underactuated rigid spacecraft (Example 11.3 on page 289).

There are two ways to enlarge the class of feedback laws in order to recover stabilizability properties. The first one is the use of discontinuous feedback laws. The second one is the use of time-varying feedback laws. We give only comments and references on the first method, but we give details on the second method. We also show the interest of time-varying feedback laws for output stabilization. In this case the feedback depends only on the output, which is only part of the state.

Chapter 12. In this chapter, we present important tools to construct explicit stabilizing feedback laws, namely:

1. control Lyapunov function (Section 12.1),
2. damping (Section 12.2),
3. homogeneity (Section 12.3),
4. averaging (Section 12.4),
5. backstepping (Section 12.5),
6. forwarding (Section 12.6),
7. transverse functions (Section 12.7).

These methods are illustrated on various control systems, in particular the stabilization of the attitude of an underactuated rigid spacecraft and to satellite orbit transfer by means of electric propulsion.

Chapter 13. In this chapter, we give examples of how some tools introduced for the case of *finite-dimensional* control systems can be used for stabilizing some partial differential equations. We treat four examples:
1. rapid exponential stabilization by means of Gramians, for linear time-reversible partial differential equations,
2. stabilization of a rotating body-beam without damping,
3. stabilization of the Euler equations of incompressible fluids,
4. stabilization of hyperbolic systems.

CHAPTER 10

Linear control systems in finite dimension and applications to nonlinear control systems

This chapter is mainly concerned with the stabilization of *finite-dimensional linear* control systems. In Section 10.1, we recall and prove the classical pole-shifting theorem (Theorem 10.1). A consequence of this theorem is that every controllable linear system can be stabilized by means of linear feedback laws (Corollary 10.12 on page 281). As we will see in Section 10.2, this also implies that, if the linearized control system at an equilibrium of a nonlinear control system is controllable, then this equilibrium can be stabilized by smooth feedback laws (Theorem 10.14 on page 281). We give an application to the stabilization of the attitude of a rigid spacecraft (Example 10.15 on page 282). Finally, in Section 10.3, we present a method to construct stabilizing feedback laws from the Gramian for linear control systems. This method will turn out to be quite useful for the stabilization of linear partial differential equations (see Section 13.1).

10.1. Pole-shifting theorem

In this section we consider the linear control system

(10.1) $$\dot{x} = Ax + Bu,$$

where the state is $x \in \mathbb{R}^n$, the control is $u \in \mathbb{R}^m$, and where $A \in \mathcal{L}(\mathbb{R}^n; \mathbb{R}^n) \simeq \mathcal{M}_{n,n}(\mathbb{R})$ and $B \in \mathcal{L}(\mathbb{R}^m; \mathbb{R}^n) \simeq \mathcal{M}_{n,m}(\mathbb{R})$ are given.

Let us denote by $\mathcal{P}_n$ the set of polynomials of degree n in z such that the coefficients are real numbers and such that the coefficient of z^n is 1. For a matrix $M \in \mathcal{M}_{n,n}(\mathbb{C})$, let us recall that P_M denotes the characteristic polynomial of M:

$$P_M(z) := \det(zI - M).$$

Then one has the classical pole-shifting theorem (see, e.g., [**290**, Theorem 3.1, page 198] or [**458**, Theorem 13, page 186]).

THEOREM 10.1. *Let us assume that the linear control system (10.1) is controllable. Then*

$$\{P_{A+BK}; K \in \mathcal{M}_{m,n}(\mathbb{R})\} = \mathcal{P}_n.$$

Theorem 10.1 is due to Murray Wonham [**504**]. See [**458**, Section 5.10, page 256] for historical comments.

Proof of Theorem 10.1. Let us first treat the case where the control is scalar, i.e., the case where $m = 1$. The starting point is the following lemma, which has its own interest.

LEMMA 10.2 (Phase variable canonical form or controller form). *Let us assume that $m = 1$ and that the linear control system (10.1) is controllable. Let us denote* by $\alpha_1, \ldots, \alpha_n$ the n real numbers such that

$$(10.2) \qquad P_A(z) = z^n - \sum_{i=0}^{n-1} \alpha_{i+1} z^i.$$

Let

$$\tilde{A} := \begin{pmatrix} 0 & 1 & 0 & \cdots & 0 \\ 0 & 0 & 1 & \cdots & 0 \\ \vdots & \vdots & \vdots & \ddots & \vdots \\ 0 & 0 & 0 & \cdots & 1 \\ \alpha_1 & \alpha_2 & \cdots & \alpha_{n-1} & \alpha_n \end{pmatrix} \in \mathcal{M}_{n,n}(\mathbb{R}), \; \tilde{B} := \begin{pmatrix} 0 \\ \vdots \\ \vdots \\ 0 \\ 1 \end{pmatrix} \in \mathcal{M}_{n,1}(\mathbb{R}).$$

(The matrix $\tilde{A}$ is called a companion matrix.) Then there exists an invertible matrix $S \in \mathcal{M}_{n,n}(\mathbb{R})$ such that

$$(10.3) \qquad S^{-1} A S = \tilde{A}, \; S^{-1} B = \tilde{B}.$$

Proof of Lemma 10.2. Let $(f_1, \ldots, f_n) \in (\mathbb{R}^n)^n$ be defined by

$$(10.4) \quad \begin{aligned} f_n &:= B, \\ f_{n-1} &:= A f_n - \alpha_n B, \\ f_{n-2} &:= A f_{n-1} - \alpha_{n-1} B = A^2 B - \alpha_n A B - \alpha_{n-1} B, \\ &\vdots \\ f_1 &:= A f_2 - \alpha_2 B = A^{n-1} B - \alpha_n A^{n-2} B \\ & \qquad - \alpha_{n-1} A^{n-3} B - \ldots - \alpha_3 A B - \alpha_2 B. \end{aligned}$$

Then

$$(10.5) \quad \begin{aligned} A f_n &= f_{n-1} + \alpha_n f_n, \\ A f_{n-1} &= f_{n-2} + \alpha_{n-1} f_n, \\ A f_{n-2} &= f_{n-3} + \alpha_{n-2} f_n, \\ &\vdots \\ A f_1 &= A^n B - \alpha_n A^{n-1} B - \ldots - \alpha_2 A B = \alpha_1 B. \end{aligned}$$

The last equality of (10.5) comes from the Cayley-Hamilton theorem and (10.2). By (10.4), an easy induction argument on k shows that

$$(10.6) \qquad A^k B \in \mathrm{Span}\, \{f_j;\, j \in \{n-k, \ldots, n\}\}, \; \forall k \in \{0, \ldots, n-1\}.$$

In particular,

$$(10.7) \qquad \mathrm{Span}\, \{A^k B;\, k \in \{0, \ldots, n-1\}\} \subset \mathrm{Span}\, \{f_j;\, j \in \{1, \ldots, n\}\}.$$

Since the control system $\dot{x} = Ax + Bu$ is controllable, the left hand side of (10.7) is, by Theorem 1.16 on page 9, equal to all of $\mathbb{R}^n$. Hence, by (10.7), $(f_1, \ldots, f_n)$ is a basis of $\mathbb{R}^n$. We define $S \in \mathcal{M}_{n,n}(\mathbb{R})$ by requiring that

$$S e_i := f_i, \; i \in \{1, \ldots, n\},$$

where $(e_1, \ldots, e_n)$ is the canonical basis of $\mathbb{R}^n$. Then S is invertible and we have (10.3). This concludes the proof of Lemma 10.2. ∎

Let us go back to the proof of Theorem 10.1 when $m = 1$. Let $P \in \mathcal{P}_n$. Let us denote by $\beta_1, \ldots, \beta_n$ the n real numbers such that

$$(10.8) \qquad P(z) = z^n - \sum_{i=0}^{n-1} \beta_{i+1} z^i.$$

Let us also denote by $\alpha_1, \ldots, \alpha_n$ the n real numbers such that

$$(10.9) \qquad P_A(z) = z^n - \sum_{i=0}^{n-1} \alpha_{i+1} z^i.$$

Let

$$\tilde{K} := (\beta_1 - \alpha_1, \ldots, \beta_n - \alpha_n) \in \mathcal{M}_{1,n}(\mathbb{R}).$$

We have

$$\tilde{A} + \tilde{B}\tilde{K} = \begin{pmatrix} 0 & 1 & 0 & \cdots & 0 \\ 0 & 0 & 1 & \cdots & 0 \\ \vdots & \vdots & \vdots & \ddots & \vdots \\ 0 & 0 & 0 & \cdots & 1 \\ \beta_1 & \beta_2 & \cdots & \beta_{n-1} & \beta_n \end{pmatrix}.$$

Then, with an easy induction argument on n, one gets

$$P_{\tilde{A}+\tilde{B}\tilde{K}}(z) = z^n - \sum_{i=0}^{n-1} \beta_{i+1} z^i = P(z).$$

Therefore, if we let $K := \tilde{K} S^{-1}$,

$$P_{A+BK} = P_{S^{-1}(A+BK)S} = P_{\tilde{A}+\tilde{B}\tilde{K}} = P,$$

which concludes the proof of Theorem 10.1 if $m = 1$. ∎

Let us now reduce the case $m > 1$ to the case $m = 1$. The key lemma for that is the following one due to Michael Heymann [**232**].

LEMMA 10.3. *Let us assume that the control system $\dot{x} = Ax + Bu$ is controllable. Then there exist $f \in \mathcal{M}_{m,1}(\mathbb{R})$ and $C \in \mathcal{M}_{m,n}(\mathbb{R})$ such that the control system*

$$\dot{x} = (A + BC)x + Bfu, \ x \in \mathbb{R}^n, \ u \in \mathbb{R},$$

is controllable.

Postponing the proof of Lemma 10.3, let us conclude the proof of Theorem 10.1. Let $P \in \mathcal{P}_n$. Let $f \in \mathcal{M}_{m,1}(\mathbb{R})$ and $C \in \mathcal{M}_{m,n}(\mathbb{R})$ be as in Lemma 10.3. Applying Theorem 10.1 to the single-input controllable control system $\dot{x} = (A+BC)x + Bfu$, there exists $K_1 \in \mathcal{M}_{1,n}(\mathbb{R})$ such that

$$P_{A+BC+BfK_1} = P.$$

Hence, if $K := C + fK_1$, then

$$P_{A+BK} = P.$$

This concludes the proof of Theorem 10.1. ∎

Proof of Lemma 10.3. We follow Matheus Hautus [218]. Let $f \in \mathcal{M}_{m,1}(\mathbb{R})$ be such that

(10.10) $$Bf \neq 0.$$

Such an f exists. Indeed, if this is not the case, $B = 0$ and the control system $\dot{x} = Ax + Bu$ is not controllable, in contradiction to the assumptions made in Lemma 10.3. (Note that (10.10) is the only property of f that will be used.) We now construct by induction a sequence $(x_i)_{i \in \{1,\ldots,k\}}$ in the following way. Let

(10.11) $$x_1 := Bf.$$

Let us assume that $(x_i)_{i \in \{1,\ldots,j\}}$ is constructed. If

(10.12) $$Ax_j + B\mathbb{R}^m \subset \text{Span } \{x_1, \ldots, x_j\},$$

we take $k := j$. If (10.12) does not hold, we choose x_{j+1} in such a way that

(10.13) $$x_{j+1} \notin \text{Span } \{x_1, \ldots, x_j\},$$
(10.14) $$x_{j+1} \in Ax_j + B\mathbb{R}^m.$$

(Of course x_{j+1} is not unique, but, since (10.12) does not hold, there exists such an x_{j+1}.) Clearly (see (10.10), (10.11) and (10.13)),

(10.15) \qquad the vectors $x_1, \ldots, x_k$ are independent and $k \leqslant n$.

Let

(10.16) $$E := \text{Span } \{x_1, \ldots, x_k\} \subset \mathbb{R}^n.$$

Let us check that

(10.17) $$E = \mathbb{R}^n \text{ and thus } k = n.$$

Since "x_{k+1} does not exist", we have, by (10.12),

(10.18) $$Ax_k + Bu \in E, \ \forall u \in \mathbb{R}^m.$$

Taking $u = 0$ in (10.18), we get $Ax_k \in E$, which, using (10.18) once more, gives us

(10.19) $$B\mathbb{R}^m \subset E.$$

Note also that, by (10.14),

$$Ax_j \in x_{j+1} + B\mathbb{R}^m, \ \forall j \in \{1, \ldots, k-1\},$$

which, together with (10.16), (10.18) and (10.19), shows that

(10.20) $$AE \subset E.$$

From (10.19) and (10.20), we get

(10.21) $$\text{Span } \{A^i Bu; \ u \in \mathbb{R}^m, \ i \in \{0, \ldots, n-1\}\} \subset E.$$

But, using the Kalman rank condition (Theorem 1.16 on page 9), the controllability of $\dot{x} = Ax + Bu$ implies that

$$\text{Span } \{A^i Bu; \ u \in \mathbb{R}^m, \ i \in \{0, \ldots, n-1\}\} = \mathbb{R}^n,$$

which, together with (10.15) and (10.21), implies (10.17).

By (10.14), for every $j \in \{1, \ldots, n-1\}$, there exists $u_j \in \mathbb{R}^m$ such that

(10.22) $$x_{j+1} = Ax_j + Bu_j.$$

Let $C \in \mathcal{M}_{m,n}(\mathbb{R})$ be such that

(10.23) $$Cx_j = u_j, \; \forall j \in \{1, \ldots, n-1\}.$$

By (10.15) such a C exists (C is not unique since Cx_n is arbitrary). We have
$$(A + BC)x_1 = Ax_1 + Bu_1 = x_2$$
and more generally, by induction on i,

(10.24) $$(A + BC)^i x_1 = x_{i+1}, \; \forall i \in \{0, \ldots, n-1\}.$$

From the Kalman rank condition (Theorem 1.16 on page 9), (10.11), (10.16) and (10.17), the control system $\dot{x} = (A + BC)x + Bfu$, $u \in \mathbb{R}$ is controllable. This concludes the proof of Lemma 10.3. ∎

EXERCISE 10.4. *Let us assume that there exists $K \in \mathcal{M}_{m,n}(\mathbb{R})$ such that $A + BK$ has only eigenvalues with strictly negative real parts. Let us also assume that there exists $\tilde{K} \in \mathcal{M}_{m,n}(\mathbb{R})$ such that $A + B\tilde{K}$ has only eigenvalues with strictly positive real parts. Prove that the linear control system (10.1) is controllable. (It has been proved by David Russell in [**425**] that this property also holds for infinite-dimensional control systems and has important applications in this framework.)*

REMARK 10.5. It follows from the proof of Theorem 10.1 that, if $m = 1$ (single-input control) the controllability of $\dot{x} = Ax + Bu$ implies that, for every $P \in \mathcal{P}_n$, there exists *one and only one* $K \in \mathcal{M}_{m,n}(\mathbb{R})$ such that $P_{A+BK} = P$. The uniqueness is no longer true for $m \geqslant 2$.

10.2. Direct applications to the stabilization of finite-dimensional control systems

Let us first recall the definition of asymptotic stability for a dynamical system.

DEFINITION 10.6. *Let Ω be an open subset of $\mathbb{R}^n$ and let $x_e \in \Omega$. Let X in $C^0(\Omega; \mathbb{R}^n)$ be such that*
$$X(x_e) = 0.$$
One says that x_e is locally asymptotically stable for $\dot{x} = X(x)$ *if there exists $\delta > 0$ such that, for every $\varepsilon > 0$, there exists $M > 0$ such that*

(10.25) $$\dot{x} = X(x) \text{ and } |x(0) - x_e| < \delta$$

imply

(10.26) $$|x(\tau) - x_e| < \varepsilon, \; \forall \tau > M.$$

If $\Omega = \mathbb{R}^n$ and if, for every $\delta > 0$ and for every $\varepsilon > 0$, there exists $M > 0$ such that (10.25) implies (10.26), then one says that x_e is globally asymptotically stable for $\dot{x} = X(x)$.

Throughout this chapter, and in particular in (10.25), every solution of $\dot{x} = X(x)$ is assumed to be a *maximal* solution of this differential equation.

Let us emphasize the fact that, since the vector field X is only continuous, the Cauchy problem $\dot{x} = X(x), x(t_0) = x_0$, where t_0 and x_0 are given, may have many maximal solutions.

REMARK 10.7. Jaroslav Kurzweil has shown in [**289**] that, even for vector fields which are only continuous, asymptotic stability is equivalent to the existence of a Lyapunov function of class C^∞; see also [**99**] by Francis Clarke, Yuri Ledyaev, and Ronald Stern.

Let us recall some classical theorems. The first one gives equivalent properties for asymptotic stability. (In fact, one often uses these properties for the definition of asymptotic stability instead of Definition 10.6 on the preceding page.)

THEOREM 10.8. *Let Ω be an open subset of $\mathbb{R}^n$ and let $x_e \in \Omega$. Let X in $C^0(\Omega; \mathbb{R}^n)$ be such that*
$$X(x_e) = 0.$$
The point x_e is locally asymptotically stable for $\dot{x} = X(x)$ if and only if the two following properties (i) and (ii) are satisfied:
(i) x_e is a stable point for $\dot{x} = X(x)$, i.e., for every $\varepsilon > 0$, there exists $\eta > 0$ such that
$$(\dot{x} = X(x) \text{ and } |x(0) - x_e| < \eta) \Rightarrow (|x(t) - x_e| < \varepsilon, \forall t \geqslant 0).$$
(ii) x_e is an attractor for $\dot{x} = X(x)$, i.e., there exists $\rho > 0$ such that
$$(\dot{x} = X(x) \text{ and } |x(0) - x_e| < \rho) \Rightarrow (\lim_{t \to +\infty} x(t) = x_e).$$
The point x_e is globally asymptotically stable for $\dot{x} = X(x)$ if and only if x_e is a stable point for $\dot{x} = X(x)$ and
$$(\dot{x} = X(x)) \Rightarrow (\lim_{t \to +\infty} x(t) = x_e).$$

The next theorem deals with the case of linear systems.

THEOREM 10.9. *(See, for example, [**273**, Theorem 3.5, Chapter 3, page 124] or [**458**, Proposition 5.5.5, Chapter 5, page 212].) Let $A \in \mathcal{L}(\mathbb{R}^n; \mathbb{R}^n)$. Then 0 is locally asymptotically stable for the linear differential equation $\dot{x} = Ax$ if and only if every eigenvalue of A has a strictly negative real part.*

The last theorem is a linear test which gives a sufficient condition and a necessary condition for local asymptotic stability.

THEOREM 10.10. *(See, for example, [**273**, Theorem 3.7, Chapter 3, pages 130–131].) Let Ω be an open subset of $\mathbb{R}^n$ and let $x_e \in \Omega$. Let $X \in C^1(\Omega; \mathbb{R}^n)$ be such that $X(x_e) = 0$. If every eigenvalue of $X'(x_e)$ has a strictly negative real part, then $x_e \in \mathbb{R}^n$ is locally asymptotically stable for $\dot{x} = X(x)$. If $X'(x_e)$ has an eigenvalue with a strictly positive real part, then x_e is not stable for $\dot{x} = X(x)$ (see (i) of Theorem 10.8 for the definition of stable point) and therefore x_e is not locally asymptotically stable for $\dot{x} = X(x)$.*

Let us now go back to our general nonlinear control system

$$(C) \qquad \dot{x} = f(x, u),$$

where the state is $x \in \mathbb{R}^n$, the control is $u \in \mathbb{R}^m$ and the function f is of class C^1 in a neighborhood of $(0,0) \in \mathbb{R}^n \times \mathbb{R}^m$ and such that $f(0,0) = 0$. We now define "asymptotic stabilizability by means of a continuous stationary feedback law".

DEFINITION 10.11. The control system (C) is *locally (resp. globally) asymptotically stabilizable by means of continuous stationary feedback laws* if there exists $u \in C^0(\mathbb{R}^n; \mathbb{R}^m)$, satisfying
$$u(0) = 0,$$
such that $0 \in \mathbb{R}^n$ is a locally (resp. globally) asymptotically stable point, for the dynamical system
$$\dot{x} = f(x, u(x)). \tag{10.27}$$
(The dynamical system (10.27) is called a closed-loop system.)

Note that we should in fact say, "The control system (C) is *locally (resp. globally) asymptotically stabilizable by means of continuous stationary feedback laws at the equilibrium* $(x_e, u_e) := (0,0) \in \mathbb{R}^n \times \mathbb{R}^m$". But, without loss of generality, performing a translation on the state and the control, we may assume that the equilibrium of interest is $(x_e, u_e) := (0,0) \in \mathbb{R}^n \times \mathbb{R}^m$. Hence we omit "at the equilibrium". In contrast, we mention "stationary" since later on (in Section 11.2) we will consider *time-varying* feedback laws.

As a direct consequence of the pole-shifting theorem (Theorem 10.1) and of Theorem 10.9, we have the following corollary.

COROLLARY 10.12. *Let us assume that the linear control system (10.1) is controllable. Then there exists a linear feedback law $x \in \mathbb{R}^n \mapsto Kx \in \mathbb{R}^m$, $K \in \mathcal{M}_{m,n}(\mathbb{R})$, such that $0 \in \mathbb{R}^n$ is globally asymptotically stable for the closed-loop system $\dot{x} = Ax + BKx$. In particular, every linear controllable system (10.1) is globally asymptotically stabilizable by means of continuous stationary feedback laws.*

REMARK 10.13. In practical situations, the choice of K is crucial in order to have good performances for the closed-loop system $\dot{x} = Ax + BKx$. In general, one desires to have *robust* feedback laws for *uncertain* linear control systems. There are many tools available to deal with this problem. Besides classical Riccati approaches, let us mention, in particular,

1. H^∞ control; see, for example, the books [36] by Tamer Başar and Pierre Bernhard, [219] by William Helton and Orlando Merino, [383] by Ian Petersen, Valery Ugrinovskii and Andrey Savkin;
2. μ analysis; see, for example, the book [339] by Uwe Mackenroth;
3. CRONE control, due to Alain Oustaloup [379].

Let us now prove the following fundamental theorem.

THEOREM 10.14. *Let us assume that the linearized control system*
$$\dot{x} = \frac{\partial f}{\partial x}(0,0)x + \frac{\partial f}{\partial u}(0,0)u \tag{10.28}$$
is controllable. Then there exists $K \in \mathcal{M}_{m,n}(\mathbb{R})$ such that $0 \in \mathbb{R}^n$ is locally asymptotically stable for the closed-loop system $\dot{x} = f(x, Kx)$. In particular, the control system (C) is locally asymptotically stabilizable by means of continuous stationary feedback laws.

Proof of Theorem 10.14. By the pole-shifting Theorem 10.1 on page 275 applied to the linear controllable system (10.28), there exists $K \in \mathcal{M}_{m,n}(\mathbb{R})$ such that

(10.29) the spectrum of $\dfrac{\partial f}{\partial x}(0,0) + \dfrac{\partial f}{\partial u}(0,0)K$ is equal to $\{-1\}$.

Let $X : x \in \mathbb{R}^n \mapsto X(x) \in \mathbb{R}^n$, be defined by
$$X(x) := f(x, Kx).$$
Then $X(0) = 0$ and, by (10.29),

(10.30) the spectrum of $X'(0)$ is equal to $\{-1\}$.

Theorem 10.10 and (10.30) imply that $0 \in \mathbb{R}^n$ is locally asymptotically stable for $\dot{x} = X(x) = f(x, Kx)$. This concludes the proof of Theorem 10.14. ∎

EXAMPLE 10.15. *Let us go back to the control of the attitude of a rigid spacecraft with control torques provided by thruster jets, a control system that we have already considered in Example 3.9 on page 128 and Example 3.35 on page 144. Hence our control system is (3.12) with $b_1, \ldots, b_m$ independent. Here we take $m = 3$. In this case, as we have seen in Example 3.9 on page 128, the linearized control system of (3.12) at the equilibrium $(0,0) \in \mathbb{R}^6 \times \mathbb{R}^3$ is controllable. Hence, by Theorem 10.14, there exists $K \in \mathcal{M}_{3,6}(\mathbb{R})$ such that $0 \in \mathbb{R}^6$ is locally asymptotically stable for the associated closed-loop system $\dot{x} = f(x, Kx)$.*

One can find many other applications of Theorem 10.14 to physical control systems in various books. Let us mention in particular the following ones:

- [**19**] by Brigitte d'Andréa-Novel and Michel Cohen de Lara,
- [**179**] by Bernard Friedland.

10.3. Gramian and stabilization

The pole-shifting theorem (Theorem 10.1 on page 275) is a quite useful method to stabilize *finite-dimensional* linear control systems. But for *infinite-dimensional* linear control systems, it cannot be used in many situations. In this section we present another method, which has been introduced independently by Dahlard Lukes [**336**] and David Kleinman [**275**] to stabilize *finite-dimensional* linear control systems, and which can be applied to handle many *infinite-dimensional* linear control systems. For applications to *infinite-dimensional* linear control systems, see, in particular, the papers [**450**] by Marshall Slemrod, [**278**] by Vilmos Komornik and [**490**] by José Urquiza, as well as Section 13.1.

We consider again the linear control system

(10.31) $$\dot{x} = Ax + Bu,$$

where the state is $x \in \mathbb{R}^n$ and the control is $u \in \mathbb{R}^m$. Let $T > 0$. Throughout this section, we assume that this linear control system is controllable on $[0, T]$. Let us recall that, as pointed in Theorem 1.16 on page 9, this property in fact does not depend on $T > 0$. By Definition 1.10 on page 6, the controllability Gramian of the linear control system (10.31) on the time interval $[0, T]$ is the $n \times n$ symmetric matrix $\mathfrak{C}$ defined by

(10.32) $$\mathfrak{C} := \int_0^T e^{(T-t)A} B B^{\text{tr}} e^{(T-t)A^{\text{tr}}} dt.$$

This symmetric matrix is nonnegative and, in fact, by Theorem 1.11 on page 6,

(10.33) $\mathfrak{C}$ is positive definite.

Let

(10.34) $$C_T := e^{-TA}\mathfrak{C}e^{-TA^{\mathrm{tr}}} = \int_0^T e^{-tA}BB^{\mathrm{tr}}e^{-tA^{\mathrm{tr}}}dt.$$

Clearly, C_T is a symmetric matrix. By (10.33) and (10.34),

(10.35) C_T is positive definite.

Let $K \in \mathcal{L}(\mathbb{R}^n; \mathbb{R}^m)$ be defined by

(10.36) $$K := -B^{\mathrm{tr}}C_T^{-1}.$$

Then one has the following theorem due to Dahlard Lukes [336, Theorem 3.1] and David Kleinman [275].

THEOREM 10.16. *There exist $\mu > 0$ and $M > 0$ such that, for every (maximal) solution of $\dot{x} = (A+BK)x$,*

(10.37) $$|x(t)| \leqslant Me^{-\mu t}|x(0)|, \forall t \geqslant 0.$$

Proof of Theorem 10.16. Since the system $\dot{x} = (A+BK)x$ is linear, it suffices to check that every (maximal) solution $x : \mathbb{R} \to \mathbb{R}^n$ of $\dot{x} = (A+BK)x$ satisfies

(10.38) $$\lim_{t\to+\infty} x(t) = 0.$$

Thus, let $x : \mathbb{R} \to \mathbb{R}^n$ be a (maximal) solution of

(10.39) $$\dot{x} = (A+BK)x.$$

Let $V : \mathbb{R}^n \to [0,+\infty)$ be defined by

(10.40) $$V(z) := z^{\mathrm{tr}} C_T^{-1} z, \forall z \in \mathbb{R}^n.$$

By (10.35),

(10.41) $$V(z) > V(0), \forall z \in \mathbb{R}^n \setminus \{0\},$$
(10.42) $$V(z) \to +\infty \text{ as } |z| \to +\infty.$$

Let $v : \mathbb{R} \to \mathbb{R}$ be defined by

(10.43) $$v(t) := V(x(t)), \forall t \in \mathbb{R}.$$

From (10.36), (10.39), (10.40) and (10.43), we have

(10.44) $$\dot{v} = -|B^{\mathrm{tr}} C_T^{-1} x|^2 + x^{\mathrm{tr}} A^{\mathrm{tr}} C_T^{-1} x + x^{\mathrm{tr}} C_T^{-1} A x - x^{\mathrm{tr}} C_T^{-1} BB^{\mathrm{tr}} C_T^{-1} x.$$

From (10.34) and an integration by parts, we get

(10.45) $$AC_T = -e^{-TA}BB^{\mathrm{tr}}e^{-TA^{\mathrm{tr}}} + BB^{\mathrm{tr}} - C_T A^{\mathrm{tr}}.$$

Multiplying (10.45) on the right and on the left by C_T^{-1}, one gets

(10.46) $$C_T^{-1} A = -C_T^{-1} e^{-TA} BB^{\mathrm{tr}} e^{-TA^{\mathrm{tr}}} C_T^{-1} + C_T^{-1} BB^{\mathrm{tr}} C_T^{-1} - A^{\mathrm{tr}} C_T^{-1}.$$

From (10.44) and (10.46), one gets

(10.47) $$\dot{v} = -|B^{\mathrm{tr}} C_T^{-1} x|^2 - |B^{\mathrm{tr}} e^{-TA^{\mathrm{tr}}} C_T^{-1} x|^2.$$

By (10.41), (10.42) and the LaSalle invariance principle (see, for example, [**458**, Lemma 5.7.8, page 226]), it suffices to check that, if

(10.48) $$\dot{v}(t) = 0, \forall t \in \mathbb{R},$$

then

(10.49) $$x(0) = 0.$$

Let us assume that (10.48) holds. Then, using (10.47),

(10.50) $$B^{\text{tr}} C_T^{-1} x(t) = 0, \forall t \in \mathbb{R},$$

(10.51) $$B^{\text{tr}} e^{-TA^{\text{tr}}} C_T^{-1} x(t) = 0, \forall t \in \mathbb{R}.$$

From (10.36), (10.39) and (10.50), we get

(10.52) $$\dot{x}(t) = Ax(t), \forall t \in \mathbb{R}.$$

Differentiating (10.50) with respect to time and using (10.52), one gets

(10.53) $$B^{\text{tr}} C_T^{-1} Ax(t) = 0, \forall t \in \mathbb{R}.$$

Using (10.46), (10.50) and (10.51), we get

(10.54) $$C_T^{-1} Ax(t) = -A^{\text{tr}} C_T^{-1} x(t), \forall t \in \mathbb{R},$$

which, together with (10.53), gives

$$B^{\text{tr}} A^{\text{tr}} C_T^{-1} x(t) = 0, \forall t \in \mathbb{R}.$$

Then, using (10.52) and (10.54), an easy induction argument shows that

$$B^{\text{tr}} (A^{\text{tr}})^i C_T^{-1} x(t) = 0, \forall t \in \mathbb{R}, \forall i \in \mathbb{N},$$

and therefore

(10.55) $$x(t)^{\text{tr}} C_T^{-1} A^i B = 0, \forall t \in \mathbb{R}, \forall i \in \{0, \ldots, n-1\}.$$

From (10.55) and the Kalman rank condition (Theorem 1.16 on page 9), one gets that $x(t)^{\text{tr}} C_T^{-1} = 0$ and therefore $x(t) = 0$, for every $t \in \mathbb{R}$. This concludes the proof of Theorem 10.16. ∎

Let us now try to construct a feedback which leads to fast decay. More precisely, let $\lambda > 0$. We want to construct a linear feedback $x \in \mathbb{R}^n \mapsto K_\lambda x \in \mathbb{R}^m$, with $K_\lambda \in \mathcal{L}(\mathbb{R}^n; \mathbb{R}^m)$ such that there exists $M > 0$ such that

(10.56) $$|x(t)| \leqslant Me^{-\lambda t}|x(0)|, \forall t \in [0, +\infty),$$

for every (maximal) solution of $\dot{x} = (A + BK_\lambda)x$. Note that the existence of such a linear map K_λ is a corollary of the pole-shifting theorem (Theorem 10.1 on page 275). However, we would like to construct such a feedback without using this theorem. The idea is the following one. Let $y := e^{\lambda t} x$. Then $\dot{x} = Ax + Bu$ is equivalent to

(10.57) $$\dot{y} = (A + \lambda \text{Id}_n) y + Bv,$$

with $v := ue^{\lambda t}$. Let us consider (10.57) as a control system where the state is $y \in \mathbb{R}^n$ and the control is $v \in \mathbb{R}^m$. The controllability of the control system $\dot{x} = Ax + Bu$ is equivalent to the controllability of the control system $\dot{y} = (A + \lambda \text{Id}_n) y + Bv$. This can be seen by using the equality $y = e^{\lambda t} x$ or by using the Kalman rank condition (see Theorem 1.16 on page 9). We apply Theorem 10.16 to the controllable control system $\dot{y} = (A + \lambda\mathrm{Id}_n)y + Bv$. Let us define $K_{\lambda,T} \in \mathcal{M}_{m,n}(\mathbb{R})$ by

$$
\begin{aligned}
K_{\lambda,T} &:= -B^{\mathrm{tr}}\left(\int_0^T e^{-t(A+\lambda\mathrm{Id}_n)} BB^{\mathrm{tr}} e^{-t(A^{\mathrm{tr}}+\lambda\mathrm{Id}_n)} dt\right)^{-1} \\
&= -B^{\mathrm{tr}}\left(\int_0^T e^{-2\lambda t} e^{-tA} BB^{\mathrm{tr}} e^{-tA^{\mathrm{tr}}} dt\right)^{-1}.
\end{aligned}
\tag{10.58}
$$

Then, by Theorem 10.16, there exists $M > 0$ such that

$$|y(t)| \leqslant M|y(0)|, \forall t \in [0, +\infty), \tag{10.59}$$

for every (maximal) solution of the closed-loop system $\dot{y} = (A + \lambda\mathrm{Id}_n + BK_{\lambda,T})y$. Now let $x : [0, +\infty) \to \mathbb{R}^n$ be a solution of the closed-loop system $\dot{x} = (A + BK_{\lambda,T})x$. Then $y : [0, +\infty) \to \mathbb{R}^n$ defined by $y(t) := e^{\lambda t}x(t)$, $t \in [0, +\infty)$ is a solution of the closed-loop system $\dot{y} = (A + \lambda\mathrm{Id}_n + BK_{\lambda,T})y$. Hence we have (10.59), which, coming back to the x variable, gives (10.56). This ends the construction of the desired feedback.

Note that the convergence to 0 can be better than the one expected by (10.56) and one can take $T = \infty$ in (10.34); this leads to a method called the Bass method by David Russell in [**427**, pages 117–118]. According to [**427**], Roger Bass introduced this method in *"Lecture notes on control and optimization presented at NASA Langley Research Center in August 1961"*. Let us explain how this method works (and how one can prove a better convergence to 0 than the one expected by (10.56), at least if $T > 0$ is large enough). For $L \in \mathcal{L}(\mathbb{R}^n; \mathbb{R}^n)$, let us denote by $\sigma(L)$ the set of eigenvalues of L and let

$$\mu(L) := \mathrm{Max}\,\{\Re\zeta; \zeta \in \sigma(L)\}.$$

(Let us recall that, for $z \in \mathbb{C}$, $\Re z$ denotes the real part of z.) Let $\lambda \in (\mu(-A), +\infty)$. Let $C_{\lambda,\infty} \in \mathcal{L}(\mathbb{R}^n; \mathbb{R}^n)$ be defined by

$$C_{\lambda,\infty} := \int_0^\infty e^{-2\lambda t} e^{-tA} BB^{\mathrm{tr}} e^{-tA^{\mathrm{tr}}} dt. \tag{10.60}$$

Note that $C_{\lambda,\infty}$ is a nonnegative symmetric matrix. Moreover,

$$x^{\mathrm{tr}} C_{\lambda,\infty} x \geqslant e^{-2\,\mathrm{Max}\,\{0,\lambda\}} x^{\mathrm{tr}} \left(\int_0^1 e^{-tA} BB^{\mathrm{tr}} e^{-tA^{\mathrm{tr}}} dt\right) x, \forall x \in \mathbb{R}^n. \tag{10.61}$$

Since the control system $\dot{x} = Ax + Bu$ is assumed to be controllable, by Theorem 1.11 on page 6 and Remark 1.12 on page 7, there exists $c > 0$ such that

$$x^{\mathrm{tr}}\left(\int_0^1 e^{-tA} BB^{\mathrm{tr}} e^{-tA^{\mathrm{tr}}} dt\right) x \geqslant c|x|^2, \forall x \in \mathbb{R}^n,$$

which, together with (10.61), implies that the symmetric matrix $C_{\lambda,\infty}$ is invertible.

We define $K_{\lambda,\infty} \in \mathcal{L}(\mathbb{R}^n; \mathbb{R}^m)$ by

$$
\begin{aligned}
K_{\lambda,\infty} &:= -B^{\mathrm{tr}}\left(\int_0^\infty e^{-t(A+\lambda\mathrm{Id}_n)} BB^{\mathrm{tr}} e^{-t(A^{\mathrm{tr}}+\lambda\mathrm{Id}_n)} dt\right)^{-1} \\
&= -B^{\mathrm{tr}}\left(\int_0^\infty e^{-2\lambda t} e^{-tA} BB^{\mathrm{tr}} e^{-tA^{\mathrm{tr}}} dt\right)^{-1} \\
&= -B^{\mathrm{tr}} C_{\lambda,\infty}^{-1}.
\end{aligned}
\tag{10.62}
$$

Then one has the following proposition, proved by José Urquiza in [**490**].

PROPOSITION 10.17. *Let $\lambda \in (\mu(-A), +\infty)$. Then*
$$\sigma(A + BK_{\lambda,\infty}) = \{-2\lambda - \zeta;\ \zeta \in \sigma(A)\}. \tag{10.63}$$

Proof of Proposition 10.17. From (10.60) and an integration by parts
$$(A + \lambda \mathrm{Id}_n)C_{\lambda,\infty} + C_{\lambda,\infty}(A + \lambda \mathrm{Id}_n)^{\mathrm{tr}} = BB^{\mathrm{tr}}. \tag{10.64}$$
(Compare with (10.45).) Let us multiply (10.64) from the right by $C_{\lambda,\infty}^{-1}$. One gets
$$(A + \lambda \mathrm{Id}_n) + C_{\lambda,\infty}(A + \lambda \mathrm{Id}_n)^{\mathrm{tr}} C_{\lambda,\infty}^{-1} = BB^{\mathrm{tr}} C_{\lambda,\infty}^{-1}. \tag{10.65}$$
From (10.62) and (10.65), one gets
$$A + BK_{\lambda,\infty} = C_{\lambda,\infty}(-A^{\mathrm{tr}} - 2\lambda \mathrm{Id}_n)C_{\lambda,\infty}^{-1}, \tag{10.66}$$
which gives (10.63). This concludes the proof of Proposition 10.17. ∎

As a consequence of (10.66), one has the following corollary which has been pointed out by José Urquiza in [**490**] (compare with (10.56)).

COROLLARY 10.18. *Assume that*
$$A^{tr} = -A. \tag{10.67}$$
Let $\lambda > 0$. Then there exists $M \in (0, +\infty)$ such that, for every (maximal) solution of $\dot{x} = (A + BK_{\lambda,\infty})x$,
$$|x(t)| \leqslant M e^{-2\lambda t}|x(0)|,\ \forall t \geqslant 0.$$

Proof of Corollary 10.18. Let $x : [0, +\infty) \to \mathbb{R}^n$ be a (maximal) solution of
$$\dot{x} = (A + BK_{\lambda,\infty})x. \tag{10.68}$$
Let $y : [0, +\infty) \to \mathbb{R}^n$ be defined by
$$y(t) := C_{\lambda,\infty}^{-1} x(t),\ t \in [0, +\infty). \tag{10.69}$$
From (10.66), (10.68) and (10.69), one gets that
$$\dot{y} = -(A^{\mathrm{tr}} + 2\lambda \mathrm{Id}_n)y. \tag{10.70}$$
From (10.67) and (10.70), we have
$$\frac{\mathrm{d}}{\mathrm{d}t}|y|^2 = -4\lambda |y|^2,$$
which implies that
$$|y(t)| = e^{-2\lambda t}|y(0)|,\ \forall t \in [0, +\infty). \tag{10.71}$$
From (10.69) and (10.71), one gets that
$$|x(t)| \leqslant \|C_{\lambda,\infty}\|_{\mathcal{L}(\mathbb{R}^n;\mathbb{R}^n)} \|C_{\lambda,\infty}^{-1}\|_{\mathcal{L}(\mathbb{R}^n;\mathbb{R}^n)} e^{-2\lambda t}|x(0)|,\ \forall t \in [0, +\infty),$$
which concludes the proof of Corollary 10.18. ∎

CHAPTER 11

Stabilization of nonlinear control systems in finite dimension

Let us recall (Corollary 10.12 on page 281) that, in finite dimension, every linear control system which is controllable can be asymptotically stabilized by means of continuous stationary feedback laws. As we have seen in Section 10.2, this result implies that, if the linearized control system of a given nonlinear control system is controllable, then the control system can be asymptotically stabilized by means of continuous stationary feedback laws; see Theorem 10.14 on page 281. A natural question is whether every controllable nonlinear system can be asymptotically stabilized by means of continuous stationary feedback laws. In 1979, Héctor Sussmann showed that the global version of this result does not hold for nonlinear control systems: In [**467**], he gave an example of a nonlinear analytic control system which is globally controllable but cannot be globally asymptotically stabilized by means of continuous stationary feedback laws. In 1983, Roger Brockett showed that the local version also does not hold even for analytic control systems: In [**74**] he gave a necessary condition (Theorem 11.1 on page 289) for local asymptotic stabilizability by means of continuous stationary feedback laws, which is not implied by local controllability even for analytic control systems. For example, as pointed out in [**74**], the analytic control system (also called the nonholonomic integrator)

(11.1) $$\dot{x}_1 = u_1, \; \dot{x}_2 = u_2, \; \dot{x}_3 = x_1 u_2 - x_2 u_1,$$

where the state is $x = (x_1, x_2, x_3)^{\mathrm{tr}} \in \mathbb{R}^3$ and the control is $u = (u_1, u_2)^{\mathrm{tr}} \in \mathbb{R}^2$, is locally and globally controllable (see Theorem 3.19 on page 135 and Theorem 3.18 on page 134) but does not satisfy the Brockett necessary condition (and therefore cannot be asymptotically stabilized by means of continuous stationary feedback laws). To overcome the problem of impossibility to stabilize many controllable systems by means of continuous stationary feedback laws, two main strategies have been proposed:

(i) asymptotic stabilization by means of discontinuous feedback laws,
(ii) asymptotic stabilization by means of continuous time-varying feedback laws.

In this chapter, we mainly consider continuous time-varying feedback laws. We only give bibliographical information on the discontinuous feedback approach in Section 11.4. Let us just mention here that the main issue for these feedbacks is the problem of robustness to (small) measurements disturbances (they are robust to (small) actuators disturbances).

For *continuous* time-varying feedback laws, let us first point out that, due to a converse to Lyapunov's second theorem proved by Jaroslav Kurzweil in [**289**] (see also [**99**]), periodic time-varying feedback laws are robust to (small) actuators and measurement disturbances. The pioneering works concerning time-varying feedback laws are due to Eduardo Sontag and Héctor Sussmann [**460**], and Claude Samson [**435**]. In [**460**], it is proved that, if the dimension of the state is 1, controllability implies asymptotic stabilizability by means of time-varying feedback laws. In [**435**], it is proved that the control system (11.1) can be asymptotically stabilized by means of time-varying feedback laws. In Sections 11.2.1 and 11.2.2, we present results showing that, in many cases, (local) controllability implies stabilizability by means of time-varying feedback laws.

In many practical situations, only part of the state—called the output—is measured and therefore state feedback laws cannot be implemented; only output feedback laws are allowed. It is well known (see, e.g., [**458**, Theorem 32, page 324]) that any linear control system which is controllable and observable can be asymptotically stabilized by means of dynamic feedback laws. Again it is natural to ask whether this result can be extended to the nonlinear case. In the nonlinear case, there are many possible definitions for observability. The weakest requirement for observability is that, given two different states, there exists a control $t \mapsto u(t)$ which leads to maps "$t \mapsto$ output at time t" which are not identical. With this definition of observability, the nonlinear control system

(11.2) $$\dot{x} = u, \, y = x^2,$$

where the state is $x \in \mathbb{R}$, the control $u \in \mathbb{R}$, and the output $y \in \mathbb{R}$, is observable. This system is also clearly controllable and asymptotically stabilizable by means of (stationary) feedback laws (e.g. $u(x) = -x$). But, see [**107**], this system cannot be asymptotically stabilized by means of continuous stationary dynamic output feedback laws. Again, as we shall see in Section 11.3, the introduction of time-varying feedback laws improves the situation. In particular, the control system (11.2) can be asymptotically stabilized by means of continuous time-varying dynamic output feedback laws. (In fact, for the control system (11.2), continuous time-varying output feedback laws are sufficient to stabilize asymptotically the control system.)

Let us also mention that the usefulness of time-varying controls for different goals has been pointed out by many authors. For example, by

1. V. Polotskiĭ [**386**] for observers to avoid peaking,
2. Jan Wang [**498**] for decentralized linear systems,
3. Dirk Aeyels and Jacques Willems [**4**] for the pole-shifting problem for linear time-invariant systems,
4. Pramod Khargonekar, Antonio Pascoal and R. Ravi [**274**], Bertina Ho-Mock-Qai and Wijesuriya Dayawansa [**235**] for simultaneous stabilization of a family of control systems,
5. Nicolas Chung Siong Fah [**157**] for Input to State Stability (ISS), a notion introduced by Eduardo Sontag in [**454**] in order to take care of robustness issues.

See also the references therein.

11.1. Obstructions to stationary feedback stabilization

As above, we continue to denote by (C) the nonlinear control system

(C) $$\dot{x} = f(x, u),$$

where $x \in \mathbb{R}^n$ is the state and $u \in \mathbb{R}^m$ is the control. We again assume that

(11.3) $$f(0,0) = 0$$

and that, unless otherwise specified, $f \in C^\infty(\mathcal{O}; \mathbb{R}^n)$, where $\mathcal{O}$ is an open subset of $\mathbb{R}^n \times \mathbb{R}^m$ containing $(0,0)$.

Let us start by recalling the following necessary condition for stabilizability due to Roger Brockett [**74**].

THEOREM 11.1. *If the control system $\dot{x} = f(x,u)$ can be locally asymptotically stabilized by means of continuous stationary feedback laws, then the image by f of every neighborhood of $(0,0) \in \mathbb{R}^n \times \mathbb{R}^m$ is a neighborhood of $0 \in \mathbb{R}^n$.*

EXAMPLE 11.2 (This example is due to Roger Brockett [**74**]). *Let us return to the nonholonomic integrator*

(11.4) $$\dot{x}_1 = u_1, \dot{x}_2 = u_2, \dot{x}_3 = x_1 u_2 - x_2 u_1.$$

It is a driftless control affine system where the state is $(x_1, x_2, x_3)^{tr} \in \mathbb{R}^3$ and the control is $(u_1, u_2)^{tr} \in \mathbb{R}^2$. As we have seen in Example 3.20 on page 135, it follows from Theorem 3.19 on page 135 that this control system is small-time locally controllable at the equilibrium $(x_e, u_e) = (0,0) \in \mathbb{R}^3 \times \mathbb{R}^2$ (see also Example 6.3 on page 190 for a proof based on the return method). However, for every $\eta \in \mathbb{R} \setminus \{0\}$, the equation

$$u_1 = 0, u_2 = 0, x_1 u_2 - x_2 u_1 = \eta,$$

where the unknown are $(x_1, x_2, u_1, u_2)^{tr} \in \mathbb{R}^4$, has no solution. Hence, the nonholonomic integrator does not satisfy Brockett's condition and therefore, by Theorem 11.1, it cannot be locally asymptotically stabilized by means of continuous stationary feedback laws.

EXAMPLE 11.3. *Let us go back to the control system of the attitude of a rigid spacecraft, already considered in Example 3.9 on page 128, Example 3.35 on page 144 and Example 10.15 on page 282. One has the following cases.*

1. *For $m = 3$, the control system (3.12) satisfies Brockett's condition. In fact, in this case, the control system (3.12) is indeed asymptotically stabilizable by means of continuous stationary feedback laws, as we have seen in Example 10.15 on page 282 (see also [**137**] and [**80**]).*
2. *For $m \in \{1, 2\}$, the control system (3.12) does not satisfy Brockett's condition (and so is not locally asymptotically stabilizable by means of continuous stationary feedback laws). Indeed, if $b \in \mathbb{R}^3 \setminus (\text{Span}\{b_1, b_2\})$, there exists no $((\omega, \eta), u)$ such that*

(11.5) $$S(\omega)\omega + u_1 b_1 + u_2 b_2 = b,$$
(11.6) $$A(\eta)\omega = 0.$$

*Indeed, (11.6) gives $\omega = 0$, which, together with (11.5), implies that $b = u_1 b_1 + u_2 b_2$; see also [**80**].*

In [**508**], Jerzy Zabczyk observed that, from a theorem due to Aleksandrovich Krasnosel'skiĭ [**283, 284**] (at least for feedback laws of class C^1; see below for feedback laws which are only continuous), one can deduce the following stronger necessary condition, that we shall call the index condition.

THEOREM 11.4. *If the control system $\dot{x} = f(x, u)$ can be locally asymptotically stabilized by means of continuous stationary feedback laws, then there exists $u \in C^0(\mathbb{R}^n; \mathbb{R}^m)$ vanishing at $0 \in \mathbb{R}^n$ such that $f(x, u(x)) \neq 0$ for $|x|$ small enough but not 0 and the index of $x \mapsto f(x, u(x))$ at 0 is equal to $(-1)^n$.*

Let us recall that the index of $x \mapsto X(x) := f(x, u(x))$ at 0 is (see, for example, [**284**, page 9])
$$\text{degree}\,(X, \{x \in \mathbb{R}^n;\, |x| < \varepsilon\}, 0)$$
where $\epsilon > 0$ is such that

(11.7) $\qquad (X(x) = 0 \text{ and } |x| \leqslant \varepsilon) \Rightarrow (x = 0).$

For the definition of degree $(X, \{x \in \mathbb{R}^n;\, |x| < \varepsilon\}, 0)$, see Appendix B. (In fact, with the notations in this appendix, we should write
$$\text{degree}\,(X|_{\{x \in \mathbb{R}^n;\, |x| \leqslant \varepsilon\}}, \{x \in \mathbb{R}^n;\, |x| < \varepsilon\}, 0)$$
instead of degree $(X, \{x \in \mathbb{R}^n;\, |x| < \varepsilon\}, 0)$. For simplicity, we use this slight abuse of notation throughout this whole section.) Note that, by the excision property of the degree (Proposition B.11 on page 388), degree $(X, \{x \in \mathbb{R}^n;\, |x| < \varepsilon\}, 0)$ is independent of $\varepsilon > 0$ provided that (11.7) holds. Let us point out that

(11.8) $\qquad \text{degree}\,(X, \{x \in \mathbb{R}^n;\, |x| < \varepsilon\}, 0) \neq 0$

implies that

(11.9) $\qquad X(\{x \in \mathbb{R}^n;\, |x| < \varepsilon\})$ is a neighborhood of $0 \in \mathbb{R}^n$.

Indeed, by a classical continuity property of the degree (Proposition B.9 on page 387), there exists $\nu > 0$ such that, for every $a \in \mathbb{R}^n$ such that $|a| \leqslant \nu$,

(11.10) $\qquad \text{degree}\,(X, \{x \in \mathbb{R}^n;\, |x| < \varepsilon\}, a)$ exists,

(11.11) $\quad \text{degree}\,(X, \{x \in \mathbb{R}^n;\, |x| < \varepsilon\}, a) = \text{degree}\,(X, \{x \in \mathbb{R}^n;\, |x| < \varepsilon\}, 0).$

From (11.8), (11.10), (11.11) and another classical property of the degree (Proposition B.10 on page 387), we get
$$\{a \in \mathbb{R}^n;\, |a| \leqslant \nu\} \subset X(\{x \in \mathbb{R}^n;\, |x| < \varepsilon\}),$$
which gives (11.9). Hence the index condition implies that the image by f of every neighborhood of $(0, 0) \in \mathbb{R}^n \times \mathbb{R}^m$ is a neighborhood of $0 \in \mathbb{R}^n$. Thus Theorem 11.4 implies Theorem 11.1.

Proof of Theorem 11.4. Since $0 \in \mathbb{R}^n$ is asymptotically stable for $\dot{x} = X(x) := f(x, u(x))$, there exists $\varepsilon_0 > 0$ such that
$$X(x) \neq 0,\, \forall x \in \mathbb{R}^n \text{ such that } 0 < |x| \leqslant \varepsilon_0.$$
Let us now give the proof of Theorem 11.4 when X is not only continuous but of class C^1 in a neighborhood of $0 \in \mathbb{R}^n$ (which is the case treated in [**283, 284**]). Then one can associate to X its flow $\Phi : \mathbb{R} \times \mathbb{R}^n \to \mathbb{R}^n$, $(t, x) \mapsto \Phi(t, x)$, defined by
$$\frac{\partial \Phi}{\partial t} = X(\Phi),\, \Phi(0, x) = x.$$
This flow is well defined in a neighborhood of $\mathbb{R} \times \{0\}$ in $\mathbb{R} \times \mathbb{R}^n$. For $\varepsilon \in (0, \varepsilon_0]$, let
$$\bar{B}_\varepsilon := \{x \in \mathbb{R}^n;\, |x| \leqslant \varepsilon\},$$

and let $H : [0,1] \times \bar{B}_\varepsilon \to \mathbb{R}^n$ be defined by

$$\begin{aligned} H(t,x) &:= \frac{\Phi\left(\frac{t}{1-t}, x\right) - x}{t}, & \forall (t,x) &\in (0,1) \times \bar{B}_\varepsilon, \\ H(0,x) &:= X(x), & \forall x &\in \bar{B}_\varepsilon, \\ H(1,x) &:= -x, & \forall x &\in \bar{B}_\varepsilon. \end{aligned}$$

Using the hypothesis that 0 is locally asymptotically stable for $\dot{x} = X(x)$, one gets that, for $\varepsilon \in (0, \varepsilon_0]$ small enough, H is continuous and does not vanish on $[0,1] \times \partial \bar{B}_\varepsilon$. We choose such an ε. From the homotopy invariance of the degree (Proposition B.8 on page 387)

(11.12) $$\text{degree}\,(H(0,\cdot), B_\varepsilon, 0) = \text{degree}\,(H(1,\cdot), B_\varepsilon, 0).$$

By a classical formula of the degree (see (B.4) on page 380),

$$\text{degree}\,(H(1,\cdot), B_\varepsilon, 0) = \text{degree}\,(-\text{Id}_n|_{\bar{B}_\varepsilon}, B_\varepsilon, 0) = (-1)^n,$$

which, together with (11.12), concludes the proof of Theorem 11.4 if X is of class C^1 in a neighborhood of $0 \in \mathbb{R}^n$.

If X is only continuous, we apply an argument introduced in [**108**]. One first uses a theorem due to Jaroslav Kurzweil [**289**] which tells us that local asymptotic stability implies the existence of a Lyapunov function of class C^∞. Hence there exist $\varepsilon_1 \in (0, \varepsilon_0]$ and $V \in C^\infty(\bar{B}_{\varepsilon_1})$ such that

(11.13) $$V(x) > V(0), \forall x \in \bar{B}_{\varepsilon_1} \setminus \{0\},$$

(11.14) $$\nabla V(x) \cdot X(x) < 0, \forall x \in \bar{B}_{\varepsilon_1} \setminus \{0\}.$$

Note that (11.14) implies that

(11.15) $$\nabla V(x) \neq 0, \forall x \in \bar{B}_{\varepsilon_1} \setminus \{0\}.$$

From (11.13) and (11.15), we get that V is a Lyapunov function for $\dot{x} = -\nabla V(x)$. Hence, $0 \in \mathbb{R}^n$ is locally asymptotically stable for $\dot{x} = -\nabla V(x)$. Therefore, by our study of the C^1 case,

(11.16) $$\text{degree}\,(-\nabla V, B_{\varepsilon_1}, 0) = (-1)^n.$$

Now let $h : [0,1] \times \bar{B}_{\varepsilon_1}$ be defined by

$$h(t,x) = -t\nabla V(x) + (1-t)X(x), \forall (t,x) \in [0,1] \times \bar{B}_{\varepsilon_1}.$$

From (11.14), one gets

$$h(t,x) \cdot \nabla V(x) < 0, \forall (t,x) \in [0,1] \times (\partial \bar{B}_{\varepsilon_1}).$$

In particular,

(11.17) $$h(t,x) \neq 0, \forall (t,x) \in [0,1] \times (\partial \bar{B}_{\varepsilon_1}).$$

Using (11.17) and once more the homotopy invariance of the degree (Proposition B.8 on page 387), we get

$$\begin{aligned} \text{degree}\,(X, B_{\varepsilon_1}, 0) &= \text{degree}\,(h(0,\cdot), B_{\varepsilon_1}, 0) \\ &= \text{degree}\,(h(1,\cdot), B_{\varepsilon_1}, 0) \\ &= \text{degree}\,(-\nabla V, B_{\varepsilon_1}, 0), \end{aligned}$$

which, together with (11.16), concludes the proof of Theorem 11.4. ∎

It turns out that the index condition is too strong for another natural stabilizability notion that we introduce now.

DEFINITION 11.5. *The control system $\dot{x} = f(x, u)$ is locally asymptotically stabilizable by means of dynamic continuous stationary feedback laws if, for some integer $p \in \mathbb{N}$, the control system*

(11.18) $$\dot{x} = f(x, u), \ \dot{y} = v \in \mathbb{R}^p,$$

where the control is $(u, v)^{\mathrm{tr}} \in \mathbb{R}^{m+p}$ with $(u, v) \in \mathbb{R}^m \times \mathbb{R}^p$, and the state is $(x, y)^{\mathrm{tr}} \in \mathbb{R}^{n+p}$ with $(x, y) \in \mathbb{R}^n \times \mathbb{R}^p$, is locally asymptotically stabilizable by means of continuous stationary feedback laws.

In Definition 11.5 and in the following, for $\alpha \in \mathbb{R}^k$ and $\beta \in \mathbb{R}^l$, $(\alpha, \beta)^{\mathrm{tr}}$ is defined by

(11.19) $$(\alpha, \beta)^{\mathrm{tr}} := \begin{pmatrix} \alpha \\ \beta \end{pmatrix} \in \mathbb{R}^{k+l}.$$

By convention, when $p = 0$, the control system (11.18) is just the control system $\dot{x} = f(x, u)$. The control system (11.18) is called a dynamic extension of the control system $\dot{x} = f(x, u)$.

Let us emphasize that for the control system (11.18), the feedback laws u and v are functions of x and y. Clearly, if the control system $\dot{x} = f(x, u)$ is locally asymptotically stabilizable by means of continuous stationary feedback laws, it is locally asymptotically stabilizable by means of dynamic continuous stationary feedback laws. But it is proved in [**128**] that the converse does not hold. Moreover, the example given in [**128**] shows that the index condition is not necessary for local asymptotic stabilizability by means of dynamic continuous stationary feedback laws. Clearly the Brockett necessary condition is still necessary for local asymptotic stabilizability by means of dynamic continuous stationary feedback laws. But this condition turns out to be not sufficient for local asymptotic stabilizability by means of dynamic continuous stationary feedback laws even if one assumes that the control system (C) is small-time locally controllable at the equilibrium $(0, 0) \in \mathbb{R}^n \times \mathbb{R}^m$ (see Definition 3.2 on page 125) and that the system is analytic. In [**102**] we proposed the following slightly stronger necessary condition.

THEOREM 11.6. *Assume that the control system $\dot{x} = f(x, u)$ can be locally asymptotically stabilized by means of dynamic continuous stationary feedback laws. Then, for every positive and small enough ε,*

(11.20) $$f_\star \left(\sigma_{n-1} \left(\{ (x, u); \ |x| + |u| \leqslant \varepsilon, f(x, u) \neq 0 \} \right) \right) = \sigma_{n-1}(\mathbb{R}^n \setminus \{0\}) (= \mathbb{Z}),$$

*where $\sigma_{n-1}(A)$ denotes the stable homotopy group of order $(n-1)$. (For a definition of stable homotopy groups and of $f_\star$, see, e.g., [**501**], Chapter XII.)*

EXERCISE 11.7. Let $n = 2$, $m = 1$, $\mathcal{O} = \mathbb{R}^2 \times \mathbb{R}$. We define f by
$$f(x, u) := (x_2^3 - 3(x_1 - u)^2 x_2, \ (x_1 - u)^3 - 3(x_1 - u)x_2^2)^{tr},$$
for every $x = (x_1, x_2)^{tr} \in \mathbb{R}^2$ and every $u \in \mathbb{R}$.
1. Prove that the control system $\dot{x} = f(x, u)$ is small-time locally controllable at $(0, 0) \in \mathbb{R}^2 \times \mathbb{R}$. (**Hint.** Use Theorem 3.32 on page 143.)
2. Check that the control system $\dot{x} = f(x, u)$ satisfies the Brockett necessary condition, i.e., that the image by f of every neighborhood of $(0, 0) \in \mathbb{R}^2 \times \mathbb{R}$ is a neighborhood of $0 \in \mathbb{R}^2$.
3. Prove that (11.20) does not hold.

11.1. OBSTRUCTIONS TO STATIONARY FEEDBACK STABILIZATION

Let us point out that the index condition implies (11.20). Moreover, (11.20) implies that a "dynamic extension" of $\dot{x} = f(x, u)$ satisfies the index condition if the system is analytic. More precisely, one has the following theorem.

THEOREM 11.8 ([**108**, Section 2 and Appendice]). *Assume that f is analytic (or continuous and subanalytic) in a neighborhood of $(0,0) \in \mathbb{R}^n \times \mathbb{R}^m$. Assume that (11.20) is satisfied. Then, if $p \geqslant 2n + 1$, the control system (11.18) satisfies the index condition.*

For a definition of a subanalytic function, see, e.g., [**233**] by Heisuke Hironaka.

Proof of Theorem 11.8. Clearly, it suffices to prove this theorem for $p = 2n + 1$. For $R > 0$ and $k \in \mathbb{N} \setminus \{0\}$, let

$$B_k(R) := \{z \in \mathbb{R}^k; |z| < R\}, \ \overline{B_k(R)} := \{z \in \mathbb{R}^k; |z| \leqslant R\}.$$

Let us assume, for the moment, that the following lemma holds.

LEMMA 11.9. *Let E be a subanalytic subset of $\mathbb{R}^k$ such that $0 \in E$ and belongs to the closure in $\mathbb{R}^k$ of $E \setminus \{0\}$. Then there exist $\varepsilon > 0$ and an homeomorphism φ from $\overline{B_k(\varepsilon)}$ into itself such that*

(11.21) $$|\varphi(z)| = |z|, \forall z \in \overline{B_k(\varepsilon)},$$

(11.22) $$\varphi(z) = z, \forall z \in \partial B_k(\varepsilon),$$

(11.23) $$\varphi(E \cap \overline{B_k(\varepsilon)}) = \{tx; t \in [0,1], x \in E \cap \overline{B_k(\varepsilon)}\}.$$

For the definition of subanalytic set, see, again, [**233**]. We apply this lemma with $k := n + m$ and, identifying $\mathbb{R}^{n+m}$ to $\mathbb{R}^n \times \mathbb{R}^m$,

$$E := \{(x,u) \in \mathbb{R}^n \times \mathbb{R}^m; f(x,u) = 0, |x|^2 + |u|^2 \leqslant \varepsilon^2\},$$

with $\varepsilon > 0$ small enough so that f is analytic on $\{(x,u) \in \mathbb{R}^n \times \mathbb{R}^m; |x|^2 + |u|^2 \leqslant 2\varepsilon^2\}$. Let $\tilde{f} : \overline{B_{n+m}(\varepsilon)} \to \mathbb{R}^n$ be defined by

(11.24) $$\tilde{f} := f \circ \varphi^{-1}.$$

Decreasing if necessary $\varepsilon > 0$ and using (11.20) we may also assume that

(11.25) $$\tilde{f}_* \left(\sigma_{n-1}(\{(x,u); |x|+|u| \leqslant \varepsilon, f(x,u) \neq 0\}) \right) = \sigma_{n-1}(\mathbb{R}^n \setminus \{0\}) (= \mathbb{Z}),$$

Let $\mathbb{S}^{2n} = \{y \in \mathbb{R}^{2n+1}; |y| = 1\}$. For $(a_1, a_2, a_3) \in \mathbb{R}^{n_1} \times \mathbb{R}^{n_2} \times \mathbb{R}^{n_3}$, we define $(a_1, a_2, a_3)^{\text{tr}}$ by

$$(a_1, a_2, a_3)^{\text{tr}} := \begin{pmatrix} a_1 \\ a_2 \\ a_3 \end{pmatrix} \in \mathbb{R}^{n_1+n_2+n_3}.$$

From (11.25) and the suspension theorem (see, e.g., [**461**, Theorem 11, Chapter 8, page 458]), there exists a map $\tilde{\theta} = (\tilde{x}, \tilde{u}, \tilde{\alpha})^{\text{tr}} \in C^0(\mathbb{S}^{2n}; \mathbb{R}^{n+m+n+1})$, with $(\tilde{x}, \tilde{u}, \tilde{\alpha}) \in C^0(\mathbb{S}^{2n}; \mathbb{R}^n \times \mathbb{R}^m \times \mathbb{R}^{n+1})$, such that

(11.26) $$\tilde{\theta}(\mathbb{S}^{2n}) \subset \left(\overline{B_{n+m}(\varepsilon)} \times \mathbb{R}^{n+1}\right)^{\text{tr}} \setminus (\tilde{f}^{-1}(0) \times \{0\})^{\text{tr}},$$

(11.27) $$\text{the degree of } g : \mathbb{S}^{2n} \to \mathbb{R}^{2n+1} \setminus \{0\} \text{ is } -1,$$

with g defined by

$$g(y) := (\tilde{f}(\tilde{x}(y), \tilde{u}(y)), \tilde{\alpha}(y))^{\text{tr}}, \forall y \in \mathbb{S}^{2n}.$$

By (11.27) and relying on the degree theory detailed in Appendix B, we mean the following. We extend $\hat{\theta} = (\tilde{x}, \tilde{u}, \tilde{\alpha})^{\text{tr}}$ to $\overline{B_{2n+1}(1)}$ by letting

(11.28) $$\tilde{\theta}(z) = |z|\theta(z/|z|) \text{ if } |z| \neq 0,$$

(11.29) $$\tilde{\theta}(0) = 0.$$

Then, we extend g to $\overline{B_{2n+1}(1)}$ by letting

(11.30) $$g(y) := (\tilde{f}(\tilde{x}(y), \tilde{u}(y)), \tilde{\alpha}(y))^{\text{tr}}, \forall y \in \overline{B_{2n+1}(1)}.$$

Property (11.27) means that

(11.31) $$\text{degree}\,(g, B_{2n+1}(1), 0) = -1.$$

Clearly, $\theta \in C^0(\overline{B_{2n+1}(1)}; \mathbb{R}^{2n+1})$ and vanishes at 0. Let r be a small positive real number. Let $u \in C^0(\mathbb{R}^n \times \mathbb{R}^{2n+1}; \mathbb{R}^m)$ and $v \in C^0(\mathbb{R}^n \times \mathbb{R}^{2n+1}; \mathbb{R}^{2n+1})$ be such that, for every $(x, y) \in \mathbb{R}^n \times \mathbb{R}^{2n+1}$ such that $|y| \leqslant r$,

(11.32) $$u(x, y) = \pi_2 \circ \varphi^{-1}(\tilde{x}(y), \tilde{u}(y)),$$

(11.33) $$v(x, y) = (\tilde{\alpha}(y), x - \pi_1 \circ \varphi^{-1}(\tilde{x}(y), \tilde{u}(y)))^{\text{tr}},$$

where π_1 and π_2 are defined by

$$\pi_1(z_1, z_2) := z_1 \in \mathbb{R}^n, \pi_2(z_1, z_2) := z_2 \in \mathbb{R}^m, \forall (z_1, z_2) \in \mathbb{R}^n \times \mathbb{R}^m.$$

Clearly, u and v vanish at $(0, 0) \in \mathbb{R}^n \times \mathbb{R}^{2n+1}$. Let $F : \overline{B_{3n+1}(r)} \to \mathbb{R}^n \times \mathbb{R}^{2n+1}$ be defined by

$$F((x, y)^{\text{tr}}) := (f(x, u(x, y)), v(x, y))^{\text{tr}},$$

for every $(x, y) \in \mathbb{R}^n \times \mathbb{R}^{2n+1}$, such that $|x|^2 + |y|^2 \leqslant r^2$. Let us check that, at least if $r > 0$ is small enough,

(11.34) $$((x, y)^{\text{tr}} \in \overline{B_{3n+1}(r)} \setminus \{0\}, x \in \mathbb{R}^n, y \in \mathbb{R}^{2n+1}) \Rightarrow (F(x, y) \neq 0).$$

Let $x \in \mathbb{R}^n$, $y \in \mathbb{R}^{2n+1}$ be such that $(x, y)^{\text{tr}} \in \overline{B_{3n+1}(r)}$ and $F((x, y)^{\text{tr}}) = 0$. Then

(11.35) $$x = \pi_1 \circ \varphi^{-1}(\tilde{x}(y), \tilde{u}(y)),$$

(11.36) $$\tilde{\alpha}(y) = 0,$$

(11.37) $$f(x, \pi_2 \circ \varphi^{-1}(\tilde{x}(y), \tilde{u}(y))) = 0.$$

From (11.35) and (11.37), one gets

(11.38) $$\tilde{f}(\tilde{x}(y), \tilde{u}(y)) = 0.$$

From (11.35), (11.21), (11.28), (11.36) and (11.38), one has

(11.39) $$y = 0.$$

From (11.29), (11.33), (11.35) and (11.39), one gets

$$x = 0,$$

which, together with (11.39), concludes the proof of (11.34).

Let us now prove that, at least if $r > 0$ is small enough (which is always assumed), then

(11.40) $$\text{degree}\,(F, B_{3n+1}(r), 0) = (-1)^{3n+1}.$$

Let $U : \overline{B_{3n+1}(r)} \to \mathbb{R}^{3n+1}$ be defined by

$$U\begin{pmatrix} x \\ y \end{pmatrix} := \begin{pmatrix} x - \pi_1 \circ \varphi^{-1}(\tilde{x}(y), \tilde{u}(y)) \\ f(x, \pi_2 \circ \varphi^{-1}(\tilde{x}(y), \tilde{u}(y))) \\ \tilde{\alpha}(y) \end{pmatrix},$$

for every $(x,y) \in \mathbb{R}^n \times \mathbb{R}^{2n+1}$ such that $|x|^2 + |y|^2 \leqslant r^2$.

Using Proposition B.12 on page 388, one gets

(11.41) $\quad$ degree $(F, B_{3n+1}(r), 0) = (-1)^n$degree $(U, B_{3n+1}(r), 0)$.

Using (11.30), (11.41), Proposition B.11 on page 388 and Proposition B.14 on page 389, one gets that

$$\text{degree}\,(F, B_{3n+1}(r), 0) = (-1)^n \text{degree}\,(g, B_{2n+1}(1), 0),$$

which, together with (11.31), gives (11.40).

It remains to prove Lemma 11.9. Let us point out that this lemma is proved in [**358**, Theorem 2.10] by John Milnor if E is a subanalytic set having an isolated singularity at $(0,0)$ (if $(0,0)$ is not a singularity of E, Lemma 11.9 obviously holds). The general case has been treated by Robert Hardt in [**217**, pages 295–296]: with the notations of [**217**], one takes E for X, $\mathbb{R}^{m+n}$ for $\mathbb{R}^m$, $[0, +\infty)$ for Y, $y \mapsto |y|^2$ for f. (In fact, [**217**] deals with semialgebraic sets, instead of subanalytic sets. However, the part of [**217**] we use holds for subanalytic sets.) This concludes the proof of Theorem 11.8. ∎

Let us end this section by an open problem:

OPEN PROBLEM 11.10. *Let us assume that f is analytic, satisfies (11.20) and that the control system $\dot{x} = f(x,u)$ is small-time locally controllable at the equilibrium $(0,0) \in \mathbb{R}^n \times \mathbb{R}^m$ (or even that $0 \in \mathbb{R}^n$ is locally continuously reachable in small time for the control system $\dot{x} = f(x,u)$; see Definition 11.20 below). Is the control system $\dot{x} = f(x,u)$ locally asymptotically stabilizable by means of dynamic continuous stationary feedback laws?*

A natural guess is that, unfortunately, a positive answer is unlikely to be true. A possible candidate for a negative answer is the control system, with $n = 3$ and $m = 1$,

$$\dot{x}_1 = x_3^2(x_1 - x_2), \ \dot{x}_2 = x_3^2(x_2 - x_3), \ \dot{x}_3 = u.$$

This system satisfies the Hermes condition $S(0)$ and so, by Sussmann's Theorem 3.29 on page 143, is small-time locally controllable at the equilibrium $(0,0) \in \mathbb{R}^3 \times \mathbb{R}$. Moreover, it satisfies the index condition (take $u = x_3 - (x_1^2 + x_2^2)$).

11.2. Time-varying feedback laws

In this section we will see that, in many cases, small-time local controllability at the equilibrium $(0,0) \in \mathbb{R}^n \times \mathbb{R}^m$ implies local asymptotic stabilizability by means of continuous time-varying feedback laws. We first treat the case of driftless control systems (Section 11.2.1). Then we consider general control systems (Section 11.2.2).

Let us start with some classical definitions. Let us first recall the definition of asymptotic stability for a time-varying dynamical system—we should in fact say uniform asymptotic stability.

DEFINITION 11.11. Let Ω be an open subset of $\mathbb{R}^n$ containing 0. Let X in $C^0(\Omega \times \mathbb{R}; \mathbb{R}^n)$ be such that
$$X(0,t) = 0, \ \forall t \in \mathbb{R}.$$
One says that 0 is *locally asymptotically stable for* $\dot{x} = X(x,t)$ if
   (i) for every $\varepsilon > 0$, there exists $\eta > 0$ such that, for every $s \in \mathbb{R}$ and for every $\tau \geqslant s$,
$$(11.42) \qquad (\dot{x} = X(x,t), |x(s)| < \eta) \Rightarrow (|x(\tau)| < \varepsilon),$$
   (ii) there exists $\delta > 0$ such that, for every $\varepsilon > 0$, there exists $M > 0$ such that, for every s in $\mathbb{R}$,
$$(11.43) \qquad \dot{x} = X(x,t) \text{ and } |x(s)| < \delta$$
   imply
$$(11.44) \qquad |x(\tau)| < \varepsilon, \ \forall \tau > s + M.$$
If $\Omega = \mathbb{R}^n$ and if, for every $\delta > 0$ and for every $\varepsilon > 0$, there exists $M > 0$ such that (11.43) implies (11.44) for every s in $\mathbb{R}$, one says that 0 is *globally asymptotically stable* for $\dot{x} = X(x,t)$.

Throughout the whole chapter, and in particular in (11.42) and (11.43), every solution of $\dot{x} = X(x,t)$ is assumed to be a *maximal* solution of this differential equation. Let us emphasize again the fact that, since the vector field X is only continuous, the Cauchy problem $\dot{x} = X(x,t), x(t_0) = x_0$, where t_0 and x_0 are given, may have many maximal solutions.

Let us recall that Jaroslav Kurzweil in [**289**] has shown that, even for time-varying vector fields which are only continuous, asymptotic stability is equivalent to the existence of a Lyapunov function of class C^∞; see also [**99**]. This Lyapunov function provides some important robustness properties with respect to (small) perturbations.

REMARK 11.12. One can easily check that, if X does not depend on time or is periodic with respect to time, then (ii) implies (i) in Definition 11.11. But this is not the case for general time-varying X.

Let us now define "asymptotic stabilizability by means of continuous time-varying feedback laws".

DEFINITION 11.13. The control system (C) is *locally (resp. globally) asymptotically stabilizable by means of continuous time-varying feedback laws* if there exists $u \in C^0(\mathbb{R}^n \times \mathbb{R}; \mathbb{R}^m)$ satisfying
$$u(0,t) = 0, \ \forall t \in \mathbb{R},$$
such that $0 \in \mathbb{R}^n$ is locally (resp. globally) asymptotically stable for the closed-loop system $\dot{x} = f(x, u(x,t))$.

11.2.1. Stabilization of driftless control affine systems. In this section, we assume that
$$f(x,u) = \sum_{i=1}^m u_i f_i(x).$$
We are going to see that the driftless control systems which satisfy the Lie algebra rank condition (see Definition 3.16 on page 134) at every $(x,0) \in \mathbb{R}^n \times \mathbb{R}^m$ with

$x \in \mathbb{R}^n \setminus \{0\}$ can be globally asymptotically stabilized by means of continuous time-varying feedback laws.

Let us recall that $\mathrm{Lie}\{f_1,\ldots,f_m\} \subset C^\infty(\mathbb{R}^n;\mathbb{R}^n)$ denotes the Lie sub-algebra of $C^\infty(\mathbb{R}^n;\mathbb{R}^n)$ generated by the vector fields $f_1,\ldots,f_m$ (see Definition 3.13 on page 133). One then has the following theorem.

THEOREM 11.14 ([**103**]). *Assume that*

(11.45) $$\{g(x);\, g \in Lie\{f_1,\ldots,f_m\}\} = \mathbb{R}^n,\; \forall x \in \mathbb{R}^n \setminus \{0\}.$$

Then, for every $T > 0$, there exists u in $C^\infty(\mathbb{R}^n \times \mathbb{R};\mathbb{R}^m)$ such that

(11.46) $$u(0,t) = 0,\; \forall t \in \mathbb{R},$$

(11.47) $$u(x, t+T) = u(x,t),\; \forall x \in \mathbb{R}^n,\; \forall t \in \mathbb{R},$$

and 0 is globally asymptotically stable for

(11.48) $$\dot{x} = f(x, u(x,t)) = \sum_{i=1}^m u_i(x,t) f_i(x).$$

REMARK 11.15. By the Rashevski-Chow theorem [**398, 94**] (Theorem 3.18 on page 134), Property (11.45) implies the global controllability of the driftless control affine system $\dot{x} = f(x,u) = \sum_{i=1}^m u_i f_i(x)$ in $\mathbb{R}^n \setminus \{0\}$, i.e., for every $x^0 \in \mathbb{R}^n \setminus \{0\}$, every $x^1 \in \mathbb{R}^n \setminus \{0\}$ and every $T > 0$, there exists $u \in L^\infty((0,T);\mathbb{R}^m)$ such that, if $\dot{x} = \sum_{i=1}^m u_i(t) f_i(x)$ and $x(0) = x^0$, then $x(T) = x^1$ (and $x(t) \in \mathbb{R}^n \setminus \{0\}$ for every $t \in [0,T]$). Let us recall that, by Theorem 3.17 on page 134, for every $a \in \mathbb{R}^n$,

$$\{g(a);\, g \in \mathrm{Lie}\{f_1,\ldots,f_m\}\} = \mathbb{R}^n,$$

is also a necessary condition for small-time local controllability at the equilibrium $(a,0) \in \mathbb{R}^n \times \mathbb{R}^m$ of the driftless control affine system $\dot{x} = f(x,u) = \sum_{i=1}^m u_i f_i(x)$ if the f_i, $1 \leqslant i \leqslant m$, are analytic.

EXAMPLE 11.16. *Let us return to the nonholonomic integrator*

(11.49) $$\dot{x}_1 = u_1,\; \dot{x}_2 = u_2,\; \dot{x}_3 = x_1 u_2 - x_2 u_1.$$

It is a driftless control affine system where the state is $(x_1,x_2,x_3)^{tr} \in \mathbb{R}^3$ and the control is $(u_1,u_2)^{tr} \in \mathbb{R}^2$. We have $m = 2$ and

$$f_1(x) := \begin{pmatrix} 1 \\ 0 \\ -x_2 \end{pmatrix},\; f_2(x) := \begin{pmatrix} 0 \\ 1 \\ x_1 \end{pmatrix}.$$

As we have seen in Example 3.20 on page 135, it follows from Theorem 3.19 on page 135 that this control system is small-time locally controllable at the equilibrium $(x_e, u_e) = (0,0) \in \mathbb{R}^3 \times \mathbb{R}^2$ and also globally controllable. (See also Example 6.3 on page 190 for a proof based on the return method. A simple scaling argument shows that the small-time local controllability proved in this example, implies global controllability.) However, as we have seen in Example 11.2 on page 289, the nonholonomic integrator does not satisfy Brockett's condition and therefore, by Theorem 11.1, it cannot be, even locally, asymptotically stabilized by means of continuous stationary feedback laws. But, as we have seen in Example 3.20 on page 135, the nonholonomic integrator satisfies the Lie algebra rank condition at every $(x,0) \in \mathbb{R}^3 \times \mathbb{R}^2$, i.e.,

$$\{g(x);\, g \in Lie\{f_1, f_2\}\} = \mathbb{R}^3,\; \forall x \in \mathbb{R}^3.$$

Hence, by Theorem 11.14, the nonholonomic integrator can be, even globally, asymptotically stabilized by means of periodic time-varying feedback laws of class C^∞.

Sketch of the proof of Theorem 11.14. Let us just briefly describe the idea of the proof: Assume that, for every positive real number T, there exists $\bar{u}$ in $C^\infty(\mathbb{R}^n \times \mathbb{R}; \mathbb{R}^m)$ satisfying (11.46) and (11.47), and such that, if $\dot{x} = f(x, \bar{u}(x,t))$, then

(11.50) $$x(T) = x(0),$$

(11.51) if $x(0) \neq 0$, then the linearized control system around the trajectory
$$t \in [0,T] \mapsto (x(t), \bar{u}(x(t), t)) \text{ is controllable on } [0,T].$$

Using (11.50) and (11.51), one easily sees that one can construct a "small" feedback v in $C^\infty(\mathbb{R}^n \times \mathbb{R}; \mathbb{R}^m)$ satisfying (11.46) and (11.47), and such that, if

(11.52) $$\dot{x} = f(x, (\bar{u} + v)(x, t))$$

and $x(0) \neq 0$, then

(11.53) $$|x(T)| < |x(0)|,$$

which implies that 0 is globally asymptotically stable for (11.48) with $u = \bar{u} + v$.

So it only remains to construct $\bar{u}$. In order to get (11.50), one just imposes on $\bar{u}$ the condition

(11.54) $$\bar{u}(x, t) = -\bar{u}(x, T - t), \forall (x, t) \in \mathbb{R}^n \times \mathbb{R},$$

which implies that $x(t) = x(T-t)$, $\forall t \in [0,T]$ for every solution of $\dot{x} = f(x, u(x,t))$ (proceed as in the proof of (6.9) above), and therefore gives (11.50). Finally, one proves that (11.51) holds for "many" $\bar{u}$'s. ∎

REMARK 11.17. The above method, that we called "return method" (see Chapter 6), can also be used to get controllability results for control systems even in infinite dimension, as we have seen in Section 6.2, Section 6.3 and in Chapter 9.

REMARK 11.18. The fact that (11.51) holds for "many" $\bar{u}$'s is related to the prior works [**452**] by Eduardo Sontag and [**206**] by Mikhael Gromov. In [**452**] Eduardo Sontag showed that, if a system is completely controllable, then any two points can be joined by means of a control law such that the linearized control system around the associated trajectory is controllable. In [**206**, Theorem page 156] Mikhael Gromov has shown that generic under-determined linear (partial) differential equations are algebraically solvable, which is related to controllability for time-varying linear control systems (see the second proof of Theorem 1.18 on page 11). In our situation the linear differential equations are not generic; only the controls are generic, but this is sufficient to get the result. Moreover, as pointed out by Eduardo Sontag in [**456**], for analytic systems, one can get (11.51) by using a result due to Héctor Sussmann on observability [**466**]. Note that the proof we give for (11.51) in [**103**] (see also [**105**]) can be used to get a C^∞-version of [**466**]; see [**106**].

REMARK 11.19. Using a method due to Jean-Baptiste Pomet [**387**], we gave in [**127**] a method to deduce a suitable v from $\bar{u}$; see Section 12.2.2 below for a description of this method.

11.2.2. Stabilization of general systems.

In this section we still denote by (C) the nonlinear control system

$$(C) \qquad \dot{x} = f(x, u),$$

where $x \in \mathbb{R}^n$ is the state, $u \in \mathbb{R}^m$ is the control. We again assume that

$$(11.55) \qquad f(0, 0) = 0$$

and that, unless otherwise specified, $f \in C^\infty(\mathcal{O}; \mathbb{R}^n)$, where $\mathcal{O}$ is a an open subset of $\mathbb{R}^n \times \mathbb{R}^m$ containing $(0,0)$.

Let us first point out that in [**460**] Eduardo Sontag and Héctor Sussmann proved that every one-dimensional state (i.e., $n = 1$) nonlinear control system which is locally (resp. globally) controllable can be locally (resp. globally) asymptotically stabilized by means of continuous time-varying feedback laws.

Let us also point out that it follows from [**467**] by Héctor Sussmann that a result similar to Theorem 11.14 does not hold for systems with a drift term; more precisely, there are analytic control systems (C) which are globally controllable, for which there is no u in $C^0(\mathbb{R}^n \times \mathbb{R}; \mathbb{R}^m)$ such that 0 is globally asymptotically stable for $\dot{x} = f(x, u(x,t))$. In fact, the proof of [**467**] requires uniqueness of the trajectories of $\dot{x} = f(x, u(x,t))$. But this can always be assumed; indeed, it follows easily from Kurzweil's result [**289**] that, if there exists u in $C^0(\mathbb{R}^n \times \mathbb{R}; \mathbb{R}^m)$ such that 0 is globally asymptotically stable for $\dot{x} = f(x, u(x,t))$, then there exists $\bar{u}$ in $C^0(\mathbb{R}^n \times \mathbb{R}; \mathbb{R}^m) \cap C^\infty((\mathbb{R}^n \setminus \{0\}) \times \mathbb{R}; \mathbb{R}^m)$ such that 0 is globally asymptotically stable for $\dot{x} = f(x, \bar{u}(x,t))$; for such a $\bar{u}$ one has uniqueness of the trajectories of $\dot{x} = f(x, \bar{u}(x,t))$.

However, we are going to see in this section that a local version of Theorem 11.14 holds for many control systems which are small-time locally controllable.

For $\varepsilon \in (0, +\infty)$, let

$$(11.56) \qquad B_\varepsilon := \{a \in \mathbb{R}^n;\ |a| < \varepsilon\},\ \bar{B}_\varepsilon := \{a \in \mathbb{R}^n;\ |a| \leqslant \varepsilon\}.$$

Let us again introduce some definitions.

DEFINITION 11.20. *The origin (of $\mathbb{R}^n$) is* locally continuously reachable in small time *(for the control system (C)) if, for every positive real number T, there exist a positive real number ε and an element u in $C^0\left(\bar{B}_\varepsilon; L^1((0,T); \mathbb{R}^m)\right)$ such that*

$$(11.57) \qquad \mathrm{Sup}\{|u(a)(t)|;\ t \in (0,T)\} \to 0 \text{ as } a \to 0,$$

$$(11.58) \qquad ((\dot{x} = f(x, u(a)(t)),\ x(0) = a) \Rightarrow (x(T) = 0)), \forall a \in \bar{B}_\varepsilon.$$

Let us prove that "many" sufficient conditions for small-time local controllability (see Definition 3.2 on page 125) at $(0,0) \in \mathbb{R}^n \times \mathbb{R}^m$ imply that the origin is locally continuously reachable in small time. This is in particular the case for the Sussmann condition (Theorems 3.29 on page 143 and 3.32 on page 143); this is in fact also the case for the Bianchini and Stefani condition [**56**, Corollary page 970], which extends Theorem 3.29 on page 143.

In order to prove this result, we first introduce the following definition.

DEFINITION 11.21. For $p \in \mathbb{N} \setminus \{0\}$, let D^p be the set of vectors ξ in $\mathbb{R}^n$ such that there exists u in $C^0([0,1]; L^1((0,1); \mathbb{R}^m))$ such that
$$|u(s)(t)| \leqslant s, \, \forall (s,t) \in [0,1] \times [0,1], \tag{11.59}$$
$$u(s)(t) = 0, \text{ if } t \geqslant s, \tag{11.60}$$
$$\psi(u(s)) = s^p \xi + o(s^p), \text{ as } s \to 0, \tag{11.61}$$
where $\psi(u(s))$ denotes the value at time 0 of the solution of $\dot{x} = f(x, u(s)(t))$, $x(1) = 0$. Let $D \subset \mathbb{R}^n$ be defined by
$$D := \bigcup_{p \geqslant 1} D^p. \tag{11.62}$$

Then one has the following proposition.

PROPOSITION 11.22 ([**105**, Lemma 3.1 and Section 5]). *Assume that*
$$0 \text{ is in the interior of } D. \tag{11.63}$$
Then $0 \in \mathbb{R}^n$ is locally continuously reachable in small time (for the control system (C)).

Proof of Proposition 11.22. We follow a method due to Matthias Kawski [**270**] (see also [**227**] by Henry Hermes). In [**270**, Appendix], Matthias Kawski proved (under the assumption of Proposition 11.22) the existence of $u : B_\varepsilon \to L^1((0,T); \mathbb{R}^m)$ satisfying (11.57) and (11.58). His u is not continuous. However, let us show how to slightly modify his proof in order to have a continuous u. Let $(e_i)_{1 \leqslant i \leqslant n}$ be the usual canonical basis of $\mathbb{R}^n$ and let $e_i = -e_{i-n}$ for i in $\{n+1, \ldots, 2n\}$. By (11.63) and noticing that, for every $l \in \mathbb{N} \setminus \{0\}$, $D^l \subset D^{l+1}$, we may assume, possibly after a change of scale on x, that, for some $p \geqslant 1$, $e_i \in D^p$ for every i in $\{1, 2, \ldots, 2n\}$. For i in $\{1, 2, \ldots, 2n\}$, let u_i be as in Definition 11.21 with $\xi = e_i$. For a in $\mathbb{R}^n$ we write $a = \sum_{i=1}^{2n} a_i e_i$ with $a_i \geqslant 0$ for every i in $\{1, 2, \ldots, 2n\}$ and $a_i a_{i+n} = 0$ for every i in $\{1, \ldots, n\}$. Let $\mu(a) = \sum_{i=1}^{2n} a_i^{1/p} e_i$. For a in $\mathbb{R}^n$ with $|a| \leqslant 1$, let $u_a : [0, \mu(a)) \to \mathbb{R}^m$ be defined by
$$u_a(t) = u_j(a_j^{1/p})(t - t_j) \text{ if } t_j := \sum_{i=1}^{j-1} a_i^{1/p} \leqslant t < \sum_{i=1}^{j} a_i^{1/p}. \tag{11.64}$$
From Gronwall's lemma, (11.59), (11.60), (11.61) and (11.64), we get
$$\mu(x_a^1) = o(\mu(a)) \text{ as } a \to 0, \tag{11.65}$$
where $x_a^1 := \zeta_a(\mu(a))$, $\zeta_a : [0, \mu(a)]$ being the solution of the Cauchy problem
$$\dot{\zeta}_a = f(\zeta_a, u_a(t)), \, \zeta_a(0) = a. \tag{11.66}$$
Hence, for $|a|$ small enough,
$$\mu(x_a^1) \leqslant \frac{\mu(a)}{2}. \tag{11.67}$$
We now define u_a on $[\mu(a), \mu(a) + \mu(x_a^1))$ by
$$u_a(t) := u_{x_a^1}(t - \mu(a)), \, \forall t \in [\mu(a), \mu(a) + \mu(x_a^1)).$$
We have, if $|a|$ is small enough, that
$$\mu(x_a^2) \leqslant \frac{\mu(x_a^1)}{2},$$

where $x_a^2 := \zeta_a(\mu(a) + \mu(x_a^1))$, ζ_a still being the solution of the Cauchy problem (11.66), but now on $[0, \mu(a) + \mu(x_a^1)]$ (instead of $[0, \mu(a)]$). We keep going and define in this way u_a on $[0, \mu(a) + \sum_{i=1}^{\infty} \mu(x_a^i)]$ (with obvious notations). Note that, if a is small enough, $\mu(a) + \sum_{i=1}^{\infty} \mu(x_i) \leqslant 2\mu(a) \leqslant T$. We extend u_a on $[0, T]$ by $u_a(t) = 0$ if $t \in [\mu(a) + \sum_{i=1}^{\infty} \mu(x_i), T]$. Then $u(a) := u_a$ satisfies all the required properties required in Definition 11.20 on page 299. This concludes the proof of Proposition 11.22. ∎

Our next proposition is the following.

PROPOSITION 11.23. *Assume that the control system (C) satisfies the Sussmann condition $S(\theta)$ (see Definition 3.31 on page 143) for some $\theta \in [0, 1]$. Then (11.63) holds.*

The proof of Proposition 11.23 follows directly from the proof of [**469**, Theorem 7.3] by Héctor Sussmann (see also the proof of Theorem 3.32 on page 143). We omit it.

The following theorem is a consequence of Proposition 11.22 and Proposition 11.23.

THEOREM 11.24. *Assume that the control system (C) (see page 299) satisfies the Sussmann condition $S(\theta)$ (see Definition 3.31 on page 143) for some $\theta \in [0, 1]$. Then $0 \in \mathbb{R}^n$ is locally continuously reachable in small time (for the control system (C)).*

In fact, the sufficient conditions for small-time local controllability we mentioned earlier on page 143 (namely [**5**] by Andrei Agrachev, [**6, 7**] by Andrei Agrachev and Revaz Gamkrelidze, [**55**] by Rosa Maria Bianchini, and [**56**] by Rosa Maria Bianchini and Gianna Stefani) all implied (11.63). In particular, the hypothesis of [**56**, Corollary p. 970], which is weaker than the hypothesis of Proposition 11.23, also implies (11.63). It sounds natural to conjecture a positive answer to the question raised in the following open problem.

OPEN PROBLEM 11.25. *Assume that the map f is analytic in an open neighborhood of $(0, 0) \in \mathbb{R}^n \times \mathbb{R}^m$ and that the control system (C) is small-time locally controllable at the equilibrium $(0, 0) \in \mathbb{R}^n \times \mathbb{R}^m$. Is $0 \in \mathbb{R}^n$ locally continuously reachable in small time (for the control system (C))?*

Concerning the converse, one has the following theorem.

THEOREM 11.26. *Let us assume that 0 is locally continuously reachable in small-time for the control system (C). Then the control system (C) is small-time locally controllable at $(0, 0) \in \mathbb{R}^n \times \mathbb{R}^m$.*

Proof of Theorem 11.26. We assume that 0 is locally continuously reachable in small time (for the control system (C)). Let $\varepsilon > 0$. Then there exist a real number $\rho > 0$ and a map u in $C^0\left(\bar{B}_\rho; L^1\left((0, \varepsilon); \mathbb{R}^m\right)\right)$ such that

(11.68) $$|u(a)(t)| \leqslant \varepsilon, \, \forall a \in \bar{B}_\rho, \, \forall t \in [0, \varepsilon],$$

(11.69) $$(\dot{x} = f(x, u(a)(t)), |a| \leqslant \rho, x(0) = a) \Rightarrow (x(\varepsilon) = 0).$$

Let $\mu > 0$. Let Φ be the map
$$\begin{array}{rcl} \bar{B}_\rho \times \bar{B}_\mu & \to & \mathbb{R}^n \\ (a, x^1) & \mapsto & x(0), \end{array}$$

where $x : [0, \varepsilon] \to \mathbb{R}^n$ is the solution of the Cauchy problem
$$\dot{x} = f(x, u(a)(t)), \ x(\varepsilon) = x^1.$$
By (11.69), there exists $\mu > 0$ such that Φ is well defined. We fix such a μ. Then Φ is continuous and, still by (11.69),

(11.70) $$\Phi(a, 0) = a, \ \forall a \in \bar{B}_\rho.$$

Hence, degree$(\Phi, B_\rho, 0)$ is well defined and

(11.71) $$\text{degree}\,(\Phi(\cdot, 0), B_\rho, 0) = 1.$$

For the definition of the degree, see Appendix B. By the continuity of the degree (Property (iii) of Theorem B.1 on page 379 and Proposition B.9 on page 387), there exists $\eta \in (0, \mu)$ such that, for every $x^0 \in \bar{B}_\eta$ and for every $x^1 \in \bar{B}_\eta$,

(11.72) $$\text{degree}\,(\Phi(\cdot, x^1), B_\rho, x^0) = 1.$$

By a classical property of the degree theory (Proposition B.10 on page 387), (11.72) implies the existence of $a \in B_\rho$ such that

(11.73) $$\Phi(a, x^1) = x^0.$$

But (11.73) just means that
$$(\dot{x} = f(x, u(a)(t)),\ x(0) = x^0) \Rightarrow (x(\varepsilon) = x^1).$$

This concludes the proof of Theorem 11.26. ∎

Our next definition follows.

DEFINITION 11.27. *The control system* (C) *is locally stabilizable in small time by means of continuous periodic time-varying feedback laws if, for every positive real number* T, *there exist* ε *in* $(0, +\infty)$ *and* u *in* $C^0(\mathbb{R}^n \times \mathbb{R}; \mathbb{R}^m)$ *such that*

(11.74) $$u(0, t) = 0, \ \forall t \in \mathbb{R},$$

(11.75) $$u(x, t + T) = u(x, t), \ \forall t \in \mathbb{R},$$

(11.76) $$((\dot{x} = f(x, u(x, t)) \text{ and } x(s) = 0) \Rightarrow (x(\tau) = 0, \ \forall \tau \geqslant s)),\ \forall s \in \mathbb{R},$$

(11.77) $$(\dot{x} = f(x, u(x, t)) \text{ and } |x(s)| \leqslant \varepsilon) \Rightarrow (x(\tau) = 0, \ \forall \tau \geqslant s + T)),\ \forall s \in \mathbb{R}.$$

If u *can be chosen to be of class* C^∞ *on* $(\mathbb{R}^n \setminus \{0\}) \times \mathbb{R}$, *one says that the control system* (C) *is locally stabilizable in small time by means of almost smooth periodic time-varying feedback laws*

Note that (11.75), (11.76), and (11.77) imply that $0 \in \mathbb{R}^n$ is locally asymptotically stable for $\dot{x} = f(x, u(x, t))$; see [109, Lemma 2.15] for a proof. Note that, if (C) is locally stabilizable in small time by means of almost smooth periodic time-varying feedback laws, then $0 \in \mathbb{R}^n$ is locally continuously reachable for (C). The main result of this section is that the converse holds if $n \notin \{2, 3\}$ and if (C) satisfies the Lie algebra rank condition at $(0, 0) \in \mathbb{R}^n \times \mathbb{R}^m$:

THEOREM 11.28 ([109] for $n \geqslant 4$ and [110] for $n = 1$). *Assume that* $0 \in \mathbb{R}^n$ *is locally continuously reachable in small time for the control system* (C), *that* (C) *satisfies the Lie algebra rank condition at* $(0, 0) \in \mathbb{R}^n \times \mathbb{R}^m$ *(see Definition 3.16 on page 134) and that*

(11.78) $$n \notin \{2, 3\}.$$

Then (C) is locally stabilizable in small time by means of almost smooth periodic time-varying feedback laws.

We conjecture a positive answer to the following open problem.

OPEN PROBLEM 11.29. *Can Assumption (11.78) be removed in Theorem 11.28?*

Sketch of the proof of Theorem 11.28 for n ⩾ 4. Let I be an interval of $\mathbb{R}$. By a smooth trajectory of the control system (C) on I, we mean a $(\gamma, u) \in C^\infty(I; \mathbb{R}^n \times \mathbb{R}^m)$ satisfying $\dot{\gamma}(t) = f(\gamma(t), u(t))$ for every t in I. The linearized control system around (γ, u) is (see Definition 3.5 on page 127)

$$\dot{\xi} = A(t)\xi + B(t)w,$$

where the state is $\xi \in \mathbb{R}^n$, the control is $w \in \mathbb{R}^m$, and

$$A(t) = \frac{\partial f}{\partial x}(\gamma(t), u(t)) \in \mathcal{L}(\mathbb{R}^n; \mathbb{R}^n), \ B(t) = \frac{\partial f}{\partial u}(\gamma(t), u(t)) \in \mathcal{L}(\mathbb{R}^m; \mathbb{R}^n), \ \forall t \in I.$$

We first introduce the following definition.

DEFINITION 11.30. *The trajectory (γ, u) is supple on $S \subset I$ if, for every s in S,*

(11.79) $$\mathrm{Span}\left\{\left(\frac{\mathrm{d}}{\mathrm{d}t} - A(t)\right)^i B(t)|_{t=s} w \ ; \ w \in \mathbb{R}^m, i \geqslant 0\right\} = \mathbb{R}^n.$$

In (11.79), we use the classical convention

$$\left(\frac{\mathrm{d}}{\mathrm{d}t} - A(t)\right)^0 B(t) = B(t).$$

Let us recall that, by Theorem 1.18 on page 11, (11.79) implies that the linearized control system around (γ, u) is controllable on every interval $[T_0, T_1]$, with $T_0 < T_1$, which meets S. Let T be a positive real number. For u in $C^0(\mathbb{R}^n \times [0, T]; \mathbb{R}^m)$ and a in $\mathbb{R}^n$, let $t \mapsto x(a, t; u)$ be the maximal solution of

$$\frac{\partial x}{\partial t} = f(x, u(a, t)), \ x(a, 0; u) = a.$$

Also, let C^* be the set of $u \in C^0(\mathbb{R}^n \times [0, T]; \mathbb{R}^m)$ of class C^∞ on $(\mathbb{R}^n \setminus \{0\}) \times [0, T]$ and vanishing on $\{0\} \times [0, T]$. For simplicity, in this sketch of proof, we omit some details which are important to take care of the uniqueness property (11.76) (note that without (11.76) one does not have stability).

Step 1. Using the fact that $0 \in \mathbb{R}^n$ is locally continuously reachable in small time for the control system (C), that (C) satisfies the Lie algebra rank condition at $(0, 0) \in \mathbb{R}^n \times \mathbb{R}^m$ and [**105**] or [**106**], one proves that there exist ε_1 in $(0, +\infty)$ and u_1 in C^*, vanishing on $\mathbb{R}^n \times \{T\}$, such that

$$|a| \leqslant \varepsilon_1 \Rightarrow x(a, T; u_1) = 0,$$
$$0 < |a| \leqslant \varepsilon_1 \Rightarrow (x(a, \cdot; u_1), u_1(a, \cdot)) \text{ is supple on } [0, T].$$

Step 2. Let Γ be a 1-dimensional closed submanifold of $\mathbb{R}^n \setminus \{0\}$ such that

$$\Gamma \subset \{x \in \mathbb{R}^n; 0 < |x| < \varepsilon_1\}.$$

Perturbing in a suitable way u_1, one obtains a map $u_2 \in C^*$, vanishing on $\mathbb{R}^n \times \{T\}$, such that

$$|a| \leqslant \varepsilon_1 \Rightarrow x(a, T; u_2) = 0,$$
$$0 < |a| \leqslant \varepsilon_1 \Rightarrow (x(a, \cdot; u_2), u_2(a, \cdot)) \text{ is supple on } [0, T],$$
$$a \in \Gamma \mapsto x(t, a; u_2) \text{ is an embedding of } \Gamma \text{ into } \mathbb{R}^n \setminus \{0\}, \forall t \in [0, T).$$

Here one uses the assumption $n \geqslant 4$ and one proceeds as in the classical proof of the Whitney embedding theorem (see e.g. [**202**, Chapter II, Section 5]). Let us emphasize that it is only in this step that this assumption is used.

Step 3. From Step 2, one deduces the existence of u_3^* in C^*, vanishing on $\mathbb{R}^n \times \{T\}$, and of an open neighborhood $\mathcal{N}^*$ of Γ in $\mathbb{R}^n \setminus \{0\}$ such that

(11.80) $$a \in \mathcal{N}^* \Rightarrow x(a, T; u_3^*) = 0,$$

$a \in \mathcal{N}^* \mapsto x(a, t; u_3^*)$ is an embedding of $\mathcal{N}^*$ into $\mathbb{R}^n \setminus \{0\}, \forall t \in [0, T)$.

This embedding property allows us to transform the open-loop control u_3^* into a feedback law u_3 on $\{(x(a, t; u_3^*), t); a \in \mathcal{N}^*, t \in [0, T)\}$. So (see in particular (11.80) and note that u_3^* vanishes on $\mathbb{R}^n \times \{T\}$) there exist u_3 in C^* and an open neighborhood $\mathcal{N}$ of Γ in $\mathbb{R}^n \setminus \{0\}$ such that

$$(x(0) \in \mathcal{N} \text{ and } \dot{x} = f(x, u_3(x, t))) \Rightarrow (x(T) = 0).$$

One can also impose that, for every τ in $[0, T]$,

$$(\dot{x} = f(x, u_3(x, t)) \text{ and } x(\tau) = 0) \Rightarrow (x(t) = 0, \forall t \in [\tau, T]).$$

Step 4. In this last step, one shows the existence of a closed submanifold Γ of $\mathbb{R}^n \setminus \{0\}$ of dimension 1, contained in the set $\{x \in \mathbb{R}^n; 0 < |x| < \varepsilon_1\}$, such that, for every neighborhood $\mathcal{N}$ of Γ in $\mathbb{R}^n \setminus \{0\}$, there exists a time-varying feedback law u_4 in C^* such that, for some ε_4 in $(0, +\infty)$,

$$(\dot{x} = f(x, u_4(x, t)) \text{ and } |x(0)| < \varepsilon_4) \Rightarrow (x(T) \in \mathcal{N} \cup \{0\}),$$
$$((\dot{x} = f(x, u_4(x, t)) \text{ and } x(\tau) = 0) \Rightarrow (x(t) = 0, \forall t \in [\tau, T])), \forall \tau \in [0, T].$$

Finally, let $u : \mathbb{R}^n \times \mathbb{R} \to \mathbb{R}^m$ be equal to u_4 on $\mathbb{R}^n \times [0, T]$, $2T$-periodic with respect to time, and such that $u(x, t) = u_3(x, t - T)$ for every (x, t) in $\mathbb{R}^n \times (T, 2T)$. Then u vanishes on $\{0\} \times \mathbb{R}$, is continuous on $\mathbb{R}^n \times (\mathbb{R} \setminus \mathbb{Z}T)$, of class C^∞ on $(\mathbb{R}^n \setminus \{0\}) \times (\mathbb{R} \setminus \mathbb{Z}T)$, and satisfies

$$(\dot{x} = f(x, u(x, t)) \text{ and } |x(0)| < \varepsilon_4) \Rightarrow (x(2T) = 0),$$
$$(\dot{x} = f(x, u(x, t)) \text{ and } x(\tau) = 0) \Rightarrow (x(t) = 0, \forall t \geqslant \tau), \forall \tau \in \mathbb{R},$$

which implies (see [**109**]) that (11.77) holds, with $4T$ instead of T and $\varepsilon > 0$ small enough, and that $0 \in \mathbb{R}^n$ is locally asymptotically stable for the system $\dot{x} = f(x, u(x, t))$. Since T is arbitrary, Theorem 11.28 is proved (modulo a problem of regularity of u at (x, t) in $\mathbb{R}^n \times \mathbb{Z}T$, which is fixed in [**109**]). ∎

EXAMPLE 11.31. *Let us again go back to control system (3.12) of the attitude of a rigid spacecraft, already considered in Example 3.9 on page 128, Example 3.35 on page 144, Example 10.15 on page 282 and Example 11.3 on page 289. We assume that $m = 2$, and that (3.81) holds, which is generically satisfied. Let us* *recall (see Example 3.35 on page 144) that, under this assumption, El-Yazid Keraï checked in* [**272**] *that the control system (3.12) satisfies Sussmann's condition S(1) which, by Theorem 11.24 on page 301, implies that $0 \in \mathbb{R}^6$ is locally continuously reachable for the control system (3.12). Hence, by Theorem 11.28 on page 302, for every $T > 0$, there exists a T-periodic continuous time-varying feedback law which locally asymptotically stabilizes the control system (3.12). For the special case where the torque actions are exerted on the principal axis of the inertia matrix of the spacecraft, the construction of such feedback laws was done independently by Pascal Morin, Claude Samson, Jean-Baptiste Pomet and Zhong-Ping Jiang in* [**367**]*, and by Gregory Walsh, Richard Montgomery and Shankar Sastry in* [**497**]*. The general case has been treated in* [**126**]*; see Example 12.18 on page 330. Simpler feedback laws have been proposed by Pascal Morin and Claude Samson in* [**364**]*. In Sections 12.4 and 12.5, we explain how the feedback laws of* [**364**] *are constructed.*

Many authors have constructed explicit time-varying stabilizing feedback laws for other control systems. Let us mention, in particular:

- Lotfi Beji, Azgal Abichou and Yasmina Bestaoui for their paper [**46**] dealing with the stabilization of an under-actuated autonomous airship.
- Zhong-Ping Jiang and Henk Nijmeijer for their paper [**256**] dealing with the tracking of nonholonomic systems in chained form.
- Naomi Leonard and P.S. Krishnaprasad for their paper [**312**] explaining how to construct explicit time-varying stabilizing feedback laws for driftless, left-invariant systems on Lie groups, with an illustration for autonomous underwater vehicles (see also [**311**]).
- Sonia Martínez, Jorge Cortés and Francesco Bullo for their paper [**345**] where they show how to use averaging techniques in order to construct stabilizing time-varying feedback (see also Section 12.4), with an application to a planar vertical takeoff and landing aircraft (PVTOL).
- Frédéric Mazenc, Kristin Pettersen and Henk Nijmeijer for their paper [**348**] dealing with an underactuated surface vessel.
- Robert M'Closkey and Richard Murray for their paper [**351**] dealing with the exponential stabilization of driftless systems using time-varying homogeneous feedback laws.
- Pascal Morin, Jean-Baptiste Pomet and Claude Samson for their paper [**363**], in which they use oscillatory control inputs which approximate motion in the direction of iterated Lie brackets in order to stabilize driftless systems.
- Abdelhamid Tayebi and Ahmed Rachid for their paper [**474**] dealing with the parking problem of a wheeled mobile robot.
- Bernard Thuilot, Brigitte d'Andréa-Novel, and Alain Micaelli for their paper [**480**] dealing with mobile robots equipped with several steering wheels.

11.3. Output feedback stabilization

In this section, only part of the state (called the output) is measured. Let us denote by $(\tilde{C})$ the control system

(11.81) $\qquad (\tilde{C}): \quad \dot{x} = f(x, u), \; y = h(x),$

where $x \in \mathbb{R}^n$ is the state, $u \in \mathbb{R}^m$ is the control, and $y \in \mathbb{R}^p$ is the output. Again, $f \in C^\infty(\mathbb{R}^n \times \mathbb{R}^m; \mathbb{R}^n)$ and satisfies (11.3). We also assume that $h \in C^\infty(\mathbb{R}^n; \mathbb{R}^p)$

and satisfies
$$h(0) = 0. \tag{11.82}$$

In order to state the main result of this section, we first introduce some definitions.

DEFINITION 11.32. The control system $(\tilde{C})$ is said to be *locally stabilizable in small time by means of continuous periodic time-varying output feedback laws* if, for every positive real number T, there exist ε in $(0, +\infty)$ and u in $C^0(\mathbb{R}^n \times \mathbb{R}; \mathbb{R}^m)$ such that (11.74), (11.75), (11.76), (11.77) hold and such that
$$u(x, t) = \bar{u}(h(x), t) \tag{11.83}$$
for some $\bar{u}$ in $C^0(\mathbb{R}^p \times \mathbb{R}; \mathbb{R}^n)$.

Our next definition concerns dynamic stabilizability.

DEFINITION 11.33. The control system $(\tilde{C})$ *is locally stabilizable in small time by means of dynamic continuous periodic time-varying state (resp. output) feedback laws* if, for some integer $k \geqslant 0$, the control system
$$\dot{x} = f(x, u), \; \dot{z} = v, \; \tilde{h}(x, z) = (h(x), z), \tag{11.84}$$
where the state is $(x, z) \in \mathbb{R}^n \times \mathbb{R}^k$, the control $(u, v) \in \mathbb{R}^m \times \mathbb{R}^k$, and the output $\tilde{h}(x, z) \in \mathbb{R}^p \times \mathbb{R}^k$, is locally stabilizable in small time by means of continuous periodic time-varying state (resp. output) feedback laws.

In the above definition, the control system (11.84) with $k = 0$ is, by convention, the control system $(\tilde{C})$. Note that "locally stabilizable in small time by means of continuous periodic time-varying state feedback laws" is simply what we called "locally stabilizable in small time by means of continuous periodic time-varying feedback laws" in Definition 11.27 on page 302.

For our last definition, one needs to introduce some notations. For α in $\mathbb{N}^m$ and $\bar{u}$ in $\mathbb{R}^m$, let $f_{\bar{u}}^\alpha$ in $C^\infty(\mathbb{R}^n; \mathbb{R}^n)$ be defined by
$$f_{\bar{u}}^\alpha(x) := \frac{\partial^{|\alpha|} f}{\partial u^\alpha}(x, \bar{u}), \; \forall x \in \mathbb{R}^n. \tag{11.85}$$

Let $O(\tilde{C})$ be the vector subspace of $C^\infty(\mathbb{R}^n \times \mathbb{R}^m; \mathbb{R}^p)$ spanned by the maps ω such that, for some integer $r \geqslant 0$ (depending on ω) and for some sequence $\alpha_1, ..., \alpha_r$ of r multi-indices in $\mathbb{N}^m$ (also depending on ω), we have, for every $x \in \mathbb{R}^n$ and for every $u \in \mathbb{R}^m$,
$$\omega(x, u) = L_{f_u^{\alpha_1}} ... L_{f_u^{\alpha_r}} h(x). \tag{11.86}$$

Let us recall that in (11.86) and in the following, for $X = (X_1, \ldots, X_n)^{\text{tr}} : \mathbb{R}^n \to \mathbb{R}^n$ and $V : \mathbb{R}^n \to \mathbb{R}$, $L_X V : \mathbb{R}^n \to \mathbb{R}$ denotes the (Lie) derivative of V in the direction of X as defined by (3.91). By convention, if $r = 0$, the right hand side of (11.86) is $h(x)$. With this notation, our last definition is the following.

DEFINITION 11.34. The control system $(\tilde{C})$ is *locally Lie null-observable at the equilibrium* $(0, 0) \in \mathbb{R}^n \times \mathbb{R}^m$ if there exists a positive real number $\bar{\varepsilon}$ such that the following two properties hold.

(i) For every a in $\mathbb{R}^n \setminus \{0\}$ such that $|a| < \bar{\varepsilon}$, there exists $q \in \mathbb{N}$ such that
$$L_{f_0}^q h(a) \neq 0, \tag{11.87}$$

with $f_0(x) = f(x,0)$ and the usual convention $L_{f_0}^0 h = h$.

(ii) For every $(a_1, a_2) \in (\mathbb{R}^n \setminus \{0\})^2$ with $a_1 \neq a_2, |a_1| < \bar{\varepsilon}$, and $|a_2| < \bar{\varepsilon}$, and for every u in $\mathbb{R}^m$ with $|u| < \bar{\varepsilon}$, there exists ω in $O(\tilde{C})$ such that
$$\omega(a_1, u) \neq \omega(a_2, u). \tag{11.88}$$

In (11.87) and in the following, for $k \in \mathbb{N}$, $X = (X_1, \ldots, X_n)^{\text{tr}} : \mathbb{R}^n \to \mathbb{R}^n$ and $V : \mathbb{R}^n \to \mathbb{R}^p$, $L_X^k V : \mathbb{R}^n \to \mathbb{R}^p$ denotes the iterated (Lie) derivative of V in the direction of X. It is defined by induction on k by
$$L_X^0 V := V,\ L_X^{k+1} V := L_X(L_X^k V). \tag{11.89}$$

REMARK 11.35. Note that (i) implies the following property:

(i)* for every $a \neq 0$ in $B_{\bar{\varepsilon}} := \{x \in \mathbb{R}^m, |x| < \bar{\varepsilon}\}$, there exists a positive real number τ such that
$$x(\tau) \text{ exists and } h(x(\tau)) \neq 0,$$
where x is the (maximal) solution of $\dot{x} = f(x, 0), x(0) = a$. Moreover, if f and g are analytic, (i)* implies (i). The reason for "null" in "null-observable" comes from condition (i) or (i)*; roughly speaking we want to be able to distinguish from 0 every a in $B_{\bar{\varepsilon}} \setminus \{0\}$ by using the control law which vanishes identically—$u \equiv 0$.

REMARK 11.36. When f is affine with respect to u, i.e., $f(x, u) = f_0(x) + \sum_{i=1}^m u_i f_i(x)$ with $f_1, \ldots, f_m$ in $C^\infty(\mathbb{R}^n; \mathbb{R}^n)$, then a slightly simpler version of (ii) can be given. Let $\tilde{O}(\tilde{C})$ be the observation space (see e.g. [**222**] or [**458**, Remark 6.4.2]), i.e., the set of maps $\tilde{\omega}$ in $C^\infty(\mathbb{R}^n; \mathbb{R}^p)$ such that, for some integer $r \geqslant 0$ (depending on $\tilde{\omega}$) and for some sequence $i_1, \ldots, i_r$ of integers in $[0, m]$,
$$\tilde{\omega}(x) = L_{f_{i_1}} \ldots L_{f_{i_r}} h(x),\ \forall x \in \mathbb{R}^n, \tag{11.90}$$

with the convention that, if $r = 0$, the right hand side of (11.90) is $h(x)$. Then (ii) is equivalent to
$$((a_1, a_2) \in B_{\bar{\varepsilon}}^2,\quad \tilde{\omega}(a_1) = \tilde{\omega}(a_2),\ \forall \tilde{\omega} \in \tilde{O}(\tilde{C})) \Rightarrow (a_1 = a_2). \tag{11.91}$$

Finally, let us remark that if f is a polynomial with respect to u or if f and g are analytic, then (ii) is equivalent to

(ii)* for every $(a_1, a_2) \in \mathbb{R}^n \setminus \{0\}$ with $a_1 \neq a_2, |a_1| < \bar{\varepsilon}$ and $|a_2| < \bar{\varepsilon}$, there exists u in $\mathbb{R}^m$ and ω in $O(\tilde{C})$ such that (11.88) holds. Indeed, in these cases, the subspace of $\mathbb{R}^p$ spanned by $\{\omega(x, u); \omega \in O(\tilde{C})\}$ does not depend on u; it is the observation space of $(\tilde{C})$ evaluated at x, as defined for example in [**222**].

With these definitions we have the following theorem, proved in [**107**].

THEOREM 11.37. *Assume that the origin (of $\mathbb{R}^n$) is locally continuously reachable (for (C)) in small time (see Definition 11.20). Assume that $(\tilde{C})$ is locally Lie null-observable at the equilibrium $(0,0) \in \mathbb{R}^n \times \mathbb{R}^m$. Then $(\tilde{C})$ is locally stabilizable in small time by means of dynamic continuous periodic time-varying output feedback laws.*

Sketch of the proof of Theorem 11.37. We assume that the assumptions of Theorem 11.37 are satisfied. Let T be a positive real number. The proof of Theorem 11.37 is divided into four steps.

Step 1. Using the assumption that the system (C) is locally Lie null-observable at the equilibrium $(0,0) \in \mathbb{R}^n \times \mathbb{R}^m$ and [**106**], one proves that there exist u^* in $C^\infty(\mathbb{R}^p \times [0,T]; \mathbb{R}^m)$ and a positive real number ε^* such that

$$u^*(y,T) = u^*(y,0) = 0, \quad \forall y \in \mathbb{R}^p,$$
(11.92)
$$u^*(0,t) = 0, \quad \forall t \in [0,T],$$

and, for every (a_1, a_2) in $B_{\varepsilon^*}^2$, for every s in $(0,T)$,

(11.93) $\quad (h_{a_1}^{(i)}(s) = h_{a_2}^{(i)}(s), \ \forall i \in \mathbb{N}) \Rightarrow (a_1 = a_2),$

where $h_a(s) = h(x^*(a,s))$ with $x^* : B_{\varepsilon^*} \times [0,T] \to \mathbb{R}^n$, $(a,t) \mapsto x^*(a,t)$, defined by

$$\frac{\partial x^*}{\partial t} = f(x^*, u^*(h(x^*), t)), \ x^*(a, 0) = a.$$

Let us note that in [**350**] a similar u^* was introduced by Frédéric Mazenc and Laurent Praly, but it was taken depending only on time and so (11.92), which is important to get stability, was not satisfied in general. In this step we do not use the local continuous reachability in small time.

Step 2. Let $q = 2n+1$. In this step, using (11.93), one proves the existence of $(q+1)$ real numbers $0 < t_0 < t_1 ... < t_q < T$ such that the map $K : B_{\varepsilon^*} \to (\mathbb{R}^p)^q$ defined by

(11.94) $\quad K(a) := \left(\int_{t_0}^{t_1} (s - t_0)(t_1 - s) h_a(s) ds, ..., \int_{t_0}^{t_q} (s - t_0)(t_q - s) h_a(s) ds \right)$

is one-to-one. Thus, there exists a map $\theta : (\mathbb{R}^p)^q \to \mathbb{R}^n$ such that

(11.95) $\quad \theta \circ K(a) = x^*(a, T), \ \forall a \in B_{\varepsilon^*/2}.$

Step 3. In this step, one proves the existence of $\bar{u}$ in $C^0(\mathbb{R}^n \times [0,T]; \mathbb{R}^m)$ and $\bar{\varepsilon}$ in $(0, +\infty)$ such that

(11.96) $\quad \bar{u} = 0$ on $(\mathbb{R}^n \times \{0, T\}) \cup (\{0\} \times [0, T]),$

(11.97) $\quad (\dot{x} = f(x, \bar{u}(x(0), t))$ and $|x(0)| < \bar{\varepsilon}) \Rightarrow (x(T) = 0).$

Property (11.97) means that $\bar{u}$ is a "dead-beat" open-loop control. In this last step, we use the small-time local reachability assumption on (C), but do not use the Lie null-observability assumption.

Step 4. Using the above three steps, let us conclude the proof of Theorem 11.37. The dynamic extension of system (C) that we consider is

(11.98) $\quad \dot{x} = f(x, u), \ \dot{z} = v = (v_1, ..., v_q, v_{q+1}) \in \mathbb{R}^p \times ... \times \mathbb{R}^p \times \mathbb{R}^n \simeq \mathbb{R}^{pq+n},$

with $z_1 = (z_1, ..., z_q, z_{q+1}) \in \mathbb{R}^p \times ... \times \mathbb{R}^p \times \mathbb{R}^n \simeq \mathbb{R}^{pq+n}$. For this system the output is $\tilde{h}(x, z) = (h(x), z) \in \mathbb{R}^p \times \mathbb{R}^{pq+n}$. For $s \in \mathbb{R}$, let $s^+ = \max(s, 0)$ and let $\text{Sign}(s) = 1$ if $s > 0$, 0 if $s = 0$, -1 if $s < 0$. Finally, for r in $\mathbb{N} \setminus \{0\}$ and $b = (b_1, ..., b_r)$ in $\mathbb{R}^r$, let

(11.99) $\quad b^{1/3} := (|b_1|^{1/3} \text{Sign}(b_1), ..., |b_r|^{1/3} \text{Sign}(b_r)).$

We now define $u : \mathbb{R}^p \times \mathbb{R}^{pq+n} \times \mathbb{R} \to \mathbb{R}^m$ and $v : \mathbb{R}^p \times \mathbb{R}^{pq+n} \times \mathbb{R} \to \mathbb{R}^{pq+n}$ by requiring, for (y,z) in $\mathbb{R}^p \times \mathbb{R}^{pq+n}$ and for every i in $\{1, ..., q\}$,

(11.100) $\quad u(y, z, t) := u^*(y, t), \ \forall t \in [0, T],$

(11.101) $\quad v_i(y, z, t) := -t(t_0 - t)^+ z_i^{1/3} + (t - t_0)^+ (t_i - t)^+ y, \ \forall t \in [0, T],$

$$
\begin{align}
v_{q+1}(y,z,t) &:= -t(t_q-t)^+ z_{q+1}^{1/3} \tag{11.102}\\
&\quad + 6\frac{(T-t)^+(t-t_q)^+}{(T-t_q)^3}\theta(z_1,\ldots,z_q),\ \forall t\in[0,T],\notag
\end{align}
$$

$$u(y,z,t):=\bar{u}(z_{q+1},t-T),\ \forall t\in[T,2T], \tag{11.103}$$

$$v(y,z,t):=0,\ \forall t\in[T,2T], \tag{11.104}$$

$$u(y,z,t)=u(y,z,t+2T),\ \forall t\in\mathbb{R}, \tag{11.105}$$

$$v(y,z,t)=v(y,z,t+2T),\ \forall t\in\mathbb{R}. \tag{11.106}$$

One easily sees that u and v are continuous and vanish on $\{(0,0)\}\times\mathbb{R}$. Let (x,z) be any maximal solution of the closed-loop system

$$\dot{x}=f(x,u(\tilde{h}(x,z),t)),\quad \dot{z}=v(\tilde{h}(x,z),t). \tag{11.107}$$

Then one easily checks that, if $|x(0)|+|z(0)|$ is small enough, then

$$z_i(t_0)=0,\ \forall i\in\{1,\ldots,q\}, \tag{11.108}$$

$$(z_1(t),\ldots,z_q(t))=K(x(0)),\ \forall t\in[t_q,T], \tag{11.109}$$

$$z_{q+1}(t_q)=0, \tag{11.110}$$

$$z_{q+1}(T)=\theta\circ K(x(0))=x(T), \tag{11.111}$$

$$x(t)=0,\ \forall t\in[2T,3T], \tag{11.112}$$

$$z(2T+t_q)=0. \tag{11.113}$$

Equalities (11.108) (resp. (11.110)) are proved by computing explicitly, for $i\in\{1,\ldots,q\}$, z_i on $[0,t_0]$ (resp. z_{q+1} on $[0,t_q]$) and by seeing that this explicit solution reaches 0 before time t_0 (resp. t_q) and by pointing out that if, for some s in $[0,t_0]$ (resp. $[0,t_q]$), $z_i(s)=0$ (resp. $z_{q+1}(s)=0$), then $z_i=0$ on $[s,t_0]$ (resp. $z_{q+1}=0$ on $[s,t_q]$); note that $z_i\dot{z}_i\leqslant 0$ on $[0,t_0]$ (resp. $z_{q+1}\dot{z}_{q+1}\leqslant 0$ on $[0,t_q]$).

Moreover, one also has, for every s in $\mathbb{R}$ and every $t\geqslant s$,

$$((x(s),z(s))=(0,0))\Rightarrow((x(t),z(t))=(0,0)). \tag{11.114}$$

Indeed, first note that, without loss of generality, we may assume that $s\in[0,2T]$ and $t\in[0,2T]$. If $s\in[0,T]$, then, since u^* is of class C^∞, we get, using (11.92), that $x(t)=0,\ \forall t\in[s,T]$ and then, using (11.82) and (11.101), we get that, for every $i\in\{1,\ldots,q\}$, $z_i\dot{z}_i\leqslant 0$ on $[s,T]$ and so z_i also vanishes on $[s,T]$; this, together with (11.102) and $\theta(0)=0$ (see (11.94) and (11.95)), implies that $z_{q+1}=0$ also on $[s,T]$. Hence we may assume that $s\in[T,2T]$. But, in this case, using (11.104), we get that $z=0$ on $[s,2T]$ and, from (11.96) and (11.103), we get that $x=0$ also on $[s,2T]$.

From (11.112), (11.113), and (11.114), we get (see Lemma 2.15 in [**109**]) the existence of ε in $(0,+\infty)$ such that, for every s in $\mathbb{R}$ and every maximal solution (x,z) of $\dot{x}=f(x,u(\tilde{h}(x,z),t))$, $\dot{z}=v(\tilde{h}(x,z),t)$, we have

$$(|x(s)|+|y(s)|\leqslant\varepsilon)\Rightarrow((x(t),z(t))=(0,0),\ \forall t\geqslant s+5T).$$

Since T is arbitrary, Theorem 11.37 is proved. ∎

REMARK 11.38. Roughly speaking the strategy to prove Theorem 11.37 is the following one.

(i) During the time interval $[0, T]$, one "excites" system (C) by means of $u^*(y, t)$ in order to be able to deduce from the observation during this interval of time, the state at time T: At time T we have $z_{q+1} = x$.
(ii) During the time interval $[T, 2T]$, z_{q+1} does not move and one uses the deadbeat open-loop $\bar{u}$ but transforms it into an output feedback by using in its argument z_{q+1} instead of the value of x at time T (this step has been used previously in the proof of Theorem 1.7 of [**105**]).

This method has been previously used by Eduardo Sontag in [**451**], Rogelio Lozano in [**335**], Frédéric Mazenc and Laurent Praly in [**350**]. A related idea is also used in Section 3 of [**105**], where we first recover initial data from the state. Moreover, as in [**451**] and [**350**], our proof relies on the existence of an output feedback which distinguishes every pair of distinct states (see [**466**] for analytic systems and [**106**] for C^∞ systems).

REMARK 11.39. It is established by Frédéric Mazenc and Laurent Praly in [**350**] that "distinguishability" with a universal time-varying control, global stabilizability by state feedback, and observability of blow-up are sufficient conditions for the existence of a time-varying dynamic output feedback (of infinite dimension and in a more general sense than the one considered in Definition 11.33) guaranteeing boundedness and convergence of all the solutions defined at time $t = 0$. The methods developed in [**350**] can be directly applied to our situation. In this case, Theorem 11.37 still gives two improvements of the results of [**350**]: We get that 0 is asymptotically stable for the closed-loop system, instead of only attractor for time 0, and our dynamic extension is of finite dimension, instead of infinite dimension.

REMARK 11.40. If $(\tilde{C})$ is locally stabilizable in small time by means of dynamic continuous periodic time-varying output feedback laws, then the origin (of $\mathbb{R}^n$) is locally continuously reachable (for $(\tilde{C})$) in small time (use Lemma 3.5 in [**107**]) and, if moreover f and h are analytic, then $(\tilde{C})$ is locally Lie null-observable; see [**107**, Proposition 4.3].

REMARK 11.41. There are linear control systems which are controllable and observable but which cannot be locally asymptotically stabilized by means of continuous time-varying feedback laws. This is, for example, the case for the controllable and observable linear system, with $n = 2$, $m = 1$, and $p = 1$,
$$\dot{x}_1 = x_2, \ \dot{x}_2 = u, \ y = x_1.$$
Indeed, assume that this system can be locally asymptotically stabilized by means of a continuous time-varying output feedback law $u : \mathbb{R} \times \mathbb{R} \to \mathbb{R}$. Then there exist $r > 0$ and $\tau > 0$ such that, if $\dot{x}_1 = x_2$, $\dot{x}_2 = u(x_1, t)$,

(11.115) $\quad (x_1(0)^2 + x_2(0)^2 \leqslant r^2) \Rightarrow (x_1(\tau)^2 + x_2(\tau)^2 \leqslant r^2/5)$.

Let $(u^n; n \in \mathbb{N})$ be a sequence of functions from $\mathbb{R} \times \mathbb{R}$ into $\mathbb{R}$ of class C^∞ which converges uniformly to u on each compact subset of $\mathbb{R} \times \mathbb{R}$. Then, for n large enough, $\dot{x}_1 = x_2$, $\dot{x}_2 = u^n(x_1, t)$ imply

(11.116) $\quad (x_1(0)^2 + x_2(0)^2 \leqslant r^2) \Rightarrow (x_1(\tau)^2 + x_2(\tau)^2 \leqslant r^2/4)$.

However, since the time-varying vector field X on $\mathbb{R}^2$ defined by
$$X_1(x_1,x_2,t)=x_2,\; X_2(x_1,x_2,t)=u^n(x_1,t),\; \forall (x_1,r_2,t)\in\mathbb{R}^3,$$
has a divergence equal to 0, the flow associated with X preserves area, which is a contradiction to (11.116).

11.4. Discontinuous feedback laws

In this section, we just briefly describe some results on discontinuous feedback laws.

The pioneering work on this type of feedback laws is [**467**] by Héctor Sussmann. It is proved in [**467**] that any controllable analytic system can be asymptotically stabilized by means of piecewise analytic feedback laws.

One of the key questions for discontinuous feedback laws is what is the relevant definition of a solution of the closed-loop system. In [**467**], this question is solved by specifying an "exit rule" on the singular set. However, it is not completely clear how to implement this exit rule (but see Remark 11.42 on the following page for this problem), which is important in order to analyze the robustness.

One possibility, as introduced and justified by Henry Hermes in [**224**] (see also [**131, 308**]), is to consider that the solutions of the closed-loop systems are the solutions in the sense of Alexey Filippov [**170**]. Then we prove in [**131**] that a control system which can be stabilized by means of a discontinuous feedback law can be stabilized by means of continuous periodic time-varying feedback laws. Moreover, we also prove in [**131**] that, if the system is control affine, it can be stabilized by means of continuous stationary feedback laws. See also the paper [**429**] by Eugene Ryan. In particular, the control system (11.1) cannot be stabilized by means of discontinuous feedback laws if one considers Filippov's solutions of the closed-loop system.

Another interesting possibility is to consider "Euler" solutions; see [**98**] for a definition. This is a quite natural notion for control systems since it corresponds to the idea that one uses the same control during small intervals of time. With this type of solutions, Francis Clarke, Yuri Ledyaev, Eduardo Sontag and Andrei Subbotin proved in [**98**] that controllability (or even asymptotic controllability) implies the existence of stabilizing discontinuous feedback laws. Their feedback laws are robust to (small) actuator disturbances.

The original proof of [**98**] has been simplified by Ludovic Rifford in [**403**]. In this paper he shows that asymptotic controllability implies the existence of a *semiconcave* Lyapunov function. This type of regularity allows us to construct directly stabilizing feedback laws. His feedback laws have some new interesting regularity properties and the set of singularities can be made repulsive for the closed-loop system. In [**404**] Ludovic Rifford has strongly improved his regularity result in the case of driftless control affine systems.

The previous papers [**98, 403**] use the existence of a control Lyapunov function as a major step to construct discontinuous stabilizing feedback laws. A completely different approach for the proof of the existence of these discontinuous feedback laws was proposed by Fabio Ancona and Alberto Bressan in [**12**]. They allow us to consider not only the Euler solutions, but also the usual Carathéodory solutions. These solutions are also allowed in the paper [**403**].

However, there is a problem of robustness of all these discontinuous feedback laws with respect to measurement disturbances: using a result in [**99**] due to Francis Clarke, Yuri Ledyaev and Ronald Stern, Yuri Ledyaev and Eduardo Sontag proved in [**308**] that these feedback laws are in general (e.g. for the control system (11.1)) not robust to arbitrarily small measurement disturbances. In [**308**] Yuri Ledyaev and Eduardo Sontag introduced a new class of "dynamic and hybrid" discontinuous feedback laws and showed that controllability (or even asymptotic controllability) implies the existence of stabilizing discontinuous feedback laws in this class which are robust to (small) actuators and measurement disturbances.

In order to get some robustness to measurement disturbances for the discontinuous feedback laws of the type in [**98**], it has been independently proved by Eduardo Sontag in [**459**] and by Francis Clarke, Yuri Ledyaev, Ludovic Rifford and Ronald Stern in [**97**] that it is interesting to impose a lower bound on the intervals of time on which the same control is applied. For the robustness of the feedback laws constructed by Fabio Ancona and Alberto Bressan in [**12**], we refer to their papers [**13, 14**].

To take care of this problem of robustness, Christophe Prieur proposed in [**393**] a new type of feedback laws, namely hybrid feedback laws. He adds discrete variables to the continuous control system which produce hysteresis between controllers. An explicit example of these feedback laws is given for the nonholonomic integrator (11.1) by Christophe Prieur and Emmanuel Trélat in [**394**] and is generalized for control affine driftless systems in [**395**].

Classical discontinuous feedback laws are provided by the so-called sliding mode control in engineering literature. These feedback laws are smooth outside attracting smooth hypersurfaces. We refer to [**491**] by Vadim Utkin for this approach. For the case of the nonholonomic integrator (11.1), see, in particular, [**58**] by Anthony Bloch and Sergey Drakunov. Again, it would be important to evaluate the robustness of this type of feedback laws with respect to measurement disturbances.

REMARK 11.42. It would be interesting to know if one can in some sense "implement" (a good enough approximation of) Sussmann's exit rule (see [**467**]) by means of Sontag and Ledyaev's "dynamic-hybrid" strategy. Another possibility to implement this exit rule could be by means of hysteresis feedback laws introduced by Christophe Prieur in [**393**].

CHAPTER 12

Feedback design tools

In this chapter we give some tools to design stabilizing feedback laws for our control system (C) (i.e., the control system $\dot{x} = f(x, u)$, see page 288) and present some applications. The tools we want to describe are:

1. control Lyapunov function (Section 12.1),
2. damping (Section 12.2),
3. homogeneity (Section 12.3),
4. averaging (Section 12.4),
5. backstepping (Section 12.5),
6. forwarding (Section 12.6),
7. transverse functions (Section 12.7).

There are in fact plenty of other powerful methods; e.g., frequency-domain methods, zero-dynamics, center manifolds, adaptive control, etc. See, for example, [17, 23, 130, 149, 178, 247, 288, 376] and the references therein.

12.1. Control Lyapunov function

A basic tool to study the asymptotic stability of an equilibrium point is the Lyapunov function (see, in particular, the book [24] by Andrea Bacciotti and Lionel Rosier). In the case of a control system, the control is at our disposal, so there are more "chances" that a given function could be a Lyapunov function for a suitable choice of feedback laws. Hence Lyapunov functions are even more useful for the stabilization of control systems than for dynamical systems without control.

For simplicity, we restrict our attention to global asymptotic stabilization and assume that $\mathcal{O} = \mathbb{R}^n \times \mathbb{R}^m$. The definitions and theorems of this section can be easily adapted to treat local asymptotic stabilization and smaller sets $\mathcal{O}$.

In the framework of control systems, the Lyapunov function approach leads to the following definition, due to Zvi Artstein [20].

DEFINITION 12.1. A function $V \in C^1(\mathbb{R}^n; \mathbb{R})$ is a control Lyapunov function for the control system (C) if

$$V(x) \to +\infty, \text{ as } |x| \to +\infty,$$
$$V(x) > 0, \forall x \in \mathbb{R}^n \setminus \{0\},$$
$$\forall x \in \mathbb{R}^n \setminus \{0\}, \exists u \in \mathbb{R}^m \text{ s.t. } f(x, u) \cdot \nabla V(x) < 0.$$

Moreover, V satisfies the *small control property* if, for every strictly positive real number ε, there exists a strictly positive real number η such that, for every $x \in \mathbb{R}^n$ with $0 < |x| < \eta$, there exists $u \in \mathbb{R}^m$ satisfying $|u| < \varepsilon$ and $f(x, u) \cdot \nabla V(x) < 0$.

With this definition, one has the following theorem due to Zvi Artstein [20].

THEOREM 12.2. *If the control system (C) is globally asymptotically stabilizable by means of continuous stationary feedback laws, then it admits a control Lyapunov function satisfying the small control property. If the control system (C) admits a control Lyapunov function satisfying the small control property, then it can be globally asymptotically stabilized by means of*

1. *continuous stationary feedback laws if the control system (C) is control affine (see Definition 3.12 on page 131),*
2. *relaxed controls (see* [**20**] *for a definition) for general f.*

Instead of relaxed controls in the last statement, one can use continuous periodic time-varying feedback laws. Indeed, one has the following theorem, proved in [**131**].

THEOREM 12.3. *The control system (C) can be globally asymptotically stabilized by means of continuous periodic time-varying feedback laws if it admits a control Lyapunov function satisfying the small control property.*

Let us point out that, even in the case of control affine systems, Artstein's proof of Theorem 12.2 relies on partitions of unity and so may lead to complicate stabilizing feedback laws. Explicit and simple feedback laws are given by Eduardo Sontag in [**455**]. He proves:

THEOREM 12.4. *Assume that V is a control Lyapunov function satisfying the small control property for the control system (C). Assume that (C) is control affine, i.e., (see Definition 3.12 on page 131)*

$$f(x,u) = f_0(x) + \sum_{i=1}^{m} u_i f_i(x), \; \forall (x,u) \in \mathbb{R}^n \times \mathbb{R}^m,$$

for some $f_0, \ldots, f_m$ *in* $C^\infty(\mathbb{R}^n; \mathbb{R}^n)$. *Then* $u = (u_1, \ldots, u_m)^{tr} : \mathbb{R}^n \to \mathbb{R}^m$ *defined by*

$$(12.1) \quad u_i(x) := -\varphi \left(f_0(x) \cdot \nabla V(x), \sum_{j=1}^{m} (f_j(x) \cdot \nabla V(x))^2 \right) f_i(x) \cdot \nabla V(x),$$

with

$$(12.2) \quad \varphi(a,b) = \begin{cases} \frac{a + \sqrt{a^2 + b^2}}{b} & \text{if } b \neq 0, \\ 0 & \text{if } b = 0, \end{cases}$$

is continuous, vanishes at $0 \in \mathbb{R}^n$ *and globally asymptotically stabilizes the control system (C).*

OPEN PROBLEM 12.5. *For systems which are not control affine, find some explicit formulas for globally asymptotically stabilizing continuous periodic time-varying feedback laws, given a control Lyapunov function satisfying the small control property. (By Theorem 12.3, such feedback laws exist.)*

12.2. Damping feedback laws

Control Lyapunov function is a very powerful tool used to design stabilizing feedback laws. But one needs to guess candidates for such functions in order to apply Sontag's formula (12.1). For mechanical systems at least, a natural candidate for a control Lyapunov function is given by the total energy, i.e., the sum of potential and kinetic energies; but, in general, it does not work.

EXAMPLE 12.6. *Consider the classical spring-mass system. The control system is*

(12.3) $$\dot{x}_1 = x_2, \ \dot{x}_2 = -\frac{k}{m}x_1 + \frac{u}{m},$$

where m is the mass of the point attached to the spring, x_1 is the displacement of the mass (on a line), x_2 is the speed of the mass, k is the spring constant, and u is the force applied to the mass. The state is $(x_1, x_2)^{tr} \in \mathbb{R}^2$ and the control is $u \in \mathbb{R}$. The total energy E of the system is

$$E = \frac{1}{2}(kx_1^2 + mx_2^2).$$

The control system can be written in the form

(12.4) $$\dot{x} = f_0(x) + uf_1(x), \ x = (x_1, x_2)^{tr} \in \mathbb{R}^2, \ u \in \mathbb{R},$$

with

$$f_0(x) := \begin{pmatrix} x_2 \\ -\frac{k}{m}x_1 \end{pmatrix}, \ f_1(x) := \begin{pmatrix} 0 \\ \frac{1}{m} \end{pmatrix}, \ \forall x = (x_1, x_2)^{tr} \in \mathbb{R}^2.$$

One has

$$f_0(x) \cdot \nabla E(x) = 0, \ \forall x = (x_1, x_2)^{tr} \in \mathbb{R}^2,$$
$$f_1(x) \cdot \nabla E(x) = x_2, \ \forall x = (x_1, x_2)^{tr} \in \mathbb{R}^2.$$

Hence, if $x_2 = 0$, there exists no u such that $(f_0(x) + uf_1(x)) \cdot \nabla E(x) < 0$. Therefore, the total energy is not a control Lyapunov function. But one has

$$(f_0(x) + uf_1(x)) \cdot \nabla E(x) = uf_1(x) \cdot \nabla E(x) = ux_2.$$

Hence, it is tempting to consider the feedback law

(12.5) $$u(x) := -\nu \nabla E(x) \cdot f_1(x) = -\nu x_2,$$

where $\nu > 0$ is fixed. With this feedback law, the closed-loop system is

(12.6) $$\begin{aligned} \dot{x}_1 &= x_2, \\ \dot{x}_2 &= -\frac{k}{m}x_1 - \frac{\nu}{m}x_2, \end{aligned}$$

which is the dynamics of a spring-mass-dashpot system. In other words, the feedback law adds some damping to the spring-mass system. With this feedback law,

$$\nabla E(x) \cdot (f_0(x) + u(x)f_1(x)) \leqslant 0,$$

so that $(0,0) \in \mathbb{R}^2$ is stable for the closed-loop system. In fact, $(0,0)$ is globally asymptotically stable for this system. Indeed, if a trajectory $x(t), t \in \mathbb{R}$, of the closed-loop system is such that $E(x(t))$ does not depend on time, then

(12.7) $$x_2(t) = 0, \ \forall t \in \mathbb{R}.$$

Derivating (12.7) with respect to time and using (12.6), one gets

$$x_1(t) = 0, \ \forall t \in \mathbb{R},$$

*which, together with (12.7) and LaSalle's invariance principle (see, for example, [**458**, Lemma 5.7.8, page 226]), proves that $(0,0)^{tr} \in \mathbb{R}^2$ is globally asymptotically stable for the closed-loop system.*

The previous example can be generalized in the following way. We assume that the control system (C) is control affine. Hence

$$f(x,u) = f_0(x) + \sum_{i=1}^{m} u_i f_i(x), \, \forall (x,u) \in \mathbb{R}^n \times \mathbb{R}^m,$$

for some $f_0, \ldots, f_m$ in $C^\infty(\mathbb{R}^n; \mathbb{R}^n)$. Let $V \in C^\infty(\mathbb{R}^n; \mathbb{R})$ be such that

(12.8) $\qquad V(x) \to +\infty, \text{ as } |x| \to +\infty,$

(12.9) $\qquad V(x) > V(0), \forall x \in \mathbb{R}^n \setminus \{0\},$

(12.10) $\qquad f_0 \cdot \nabla V \leqslant 0 \text{ in } \mathbb{R}^n.$

Then

$$f \cdot \nabla V \leqslant \sum_{i=1}^{m} u_i (f_i \cdot \nabla V).$$

Hence it is tempting to consider the feedback law $u = (u_1, \ldots, u_m)^{\mathrm{tr}}$ defined by

(12.11) $\qquad u_i = -f_i \cdot \nabla V, \, \forall i \in \{1, \ldots, m\}.$

With this feedback law, one has

$$f(x, u(x)) \cdot \nabla V(x) = f_0(x) \cdot \nabla V(x) - \sum_{i=1}^{m} (f_i(x) \cdot \nabla V(x))^2 \leqslant 0.$$

Therefore, $0 \in \mathbb{R}^n$ is a stable point for the closed-loop system $\dot{x} = f(x, u(x))$, that is, for every $\varepsilon > 0$, there exists $\eta > 0$ such that

$$(\dot{x} = f(x, u(x)), |x(0)| \leqslant \eta) \Rightarrow (|x(t)| \leqslant \varepsilon, \forall t \geqslant 0);$$

(see (i) of Theorem 10.8 on page 280). Moreover, by LaSalle's invariance principle (see, for example, [**458**, Lemma 5.7.8, page 226]) $0 \in \mathbb{R}^n$ is globally asymptotically stable if the following property $(\mathcal{P})$ holds.

$(\mathcal{P})$ For every $x \in C^\infty(\mathbb{R}; \mathbb{R}^n)$ such that

(12.12) $\qquad \dot{x}(t) = f_0(x(t)), \forall t \in \mathbb{R},$

(12.13) $\qquad f_i(x(t)) \cdot \nabla V(x(t)) = 0, \forall t \in \mathbb{R}, \forall i \in \{0, \ldots, m\},$

one has

(12.14) $\qquad x(t) = 0, \forall t \in \mathbb{R}.$

This method has been introduced independently by David Jacobson in [**250**] and by Velimir Jurdjevic and John Quinn [**261**]. There are many sufficient conditions available for property $(\mathcal{P})$. Let us give, for example, the following condition, due to Velimir Jurdjevic and John Quinn [**261**]; see also [**310**] by Kyun Lee and Aristotle Arapostathis for a more general condition.

THEOREM 12.7. *With the above notations and properties, assume that, for every* $x \in \mathbb{R}^n \setminus \{0\}$,

(12.15) $\qquad \mathrm{Span}\{f_0(x), \mathrm{ad}_{f_0}^k f_i(x), i = 1, \ldots, m, k \in \mathbb{N}\} = \mathbb{R}^n,$

(12.16) $\qquad \nabla V(x) \neq 0.$

Then property $(\mathcal{P})$ is satisfied. In particular, the feedback law defined by (12.11) globally asymptotically stabilizes the control system (C).

(Let us recall that $\mathrm{ad}_{f_0}^k f_i$ is defined in Definition 3.11 on page 130.)

Proof of Theorem 12.7. Let us start by recalling a classical property of the Lie derivative (defined on page 146): for every $X \in C^\infty(\mathbb{R}^n; \mathbb{R}^n)$, every $Y \in C^\infty(\mathbb{R}^n; \mathbb{R}^n)$, and every $W \in C^\infty(\mathbb{R}^n)$,

$$L_X(L_Y W) - L_Y(L_X W) = L_{[X,Y]} W. \tag{12.17}$$

(We already used this property; see (3.65).) Let $x \in C^\infty(\mathbb{R}; \mathbb{R}^n)$ be such that (12.12) and (12.13) hold. For $i \in \{0, \ldots, m\}$, let $A_i \in C^\infty(\mathbb{R}^n)$ and $\alpha_i \in C^\infty(\mathbb{R})$ be defined by

$$A_i(y) := f_i(y) \cdot \nabla V(y) = (L_{f_i} V)(y), \; \forall y \in \mathbb{R}^n, \tag{12.18}$$

$$\alpha_i(t) := A_i(x(t)), \; \forall t \in \mathbb{R}. \tag{12.19}$$

Using (12.12), (12.17), (12.18) and (12.19), one gets

$$\dot\alpha_i(t) = (L_{f_0} A_i)(x(t)) = (L_{f_0}(L_{f_i} V))(x(t)) \tag{12.20}$$
$$= (L_{[f_0, f_i]} V)(x(t)) + (L_{f_i}(L_{f_0} V))(x(t)), \; \forall t \in \mathbb{R}.$$

From (12.10), and (12.13) for $i = 0$, one sees that $L_{f_0} V$ is maximal at $x(t)$. In particular,

$$(L_{f_i}(L_{f_0} V))(x(t)) = 0, \; \forall t \in \mathbb{R}. \tag{12.21}$$

From (12.20) and (12.21), one gets

$$\dot\alpha_i(t) = (L_{[f_0, f_i]} V)(x(t)), \; \forall t \in \mathbb{R}. \tag{12.22}$$

Continuing in the same way, an easy induction argument on k shows that

$$\alpha_i^{(k)}(t) = (L_{\mathrm{ad}_{f_0}^k f_i} V)(x(t)) = \mathrm{ad}_{f_0}^k f_i(x(t)) \cdot \nabla V(x(t)), \; \forall t \in \mathbb{R}, \; \forall k \in \mathbb{N}. \tag{12.23}$$

From (12.15) and (12.23), one gets

$$(\nabla V(x(t)) = 0 \text{ or } x(t) = 0), \; \forall t \in \mathbb{R}, \tag{12.24}$$

which, together with (12.16), implies (12.14). This concludes the proof of Theorem 12.7. ∎

Let us point out that the damping method is also very useful when there are some constraints on the controls. Indeed if, for example, one wants that, for some $\varepsilon > 0$,

$$|u_i(x)| \leqslant \varepsilon, \forall i \in \{1, \ldots, m\},$$

then it suffices to replace (12.11) by

$$u_i(x) = -\sigma(f_i(x) \cdot \nabla V(x)),$$

where $\sigma \in C^0(\mathbb{R}; [-\varepsilon, \varepsilon])$ is such that $s\sigma(s) > 0$ for every $s \in \mathbb{R} \setminus \{0\}$. We give an application of this variant in the next section (Section 12.2.1).

REMARK 12.8. Note that, in general, V is not a control Lyapunov function (see Definition 12.1 on page 313) for the control system (C). However, under suitable homogeneity assumptions on the f_i's, Ludovic Faubourg and Jean-Baptiste Pomet constructed in [**163**] an explicit perturbation of V which is a control Lyapunov function for the control system (C).

REMARK 12.9. Let us show on a simple system the limitation of this damping approach. Let us consider the following control system:

(12.25) $$\dot{x}_1 = x_2, \ \dot{x}_2 = -x_1 + u.$$

This control system is the classical spring mass control system (12.3) considered in Example 12.6 on page 315, with normalized physical constants (namely $k = m = 1$). With the Lyapunov strategy used above, let $V \in C^\infty(\mathbb{R}^2; \mathbb{R})$ be defined by

$$V(x) = x_1^2 + x_2^2, \forall x = (x_1, x_2)^{\text{tr}} \in \mathbb{R}^2.$$

We have
$$\dot{V} = 2x_2 u.$$

So it is tempting to take, at least if we remain in the class of linear feedback laws,

$$u(x_1, x_2) = -\nu x_2,$$

where ν is some fixed positive real number. An a priori guess would be that, if we let ν be quite large, then we get a quite good convergence, as fast as we want. But this is completely wrong. Indeed, from now on, let us assume that $\nu > 2$. Then

$$x_1^\nu(t) := \exp\left(\frac{-\nu + \sqrt{\nu^2 - 4}}{2} t\right), \ x_2^\nu(t) := \dot{x}_1^\nu(t)$$

is a trajectory of the closed-loop system. Let $T > 0$ be given. Then, as $\nu \to \infty$,

$$x_1^\nu(0) = 1, \ x_2^\nu(0) \to 0.$$

However, still as $\nu \to \infty$,

$$x_1^\nu(T) \to 1 \neq 0.$$

Note also that the eigenvalues of the closed-loop system are

$$\lambda_1 = \frac{-\nu - \sqrt{\nu^2 - 4}}{2}, \ \lambda_2 = \frac{-\nu + \sqrt{\nu^2 - 4}}{2}.$$

We have $\lambda_1 < 0$ and $\lambda_2 < 0$, which confirms the interest of the damping. Moreover,

$$\lim_{\nu \to +\infty} \lambda_1 = -\infty.$$

However,
$$\lim_{\nu \to +\infty} \lambda_2 = 0,$$

explaining why that it is not a good idea to take $\nu \to +\infty$.

Note that, using the Kalman rank condition (Theorem 1.16 on page 9), one easily checks that the linear control system (12.25) is controllable. Hence, by the pole-shifting theorem (Theorem 10.1 on page 275), for every $\lambda > 0$, there exist $(k_1, k_2)^{\text{tr}} \in \mathbb{R}^2$ and $C > 0$ such that, for every solution $x := (x_1, x_2)^{\text{tr}}$ of the closed-loop system

$$\dot{x}_1 = x_2, \ \dot{x}_2 = k_1 x_1 + k_2 x_2,$$

one has
$$|x(t)| \leqslant C e^{-\lambda t} |x(0)|, \ \forall t \in [0, +\infty).$$

This shows an important limitation of the damping method: It can lead to a rather slow convergence compared to what can be done.

12.2.1. Orbit transfer with low-thrust systems. Electric propulsion systems for satellites are seriously considered for performing large amplitude orbit transfers. Let us recall that electric propulsion is characterized by a low-thrust acceleration level but a high specific impulse. In this section, where we follow [**129**], we are interested in a large amplitude orbit transfer of a satellite in a central gravitational field by means of an electric propulsion system.

The state of the control system is the position of the satellite (here identified to a point: we are not considering the attitude of the satellite) and the speed of the satellite. Instead of using Cartesian coordinates, one prefers to use the "orbital" coordinates. The advantage of this set of coordinates is that, in this set, the first five coordinates remain unchanged if the thrust vanishes: these coordinates characterize the Keplerian elliptic orbit. When thrust is applied, they characterize the Keplerian elliptic osculating orbit of the satellite. The last component is an angle which gives the position of the satellite on the Keplerian elliptic osculating orbit of the satellite. A usual set of orbital coordinates is

$$p := a(1 - e^2),$$
$$e_x := e \cos \tilde{\omega}, \text{ with } \tilde{\omega} = \omega + \Omega,$$
$$e_y := e \sin \tilde{\omega},$$
$$h_x := \tan \frac{i}{2} \cos \Omega,$$
$$h_y := \tan \frac{i}{2} \sin \Omega,$$
$$L := \tilde{\omega} + v,$$

where a, e, ω, Ω, i characterize the Keplerian osculating orbit:

1. a is the semi-major axis,
2. e is the eccentricity,
3. i is the inclination with respect to the equator,
4. Ω is the right ascension of the ascending node,
5. ω is the angle between the ascending node and the perigee,

and where v is the true anomaly; see, e.g., [**511**, Partie A-a].

In this set of coordinates, the equations of motion are

(12.26)
$$\begin{cases} \dfrac{dp}{dt} = 2\sqrt{\dfrac{p^3}{\mu}} \dfrac{1}{Z} S, \\ \dfrac{de_x}{dt} = \sqrt{\dfrac{p}{\mu}} \dfrac{1}{Z} \left[Z(\sin L) Q + AS - e_y (h_x \sin L - h_y \cos L) W \right], \\ \dfrac{de_y}{dt} = \sqrt{\dfrac{p}{\mu}} \dfrac{1}{Z} \left[-Z(\cos L) Q + BS - e_x (h_x \sin L - h_y \cos L) W \right], \\ \dfrac{dh_x}{dt} = \dfrac{1}{2} \sqrt{\dfrac{p}{\mu}} \dfrac{X}{Z} (\cos L) W, \\ \dfrac{dh_y}{dt} = \dfrac{1}{2} \sqrt{\dfrac{p}{\mu}} \dfrac{X}{Z} (\sin L) W, \\ \dfrac{dL}{dt} = \sqrt{\dfrac{\mu}{p^3}} Z^2 + \sqrt{\dfrac{p}{\mu}} \dfrac{1}{Z} (h_x \sin L - h_y \cos L) W, \end{cases}$$

where $\mu > 0$ is a gravitational coefficient depending on the central gravitational field, Q, S, W, are the radial, orthoradial, and normal components of the thrust delivered by the electric propulsion systems, and where

$$Z := 1 + e_x \cos L + e_y \sin L, \tag{12.27}$$

$$A := e_x + (1 + Z) \cos L, \tag{12.28}$$

$$B := e_y + (1 + Z) \sin L, \tag{12.29}$$

$$X := 1 + h_x^2 + h_y^2. \tag{12.30}$$

We study the case, useful in applications, where

$$Q = 0, \tag{12.31}$$

and, for some $\varepsilon > 0$,

$$|S| \leqslant \varepsilon \text{ and } |W| \leqslant \varepsilon.$$

Note that ε is small, since the thrust acceleration level is low.

In this section we give feedback laws, based on the damping approach, which (globally) asymptotically stabilize a given Keplerian elliptic orbit characterized by the coordinates $\bar{p}, \bar{e}_x, \bar{e}_y, \bar{h}_x, \bar{h}_y$.

REMARK 12.10. We are not interested in the position (given by $L(t)$) at time t of the satellite on the desired Keplerian elliptic orbit. For papers taking into account this position, see [**92, 129**], which use forwarding techniques developed by Frédéric Mazenc and Laurent Praly in [**349**], by Mrdjan Janković, Rodolphe Sepulchre and Petar Kokotović in [**254**], and in [**442**, Section 6.2]; see also Section 12.6.

In order to simplify the notations (this is not essential for the method), we restrict our attention to the case where the desired final orbit is geostationary, that is,

$$\bar{e}_x = \bar{e}_y = \bar{h}_x = \bar{h}_y = 0.$$

Let

$$\mathcal{A} = (0, +\infty) \times \mathcal{B}_1 \times \mathbb{R}^2 \times \mathbb{R},$$

where

$$\mathcal{B}_1 = \{e = (e_x, e_y) \in \mathbb{R}^2; e_x^2 + e_y^2 < 1\}. \tag{12.32}$$

With these notations, one requires that the state $(p, e_x, e_y, h_x, h_y, L)$ belongs to $\mathcal{A}$. We are looking for two maps

$$\begin{array}{rccc} S: & \mathcal{A} & \to & [-\varepsilon, \varepsilon] \\ & (p, e_x, e_y, h_x, h_y, L) & \mapsto & S(p, e_x, e_y, h_x, h_y, L), \end{array}$$

and

$$\begin{array}{rccc} W: & \mathcal{A} & \to & [-\varepsilon, \varepsilon] \\ & (p, e_x, e_y, h_x, h_y, L) & \mapsto & W(p, e_x, e_y, h_x, h_y, L), \end{array}$$

such that $(\bar{p}, 0, 0, 0, 0) \in \mathbb{R}^5$ is globally asymptotically stable for the closed-loop system (see (12.26) and (12.31))

(12.33)
$$\begin{cases} \dfrac{dp}{dt} = 2\sqrt{\dfrac{p^3}{\mu}\dfrac{1}{Z}} S, \\ \dfrac{de_x}{dt} = \sqrt{\dfrac{p}{\mu}\dfrac{1}{Z}} \left[AS - e_y(h_x \sin L - h_y \cos L)W \right], \\ \dfrac{de_y}{dt} = \sqrt{\dfrac{p}{\mu}\dfrac{1}{Z}} \left[BS - e_x(h_x \sin L - h_y \cos L)W \right], \\ \dfrac{dh_x}{dt} = \dfrac{1}{2}\sqrt{\dfrac{p}{\mu}}\dfrac{X}{Z}(\cos L)W, \\ \dfrac{dh_y}{dt} = \dfrac{1}{2}\sqrt{\dfrac{p}{\mu}}\dfrac{X}{Z}(\sin L)W, \\ \dfrac{dL}{dt} = \sqrt{\dfrac{\mu}{p^3}}Z^2 + \sqrt{\dfrac{p}{\mu}\dfrac{1}{Z}}(h_x \sin L - h_y \cos L)W, \end{cases}$$

with $(p, e_x, e_y, h_x, h_y, L) \in \mathcal{A}$.

Note that $\mathcal{A} \neq \mathbb{R}^6$ and that we are interested only in the first five variables. So one needs to specify what we mean by "$(\bar{p}, 0, 0, 0, 0)$ is globally uniformly asymptotically stable for the closed-loop system". Various natural definitions are possible. We take the one which seems to be the strongest natural one, namely we require:

- *Uniform stability*, that is, for every $\varepsilon_0 > 0$, there exists $\varepsilon_1 > 0$ such that every solution of (12.33) defined at time 0 and satisfying

$$|p(0) - \bar{p}| + |e_x(0)| + |e_y(0)| + |h_x(0)| + |h_y(0)| < \varepsilon_1,$$

is defined for every time $t \geqslant 0$ and satisfies, for every $t \geqslant 0$,

$$|p(t) - \bar{p}| + |e_x(t)| + |e_y(t)| + |h_x(t)| + |h_y(t)| < \varepsilon_0.$$

- *Uniform global attractivity*, that is, for every $M > 0$ and for every $\eta > 0$, there exists $T > 0$ such that every solution of (12.33), defined at time 0 and satisfying

$$\frac{1}{p(0)} + p(0) + \frac{1}{1 - (e_x(0)^2 + e_y(0)^2)} + |h_x(0)| + |h_y(0)| < M,$$

is defined for every time $t \geqslant 0$ and satisfies, for every time $t \geqslant T$,

$$|p(t) - \bar{p}| + |e_x(t)| + |e_y(t)| + |h_x(t)| + |h_y(t)| < \eta.$$

We start with a change of "time", already used by Sophie Geffroy in [**190**]. One describes the evolution of (p, e_x, e_y, h_x, h_y) as a function of L instead of t. Then system (12.33) reads

$$
(12.34) \quad \begin{cases} \dfrac{dp}{dL} = 2KpS, \\ \dfrac{de_x}{dL} = K[AS - e_y(h_x \sin L - h_y \cos L)W], \\ \dfrac{de_y}{dL} = K[BS - e_x(h_x \sin L - h_y \cos L)W], \\ \dfrac{dh_x}{dL} = \dfrac{K}{2} X (\cos L) W, \\ \dfrac{dh_y}{dL} = \dfrac{K}{2} X (\sin L) W, \\ \dfrac{dt}{dL} = K \sqrt{\dfrac{\mu}{p}} Z, \end{cases}
$$

with

$$
(12.35) \qquad K = \left[\dfrac{\mu}{p^2} Z^3 + (h_x \sin L - h_y \cos L) W \right]^{-1}.
$$

Let V be a function of class C^1 from $(0, \infty) \times \mathcal{B}_1 \times \mathbb{R}^2$ into $[0, \infty)$ such that

$$(12.36) \qquad V(p, e_x, e_y, h_x, h_y) = 0 \;\Leftrightarrow\; (p, e_x, e_y, h_x, h_y) = (\bar{p}, 0, 0, 0, 0),$$

$(12.37)\; V(p, e_x, e_y, h_x, h_y) \to +\infty$ as $(p, e_x, e_y, h_x, h_y) \to \partial((0, +\infty) \times \mathcal{B}_1 \times \mathbb{R}^2)$.

In (12.37), the boundary $\partial((0, +\infty) \times \mathcal{B}_1 \times \mathbb{R}^2)$ is taken in the set $[0, +\infty] \times \bar{\mathcal{B}}_1 \times [-\infty, +\infty]^2$. Therefore condition (12.37) is equivalent to the following condition: For every $M > 0$, there exists a compact set $\mathcal{K}$ included in $(0, +\infty) \times \mathcal{B}_1 \times \mathbb{R}^2$ such that

$$\bigl((p, e_x, e_y, h_x, h_y) \in ((0, +\infty) \times \mathcal{B}_1 \times \mathbb{R}^2) \setminus \mathcal{K}\bigr) \Rightarrow (V(p, e_x, e_y, h_x, h_y) \geqslant M).$$

(One can take, for example,

$$(12.38) \qquad V(p, e_x, e_y, h_x, h_y) = \dfrac{1}{2} \left(\dfrac{(p - \bar{p})^2}{p} + \dfrac{e^2}{1 - e^2} + h^2 \right),$$

with $e^2 = e_x^2 + e_y^2$ and $h^2 = h_x^2 + h_y^2$.)

The time derivative of V along a trajectory of (12.34) is given by

$$\dot{V} = K(\alpha S + \beta W),$$

with

$$(12.39) \qquad \alpha := 2p \dfrac{\partial V}{\partial p} + A \dfrac{\partial V}{\partial e_x} + B \dfrac{\partial V}{\partial e_y},$$

$$(12.40) \qquad \beta := (h_y \cos L - h_x \sin L) \left(e_y \dfrac{\partial V}{\partial e_x} + e_x \dfrac{\partial V}{\partial e_y} \right) \\ + \dfrac{1}{2} X \left((\cos L) \dfrac{\partial V}{\partial h_x} + (\sin L) \dfrac{\partial V}{\partial h_y} \right).$$

Following the damping method, one defines

$$(12.41) \qquad S := -\sigma_1(\alpha),$$
$$(12.42) \qquad W := -\sigma_2(\beta) \sigma_3(p, e_x, e_y, h_x, h_y),$$

where $\sigma_1 : \mathbb{R} \to \mathbb{R}, \sigma_2 : \mathbb{R} \to \mathbb{R}$ and $\sigma_3 : (0,+\infty) \times \mathcal{B}_1 \times \mathbb{R}^2 \to (0,1]$ are continuous functions such that

(12.43) $$\sigma_1(s)s > 0, \quad \forall s \in \mathbb{R} \setminus \{0\},$$
(12.44) $$\sigma_2(s)s > 0, \quad \forall s \in \mathbb{R} \setminus \{0\},$$
(12.45) $$\|\sigma_1\|_{L^\infty(\mathbb{R})} < \varepsilon,$$
(12.46) $$\|\sigma_2\|_{L^\infty(\mathbb{R})} < \varepsilon,$$

(12.47) $\sigma_3(p, e_x, e_y, h_x, h_y)$
$$\leqslant \frac{1}{1+\varepsilon} \frac{\mu}{p^2} \frac{(1-|e|)^3}{|h|}, \quad \forall (p, e_x, e_y, h_x, h_y) \in (0,+\infty) \times \mathcal{B}_1 \times (\mathbb{R}^2 \setminus \{0\}).$$

The reason for using σ_3 is to ensure the existence of K defined by (12.35). Indeed from (12.27), (12.34), (12.42), (12.46) and (12.47), one gets, for every $L \in \mathbb{R}$,
$$\frac{\mu Z^3}{p^2} + (h_x \sin L - h_y \cos L)W > 0$$
on $(0,\infty) \times \mathcal{B}_1 \times \mathbb{R}^2$ and therefore K (see (12.35)) is well defined for every
$$(p, e_x, e_y, h_x, h_y, L) \in (0,+\infty) \times \mathcal{B}_1 \times \mathbb{R}^2 \times \mathbb{R}.$$

One has

(12.48) $$\|S\|_{L^\infty((0,+\infty)\times\mathcal{B}_1\times\mathbb{R}^2\times\mathbb{R})} < \varepsilon,$$
(12.49) $$\|W\|_{L^\infty((0,+\infty)\times\mathcal{B}_1\times\mathbb{R}^2\times\mathbb{R})} < \varepsilon,$$
$$\dot{V} \leqslant 0 \text{ and } (\dot{V} = 0 \Leftrightarrow \alpha = \beta = 0).$$

Since the closed-loop system (12.34) is L-varying but *periodic* with respect to L, one may apply LaSalle's invariance principle (see, for example, [**458**, Lemma 5.7.8, page 226]): in order to prove that $(\bar{p},0,0,0,0)$ is globally asymptotically stable on $(0,+\infty) \times \mathcal{B}_1 \times \mathbb{R}^2$ for the closed-loop system (12.34), it suffices to check that every trajectory of (12.34) such that $\alpha = \beta = 0$ is identically equal to $(\bar{p},0,0,0,0)$. For such a trajectory, one has (see in particular (12.34), (12.39), (12.40), (12.41), (12.42), (12.43), (12.44) and (12.47))

(12.50) $$\frac{dp}{dL} = 0, \ \frac{de_x}{dL} = 0, \ \frac{de_y}{dL} = 0, \ \frac{dh_x}{dL} = 0, \ \frac{dh_y}{dL} = 0,$$
(12.51) $$2p\frac{\partial V}{\partial p} + A\frac{\partial V}{\partial e_x} + B\frac{\partial V}{\partial e_y} = 0,$$

(12.52) $$(h_y \cos L - h_x \sin L)\left(e_y \frac{\partial V}{\partial e_x} + e_x \frac{\partial V}{\partial e_y}\right)$$
$$+ \frac{1}{2}X\left((\cos L)\frac{\partial V}{\partial h_x} + (\sin L)\frac{\partial V}{\partial h_y}\right) = 0.$$

Hence p, e_x, e_y, h_x and h_y are constant. By (12.28) and (12.29), the left hand side of (12.51) is a linear combination of the functions $\cos L, \sin L, \cos^2 L, \sin L \cos L$ and the constant functions. These functions are linearly independent, so that

$$2p\frac{\partial V}{\partial p} + e_x\frac{\partial V}{\partial e_x} + 2e_y\frac{\partial V}{\partial e_y} = 0, \quad \frac{\partial V}{\partial e_x} = 0, \quad \frac{\partial V}{\partial e_y} = 0,$$

and therefore
$$\frac{\partial V}{\partial p} = \frac{\partial V}{\partial e_x} = \frac{\partial V}{\partial e_y} = 0,$$
which, together with (12.30) and (12.52), gives
$$\frac{\partial V}{\partial h_x} = \frac{\partial V}{\partial h_y} = 0.$$
Hence it suffices to impose on V that

(12.53) $\quad (\nabla V(p, e_x, e_y, h_x, h_y) = 0) \Rightarrow ((p, e_x, e_y, h_x, h_y) = (\bar{p}, 0, 0, 0, 0)).$

Note that, if V is given by (12.38), then V satisfies (12.53).

REMARK 12.11. It is interesting to compare the feedback constructed here to the open-loop optimal control for the minimal time problem (reach $(\bar{p}, 0, 0, 0, 0)$ in a minimal time with the constraint $|u(t)| \leqslant M$). Numerical experiments show that the use of the previous feedback laws (with suitable saturations σ_i, $i \in \{1, 2, 3\}$) gives trajectories which are nearly optimal if the state is not too close to $(\bar{p}, 0, 0, 0, 0)$. Note that our feedback laws are quite easy to compute compared to the optimal trajectory and provide already good robustness properties compared to the open-loop optimal trajectory (the optimal trajectory in a closed-loop form being, at least for the moment, out of reach numerically). However, when one is close to the desired target, our feedback laws are very far from being optimal. When one is close to the desired target, it is much better to linearize around the desired target and apply a standard Linear-Quadratic strategy. This problem is strongly related to Remark 12.9 on page 318.

REMARK 12.12. A related approach has been used for a Schrödinger control system by Mazyar Mirrahimi, Pierre Rouchon and Gabriel Turinici in [**360**]; see also [**44**] for degenerate cases. Note that, in the framework of a Schrödinger control system, the damping feedback approach is used to construct open-loop controls (see also Remark 12.11): for a Schrödinger control system, there are few quantities which can be measured and every measure modifies the state.

12.2.2. Damping feedback and driftless system. Throughout this section, we again assume that (C) is a driftless control affine system, i.e.,
$$f(x, u) = \sum_{i=1}^{m} u_i f_i(x).$$
We also assume $\mathcal{O} = \mathbb{R}^n \times \mathbb{R}^m$, that the f_i's are of class C^∞ on $\mathbb{R}^n$ and that the Lie algebra rank conditions (11.45) hold. Then Theorem 11.14 on page 297 asserts that, for every $T > 0$, the control system (C) is globally asymptotically stabilizable by means of T-periodic time-varying feedback laws. Let us recall that the main ingredient of the proof is the existence of $\bar{u}$ in $C^\infty(\mathbb{R}^n \times \mathbb{R}; \mathbb{R}^m)$ vanishing on $\{0\} \times \mathbb{R}$, T-periodic with respect to time and such that, if $\dot{x} = f(x, \bar{u}(x, t))$, then

(12.54) $\qquad\qquad\qquad x(T) = x(0),$

(12.55) If $x(0) \neq 0$, the linearized control system around the trajectory
$$t \in [0, T] \mapsto (x(t), \bar{u}(x(t), t)) \text{ is controllable on } [0, T].$$

In this section, we want to explain how the damping method allows us to construct from this $\bar{u}$ a T-periodic time-varying feedback law u which globally asymptotically stabilizes the control system (C). We follow, with slight modifications, [**127**], which is directly inspired from [**387**] by Jean-Baptiste Pomet. Let $W \in C^\infty(\mathbb{R}^n; \mathbb{R})$ be any function such that

$$W(x) \to +\infty \text{ as } |x| \to +\infty,$$
$$W(x) > W(0), \forall x \in \mathbb{R}^n \setminus \{0\},$$
(12.56) $$\nabla W(x) \neq 0, \forall x \in \mathbb{R}^n \setminus \{0\}.$$

One can take, for example, $W(x) = |x|^2$. Let $X(x,t) = \sum_{i=1}^m \bar{u}_i(x,t)f_i(x)$ and let $\Phi : \mathbb{R}^n \times \mathbb{R} \times \mathbb{R} \to \mathbb{R}^n$, $(x,t,s) \mapsto \Phi(x,t,s)$ be the flow associated to the time-varying vector field X, i.e.,

(12.57) $$\frac{\partial \Phi}{\partial t} = X(\Phi, t),$$

(12.58) $$\Phi(x, s, s) = x, \forall x \in \mathbb{R}^n, \forall s \in \mathbb{R}.$$

It readily follows from this definition that one has the following classical formula

(12.59) $$\Phi(x, t_3, t_1) = \Phi(\Phi(x, t_2, t_1), t_3, t_2), \forall x \in \mathbb{R}^n, \forall (t_1, t_2, t_3) \in \mathbb{R}^3;$$

compare with (1.8). Note that by (12.54),

(12.60) $$\Phi(x, 0, T) = \Phi(x, 0, 0) = x, \forall x \in \mathbb{R}^n.$$

Let us now define $V \in C^\infty(\mathbb{R}^n \times \mathbb{R}; \mathbb{R})$ by

(12.61) $$V(x, t) = W(\Phi(x, 0, t)).$$

By (12.60), V is T-periodic with respect to time and one easily checks that

$$V(x, t) > V(0, t) = W(0), \forall (x, t) \in (\mathbb{R}^n \setminus \{0\}) \times \mathbb{R},$$
$$\lim_{|x| \to +\infty} \text{Min}\{V(x, t); t \in \mathbb{R}\} = +\infty.$$

From (12.59) with $t_1 := t$, $t_2 := \tau + t$ and $t_3 := 0$, and (12.61), we have

(12.62) $$V(\Phi(x, \tau + t, t), \tau + t) = W(\Phi(x, 0, t)), \forall x \in \mathbb{R}^n, \forall (t, \tau) \in \mathbb{R}^2.$$

Differentiating (12.62) with respect to τ and then letting $\tau = 0$, we get, using (12.57) and (12.58),

(12.63) $$\frac{\partial V}{\partial t} + X \cdot \nabla V = 0.$$

From (12.63), along the trajectories of $\dot{x} = \sum_{i=1}^m (\bar{u}_i + v_i)f_i(x)$, the time derivative $\dot{V}$ of V is

$$\dot{V} = \frac{\partial V}{\partial t} + (\sum_{i=1}^m (\bar{u}_i + v_i)f_i) \cdot \nabla V$$

(12.64) $$= \sum_{i=1}^m v_i(f_i \cdot \nabla V).$$

Hence, as above, one takes $v_i(x,t) = -f_i(x) \cdot \nabla V(x,t)$, which, together with (12.64), gives

$$\dot{V} = -\sum_{i=1}^m (f_i(x) \cdot \nabla V(x,t))^2.$$

One has the following proposition.

PROPOSITION 12.13. *The feedback law $u := \bar{u} + v$ globally asymptotically stabilizes the control system (C).*

Proof of Proposition 12.13. By the time-periodic version of the LaSalle invariance principle (see, for example, [**414**, Theorem 1.3, Chapter II, pages 50–51]), in order to prove that $u = \bar{u} + v$ globally asymptotically stabilizes the control system (C), it suffices to check that every solution $\tilde{x} : \mathbb{R} \to \mathbb{R}^n$ of $\dot{\tilde{x}} = X(\tilde{x}, t)$ such that

(12.65) $$f_i(\tilde{x}(t)) \cdot \nabla V(\tilde{x}(t), t) = 0, \ \forall t \in \mathbb{R}, \ \forall i \in \{1, \ldots, m\},$$

satisfies

(12.66) $$\tilde{x}(0) = 0.$$

Let $\tilde{u} \in C^\infty([0, T]; \mathbb{R}^m)$ be defined by

$$\tilde{u}(t) := \bar{u}(\tilde{x}(t), t), \ \forall t \in \mathbb{R}.$$

Let $A \in C^\infty([0, T]; \mathcal{L}(\mathbb{R}^n; \mathbb{R}^n))$ and $B \in C^\infty([0, T]; \mathcal{L}(\mathbb{R}^m; \mathbb{R}^n))$ be defined by

(12.67) $$A(t) := \frac{\partial f}{\partial x}(\tilde{x}(t), \tilde{u}(t)), \ \forall t \in [0, T],$$

(12.68) $$B(t) := \frac{\partial f}{\partial u}(\tilde{x}(t), \tilde{u}(t)), \ \forall t \in [0, T].$$

The map $t \in [0, T] \mapsto (\tilde{x}(t), \tilde{u}(t)) \in \mathbb{R}^n \times \mathbb{R}^m$ is a trajectory of the control system (C) and the linearized control system around this trajectory is the time-varying linear control system

(12.69) $$\dot{x} = A(t)x + B(t)u, \ t \in [0, T],$$

where the state is $x \in \mathbb{R}^n$ and the control is $u \in \mathbb{R}^m$. We assume that (12.66) does not hold. Then, by (12.55),

(12.70) the linear control system (12.69) is controllable.

Let $\tilde{A} \in C^\infty([0, T]; \mathcal{L}(\mathbb{R}^n; \mathbb{R}^n))$ be defined by

(12.71) $$\tilde{A}(t) := \frac{\partial X}{\partial x}(\tilde{x}(t), t), \ \forall t \in [0, T].$$

We have

$$\tilde{A}(t)\xi = \frac{\partial f}{\partial x}(\tilde{x}(t), \tilde{u}(t))\xi + \sum_{i=1}^m \frac{\partial \bar{u}_i}{\partial x}(\tilde{x}(t), t)\xi f_i(\tilde{x}(t)), \ \forall t \in [0, T], \ \forall \xi \in \mathbb{R}^n.$$

In particular, there exists $K \in C^\infty([0, T]; \mathcal{L}(\mathbb{R}^n; \mathbb{R}^m))$ such that

(12.72) $$\tilde{A}(t) = A(t) + B(t)K(t), \ \forall t \in [0, T].$$

Let us consider the time-varying linear control system

(12.73) $$\dot{x} = \tilde{A}(t)x + B(t)u, \ t \in [0, T],$$

where the state is $x \in \mathbb{R}^n$ and the control is $u \in \mathbb{R}^m$. Note that, if $t \in [0, T] \mapsto (x(t), u(t)) \in \mathbb{R}^n \times \mathbb{R}^m$ is a trajectory of the linear control system (12.69), then $t \in [0, T] \mapsto (x(t), u(t) - K(t)x(t)) \in \mathbb{R}^n \times \mathbb{R}^m$ is a trajectory of the linear control system (12.73). Hence, using also (12.70), we get that

(12.74) the linear control system (12.73) is controllable.

Let $\tilde{R} : [0,T] \times [0,T] \to \mathcal{L}(\mathbb{R}^n;\mathbb{R}^n)$, $(t,s) \mapsto \tilde{R}(t,s)$, be the resolvent of the time-varying linear differential equation $\dot{y} = \tilde{A}(t)y$, $t \in [0,T]$, i.e.,

$$\frac{\partial \tilde{R}}{\partial t} = \tilde{A}(t)\tilde{R} \text{ on } [0,T] \times [0,T],$$

$$\tilde{R}(s,s)x = x, \forall (s,x) \in [0,T] \times \mathbb{R}^n;$$

see Definition 1.4 on page 4. By (12.61) one has

(12.75) $\quad (\nabla V(\tilde{x}(t),t))^{\text{tr}} = \nabla W(\tilde{x}(0))^{\text{tr}} \dfrac{\partial \Phi}{\partial x}(\tilde{x}(t),0,t), \forall t \in [0,T].$

Let us assume, for the moment, that

(12.76) $\quad \dfrac{\partial \Phi}{\partial x}(\tilde{x}(t),0,t) = \tilde{R}(0,t), \forall t \in [0,T].$

From (12.65), (12.68), (12.75) and (12.76), we get

$$(\nabla W(\bar{x}(0)))^{\text{tr}} \tilde{R}(0,t) B(t) = 0, \forall t \in \mathbb{R}.$$

In particular,

$$(\nabla W(\tilde{x}(0)))^{\text{tr}} \left(\int_0^T \tilde{R}(0,t)B(t)B(t)^{\text{tr}}\tilde{R}(0,t)^{\text{tr}} dt \right) (\nabla W(\tilde{x}(0))) = 0,$$

which, by (12.56), shows that the nonnegative symmetric matrix

$$\tilde{C} := \int_0^T \tilde{R}(0,t)B(t)B(t)^{\text{tr}}\tilde{R}(0,t)^{\text{tr}} dt$$

is not invertible. But, by Theorem 1.11 on page 6, the time-varying linear control system (12.73) is controllable on $[0,T]$ (if and) only if $\tilde{R}(T,0)\tilde{C}\tilde{R}(T,0)^{\text{tr}}$ is invertible. Hence, using (12.74), one gets a contradiction.

It remains to check that (12.76) holds. Using (12.59) with $t_1 := t$, $t_2 = 0$, and $t_3 := t$ and (12.58) with $s = t$, we get

(12.77) $\quad \Phi(\Phi(x,0,t),t,0) = \Phi(x,t,t) = x, \forall x \in \mathbb{R}^n, \forall t \in \mathbb{R}.$

Differentiating (12.77) with respect to x and then letting $x := \tilde{x}(t)$, one has

(12.78) $\quad \dfrac{\partial \Phi}{\partial x}(\tilde{x}(0),t,0) \circ \dfrac{\partial \Phi}{\partial x}(\tilde{x}(t),0,t) = \text{Id}_n, \forall t \in [0,T],$

where, as above, Id_n denotes the identity map of $\mathbb{R}^n$. Let $H \in C^\infty([0,T];\mathcal{L}(\mathbb{R}^n;\mathbb{R}^n))$ be defined by

(12.79) $\quad H(t) := \dfrac{\partial \Phi}{\partial x}(\tilde{x}(0),t,0), \forall t \in [0,T].$

Differentiating (12.57) with respect to x and using (12.71), one has, for every $t \in [0,T]$,

(12.80)
$$\begin{aligned}
\dot{H}(t) &= \dfrac{\partial X}{\partial x}(\Phi(\tilde{x}(0),t,0),t)H(t) \\
&= \dfrac{\partial X}{\partial x}(\tilde{x}(t),t)H(t) \\
&= \tilde{A}(t)H(t).
\end{aligned}$$

Differentiating (12.58) with respect to x and then choosing $x := \tilde{x}(0)$ and $s := 0$, one gets

(12.81) $$H(0) = \mathrm{Id}_n.$$

From (12.80) and (12.81), one gets

(12.82) $$H(t) = \tilde{R}(t, 0),\ \forall t \in [0, T].$$

Equation (12.76) follows from property (1.9) of the resolvent, (12.78), (12.79) and (12.82). This concludes the proof of Proposition 12.13. ∎

12.3. Homogeneity

Let us start by recalling the argument already used in the proof of Theorem 10.14 on page 281. Consider the linearized control system of (C) around $(0,0)$, i.e., the linear control system

$$\dot{x} = \frac{\partial f}{\partial x}(0,0)x + \frac{\partial f}{\partial u}(0,0)u,$$

where $x \in \mathbb{R}^n$ is the state and $u \in \mathbb{R}^m$ is the control. Assume that this linear control system is asymptotically stabilizable by means of a linear feedback law $u(x) = Kx$ with $K \in \mathcal{L}(\mathbb{R}^n; \mathbb{R}^m)$. By Theorem 10.9 on page 280 and Theorem 10.10 on page 280, this feedback law locally asymptotically stabilizes the control system (C).

The idea of "homogeneity" is a generalization of the above procedure: one wants to deduce the asymptotic stabilizability of the control system (C) from the asymptotic stabilizability of a control system which is "simpler" than the control system (C).

Let us now give the definition of a homogeneous vector field. Since we are going to give an application to periodic time-varying feedback laws, the vector fields we consider depend on time and are T-periodic with respect to time. The vector fields are also assumed to be continuous. Let $r = (r_1, \ldots, r_n) \in (0, +\infty)^n$. One has the following definition (see Lionel Rosier's Ph.D. thesis [**406**, Chapitre 3] for various generalizations).

DEFINITION 12.14. *The vector field $X = (X_1, \ldots, X_n)^{\mathrm{tr}}$ is r-homogeneous of degree 0, if, for every $\varepsilon > 0$, every $x \in \mathbb{R}^n$, every $i \in \{1, \ldots, n\}$, and every $t \in \mathbb{R}$,*

$$X_i(\varepsilon^{r_1} x_1, \ldots, \varepsilon^{r_n} x_n, t) = \varepsilon^{r_i} X_i(x_1, \ldots, x_n, t).$$

Since the degree of homogeneity will always be 0 in this section, we shall omit "of degree 0".

EXAMPLE 12.15. *1. (Linear systems.) The vector field $X(x, t) = A(t)x$ with $A(t) \in \mathcal{L}(\mathbb{R}^n; \mathbb{R}^n)$ is $(1, \ldots, 1)$-homogeneous.*
2. Take $n = 2$ and $X(x_1, x_2) = (x_1 - x_2^3, x_2)$. Then X is $(3,1)$-homogeneous.

For applications to feedback stabilization, the key theorem is:

THEOREM 12.16. *Let us assume that*

(12.83) $$X = Y + R,$$

where Y and R are continuous T-periodic time-varying vector fields such that Y is r-homogeneous and, for some $\rho > 0$, $\eta > 0$ and $M > 0$, one has, for every $i \in \{1,\ldots,n\}$, every $\varepsilon \in (0,1]$, and every $x = (x_1,\ldots,x_n) \in \mathbb{R}^n$ with $|x| \leqslant \rho$,

$$|R_i(\varepsilon^{r_1}x_1,\ldots,\varepsilon^{r_n}x_n,t)| \leqslant M\varepsilon^{r_i+\eta}. \tag{12.84}$$

If 0 is locally (or equivalently globally) asymptotically stable for $\dot{x} = Y(x,t)$, then it is locally asymptotically stable for $\dot{x} = X(x,t)$.

This theorem has been proved by Henry Hermes in [**229**] when one has uniqueness of the trajectories of $\dot{x} = Y(x)$, and in the general case by Lionel Rosier in [**405**]. In fact, [**405**], as well as [**229**], deals with the case of stationary vector fields. But the proof of [**405**] can be easily extended to the case of periodic time-varying vector fields. Let us briefly sketch the arguments. One first shows that Theorem 12.16 is a corollary of the following theorem, which has its own interest and goes back to José Massera [**346**] when the vector field is of class C^1.

THEOREM 12.17 ([**405**] by Lionel Rosier and [**388**] by Jean-Baptiste Pomet and Claude Samson). *Let Y be a continuous T-periodic time-varying vector field which is r-homogeneous. We assume that 0 is locally (=globally) asymptotically stable for $\dot{x} = Y(x,t)$. Let p be a positive integer and let*

$$k \in (p\,Max\{r_i,\ 1 \leqslant i \leqslant n\}, +\infty).$$

Then there exists a function $V \in C^\infty((\mathbb{R}^n \setminus \{0\}) \times \mathbb{R}; \mathbb{R}) \cap C^p(\mathbb{R}^n \times \mathbb{R}; \mathbb{R})$ such that

$$\tag{12.85}\begin{aligned} V(x,t) > V(0,t) &= 0,\ \forall (x,t) \in (\mathbb{R}^n \setminus \{0\}) \times \mathbb{R},\\ V(x,t+T) &= V(x,t),\ \forall (x,t) \in \mathbb{R}^n \times \mathbb{R},\\ \lim_{|x|\to+\infty} \operatorname{Min}\ \{V(x,t); t \in \mathbb{R}\} &= +\infty, \end{aligned}$$

$$\frac{\partial V}{\partial t} + Y \cdot \nabla V < 0 \text{ in } (\mathbb{R}^n \setminus \{0\}) \times \mathbb{R}, \tag{12.86}$$

$$V(\varepsilon^{r_1}x_1,\ldots,\varepsilon^{r_n}x_n,t) = \varepsilon^k V(x_1,\ldots,x_n,t), \tag{12.87}$$
$$\forall (\varepsilon,x,t) \in (0,+\infty) \times \mathbb{R}^n \times \mathbb{R}.$$

Proof of Theorem 12.16. Let us deduce, as in [**405**], Theorem 12.16 from Theorem 12.17. For $x = (x_1,\ldots,x_n) \in \mathbb{R}^n$ and $\varepsilon > 0$, let

$$\delta_\varepsilon^r(x) = (\varepsilon^{r_1}x_1,\ldots,\varepsilon^{r_n}x_n). \tag{12.88}$$

Let V be as in Theorem 12.17 with $p = 1$. From (12.86), there exists $\nu > 0$ such that

$$\left(\frac{\partial V}{\partial t} + Y \cdot \nabla V\right)(x,t) \leqslant -\nu, \tag{12.89}$$

for every $t \in [0,T]$ and every $x \in \mathbb{R}^n$ such that $|x_1|^{1/r_1} + \ldots + |x_n|^{1/r_n} = 1$. From (12.87) and the assumption that Y is r-homogeneous, we get that

$$\left(\frac{\partial V}{\partial t} + Y \cdot \nabla V\right)(\delta_\varepsilon^r(x),t) \tag{12.90}$$
$$= \varepsilon^k \left(\frac{\partial V}{\partial t} + Y \cdot \nabla V\right)(x,t),\ \forall (\varepsilon,x,t) \in (0,+\infty) \times \mathbb{R}^n \times \mathbb{R}.$$

From (12.89) and (12.90), we get

(12.91)
$$\left(\frac{\partial V}{\partial t} + Y \cdot \nabla V\right)(x,t) \leqslant -\nu\left(|x_1|^{1/r_1} + \ldots + |x_n|^{1/r_n}\right)^k, \forall (x,t) \in \mathbb{R}^n \times [0,T].$$

Using (12.84) and (12.87), similar computations show the existence of $\bar{C} > 0$ such that

(12.92)
$$(R \cdot \nabla V)(x,t) \leqslant \bar{C}\left(|x_1|^{1/r_1} + \ldots + |x_n|^{1/r_n}\right)^{\eta+k},$$

for every $t \in [0,T]$ and every $x \in \mathbb{R}^n$ with $|x| \leqslant \rho$. From (12.83), (12.91) and (12.92), we get the existence of $\rho' > 0$ such that

$$\left(\frac{\partial V}{\partial t} + X \cdot \nabla V\right)(x,t) < 0, \forall t \in [0,T], \forall x \in \mathbb{R}^n \text{ with } 0 < |x| \leqslant \rho',$$

which concludes the proof of Theorem 12.16. ∎

Sketch of the proof of Theorem 12.17. We follow the proof given in [**405**] by Lionel Rosier for stationary vector fields and extended by Jean-Baptiste Pomet and Claude Samson in [**388**] to the case of time-varying vector fields. By Kurzweil's theorem [**289**], there exists $W \in C^\infty(\mathbb{R}^n \times \mathbb{R}; \mathbb{R})$ such that

$$W(x,t) > W(0,t) = 0, \forall (x,t) \in (\mathbb{R}^n \setminus \{0\}) \times \mathbb{R},$$
$$W(x,t+T) = W(x,t), \forall (x,t) \in \mathbb{R}^n \times \mathbb{R},$$
$$\lim_{|x| \to +\infty} \text{Min}\{W(x,t); t \in \mathbb{R}\} = +\infty,$$
$$\frac{\partial W}{\partial t} + Y \cdot \nabla W < 0 \text{ in } (\mathbb{R}^n \setminus \{0\}) \times \mathbb{R}.$$

Let $a \in C^\infty(\mathbb{R}; \mathbb{R})$ be such that $a' \geqslant 0$, $a = 0$ in $(-\infty, 1]$ and $a = 1$ in $[2, +\infty)$. Then one can verify that V, defined by

$$V(x,t) := \int_0^{+\infty} \frac{1}{s^{k+1}} a(W(s^{r_1} x_1, \ldots, s^{r_n} x_n, t)) ds, \forall (x,t) \in \mathbb{R}^n \times \mathbb{R},$$

satisfies all required conditions. This concludes the sketch of proof of Theorem 12.17. ∎

EXAMPLE 12.18. *Following* [**126**], *let us give an application to the construction of explicit time-varying feedback laws stabilizing asymptotically the attitude of a rigid body spacecraft, a control system already considered in Example 3.9 on page 128, Example 3.35 on page 144, Example 10.15 on page 282, Example 11.3 on page 289 and Example 11.31 on page 304. So the control system is given by (3.12) with $b_1, \ldots, b_m$ independent and $1 \leqslant m \leqslant 3$. Here we assume that $m = 2$. Without loss of generality, we may assume that*

$$\{v_1 b_1 + v_2 b_2; (v_1, v_2) \in \mathbb{R}^2\} = \{0\} \times \mathbb{R}^2.$$

So, after a change of control variables, (3.12) can be replaced by

(12.93) $\dot{\omega}_1 = Q(\omega) + \omega_1 L_1(\omega), \quad \dot{\omega}_2 = V_1, \quad \dot{\omega}_3 = V_2, \quad \dot{\eta} = A(\eta)\omega,$

with $L_1\omega = D_1\omega_1 + E_1\omega_2 + F_1\omega_3$, $Q(\omega) = A\omega_2^2 + B\omega_2\omega_3 + C\omega_3^2$. For system (12.93), the controls are V_1 and V_2. It is proved in [**272**] that Q changes sign if and only if the control system (3.12) satisfies the Lie algebra rank condition at $(0,0) \in \mathbb{R}^6 \times \mathbb{R}^2$ which, by Theorem 3.17 on page 134, is a necessary condition for small-time local controllability at $(0,0) \in \mathbb{R}^6 \times \mathbb{R}^2$. From now on we assume that Q changes sign (this is a generic situation). Hence, after a suitable change of coordinates of the form

$$(12.94) \quad \omega = P\tilde{\omega} = \begin{pmatrix} 1 & 0 & 0 \\ 0 & a_p & b_p \\ 0 & c_p & d_p \end{pmatrix} \tilde{\omega},$$

system (12.93) can be written as

$$(12.95) \quad \dot{\tilde{\omega}}_1 = \tilde{\omega}_2\tilde{\omega}_3 + \tilde{\omega}_1 L_2(\tilde{\omega}), \quad \dot{\tilde{\omega}}_2 = u_1, \quad \dot{\tilde{\omega}}_3 = u_2, \quad \dot{\eta} = A(\eta)P\tilde{\omega},$$

with $L_2\tilde{\omega} = D_2\tilde{\omega}_1 + E_2\tilde{\omega}_2 + F_2\tilde{\omega}_3$. Let $c = \det P$. We can always choose P so that $c > 0$. Let

$$x_1 = \tilde{\omega}_1, \; x_5 = \tilde{\omega}_2, \; x_6 = \tilde{\omega}_3, \; x_3 = \frac{1}{c}(d_p\theta - b_p\psi),$$

$$x_4 = \frac{1}{c}(-c_p\theta + a_p\psi), \; x_2 = \phi - \frac{b_p d_p}{2}x_4^2 - \frac{a_p c_p}{2}x_3^2 - b_p c_p x_3 x_4.$$

In these coordinates, our system can be written as

$$(12.96) \quad \begin{cases} \dot{x}_1 = x_5 x_6 + R_1(x), \; \dot{x}_2 = x_1 + c x_3 x_6 + R_2(x), \\ \dot{x}_3 = x_5 + R_3(x), \; \dot{x}_4 = x_6 + R_4(x), \; \dot{x}_5 = u_1, \; \dot{x}_6 = u_2, \end{cases}$$

where R_1, R_2, R_3, and R_4 are analytic functions defined on a neighborhood of 0 such that, for a suitable positive constant C, one has, for every x in $\mathbb{R}^6$ with $|x|$ small enough,

$$|R_1(x)| + |R_3(x)| + |R_4(x)| \leqslant C \left(|x_1| + |x_2| + |x_3|^2 + |x_4|^2 + |x_5|^2 + |x_6|^2\right)^{3/2},$$

$$|R_2(x)| \leqslant C \left(|x_1| + |x_2| + |x_3|^2 + |x_4|^2 + |x_5|^2 + |x_6|^2\right)^2,$$

Hence our control system can be written as

$$(12.97) \quad \dot{x} = f(x,u) = X(x) + R(x) + uY,$$

where $x = (x_1, \ldots, x_6)^{tr} \in \mathbb{R}^6$ is the state, $u = (u_1, u_2)^{tr} \in \mathbb{R}^2$ is the control,
(12.98)
$$uY = u_1 Y_1 + u_2 Y_2 = (0,0,0,0,u_1,u_2), \quad X(x) = (x_5 x_6, x_1 + c x_3 x_6, x_5, x_6, 0, 0),$$

where $c \in (0, +\infty)$ is a constant, and $R := (R_1, \ldots, R_6)$ is a perturbation term in the following sense. Note that X is $(2,2,1,1,1,1)$-homogeneous and that, for a suitable constant $C > 0$, the vector field R satisfies, for every ε in $(0,1)$ and for every $x = (x_1, \ldots, x_6)$ in $\mathbb{R}^6$ with $|x| \leqslant 1$,

$$|R_i(\delta_\varepsilon(x))| \leqslant C_0 \varepsilon^{1+r_i}, \; \forall i \in \{1, \ldots, 6\},$$

with

$$\delta_\varepsilon(x) := (\varepsilon^2 x_1, \varepsilon^2 x_2, \varepsilon x_3, \varepsilon x_4, \varepsilon x_5, \varepsilon x_6).$$

Keeping in mind this $(2,2,1,1,1,1)$-homogeneity and Theorem 12.16, it is natural to consider time-varying feedback laws u which have the following property:

$$(12.99) \quad u(\delta_\varepsilon(x), t) = \varepsilon u(x,t), \; \forall x \in \mathbb{R}^6, \; \forall t \in \mathbb{R}.$$

Indeed, assume that u is a periodic time-varying feedback law satisfying (12.99) which locally asymptotically stabilizes the control system

$$\dot{x} = X(x) + uY. \tag{12.100}$$

(By homogeneity, this local asymptotic stability is equivalent to global asymptotic stability.) Then, from Theorem 12.16, u locally asymptotically stabilizes the control system (12.97). We shall give in sections 12.4 and 12.5 a method, due to Pascal Morin and Claude Samson [**364**], *to construct a periodic time-varying feedback law u satisfying (12.99) which locally (=globally) asymptotically stabilizes control system (12.100); see also* [**126**] *for another method to construct such feedback laws.*

REMARK 12.19. Let us point out that, as shown by Lionel Rosier in [**406**, Part A, Chapter II] and by Rodolphe Sepulchre and Dirk Aeyels in [**441**], there are homogeneous control systems which can be asymptotically stabilized by means of continuous stationary feedback laws and cannot be asymptotically stabilized by means of continuous homogeneous stationary feedback laws. An obstruction to the asymptotic stabilization by means of continuous homogeneous stationary feedback laws is given in [**113**, Proposition 3.6, page 374]. Let us also mention that homogeneity can be used for stabilization in finite time; see [**54**] by Sanjay Bhat and Dennis Bernstein.

REMARK 12.20. For more information on the homogeneity techniques in control theory, see, in particular, [**11**] by Fabio Ancona, [**24**, Section 5.3] by Andrea Bacciotti and Lionel Rosier, [**143**] by Wijesuriya Dayawansa, Clyde Martin and Sandra Samelson, [**229, 230**] by Henry Hermes, [**255**] by Hamadi Jerbi, [**268**] by Matthias Kawski, [**493**] by Marilena Vendittelli, Giuseppe Oriolo, Frédéric Jean, and Jean-Paul Laumond, and the references therein.

12.4. Averaging

Let us start with the following classical result (see, e.g., [**273**, Theorem 7.4, page 417]).

THEOREM 12.21. *Let X be a T-periodic time-varying vector field of class C^2. Assume that the origin is locally exponentially asymptotically stable for the "averaged" system*

$$\dot{x} = \frac{1}{T} \int_0^T X(x,t) dt. \tag{12.101}$$

Then there exists $\varepsilon_0 > 0$ such that, for every $\varepsilon \in (0, \varepsilon_0]$, the origin is locally asymptotically stable for $\dot{x} = X(x, t/\varepsilon)$.

By "locally exponentially asymptotically stable", one means the existence of $(r, C, \lambda) \in (0, +\infty)^3$ such that $|x(t)| \leqslant C|x(0)| \exp(-\lambda t)$ for every solution x of the averaged system (12.101) satisfying $|x(0)| \leqslant r$. This is equivalent (see, e.g., [**273**, Theorem 4.4, page 179]) to the following property: the origin is asymptotically stable for the linear system

$$\dot{y} = \frac{1}{T} \left(\int_0^T \frac{\partial X}{\partial x}(0, t) dt \right) y.$$

12.4. AVERAGING

In the case of homogeneous vector fields, this theorem has been improved by Robert M'Closkey and Richard Murray in [**351**]. They have proved the following theorem.

THEOREM 12.22. *Let X be a continuous T-periodic time-varying feedback law which is r-homogeneous (of degree 0). Assume that the origin is locally (=globally) asymptotically stable for the averaged system (12.101). Then there exists $\varepsilon_0 > 0$ such that, for every $\varepsilon \in (0, \varepsilon_0]$, the origin is locally asymptotically stable for $\dot{x} = X(x, t/\varepsilon)$.*

Pascal Morin and Claude Samson gave in [**364**] a proof of this theorem which provides us with a value of ε_0 if X has the following form:

$$X(x,t) = f_0(x) + \sum_{i=1}^{m} g_i(t) f_i(x).$$

EXAMPLE 12.23. *Following [**364**], let us give an application of this theorem to the construction of explicit time-varying feedback laws stabilizing asymptotically the attitude of a rigid spacecraft with two controls. In Example 12.18 on page 330, we reduced this problem to the construction of a periodic time-varying feedback law u satisfying (12.99) which locally (=globally) asymptotically stabilizes the control system (12.100). Now the strategy is to construct a periodic time-varying feedback law with the "proper homogeneity" which globally asymptotically stabilizes the control system*

$$(12.102) \qquad \dot{x}_1 = \bar{x}_5 \bar{x}_6, \ \dot{x}_2 = x_1 + cx_3 \bar{x}_6, \ \dot{x}_3 = \bar{x}_5, \ \dot{x}_4 = \bar{x}_6,$$

where the state is $(x_1, x_2, x_3, x_4) \in \mathbb{R}^4$ and the control is $(\bar{x}_5, \bar{x}_6) \in \mathbb{R}^2$. By "proper homogeneity" we mean that, for every t in $\mathbb{R}$, every (x_1, x_2, x_3, x_4) in $\mathbb{R}^4$, every ε in $(0, +\infty)$, and $i = 5$ or 6,

$$(12.103) \qquad \bar{x}_i(\varepsilon^2 x_1, \varepsilon^2 x_2, \varepsilon x_3, \varepsilon x_4, t) = \varepsilon \bar{x}_i(x_1, x_2, x_3, x_4, t).$$

Using the backstepping method explained in the following section, we shall see, in Example 12.26 on page 337, how to deduce, from such a feedback law $(\bar{x}_5, \bar{x}_6)$, a feedback law $u : \mathbb{R}^6 \times \mathbb{R} \to \mathbb{R}^2$, $(x_1, x_2, x_3, x_4, x_5, x_6, t) \to \tilde{u}(x_1, x_2, x_3, x_4, x_5, x_6, t)$ which is periodic in time, has a good homogeneity, and globally asymptotically stabilizes the control system obtained from the control system (12.102) by adding an integrator on x_5 and on x_6, i.e., the control system (12.100).

For $x = (x_1, x_2, x_3, x_4) \in \mathbb{R}^4$, let $\rho = \rho(x) = (x_1^2 + x_2^2 + x_3^4 + x_4^4)^{1/4}$. Let $\bar{x}_5 \in C^0(\mathbb{R}^4 \times \mathbb{R}; \mathbb{R})$ and $\bar{x}_6 \in C^0(\mathbb{R}^4 \times \mathbb{R}; \mathbb{R})$ be defined by

$$(12.104) \qquad \bar{x}_5 = -x_3 - \rho \sin \frac{t}{\varepsilon}, \ \bar{x}_6 = -x_4 + \frac{2}{\rho}(x_1 + x_2) \sin \frac{t}{\varepsilon}.$$

Then the closed-loop system (12.102)-(12.104) is $(2,2,1,1)$-homogeneous and its corresponding averaged system is

$$(12.105) \qquad \dot{x}_1 = -x_1 - x_2, \ \dot{x}_2 = x_1 - cx_3 x_4, \ \dot{x}_3 = -x_3, \ \dot{x}_4 = -x_4.$$

Using Theorem 10.10 on page 280, one gets that $0 \in \mathbb{R}^4$ is locally asymptotically stable for system (12.105). Then, by Theorem 12.22, the feedback law (12.104) locally asymptotically stabilizes the control system (12.102) if ε is small enough.

12.5. Backstepping

For the backstepping method, we are interested in a control system (C) having the following structure:

$$\dot{x}_1 = f_1(x_1, x_2), \tag{12.106}$$
$$\dot{x}_2 = u, \tag{12.107}$$

where the state is $x = (x_1, x_2)^{\text{tr}} \in \mathbb{R}^{n_1+m} = \mathbb{R}^n$ with $(x_1, x_2) \in \mathbb{R}^{n_1} \times \mathbb{R}^m$ and the control is $u \in \mathbb{R}^m$ (see (11.19) for the definition of $(x_1, x_2)^{\text{tr}}$). The key theorem for backstepping is the following one.

THEOREM 12.24. *Assume that $f_1 \in C^1(\mathbb{R}^{n_1} \times \mathbb{R}^m; \mathbb{R}^{n_1})$ and that the control system*

$$\dot{x}_1 = f_1(x_1, v), \tag{12.108}$$

where the state is $x_1 \in \mathbb{R}^{n_1}$ and the control $v \in \mathbb{R}^m$, can be globally asymptotically stabilized by means of a stationary feedback law of class C^1. Then the control system (12.106)-(12.107) can be globally asymptotically stabilized by means of a continuous stationary feedback law.

A similar theorem holds for time-varying feedback laws and local asymptotic stabilization. Theorem 12.24 has been proved independently by Christopher Byrnes and Alberto Isidori in [**79**], Daniel Koditschek in [**276**] and John Tsinias in [**485**]. A local version of Theorem 12.24 has been known for a long time; see, e.g., [**494**] by Mathukumalli Vidyasagar.

Proof of Theorem 12.24. We give the proof of [**79, 276, 485**]; for a different method, see [**453**] by Eduardo Sontag. Let $v \in C^1(\mathbb{R}^{n_1}; \mathbb{R}^m)$ be a feedback law which globally asymptotically stabilizes $0 \in \mathbb{R}^{n_1}$ for the control system (12.108). Then, by the converse of the second Lyapunov theorem (due to José Massera [**346**]), there exists a Lyapunov function of class C^∞ for the closed-loop system $\dot{x}_1 = f_1(x_1, v(x_1))$, that is, there exists a function $V \in C^\infty(\mathbb{R}^{n_1}; \mathbb{R})$ such that

$$f_1(x_1, v(x_1)) \cdot \nabla V(x_1) < 0, \, \forall x_1 \in \mathbb{R}^{n_1} \setminus \{0\}, \tag{12.109}$$
$$V(x_1) \to +\infty \text{ as } |x_1| \to +\infty,$$
$$V(x_1) > V(0), \, \forall x_1 \in \mathbb{R}^{n_1} \setminus \{0\}.$$

A natural candidate for a control Lyapunov function (see Section 12.1) for the control system (12.106)-(12.107) is

$$W((x_1, x_2)^{\text{tr}}) := V(x_1) + \frac{1}{2}|x_2 - v(x_1)|^2, \, \forall (x_1, x_2) \in \mathbb{R}^{n_1} \times \mathbb{R}^m. \tag{12.110}$$

Indeed, one has, for such a W,

$$W((x_1, x_2)^{\text{tr}}) \to +\infty \text{ as } |x_1| + |x_2| \to +\infty, \tag{12.111}$$
$$W((x_1, x_2)^{\text{tr}}) > W(0, 0), \, \forall (x_1, x_2) \in (\mathbb{R}^{n_1} \times \mathbb{R}^m) \setminus \{(0, 0)\}. \tag{12.112}$$

Moreover, if one computes the time-derivative $\dot{W}$ of W along the trajectories of (12.106)-(12.107), one gets

$$\dot{W} = f_1(x_1, x_2) \cdot \nabla V(x_1) - (x_2 - v(x_1)) \cdot (v'(x_1) f_1(x_1, x_2) - u). \tag{12.113}$$

Since f_1 is of class C^1, there exists $G \in C^0(\mathbb{R}^{n_1} \times \mathbb{R}^m \times \mathbb{R}^m; \mathcal{L}(\mathbb{R}^m, \mathbb{R}^{n_1}))$ such that
(12.114)
$$f_1(x_1, x_2) - f_1(x_1, y) = G(x_1, x_2, y)(x_2 - y), \, \forall (x_1, x_2, y) \in \mathbb{R}^{n_1} \times \mathbb{R}^m \times \mathbb{R}^m.$$
By (12.113) and (12.114),
$$\dot{W} = f_1(x_1, v(x_1)) \cdot \nabla V(x_1)$$
$$+ \left[u^{\text{tr}} - (v'(x_1) f_1(x_1, x_2))^{\text{tr}} + (\nabla V(x_1))^{\text{tr}} G(x_1, x_2, v(x_1))\right](x_2 - v(x_1)).$$
Hence, if one takes as a feedback law for the control system (12.106)-(12.107)
(12.115) $\quad u := v'(x_1) f_1(x_1, x_2) - G(x_1, x_2, v(x_1))^{\text{tr}} \nabla V(x_1) - (x_2 - v(x_1)),$
one gets
$$\dot{W} = f_1(x_1, v(x_1)) \cdot \nabla V(x_1) - |x_2 - v(x_1)|^2$$
which, together with (12.109), gives
$$\dot{W}((x_1, x_2)^{\text{tr}}) < 0, \, \forall (x_1, x_2) \in (\mathbb{R}^{n_1} \times \mathbb{R}^m) \setminus \{(0,0)\}.$$
Hence, the feedback law (12.115) globally asymptotically stabilizes the control system (12.106)-(12.107). This concludes the proof of Theorem 12.24. ∎

Let us point out that this proof uses the C^1-regularity of f_1 and v. In fact, one knows that Theorem 12.24 does not hold in the following cases:
- The map f_1 is only continuous; see [**131**, Remark 3.2]).
- One replaces "stationary" by "periodic time-varying" and the feedback law which asymptotically stabilizes the control system (12.108) is only assumed to be continuous; see [**131**, Proposition 3.1].

One does not know any counterexample to Theorem 12.24 when the feedback laws which asymptotically stabilize the control system (12.108) are only continuous. But it seems reasonable to conjecture that such counterexamples exist. It would be more interesting to know if there exists a counterexample such that the control system (12.108) satisfies the Hermes condition $S(0)$ (see page 143). Let us recall (see Proposition 3.30 on page 143) that, if the control system (12.108) satisfies the Hermes condition, then the control system (12.106)-(12.107) also satisfies the Hermes condition, and so, by Theorem 3.29, is small-time locally controllable at $(0,0) \in \mathbb{R}^n \times \mathbb{R}^m$.

12.5.1. Desingularization. In some cases where v is not of class C^1, one can use a "desingularization" technique introduced in [**392**]. Instead of giving the method in its full generality (see [**392**] for a precise general statement), let us explain it on a simple example. We take $n_1 = m = 1$ and $f_1(x_1, x_2) = x_1 - 2x_2^3$, so the control system (12.108) is
(12.116) $\quad\quad\quad\quad\quad\quad\quad \dot{x}_1 = x_1 - 2v^3,$
where the state is $x_1 \in \mathbb{R}$ and the control is $v \in \mathbb{R}$. Clearly the feedback law $v(x_1) := x_1^{1/3} := |x_1|^{1/3} \text{Sign}(x_1)$ globally asymptotically stabilizes the control system (12.116). Matthias Kawski gave in [**269**] an explicit continuous stationary feedback law which asymptotically stabilizes the control system (12.106)-(12.107) (and other control systems in the plane). In our situation, this control system is
(12.117) $\quad\quad\quad\quad\quad\quad \dot{x}_1 = x_1 - 2x_2^3, \, \dot{x}_2 = u,$

where the state is $x := (x_1, x_2)^{\mathrm{tr}} \in \mathbb{R}^2$ and the control is $u \in \mathbb{R}$. Note that the control systems (12.116) and (12.117) cannot be stabilized by means of feedback laws of class C^1 (see also [**142**] by Wijesuriya Dayawansa and Clyde Martin, for less regularity). Moreover, the construction of a stabilizing feedback law u given in the proof of Theorem 12.24 leads to a feedback law which is not locally bounded.

Let us explain how the desingularization technique of [**392**] works on this example (the Kawski construction given in [**269**] is different). Let $V \in C^\infty(\mathbb{R}; \mathbb{R})$ be defined by

$$V(x_1) = \frac{1}{2}x_1^2, \ \forall x_1 \in \mathbb{R}.$$

Let us first point out that the reason for the term $(1/2)|x_2 - v(x_1)|^2$ in the control Lyapunov function (12.110) is to penalize $x_2 \neq v(x_1)$. But, in our case, $x_2 = v(x_1)$ is equivalent to $x_2^3 = x_1$. So a natural idea is to replace the definition of the control Lyapunov function (12.110) by

$$W((x_1, x_2)^{\mathrm{tr}}) := V(x_1) + \int_{x_1^{1/3}}^{x_2} (s^3 - x_1) ds$$

$$= \frac{1}{2}x_1^2 + \frac{1}{4}x_2^4 - x_1 x_2 + \frac{3}{4}|x_1|^{4/3}.$$

With this W, one again has (12.111) and (12.112). Moreover, one now gets

$$\dot{W} = -x_1^2 + (x_2 - x_1^{1/3})[(x_2^2 + x_1^{1/3} x_2 + |x_1|^{2/3})(u - 2(x_1 - x_2 + x_1^{1/3})) + x_1].$$

Hence, if one takes for u the continuous function defined by

$$u((x_1, x_2)^{\mathrm{tr}}) := 2(x_1 - x_2 + x_1^{1/3}) - \frac{x_1}{x_2^2 + x_1^{1/3} x_2 + |x_1|^{2/3}} - (x_2 - x_1^{1/3}),$$

if $(x_1, x_2) \neq (0,0)$ and

$$u((x_1, x_2)^{\mathrm{tr}}) := 0 \text{ if } (x_1, x_2) = (0,0),$$

one gets $\dot{W} = -x_1^2 - (x_2 - x_1^{1/3})^2(x_2^2 + x_1^{1/3} x_2 + |x_1|^{2/3}) < 0$ for $(x_1, x_2) \neq (0,0)$. Hence the feedback law u globally asymptotically stabilizes the control system (12.116).

One can find in Section 13.2 an application of this desingularization technique to the stabilization of a nonlinear partial differential equation.

12.5.2. Backstepping and homogeneity. Note that, in order to construct u as in the proof of Theorem 12.24, one does not only need to know v; one also needs to know a Lyapunov function V. In many situations this is a difficult task. Lionel Rosier in [**406**, Chapitre VI] (see also [**24**, Proposition 2.19 page 80 and Section 2.6]) and Pascal Morin and Claude Samson in [**364**] have exhibited interesting situations where one does not need to know V. Let us briefly describe Morin and Samson's situation. It concerns homogeneous control systems, with homogeneous feedback laws, a case already considered in [**128**] but where the method introduced does not lead to explicit feedback laws. Pascal Morin and Claude Samson have proved in [**364**] the following theorem.

THEOREM 12.25. *Let $T > 0$. Assume that there exists a T-periodic time-varying feedback law v of class C^1 on $(\mathbb{R}^{n_1} \times \mathbb{R}) \setminus (\{0\} \times \mathbb{R})$ which globally asymptotically stabilizes the control system (12.108). Assume also that there exist $r =*

$(r_1, \ldots, r_m) \in (0, +\infty)^m$ and $q > 0$ such that the closed-loop vector field $(x_1, t) \mapsto f_1(x_1, v(x_1, t))$ is r-homogeneous (of degree 0) and that, with the notation of (12.88),

$$v(\delta_\varepsilon^r(x_1), t) = \varepsilon^q v(x_1, t), \forall (\varepsilon, x_1, t) \in (0, +\infty) \times \mathbb{R}^{n_1} \times \mathbb{R}.$$

Then, for $K > 0$ large enough, the feedback law $u := -K(x_2 - v(x_1, t))$ globally asymptotically stabilizes the control system (12.106)-(12.107).

EXAMPLE 12.26. *(This example is due to Pascal Morin and Claude Samson* [**364**]*.) Let us go back again to the stabilization problem of the attitude of a rigid spacecraft, already considered in Examples 3.9 on page 128, 11.3 on page 289, 11.31 on page 304, 12.18 on page 330 and 12.23 on page 333. It follows from these examples and Theorem 12.25 that the feedback law*

$$u_1 := -K(x_5 - \bar{x}_5(x_1, x_2, x_3, x_4, t)), \; u_2 := -K(x_6 - \bar{x}_6(x_1, x_2, x_3, x_4, t)),$$

where $\bar{x}_5$ and $\bar{x}_6$ are defined by (12.103) and $K > 0$ is large enough, locally asymptotically stabilizes the control system (12.97), i.e., the attitude of the rigid spacecraft.

12.6. Forwarding

The pioneering works for forwarding techniques are [**477, 478**] by Andrew Teel, [**349**] by Frédéric Mazenc and Laurent Praly (see also [**391**] by Laurent Praly for a more tutorial presentation) and [**254**] by Mrdjan Janković, Rodolphe Sepulchre and Petar Kokotović. There are essentially two types of approaches for these techniques:

1. An approach based on the asymptotic behavior of interconnected systems. This approach has been introduced by Andrew Teel in [**477, 478**].
2. An approach based on Lyapunov functions. This approach has been introduced independently by Frédéric Mazenc and Laurent Praly in [**349**], and by Mrdjan Janković, Rodolphe Sepulchre and Petar Kokotović in [**254**]. See also [**442**, Section 6.2] by Rodolphe Sepulchre, Mrdjan Janković and Petar Kokotović for a synthesis.

Here we follow the Lyapunov approach and more precisely [**349**]. We do not try to give the more general framework but only explain the main ideas of [**349**].

Let us consider the following control system

(12.118) $$\dot{x} = f(x, u), \; \dot{y} = g(x, u).$$

For this control system, the state is $(x, y)^{\text{tr}} \in \mathbb{R}^{n+p}$, with $(x, y) \in \mathbb{R}^n \times \mathbb{R}^p$, and the control is $u \in \mathbb{R}^m$ (see (11.19) for the definition of $(x, y)^{\text{tr}}$). We assume that f and g are of class C^1 and satisfy

(12.119) $$f(0, 0) = 0, \; g(0, 0) = 0.$$

Let us assume that there exists a feedback $\bar{u} : \mathbb{R}^n \to \mathbb{R}^m$ of class C^1 such that

(12.120) $\quad 0 \in \mathbb{R}^n$ is globally asymptotically stable for $\dot{x} = f(x, \bar{u}(x))$.

Let $V : \mathbb{R}^n \to \mathbb{R}$ be a smooth Lyapunov function for $\dot{x} = f(x, \bar{u}(x))$:

(12.121) $$\lim_{|x| \to +\infty} V(x) = +\infty,$$

(12.122) $$V(x) > V(0), \forall x \in \mathbb{R}^n \setminus \{0\},$$

(12.123) $$\nabla V(x) \cdot f(x, \bar{u}(x)) < 0, \forall x \in \mathbb{R}^n \setminus \{0\}.$$

Let $(x, y) \in \mathbb{R}^n \times \mathbb{R}^p$. Let us consider the following Cauchy problem:
$$\dot{X} = f(X, \bar{u}(X)), \; X(0) = x,$$
$$\dot{Y} = g(X, \bar{u}(X)), \; Y(0) = y.$$

Note that, by (12.120),

(12.124) $\qquad X(t) \to 0$ as $t \to +\infty.$

If the convergence in (12.124) is fast enough (for example exponential), $Y(t)$ converges as $t \to +\infty$. Let us assume that this is indeed the case and let

(12.125) $\qquad \varphi(x, y) := \lim_{t \to +\infty} Y(t).$

Let $W : \mathbb{R}^{n+p} \to \mathbb{R}$ be defined by

(12.126) $\qquad W((x, y)^{\mathrm{tr}}) := V(x) + \frac{1}{2} |\varphi(x, y)|^2, \; \forall (x, y) \in \mathbb{R}^n \times \mathbb{R}^p.$

Note that

(12.127) $\qquad W((x, y)^{\mathrm{tr}}) > W(0), \; \forall (x, y) \in \mathbb{R}^n \times \mathbb{R}^p$ such that $(x, y) \neq (0, 0).$

Moreover, by (12.123), $t \in [0, +\infty) \mapsto V(X(t)) \in \mathbb{R}$ is non-increasing and in fact decreasing if $X(0) \neq 0$. Furthermore, it follows directly from the definition of φ that

$$\varphi(X(t), Y(t)) \text{ does not depend on } t \in [0, +\infty).$$

Hence W can be expected to be a control Lyapunov function for our control system (12.118) and taking $u = \bar{u} + v$ with v, which could be small but well chosen, one can expect to stabilize the control system (12.118).

Let us show on a very simple example how this method works. We consider the control system

(12.128) $\qquad \dot{x} = u, \; \dot{y} = x,$

where the state is $(x, y)^{\mathrm{tr}} \in \mathbb{R}^2$, $(x, y) \in \mathbb{R} \times \mathbb{R}$, and the control is $u \in \mathbb{R}$. Let us just define $\bar{u} : \mathbb{R} \to \mathbb{R}$ by

$$\bar{u}(x) = -x, \; \forall x \in \mathbb{R}.$$

Then one has (12.120). Concerning $V : \mathbb{R} \to \mathbb{R}$, let us, for example, take

(12.129) $\qquad V(x) = \frac{1}{2} x^2, \; \forall x \in \mathbb{R}.$

One has, with the above notations,

(12.130) $\qquad X(t) = e^{-t} x, \; Y(t) = (1 - e^{-t}) x + y, \; \forall t \in \mathbb{R}.$

From (12.125) and (12.130), one gets

(12.131) $\qquad \varphi(x, y) = x + y, \; \forall (x, y) \in \mathbb{R} \times \mathbb{R}.$

From (12.126), (12.129) and (12.131), one has

(12.132) $\qquad W((x, y)^{\mathrm{tr}}) = \frac{1}{2} x^2 + \frac{1}{2} (x + y)^2.$

Note that, as seen above (see (12.127)),

(12.133) $\qquad W((x, y)^{\mathrm{tr}}) > W(0), \; \forall (x, y) \in \mathbb{R} \times \mathbb{R}$ such that $(x, y) \neq (0, 0).$

Moreover,

(12.134) $$\lim_{|x|+|y|\to+\infty} W((x,y)^{\mathrm{tr}}) = +\infty.$$

Let us now compute the time derivative $\dot{W}$ of W along the trajectory of

$$\dot{x} = \bar{u}(x) + v, \; \dot{y} = x.$$

One gets

$$\dot{W} = -x^2 + v(2x+y).$$

Hence it is tempting to take

$$v((x,y)^{\mathrm{tr}}) := -(2x+y), \; \forall (x,y) \in \mathbb{R} \times \mathbb{R}.$$

With such a v,

(12.135) $$\dot{W} = -x^2 - (2x+y)^2 < 0 \; \forall (x,y) \in (\mathbb{R} \times \mathbb{R}) \setminus \{(0,0)\}.$$

From (12.133), (12.134) and (12.135), it follows that $0 \in \mathbb{R}^2$ is globally asymptotically stable for the closed-loop system

$$\dot{x} = \bar{u}(x) + v((x,y)^{\mathrm{tr}}), \; \dot{y} = x.$$

One of the numerous interests of forwarding is to allow us to construct, under suitable assumptions, stabilizing feedback, which can be quite small. This has been used for example in [**92, 129**] to perform large amplitude orbit transfer of a satellite in a central gravitational field by means of an electric propulsion. (Let us recall that electric propulsion is characterized by a low-thrust acceleration level but a high specific impulse.) See also Section 12.2.1 if one does not care about the position at time t of the satellite on the desired final Keplerian elliptic orbit.

Let us explain how this can be done on our simple control system (12.128). Let $\delta > 0$. Let $\sigma \in C^0(\mathbb{R}; \mathbb{R})$ be defined by

(12.136) $$\sigma(x) = x, \; \forall x \in [-\delta, \delta],$$
(12.137) $$\sigma(x) = \delta, \; \forall x \in [\delta, +\infty),$$
(12.138) $$\sigma(x) = -\delta, \; \forall x \in (-\infty, -\delta].$$

We take

(12.139) $$\bar{u} := -\sigma.$$

Then $\bar{u}$ is globally Lipschitz and one has (12.120). Concerning $V : \mathbb{R} \to \mathbb{R}$, we keep the V defined in (12.129). Straightforward computations lead to the following expressions for $X(t)$:

$$X(t) = xe^{-t} \text{ for } t \geqslant 0 \text{ and } x \in [-\delta, \delta],$$
$$X(t) = x - \delta t \text{ for } t \in [0, (x-\delta)/\delta] \text{ and } x \in [\delta, +\infty),$$
$$X(t) = \delta e^{-t+((x-\delta)/\delta)} \text{ for } t \in [(x-\delta)/\delta, +\infty) \text{ and } x \in [\delta, +\infty),$$
$$X(t) = x + \delta t \text{ for } t \in [0, -(x+\delta)/\delta] \text{ and } x \in (-\infty, -\delta],$$
$$X(t) = -\delta e^{-t-((x+\delta)/\delta)} \text{ for } t \in [-(x+\delta)/\delta, +\infty) \text{ and } x \in (-\infty, -\delta].$$

Using these expressions of $X(t)$, one can compute $Y(t)$ and one finally arrives at the following expressions for $\varphi(x,y)$:

$$\varphi(x,y) = x + y, \ \forall x \in [-\delta, \delta], \ \forall y \in \mathbb{R},$$

$$\varphi(x,y) = \delta + \frac{x^2 - \delta^2}{2\delta} + y, \ \forall x \in [\delta, +\infty), \ \forall y \in \mathbb{R},$$

$$\varphi(x,y) = -\delta - \frac{x^2 - \delta^2}{2\delta} + y, \ \forall x \in (-\infty, -\delta], \ \forall y \in \mathbb{R}.$$

Note that φ is of class C^1. One now gets, still with the above notations,

$$\dot{W} = -x^2 + v(2x + y), \ \forall x \in [-\delta, \delta], \ \forall y \in \mathbb{R},$$

$$\dot{W} = -x\delta + v\frac{x}{\delta}\left(2\delta + y + \frac{x^2 - \delta^2}{2\delta}\right), \ \forall x \in [\delta, +\infty), \ \forall y \in \mathbb{R},$$

$$\dot{W} = x\delta - v\frac{x}{\delta}\left(-2\delta + y - \frac{x^2 - \delta^2}{2\delta}\right), \ \forall x \in (-\infty, -\delta], \ \forall y \in \mathbb{R}.$$

Let us point out that (12.133) and (12.134) still hold. We define $v : \mathbb{R}^2 \to \mathbb{R}$ by

(12.140) $\quad v((x,y)^{\mathrm{tr}}) = -\sigma(2x + y), \ \forall x \in [-\delta, \delta], \ \forall y \in \mathbb{R},$

(12.141) $\quad v((x,y)^{\mathrm{tr}}) = -\sigma\left(\frac{x}{\delta}\left(2\delta + y + \frac{x^2 - \delta^2}{2\delta}\right)\right), \ \forall x \in [\delta, +\infty), \ \forall y \in \mathbb{R},$

(12.142) $\quad v((x,y)^{\mathrm{tr}}) = \sigma\left(\frac{x}{\delta}\left(-2\delta + y - \frac{x^2 - \delta^2}{2\delta}\right)\right), \ \forall x \in (-\infty, -\delta], \ \forall y \in \mathbb{R}.$

Then v is locally Lipschitz, vanishes at $0 \in \mathbb{R}^2$ and

(12.143) $\quad \dot{W}((x,y)^{\mathrm{tr}}) < 0, \ \forall (x,y) \in (\mathbb{R} \times \mathbb{R}) \setminus \{(0,0)\}.$

From (12.133), (12.134) and (12.143), it follows that $(0,0) \in \mathbb{R}^2$ is globally asymptotically stable for the closed-loop system

$$\dot{x} = u((x,y)^{\mathrm{tr}}), \ \dot{y} = x,$$

where the feedback law $u : \mathbb{R}^2 \to \mathbb{R}$ is defined by

(12.144) $\quad u((x,y)^{\mathrm{tr}}) := \bar{u}(x) + v((x,y)^{\mathrm{tr}}), \ \forall (x,y) \in \mathbb{R} \times \mathbb{R}.$

From (12.136) to (12.144),

$$|u((x,y)^{\mathrm{tr}})| \leqslant 2\delta, \ \forall (x,y) \in \mathbb{R} \times \mathbb{R}.$$

Hence, using the forwarding technique, we succeeded to globally asymptotically stabilize $0 \in \mathbb{R}^2$ for the control system (12.128) by means of feedback laws with L^∞-norm as small as we want (but not 0).

12.7. Transverse functions

This section is borrowed from the paper [**365**] by Pascal Morin and Claude Samson. It gives a new simple and powerful tool to construct explicit feedback laws which almost lead to asymptotic stabilization. This method concerns, at least for the moment, control affine systems without drift,

(12.145) $$\dot{x} = \sum_{i=1}^{m} u_i f_i(x),$$

where the state is $x = (x_1, x_2, \ldots, x_n)^{\text{tr}} \in \Omega$, $u = (u_1, u_2, \ldots, u_m)^{\text{tr}} \in \mathbb{R}^m$. The set Ω is an open subset of $\mathbb{R}^n$ containing 0, the vector fields f_i, $i \in \{1, 2, \ldots, m\}$, are assumed to be of class C^∞ on Ω, and one assumes that $m < n$. One has the following theorem, due to Pascal Morin and Claude Samson.

THEOREM 12.27 ([**365**, Theorem 1]). *Let $\mathbb{T} := \mathbb{R}/2\pi\mathbb{Z}$ be the one-dimensional torus. Let us assume that the control system (12.145) satisfies the Lie algebra rank condition at the equilibrium $(x_e, u_e) := (0, 0) \in \mathbb{R}^n \times \mathbb{R}^m$ (see Definition 3.16 on page 134). Then there exists an integer $\bar{n} \geqslant n$ such that, for every open neighborhood $\tilde{\Omega} \subset \Omega$ of $0 \in \mathbb{R}^n$, there exists a function $F \in C^\infty(\mathbb{T}^{\bar{n}-m}; \tilde{\Omega})$ such that*

(12.146) $\text{Span } \{f_1(F(\theta)), f_2(F(\theta)), \ldots, f_m(F(\theta))\} + \text{Im } F'(\theta) = \mathbb{R}^n, \forall \theta \in \mathbb{T}^{\bar{n}-m}.$

For the proof, we refer to [**365**, Theorem 1]. Let us just give an example. We consider the control system

(12.147) $\quad\quad\quad\quad \dot{x}_1 = u_1, \; \dot{x}_2 = u_2, \; \dot{x}_3 = u_1 x_2,$

where the state is $(x_1, x_2, x_3)^{\text{tr}} \in \mathbb{R}^3$ and the control is $u = (u_1, u_2)^{\text{tr}} \in \mathbb{R}^2$. So $n = 3$, $m = 2$, $\Omega = \mathbb{R}^3$ and, for every $x = (x_1, x_2, x_3)^{\text{tr}} \in \mathbb{R}^3$,

(12.148) $\quad\quad\quad\quad f_1(x) := \begin{pmatrix} 1 \\ 0 \\ x_2 \end{pmatrix}, \; f_2(x) := \begin{pmatrix} 0 \\ 1 \\ 0 \end{pmatrix}.$

Note that, $\forall \eta \in \mathbb{R} \setminus \{0\}$,

$u_1 f_1(x) + u_2 f_2(x) \neq (0, 0, \eta)^{\text{tr}}, \forall x = (x_1, x_2, x_3)^{\text{tr}} \in \mathbb{R}^3, \forall (u_1, u_2)^{\text{tr}} \in \mathbb{R}^2.$

Hence, by the Brockett Theorem 11.1 on page 289, the control system (12.147) cannot be asymptotically stabilized by means of continuous (even dynamic) stationary feedback laws. We have, for every $x = (x_1, x_2, x_3)^{\text{tr}} \in \mathbb{R}^3$,

(12.149) $\quad\quad\quad\quad [f_1, f_2](x) = \begin{pmatrix} 0 \\ 0 \\ -1 \end{pmatrix}.$

From (12.148) and (12.149), one sees that the control system (12.147) satisfies the Lie algebra rank condition at every equilibrium point $(x, 0) \in \mathbb{R}^3 \times \mathbb{R}^2$. Let us take $\bar{n} := 3$. Hence $\bar{n} - 3 = 1$. Let $\varepsilon > 0$. We define $F \in C^\infty(\mathbb{T}; \mathbb{R}^3)$ by

(12.150) $\quad\quad\quad\quad F(\theta) := \begin{pmatrix} \varepsilon \sin(\theta) \\ \varepsilon \cos(\theta) \\ \dfrac{\varepsilon^2}{4} \sin(2\theta) \end{pmatrix}, \forall \theta \in \mathbb{T}.$

Let $\tilde{\Omega} \subset \mathbb{R}^3$ be an open neighborhood of $0 \in \mathbb{R}^3$. If $\varepsilon > 0$ is small enough, the image of F is included in $\tilde{\Omega}$. Let $H : \mathbb{T} \to \mathcal{M}_{3,3}(\mathbb{R})$ be defined by

$$H(\theta) := (f_1(F(\theta)), f_2(F(\theta)), -F'(\theta)).$$

One has

$$H(\theta) = \begin{pmatrix} 1 & 0 & -\varepsilon \cos(\theta) \\ 0 & 1 & \varepsilon \sin(\theta) \\ \varepsilon \cos(\theta) & 0 & -\dfrac{\varepsilon^2}{2} \cos(2\theta) \end{pmatrix}.$$

Hence
$$\det H(\theta) = \frac{\varepsilon^2}{2} \neq 0,$$
showing that (12.146) holds.

Let us now explain how such functions F can be used to construct interesting feedback laws. For simplicity we explain it on the control system (12.147) with the map F defined by (12.150). (Note that F depends on $\varepsilon > 0$.) Let $y : \mathbb{R}^3 \times \mathbb{T} \to \mathbb{R}^3$ be defined by

(12.151) $$y := \begin{pmatrix} y_1 \\ y_2 \\ y_3 \end{pmatrix} := \begin{pmatrix} x_1 - F_1(\theta) \\ x_2 - F_2(\theta) \\ x_3 - F_3(\theta) - F_1(\theta)(x_2 - F_2(\theta)) \end{pmatrix}.$$

Straightforward computations lead to

(12.152) $$\dot{y} = C(y,\theta) H(\theta) (u_1, u_2, \dot{\theta})^{\mathrm{tr}},$$

with

(12.153) $$C(y,\theta) := \begin{pmatrix} 1 & 0 & 0 \\ 0 & 1 & 0 \\ y_2 & -F_1(\theta) & 1 \end{pmatrix}, \ \forall y = (y_1, y_2, y_3)^{\mathrm{tr}} \in \mathbb{R}^3, \ \forall \theta \in \mathbb{R}.$$

Note that
$$\det C(y,\theta) = 1, \ \forall y \in \mathbb{R}^3, \ \forall \theta \in \mathbb{T}.$$

Hence the 3×3 matrix $C(y,\theta)$ is invertible for every $y \in \mathbb{R}^3$ and for every $\theta \in \mathbb{T}$. In fact, as one easily checks,

(12.154) $$C(y,\theta)^{-1} = \begin{pmatrix} 1 & 0 & 0 \\ 0 & 1 & 0 \\ -y_2 & F_1(\theta) & 1 \end{pmatrix}, \ \forall y = (y_1, y_2, y_3)^{\mathrm{tr}} \in \mathbb{R}^3, \ \forall \theta \in \mathbb{R}.$$

Let us consider the following dynamical extension (see Definition 11.5 on page 292) of the control system (12.147)

(12.155) $$\dot{x}_1 = u_1, \ \dot{x}_2 = u_2, \ \dot{x}_3 = u_1 x_2, \ \dot{\theta} = v,$$

where the state is $(x_1, x_2, x_3, \theta)^{\mathrm{tr}} \in \mathbb{R}^4$ and the control is $(u_1, u_2, v)^{\mathrm{tr}} \in \mathbb{R}^3$. Let us choose $\lambda > 0$. We consider the following feedback law for the control system (12.155):

(12.156) $$(u_1, u_2, v)^{\mathrm{tr}} := -\lambda H(\theta)^{-1} C(y,\theta)^{-1} y.$$

From (12.152) and (12.156), we get

(12.157) $$\dot{y} = -\lambda y.$$

Let us consider a solution $t \in [0, +\infty) \mapsto (x_1(t), x_2(t), x_3(t), \theta(t))^{\mathrm{tr}}$ of the closed-loop system (12.155)-(12.156) (with y defined by (12.151)). From (12.151), (12.154), (12.155), (12.156) and (12.157), one gets that

(12.158) $$\text{there exists } \theta_\infty \in \mathbb{R} \text{ such that } \lim_{t \to \infty} \theta(t) = \theta_\infty,$$

(12.159) $$\lim_{t \to \infty} x(t) = F(\theta_\infty).$$

Note that (12.150) and (12.159) imply the existence of $T > 0$ such that

(12.160) $$(t \geqslant T) \Rightarrow (|x(t)| \leqslant 2\varepsilon).$$

Moreover, it is not hard to see that for every compact $K \subset \mathbb{R}^4$, there exists $T > 0$ such that (12.160) holds for every solution $t \in [0, +\infty) \mapsto (x_1(t), x_2(t), x_3(t), \theta(t))^{\mathrm{tr}}$ of the closed-loop system (12.155)-(12.156) such that $(x_1(0), x_2(0), x_3(0), \theta(0))^{\mathrm{tr}} \in K$. Since $\varepsilon > 0$ is arbitrary, one gets what is usually called global "practical" asymptotic stabilizability.

Let us go back to the definition of y given in (12.151). Let us first point out that, if one takes, instead of (12.151),

$$(12.161) \qquad y := x - F(\theta),$$

(which could sound more natural), one gets, instead of (12.152),

$$(12.162) \qquad \dot{y} = Y(x_1, x_2, x_3, \theta, u_1, u_2, v),$$

with

$$(12.163) \qquad Y(x_1, x_2, x_3, \theta, u_1, u_2, v) := \begin{pmatrix} u_1 - \varepsilon v \cos(\theta) \\ u_2 + \varepsilon v \sin(\theta) \\ u_1 x_2 - \dfrac{\varepsilon^2}{2} v \cos(2\theta) \end{pmatrix}.$$

But now we can no longer get (12.157), no matter what we choose for the feedback $(u_1, u_2, v)^{\mathrm{tr}}$. Indeed, if we denote by Y_3 the last component of Y, then

$$Y_3(x_1, 0, x_3, \pi/4, u_1, u_2, v) = 0, \ \forall (x_1, x_3, u_1, u_2, v) \in \mathbb{R}^5.$$

Hence, the choice of y is a key point in this method. The choice given by (12.151) is dictated by the Lie group structure of the control system (12.147). Let us denote by $*$ the following product in $\mathbb{R}^3$:

$$(12.164) \qquad \begin{pmatrix} a_1 \\ a_2 \\ a_3 \end{pmatrix} * \begin{pmatrix} b_1 \\ b_2 \\ b_3 \end{pmatrix} = \begin{pmatrix} a_1 + b_1 \\ a_2 + b_2 \\ a_3 + b_3 + a_2 b_1 \end{pmatrix},$$

for every $a = (a_1, a_2, a_3)^{\mathrm{tr}} \in \mathbb{R}^3$ and for every $b = (b_1, b_2, b_3)^{\mathrm{tr}} \in \mathbb{R}^3$. One easily checks that the C^∞-manifold $\mathbb{R}^3$ equipped with $*$ is a Lie group (for a definition of a Lie group, see, for example, [**2**, 4.1.1 Definition, page 253]). This can also be proved by pointing out that

$$\begin{pmatrix} 1 & b_1 & b_3 \\ 0 & 1 & b_2 \\ 0 & 0 & 1 \end{pmatrix} \begin{pmatrix} 1 & a_1 & a_3 \\ 0 & 1 & a_2 \\ 0 & 0 & 1 \end{pmatrix} = \begin{pmatrix} 1 & a_1 + b_1 & a_3 + b_3 + a_2 b_1 \\ 0 & 1 & a_2 + b_2 \\ 0 & 0 & 1 \end{pmatrix},$$

showing that $(\mathbb{R}^3, *)$ is the group of upper triangular 3×3 real matrices with 1 on the diagonal. Note that the identity element of this Lie group is $0 \in \mathbb{R}^3$ and that the inverse a^{-1} of $a = (a_1, a_2, a_3)^{\mathrm{tr}}$ is $(-a_1, -a_2, -a_3 + a_1 a_2)^{\mathrm{tr}}$.

For $a \in \mathbb{R}^3$, let $l_a : \mathbb{R}^3 \to \mathbb{R}^3$ be defined by

$$l_a(b) = a * b, \ \forall b \in \mathbb{R}^3.$$

We recall that a vector field $X \in C^\infty(\mathbb{R}^3; \mathbb{R}^3)$ is said to be left invariant if, for every $a \in \mathbb{R}^3$,

$$((l_a)'(b)) X(b) = X(a * b), \ \forall b \in \mathbb{R}^3;$$

see, for example, [**2**, page 254]. In other words, X is preserved by left translations.

Let us check that the vector field f_1 is left invariant. For $a = (a_1, a_2, a_3)^{\mathrm{tr}} \in \mathbb{R}^3$ and for $b = (b_1, b_2, b_3)^{\mathrm{tr}} \in \mathbb{R}^3$, one has

$$(12.165) \qquad ((l_a)'(b))c = \begin{pmatrix} c_1 \\ c_2 \\ c_3 + a_2 c_1 \end{pmatrix}, \ \forall c = (c_1, c_2, c_3)^{\mathrm{tr}} \in \mathbb{R}^3.$$

Hence

$$((l_a)'(b))f_1(b) = (1, 0, b_2 + a_2)^{\mathrm{tr}} = f_1(a * b),$$

which shows that f_1 is left invariant. One similarly checks that the vector field f_2 is also left invariant.

Note that the definition of y given by (12.151) can now be rewritten as

$$(12.166) \qquad y := x * F(\theta)^{-1}.$$

Let us now show why the fact that the vector fields f_1 and f_2 are left invariant is crucial to avoid the problem that we faced when we have defined y by (12.161). For $a \in \mathbb{R}^3$, let $r_a : \mathbb{R}^3 \to \mathbb{R}^3$ (the right translation by a) be defined by

$$r_a(b) = b * a, \ \forall b \in \mathbb{R}^3.$$

From (12.166), one has

$$(12.167) \qquad x = y * F(\theta).$$

Differentiating (12.167) with respect to time, one gets, using (12.155),

$$(12.168) \qquad u_1 f_1(y * F(\theta)) + u_2 f_2(y * F(\theta)) = r'_{F(\theta)}(y)\dot{y} + l'_y(F(\theta))F'(\theta)v.$$

Since f_1 and f_2 are left invariant,

$$(12.169) \qquad f_1(y * F(\theta)) = l'_y(F(\theta))f_1(y), \ f_2(y * F(\theta)) = l'_y(F(\theta))f_2(y).$$

From (12.168) and (12.169), one gets

$$(12.170) \qquad \dot{y} = (r'_{F(\theta)}(y))^{-1}(l'_y F(\theta))H(\theta)(u_1, u_2, v)^{\mathrm{tr}},$$

or, equivalently,

$$(12.171) \qquad (u_1, u_2, v)^{\mathrm{tr}} = H(\theta)^{-1}(l'_y F(\theta))^{-1} r'_{F(\theta)}(y)\dot{y}.$$

Equality (12.171) shows that $\dot{y}$ can be chosen arbitrarily. Note that

$$l'_y(F(\theta)) = \begin{pmatrix} 1 & 0 & 0 \\ 0 & 1 & 0 \\ y_2 & 0 & 1 \end{pmatrix}, \ r'_{F(\theta)}(y) = \begin{pmatrix} 1 & 0 & 0 \\ 0 & 1 & 0 \\ 0 & F_1(\theta) & 1 \end{pmatrix},$$

which, together with (12.154), lead to

$$(12.172) \qquad (l'_y(F(\theta)))^{-1} r'_{F(\theta)}(y) = \begin{pmatrix} 1 & 0 & 0 \\ 0 & 1 & 0 \\ -y_2 & 0 & 1 \end{pmatrix} \begin{pmatrix} 1 & 0 & 0 \\ 0 & 1 & 0 \\ 0 & F_1(\theta) & 1 \end{pmatrix}$$

$$= \begin{pmatrix} 1 & 0 & 0 \\ 0 & 1 & 0 \\ -y_2 & F_1(\theta) & 1 \end{pmatrix} = C(y, \theta)^{-1}.$$

Hence, if we take, as above, $\dot{y} = -\lambda y$, one recovers (12.156) from (12.171) and (12.172).

This method to construct dynamic feedback laws which lead to global practical stability works for every driftless control affine system $\dot{x} = \sum_{i=1}^{m} u_i f_i(x)$ on a Lie group provided that the vector fields f_i's are left invariant and that the control system satisfies the Lie algebra rank condition at $(e, 0) \in \mathbb{R}^3 \times \mathbb{R}^2$, where e is the identity element of the Lie group; see [**366**, Theorem 1, page 1499 and Proposition 1, page 1500].

This method can also be used in order to (see [**366**])

1. follow curves $t \mapsto \bar{x}(t)$ which could even not be such that $t \mapsto (\bar{x}(t), \bar{u}(t))$ is a trajectory of the control system for a suitable $t \mapsto \bar{u}(t)$,
2. get practical local stabilization even if the manifold is not a Lie group or the vector fields are not left invariant but have some homogeneity properties.

CHAPTER 13

Applications to some partial differential equations

The goal of this chapter is to show that some methods presented in the previous chapters can also be useful for the stabilization of linear or nonlinear partial differential control equations. We present four applications:

1. rapid exponential stabilization by means of Gramians for linear time-reversible partial differential equations (Section 13.1),
2. stabilization of a rotating body-beam without damping (Section 13.2),
3. stabilization of the Euler equations of incompressible fluids (Section 13.3),
4. stabilization of hyperbolic systems (Section 13.4).

13.1. Gramian and rapid exponential stabilization

In this section we explain how to use for infinite-dimensional linear control systems the Gramian approach for stabilization presented in Section 10.3 in the framework of finite-dimensional linear control systems. The Gramian approach for infinite-dimensional linear control systems is due to

- Marshall Slemrod [**450**] for the case of bounded control operators (i.e., the case where the operator B on page 52 is continuous from U into $H \subset D(A^*)'$ with the notations of Section 2.3),
- Vilmos Komornik [**278**] for the case of unbounded control operators.

For the extension of the Bass method (see Proposition 10.17 on page 286 and Corollary 10.18 on page 286) to the case of infinite-dimensional linear control systems, we refer to José Urquiza [**490**].

We use the notations of Section 2.3 on abstract linear control systems. However, we need to assume that the strongly continuous semigroup $S(t)$, $t \in [0, +\infty)$, of continuous linear operators is in fact a strongly continuous *group* $S(t)$, $t \in \mathbb{R}$, of continuous linear operators.

Therefore, let H and U be two real (for simplification of notation) Hilbert spaces and let $S(t)$, $t \in \mathbb{R}$, be a strongly continuous group of continuous linear operators on H (see Definition A.12 on page 376 and Definition A.13 on page 376). Let A be the infinitesimal generator of the strongly continuous group $S(t)$, $t \in \mathbb{R}$, of continuous linear operators (see Definition A.14 on page 376). Then $S(t)^*$, $t \in \mathbb{R}$, is a strongly continuous group of continuous linear operators and the infinitesimal generator of this group is the adjoint A^* of A (see Theorem A.15 on page 376). The domain $D(A^*)$ is equipped with the usual graph norm $\|\cdot\|_{D(A^*)}$ of the unbounded operator A^*:

$$\|z\|_{D(A^*)} := \|z\|_H + \|A^*z\|_H, \ \forall z \in D(A^*).$$

Let $D(A^*)'$ be the dual of $D(A^*)$ with the pivot space H. Let $B \in \mathcal{L}(U, D(A^*)')$. We also assume the following regularity property (see (2.199)):

(13.1) $\quad \forall \tau > 0, \exists C_\tau > 0$ such that $\int_0^\tau \|B^* S(t)^* z\|_U^2 dt \leqslant C_\tau \|z\|_H^2, \forall z \in D(A^*).$

The control system we consider here is

(13.2) $$\dot{y} = Ay + Bu, \, t \in [0, T],$$

where, at time t, the control is $u(t) \in U$ and the state is $y(t) \in H$. For the definition of the solution to the Cauchy problem

$$\dot{y} = Ay + Bu, \, y(0) = y^0,$$

with given $y^0 \in H$ and given $u \in L^2(0, T)$, we refer to Definition 2.36 on page 53. For the existence and uniqueness of the solution to this Cauchy problem, we refer to Theorem 2.37 on page 53.

Let $T > 0$. Let us recall (see Theorem 2.42 on page 56) that the linear control system is exactly controllable on $[0, T]$ (see Definition 2.38 on page 55) if and only if there exists $c > 0$ such that the following observability inequality holds:

(13.3) $$\int_0^T \|B^* S(t)^* z\|_U^2 dt \geqslant c\|z\|_H^2, \forall z \in D(A^*).$$

Throughout this whole section we assume that the linear control system $\dot{y} = Ay + Bu$ is controllable on $[0, T]$. Hence (13.3) holds for some $c > 0$.

Let $\lambda > 0$. Our goal is to construct a feedback law $K : y \mapsto Ky$ such that the closed-loop system $\dot{y} = (A + BK)y$ is well-posed (in a sense to be precise) and is such that, for some $M > 0$, every (maximal) solution of $\dot{y} = (A + BK)y$ defined at time 0 satisfies

(13.4) $$\|y(t)\|_H \leqslant M e^{-\lambda t} \|y(0)\|_H, \forall t \in [0, +\infty).$$

Let

(13.5) $$T_\lambda := T + \frac{1}{2\lambda}.$$

Let $f_\lambda : [0, T_\lambda] \to [0, +\infty)$ be defined by

(13.6) $$f_\lambda(t) := \begin{cases} e^{-2\lambda t} & \text{if } 0 \leqslant t \leqslant T, \\ 2\lambda e^{-2\lambda T}(T_\lambda - t) & \text{if } T \leqslant t \leqslant T_\lambda. \end{cases}$$

Let Q_λ be the continuous symmetric bilinear form on $D(A^*)$ defined by

(13.7) $$Q_\lambda(y, z) := \int_0^{T_\lambda} f_\lambda(t)(B^* S(-t)^* y, B^* S(-t)^* z) dt, \forall (y, z) \in D(A^*)^2.$$

Using (13.1) and the fact that

$$S(-t)^* = S(T_\lambda - t)^* S(-T_\lambda)^*, \forall t \in [0, T_\lambda],$$

one gets the existence of $M_\lambda > 0$ such that

(13.8) $$|Q_\lambda(y, z)| \leqslant M_\lambda \|y\|_H \|z\|_H, \forall (y, z) \in D(A^*)^2.$$

Inequality (13.8) and the density of $D(A^*)$ allow us to extend in a unique way Q_λ to a continuous symmetric quadratic form on H. We still denote this extension by Q_λ. By the Riesz representation theorem, there exists $C_\lambda \in \mathcal{L}(H; H)$ such that

(13.9) $$Q_\lambda(y, z) = (y, C_\lambda z), \forall (y, z) \in H^2.$$

Note that

(13.10) $$S(-t)^* = S(T-t)^*S(-T)^*, \forall t \in [0, T_\lambda],$$

(13.11) $$S(-T)^*S(T)^*y = y, \forall y \in H.$$

From (13.3), (13.6), (13.7), (13.10) and (13.11), one gets

(13.12) $$Q_\lambda(y,y) \geqslant ce^{-2\lambda T}\|S(-T)^*y\|_H^2 \geqslant ce^{-2\lambda T}\|S(T)^*\|_{\mathcal{L}(H;H)}^{-2}\|y\|_H^2, \forall y \in H.$$

By the Lax-Milgram theorem, (13.9) and (13.12), $C_\lambda \in \mathcal{L}(H;H)$ is invertible in $\mathcal{L}(H;H)$.

We are now in a position to define our stabilizing feedback. Let

$$K_\lambda : C_\lambda(D(A^*)) \to U$$

be defined by

(13.13) $$K_\lambda y := -B^*C_\lambda^{-1}y, \forall y \in C_\lambda(D(A^*)).$$

One then has the following theorem, due to Vilmos Komornik [**278**].

THEOREM 13.1. *The operator $A + BK_\lambda$ is the restriction of an infinitesimal generator of a strongly continuous semigroup of continuous linear operators $S(t)$, $t \in [0, +\infty)$, on H. Moreover, there exists $M > 0$ such that*

$$\|S(t)y^0\|_H \leqslant Me^{-\lambda t}\|y^0\|_H, \forall t \in [0, +\infty), \forall y^0 \in H.$$

As stated, the meaning of this theorem is unclear since we have not defined the operator $A + BK_\lambda$ (and in particular its domain of definition)! This definition is rather involved and far beyond the scope of this book. It relies on a study of algebraic Riccati equations made by Franco Flandoli in [**171**]. See also [**172**] by Franco Flandoli, Irena Lasiecka and Roberto Triggiani, and [**484**] by Roberto Triggiani.

We do not give the full details of the proof of Theorem 13.1. Let us simply give a proof of this theorem in the case of *finite* dimensional control systems in order to see the interest of the weight f_λ (see (13.6)), a choice due to Frédéric Bourquin (see [**278**, Note added in proof, page 1611] and [**279**, page 23]). Note that, if one simply takes $T_\lambda = T$ and

(13.14) $$f_\lambda(t) := e^{-2\lambda t} \text{ if } 0 \leqslant t \leqslant T,$$

one recovers what we have done in Section 10.3; see in particular Theorem 10.16 on page 283 and (10.56). However, our proof of Theorem 10.16 on page 283 relies on the LaSalle invariance principle; see page 284; see also [**450**]. With the weight function f_λ defined by (13.6), one does not need to use this principle. We readily get an explicit strict Lyapunov function. This is especially interesting in infinite dimension and/or if one wants to deal with nonlinear control systems which are small perturbations of our linear control system.

So we assume that H and U are of finite dimension. Let $y : \mathbb{R} \to H$ be a (maximal) solution of

(13.15) $$\dot{y} = (A + BK_\lambda)y.$$

Let $V : H \to [0, +\infty)$ be defined by

(13.16) $$V(z) := z^{\text{tr}}C_\lambda^{-1}z, \forall z \in H.$$

By (13.8), (13.12) and (13.16),

(13.17) $$\frac{1}{M_\lambda}\|z\|_H^2 \leqslant V(z) \leqslant c^{-1}e^{2\lambda T}\|S(T)^*\|^2_{\mathcal{L}(H;H)}\|z\|_H^2, \forall z \in H.$$

Let $v : \mathbb{R} \to \mathbb{R}$ be defined by

(13.18) $$v(t) := V(y(t)), \forall t \in \mathbb{R}.$$

From (13.13), (13.15), (13.16) and (13.18), we get

(13.19) $\dot{v} = -|B^{\mathrm{tr}}C_\lambda^{-1}y|^2 + y^{\mathrm{tr}}A^{\mathrm{tr}}C_\lambda^{-1}y + y^{\mathrm{tr}}C_\lambda^{-1}Ay - y^{\mathrm{tr}}C_\lambda^{-1}BB^{\mathrm{tr}}C_\lambda^{-1}y.$

For two symmetric matrices Δ_1 and Δ_2 in $\mathcal{L}(H;H)$, one says that $\Delta_1 \leqslant \Delta_2$ if $z^{\mathrm{tr}}\Delta_1 z \leqslant z^{\mathrm{tr}}\Delta_2 z$ for every $z \in H$. From (13.6), we have

(13.20) $$2\lambda f_\lambda(t) \leqslant -f'_\lambda(t), \forall t \in [0, T_\lambda] \setminus \{T\}.$$

From (13.7), (13.9), (13.20) and an integration by parts, we get

$$\begin{aligned} C_\lambda &= \int_0^{T_\lambda} f_\lambda(t) e^{-tA} BB^{\mathrm{tr}} e^{-tA^{\mathrm{tr}}} dt \\ &\leqslant -\frac{1}{2\lambda}\int_0^{T_\lambda} f'_\lambda(t) e^{-tA} BB^{\mathrm{tr}} e^{-tA^{\mathrm{tr}}} dt \\ &\leqslant \frac{1}{2\lambda}\left(BB^{\mathrm{tr}} - AC_\lambda - C_\lambda A^{\mathrm{tr}}\right). \end{aligned}$$

Hence

$$AC_\lambda + C_\lambda A^{\mathrm{tr}} - BB^{\mathrm{tr}} \leqslant -2\lambda C_\lambda,$$

from which we readily get

(13.21) $$C_\lambda^{-1}A + A^{\mathrm{tr}}C_\lambda^{-1} - C_\lambda^{-1}BB^{\mathrm{tr}}C_\lambda^{-1} \leqslant -2\lambda C_\lambda^{-1}.$$

From (13.19) and (13.21), one gets

$$\dot{v} \leqslant -2\lambda v,$$

which leads to

(13.22) $$v(t) \leqslant e^{-2\lambda t}v(0), \forall t \in [0, +\infty).$$

From (13.17) and (13.22), we finally obtain

$$\|y(t)\|_H^2 \leqslant c^{-1}e^{2\lambda T}\|S(T)^*\|^2_{\mathcal{L}(H;H)} M_\lambda e^{-2\lambda t}\|y(0)\|_H^2, \forall t \in [0, +\infty).$$

This concludes the proof of Theorem 13.1 in the case of finite-dimensional control systems. ∎

Let us end this section by mentioning that this method to construct feedback laws leading to rapid stabilization has been implemented and tested by various authors, for example by:

- Frédéric Bourquin in [66] for the wave equation with Dirichlet boundary condition. This paper proposes an approximation scheme of spectral type, such that the discrete controlled solutions have an energy which decays exponentially fast with the same decay rate as in the continuous case.
- Frédéric Bourquin, Michel Joly, Manuel Collet and Louis Ratier in [67]. This paper deals with the stabilization of flexible structures. It compares various methods for getting rapid exponential stabilization and it shows practical experiments.

- Jean-Séverin Briffaut in his Ph.D. thesis [**72**] under the supervision of Frédéric Bourquin and Roland Glowinski. It is numerically observed and heuristically justified in this thesis that the exponential decay rate is two times larger than the one given in Theorem 13.1, at least if A is self-adjoint and Q_λ is defined by

$$Q_\lambda(y,z) := \int_0^\infty e^{-2\lambda t}(B^*S(-t)^*y, B^*S(-t)^*z)dt, \, \forall (y,z) \in D(A^*)^2;$$

compare with (13.7). This is confirmed theoretically by José Urquiza in [**490**]; see also Proposition 10.17 on page 286. Note that, with this new Q_λ, the definition of $A + BK_\lambda$ is now simpler. This is due to the fact that more explicit expressions are now available; see, in particular, (3.11) of [**490**].

- José Urquiza in his Ph.D. thesis [**489**] under the supervision of Frédéric Bourquin.

13.2. Stabilization of a rotating body-beam without damping

In this section we study the stabilization of a system, already considered by John Baillieul and Mark Levi in [**25**], consisting of a disk with a beam attached to its center and perpendicular to the disk's plane. The beam is confined to another plane, which is perpendicular to the disk and rotates with the disk; see Figure 1 below.

The dynamics of motion is (see [**25**] and [**26**])

$$(13.23) \quad \rho y_{tt}(t,x) + EI y_{xxxx}(t,x) + \rho B y_t(t,x)$$
$$= \rho \omega^2(t) y(t,x), \, t \in (0, +\infty), \, x \in (0, L),$$

$$(13.24) \quad y(t,0) = y_x(t,0) = y_{xx}(t,L) = y_{xxx}(t,L) = 0, \, t \in (0, +\infty),$$

$$(13.25) \quad \frac{d}{dt}\left(\omega(t)(J_d + \rho \int_0^L y^2(t,x)dx)\right) = \Gamma(t), \, t \in (0, +\infty),$$

where L is the length of the beam, ρ is the mass per unit length of the beam, EI is the flexural rigidity per unit length of the beam, $\omega(t) = \dot\theta(t)$ is the angular velocity of the disk at time t, J_d is the disk's moment of inertia, $y(t,x)$ is the beam's displacement in the rotating plane at time t with respect to the spatial variable x, By_t is the damping term and $\Gamma(t)$ is the torque control variable applied to the disk at time t (see Figure 1). It is a control system where, at time t, the state is $(y(t,\cdot), y_t(t,\cdot), \omega(t))$ and the control is $\Gamma(t)$.

If there is no damping, $B = 0$ and therefore (13.23) reads

$$(13.26) \quad \rho y_{tt}(t,x) + EI y_{xxxx}(t,x) = \rho \omega^2(t) y(t,x).$$

The asymptotic behavior of the solutions of (13.23)-(13.24)-(13.25) when there is no control (i.e., $\Gamma = 0$), but with a damping term (i.e., when $B > 0$), has been studied by John Baillieul and Mark Levi in [**25**] and by Anthony Bloch and Edriss Titi in [**59**]. Still when there is a damping term, Cheng-Zhong Xu and John Baillieul have shown in [**505**] that the feedback torque control law $\Gamma = -\nu\omega$, where ν is any positive constant, globally asymptotically stabilizes the equilibrium point $(y, y_t, \omega) = (0,0,0)$. It is easy to check that such feedback laws do not asymptotically stabilize the equilibrium point when there is no damping.

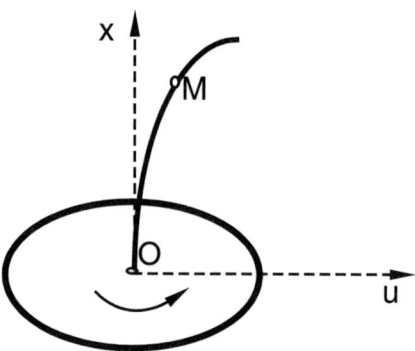

FIGURE 1. The body-beam structure

In this section, we present a result obtained in [**119**] on the stabilization problem when there is no damping. We shall see that the design tools presented in Section 12.5 will allow us to construct in this case a feedback torque control law which globally asymptotically stabilizes the origin.

Of course, by suitable scaling arguments, we may assume that $L = EI = \rho = 1$. Let
$$H = \{w = (u,v)^{\text{tr}}; u \in H^2(0,1), v \in L^2(0,1), u(0) = u_x(0) = 0\}.$$
The space H with inner product
$$((u_1, v_1)^{\text{tr}}, (u_2, v_2)^{\text{tr}})_H = \int_0^1 (u_{1xx} u_{2xx} + v_1 v_2) dx$$
is a Hilbert space. For $w \in H$, we denote by $\|w\|_H$ the norm of w and let
$$\mathcal{E}(w) = \|w\|_H^2.$$
We consider the unbounded linear operator $A : D(A) \subset H \to H$ defined by
$$D(A) := \left\{ \begin{pmatrix} u \\ v \end{pmatrix}; u \in H^4(0,1), v \in H^2(0,1), \right.$$
$$\left. u(0) = u_x(0) = u_{xx}(1) = u_{xxx}(1) = v(0) = v_x(0) = 0 \right\},$$
$$A \begin{pmatrix} u \\ v \end{pmatrix} := \begin{pmatrix} -v \\ u_{xxxx} \end{pmatrix}.$$
One easily checks that
$$\mathcal{D}(A) \text{ is dense in } H,$$
$$A \text{ is a closed operator,}$$
$$A^* = -A.$$
Hence, from the Lumer-Phillips theorem (see Theorem A.16 on page 377), A is the infinitesimal generator of a strongly continuous group $S(t)$, $t \in \mathbb{R}$, of isometries on H.

13.2. STABILIZATION OF A ROTATING BODY-BEAM WITHOUT DAMPING

With this notation, our control system (13.24)-(13.25)-(13.26) reads

$$\text{(13.27)} \qquad \frac{dw}{dt} + Aw = \omega^2 \begin{pmatrix} 0 \\ u \end{pmatrix},$$

$$\text{(13.28)} \qquad \frac{d\omega}{dt} = \gamma,$$

with

$$\text{(13.29)} \qquad \gamma := \frac{\Gamma - 2\omega \int_0^1 uv\,dx}{J_d + \int_0^1 u^2\,dx}.$$

By (13.29), we may consider γ as the control. In order to explain how we have constructed our stabilizing feedback law, let us first consider, as in the usual backstepping method (see Section 12.5 above), equation (13.27) as a control system where w is the state and ω is the control. Then natural candidates for a control Lyapunov function and a stabilizing feedback law are, respectively, $\mathcal{E}$ and $\omega = \sigma^\star(\int_0^1 uv\,dx)$, where $\sigma^\star \in C^0(\mathbb{R}, \mathbb{R})$ satisfies $\sigma^\star > 0$ on $(-\infty, 0)$ and $\sigma^\star = 0$ on $[0, +\infty)$. (This is the usual damping technique; see Section 12.2.) One can prove (see [**119**, Appendix]) that such feedback laws always give *weak* asymptotic stabilization, i.e., one gets

$$\text{(13.30)} \qquad w(t) \rightharpoonup 0 \text{ weakly in } H \text{ as } t \to +\infty,$$

instead of

$$\text{(13.31)} \qquad w(t) \to 0 \text{ in } H \text{ as } t \to +\infty.$$

But it is not clear that such feedbacks give *strong* asymptotic stabilization. It is possible to prove that one gets such stabilization for the particular case where the feedback is

$$\text{(13.32)} \qquad \omega := (\max\{0, -\int_0^1 uv\,dx\})^{\frac{1}{2}}.$$

Unfortunately, ω defined by (13.32) is not of class C^1 and so one cannot use the proof of Theorem 12.24 on page 334 to get asymptotic stabilization for the control system (13.27)-(13.28) which is obtained by adding an integrator to control system (13.27). The regularity of this ω is also not sufficient to apply the desingularization techniques introduced in [**392**]; see Section 12.5.1 above. For these reasons, we use a different control Lyapunov function and a different feedback law to asymptotically stabilize control system (13.27).

For the control Lyapunov function, we take

$$J(w) = \frac{1}{2}\left(\mathcal{E}(w) - F(\mathcal{E}(w))\int_0^1 u^2\,dx\right), \quad \forall w = (u,v)^{\text{tr}} \in H,$$

where $F \in C^3([0, +\infty); [0, +\infty))$ satisfies

$$\text{(13.33)} \qquad \sup_{s \geqslant 0} F(s) < \frac{\mu_1}{2},$$

where μ_1 is the first eigenvalue of the unbounded linear operator (d^4/dx^4) on $L^2(0,1)$ with domain

$$\text{Dom}(d^4/dx^4) = \{f;\, f \in H^4(0,1),\, f(0) = f_x(0) = f_{xx}(1) = f_{xxx}(1) = 0\}.$$

Note that this operator is self-adjoint (i.e., is equal to its adjoint) and has compact resolvent. Hence, by the definition of μ_1,

$$\int_0^1 u_{xx}^2 dx \geqslant \mu_1 \int_0^1 u^2 dx, \ \forall w \in H. \tag{13.34}$$

From (13.33) and (13.34),

$$\frac{1}{4}\mathcal{E}(w) \leqslant J(w) \leqslant \frac{1}{2}\mathcal{E}(w), \ \forall w \in H. \tag{13.35}$$

Computing the time-derivative $\dot{J}$ of J along the trajectories of (13.27), one gets

$$\dot{J} = \left(K\omega^2 - F(\mathcal{E})\right)\left(\int_0^1 uv dx\right), \tag{13.36}$$

where, for simplicity, we write $\mathcal{E}$ for $\mathcal{E}(w)$ and where

$$K(=K(w)) := 1 - F'(\mathcal{E})\int_0^1 u^2 dx. \tag{13.37}$$

Let us impose that

$$0 \leqslant F'(s)s < \mu_1 - F(s), \ \forall s \in [0, +\infty), \tag{13.38}$$

$$\exists C_4 > 0 \text{ such that } \lim_{s \to 0, s > 0} \frac{F(s)}{s} = C_4. \tag{13.39}$$

It is then natural to consider the feedback law for (13.27) vanishing at 0 and such that, on $H \setminus \{0\}$,

$$\omega = K^{-1/2}\left(F(\mathcal{E}) - \bar{\sigma}\left(\int_0^1 uv dx\right)\right)^{\frac{1}{2}}, \tag{13.40}$$

for some function $\bar{\sigma} \in C^2(\mathbb{R}; \mathbb{R})$ satisfying

$$s\bar{\sigma}(s) > 0, \ \forall s \in \mathbb{R} \setminus \{0\}, \tag{13.41}$$

$$\exists C_5 > 0 \text{ such that } \lim_{s \to 0, s \neq 0} \frac{\bar{\sigma}(s)}{s} = C_5, \tag{13.42}$$

$$\bar{\sigma}(s) < F(2\sqrt{\mu_1}s), \ \forall s > 0. \tag{13.43}$$

Note that, using (13.34), one gets that, for every $w = (u, v) \in H$,

$$\int_0^1 uv dx \leqslant \frac{1}{2}(\sqrt{\mu_1}\int_0^1 u^2 dx + \frac{1}{\sqrt{\mu_1}}\int_0^1 v^2 dx)$$

$$\leqslant \frac{1}{2\sqrt{\mu_1}}\left(\int_0^1 u_{xx}^2 dx + \int_0^1 v^2 dx\right) = \frac{1}{2\sqrt{\mu_1}}\mathcal{E}(w), \tag{13.44}$$

which, together with (13.38), (13.39) and (13.43), implies that

$$F(\mathcal{E}(w)) - \bar{\sigma}\left(\int_0^1 uv dx\right) > 0, \ \forall w \in H \setminus \{0\}. \tag{13.45}$$

From (13.34), (13.37) and (13.38), one gets

$$K(w) \geqslant 1 - \frac{F'(\mathcal{E}(w))}{\mu_1}\int_0^1 u_{xx}^2 dx \geqslant 1 - \frac{F'(\mathcal{E}(w))}{\mu_1}\mathcal{E}(w) > 0, \ \forall w \in H. \tag{13.46}$$

13.2. STABILIZATION OF A ROTATING BODY-BEAM WITHOUT DAMPING

From (13.45) and (13.46), we get that ω is well defined by (13.40) and is of class C^2 on $H \setminus \{0\}$. This regularity is sufficient to apply the desingularization technique of [**392**] (see Section 12.5.1 above). We note that (13.40) is equivalent to

$$(13.47) \qquad \omega^3 = \psi(w) := K^{-3/2}\left(F(\mathcal{E}(w)) - \bar{\sigma}\left(\int_0^1 uv dx\right)\right)^{\frac{3}{2}}$$

and therefore, following [**392**], one considers the following control Lyapunov function for control system (13.27)-(13.28):

$$V(w,\omega) = J + \int_{\psi^{\frac{1}{3}}}^{\omega}(s^3 - \psi)ds = J + \frac{1}{4}\omega^4 - \psi\omega + \frac{3}{4}\psi^{\frac{4}{3}},$$

where, for simplicity, we write J for $J(w)$ and ψ for $\psi(w)$. Then, by (13.35),

$$(13.48) \qquad V(w,\omega) \to +\infty \text{ as } \|w\|_H + |\omega| \to +\infty,$$
$$(13.49) \qquad V(w,\omega) > 0, \ \forall (w,\omega) \in H \times \mathbb{R} \setminus \{(0,0)\},$$
$$(13.50) \qquad V(0,0) = 0.$$

Moreover, if one computes the time-derivative $\dot{V}$ of V along the trajectories of (13.27)-(13.28), one gets, using in particular (13.36), that

$$(13.51) \quad \dot{V} = -\left(\int_0^1 uv dx\right)\bar{\sigma}\left(\int_0^1 uv dx\right) + (\omega - \psi^{\frac{1}{3}})[\gamma(\omega^2 + \psi^{\frac{1}{3}}\omega + \psi^{\frac{2}{3}}) + D],$$

where

$$(13.52) \qquad D := -\dot{\psi} + K(\omega + \psi^{\frac{1}{3}})\int_0^1 uv dx,$$

with
(13.53)
$$\dot{\psi} := \frac{3\psi^{\frac{1}{3}}}{2K}\left[2F'(\mathcal{E})\omega^2\int_0^1 uv dx - \sigma'\left(\int_0^1 uv dx\right)\left(\int_0^1 (v^2 - u_{xx}^2 + \omega^2 u^2)dx\right)\right]$$
$$+ \frac{3\psi}{K}\left(\int_0^1 uv dx\right)\left(\omega^2 F''(\mathcal{E})\int_0^1 u^2 dx + F'(\mathcal{E})\right).$$

Hence it is natural to define the feedback law γ by

$$(13.54) \qquad \gamma(0,0) := 0$$

and, for every $(w,\omega) \in (H \times \mathbb{R}) \setminus \{(0,0)\}$,

$$(13.55) \qquad \gamma(w,\omega) := -(\omega - \psi^{\frac{1}{3}}) - \frac{D}{\omega^2 + \psi^{\frac{1}{3}}\omega + \psi^{\frac{2}{3}}}.$$

Note that by (13.45), (13.46) and (13.47),

$$(13.56) \qquad \psi(w) > 0, \ \forall w \in H \setminus \{0\}.$$

Moreover, by (13.37), (13.39), (13.42), (13.43) and (13.44), there exists $\delta > 0$ such that

$$(13.57) \qquad \psi(w) > \delta \mathcal{E}(w)^{3/2}, \ \forall w \in H \text{ such that } \mathcal{E}(w) < \delta.$$

Using (13.37), (13.46), (13.47), (13.52), (13.53), (13.55), (13.56) and (13.57), one easily checks that γ is Lipschitz on bounded sets of $H \times \mathbb{R}$. Therefore the Cauchy problem associated to (13.27)-(13.28) has, for this feedback law γ and an initial condition given at time 0, one and only one (maximal) solution defined on an open interval containing 0. (The definition of a solution is the definition of a usual mild solution as given, for example, in [**381**, Definition 1.1, Chapter 6, page 184]. It is quite similar to the one given in Definition 2.86 on page 96. For the existence and uniqueness of (maximal) solutions, we refer, for example, to [**381**, Theorem 1.4, Chapter 6, pages 185–186].) By (13.43), (13.51), (13.52), (13.54) and (13.55), one has

$$(13.58) \quad \dot{V} = -\left(\int_0^1 uv dx\right) \bar{\sigma}\left(\int_0^1 uv dx\right) - (\omega - \psi^{\frac{1}{3}})^2(\omega^2 + \psi^{\frac{1}{3}}\omega + \psi^{\frac{2}{3}}) \leqslant 0,$$

which, together with (13.48) and a classical property of maximal solutions (see, e.g., [**381**, Theorem 1.4, Chapter 6, pages 185–186]) shows that the (maximal) solutions of (13.27)-(13.28) defined at time t_0 are defined on at least $[t_0, +\infty)$.

One has the following theorem.

THEOREM 13.2. *The feedback law γ defined by (13.54)-(13.55) globally strongly asymptotically stabilizes the equilibrium point $(0,0)$ for the control system (13.27)-(13.28), i.e.,*

(i) *for every solution of (13.27), (13.28), (13.54) and (13.55),*

$$(13.59) \quad \lim_{t \to +\infty} \|w(t)\|_H + |\omega(t)| = 0;$$

(ii) *for every $\varepsilon > 0$, there exists $\eta > 0$ such that, for every solution of (13.27), (13.28), (13.54) and (13.55),*

$$(\|w(0)\|_H + |\omega(0)| < \eta) \Rightarrow (\|w(t)\|_H + |\omega(t)| < \varepsilon, \ \forall t \geqslant 0).$$

The proof of Theorem 8.5 is given in [**119**]. Let us just mention that it is divided into two parts:

(i) First one proves that the trajectories of (13.27)-(13.28) are precompact in H for $t \geqslant 0$.
(ii) Then one concludes by means of the LaSalle invariance principle.

The main difficult point is to prove (i). More precisely, one needs to prove that the energy associated to the high frequency modes is uniformly small. In order to prove this uniform smallness a key point is that all these modes satisfy the same equation as w. Finally, an important ingredient is to get some estimates on $\int_0^1 uv dx$ for any solution of (13.27)-(13.28), which allow us to prove the uniform smallness.

13.3. Null asymptotic stabilizability of the 2-D Euler control system

In this section we address a problem of stabilization of the Euler equations. In Section 6.2.1 we considered the problem of controllability of the Euler control system of an incompressible inviscid fluid in a bounded domain. In particular, we saw that, if the controls act on an arbitrarily small open subset of the boundary which meets every connected component of this boundary, then the Euler equation is exactly controllable.

For linear control systems, the exact controllability implies in many cases the asymptotic stabilizability by means of feedback laws. As we saw in Theorem 10.1 on page 275, this is always the case in finite dimension. This also holds for important partial differential equations as shown by Marshall Slemrod in [**450**], Jacques-Louis Lions in [**326**], Irena Lasiecka and Roberto Triggiani in [**299**] and Vilmos Komornik in [**278**]; see also Section 13.1.

13.3. ASYMPTOTIC STABILIZABILITY OF THE 2-D EULER CONTROL SYSTEM

However, as we saw in Section 11.1, this is no longer true for *nonlinear* control systems, even of finite dimension. For example, the nonholonomic integrator

$$(13.60) \qquad \dot{x}_1 = u_1, \ \dot{x}_2 = u_2, \ \dot{x}_3 = x_1 u_2 - x_2 u_1,$$

where the state is $(x_1, x_2, x_3) \in \mathbb{R}^3$ and the control is $(u_1, u_2)^{\text{tr}} \in \mathbb{R}^2$, is small-time locally controllable at the equilibrium $(x_e, u_e) := (0, 0) \in \mathbb{R}^3 \times \mathbb{R}^2$ (see Example 3.20 on page 135 or Example 6.3). However, as we have seen in Example 11.2 on page 289 (an example due to Roger Brockett [74]), the control system (13.60) cannot be locally asymptotically stabilized by means of continuous stationary feedback laws. Note however that, by Theorem 11.14 on page 297, the driftless control affine system (13.60) can be globally asymptotically stabilized by means of smooth periodic time-varying feedback laws (see Example 11.16 on page 297). Let us also notice that, as for the control system (13.60), the linearized control system of the Euler equation around the origin is not controllable, as we have seen on page 195.

Therefore it is natural to ask what is the situation for the asymptotic stabilizability of the origin for the 2-D Euler equation of an incompressible inviscid fluid in a bounded domain, when the controls act on an arbitrarily small open subset of the boundary which meets every connected component of this boundary. In this section we are going to prove that the null global asymptotic stabilizability by means of feedback laws holds if the domain is simply connected.

Let Ω be a nonempty bounded connected and simply connected subset of $\mathbb{R}^2$ of class C^∞ and let Γ_0 be a nonempty open subset of the boundary $\partial\Omega$ of Ω. This set Γ_0 is the location of the control. Let y be the velocity field of the inviscid fluid contained in Ω. We assume that the fluid is incompressible, so that

$$(13.61) \qquad \text{div } y = 0.$$

Since Ω is simply connected, y is completely characterized by $\omega := \text{curl } y$ and $y \cdot n$ on $\partial\Omega$, where n denotes the unit outward normal to $\partial\Omega$. For the controllability problem, one does not really need to specify the control and the state; one considers the "Euler control system" as an under-determined system, by requiring $y \cdot n = 0$ on $\partial\Omega \setminus \Gamma_0$ instead of $y \cdot n = 0$ on $\partial\Omega$ which one requires for the uncontrolled usual Euler equation. For the stabilization problem, one needs to specify more precisely the control and the state. In this section, the state at time t is equal to $\omega(t, \cdot)$. For the control at time t, there are at least two natural possibilities:

(a) The control is $y(t, x) \cdot n(x)$ on Γ_0 and the time derivative $\partial \omega / \partial t(t, x)$ of the vorticity at the points x of Γ_0 where $y(t, x) \cdot n(x) < 0$, i.e., at the points where the fluid enters into the domain Ω.

(b) The control is $y(t, x) \cdot n(x)$ on Γ_0 and the vorticity ω at the points x of Γ_0 where $y(t, x) \cdot n(x) < 0$.

Let us point out that, by (13.61), in both cases $y \cdot n$ has to satisfy $\int_{\partial\Omega} y \cdot n = 0$. In this section, we study only case (a); for case (b), see [**114**].

Let us give stabilizing feedback laws. Let $g \in C^\infty(\partial\Omega)$ be such that

(13.62)   the support of g is included in Γ_0,

(13.63) $\Gamma_0^+ := \{g > 0\}$ and $\Gamma_0^- := \{g < 0\}$ are connected,

(13.64) $$g \neq 0,$$

(13.65) $$\overline{\Gamma_0^+} \cap \overline{\Gamma_0^-} = \emptyset,$$

(13.66) $$\int_{\partial\Omega} g = 0.$$

For every $f \in C^0(\overline{\Omega})$, we denote
$$|f|_0 = \text{Max}\{|f(x)|\,;x \in \overline{\Omega}\}.$$

Our stabilizing feedback laws are
$$y \cdot n = M\,|\omega|_0\, g \text{ on } \Gamma_0,$$
$$\frac{\partial \omega}{\partial t} = -M\,|\omega|_0\, \omega \text{ on } \Gamma_0^- \text{ if } |\omega|_0 \neq 0,$$
where $M > 0$ is large enough.

With these feedback laws, a function $\omega : I \times \overline{\Omega} \to \mathbb{R}$, where I is a time interval, is a solution of the closed-loop system Σ if

(13.67) $$\frac{\partial \omega}{\partial t} + \text{div}\,(\omega y) = 0 \text{ in } \overset{\circ}{I} \times \Omega,$$

(13.68) $$\text{div}\, y = 0 \text{ in } \overset{\circ}{I} \times \Omega,$$

(13.69) $$\text{curl}\, y = \omega \text{ in } \overset{\circ}{I} \times \Omega,$$

(13.70) $$y(t) \cdot n = M\,|\omega(t)|_0\, g \text{ on } \partial\Omega, \forall t \in I,$$

(13.71) $$\frac{\partial \omega}{\partial t} = -M\,|\omega(t)|_0\, \omega \text{ on } \{t \in \overset{\circ}{I};\, \omega(t) \neq 0\} \times \Gamma_0^-.$$

Here, $\overset{\circ}{I}$ denotes the interior of $I \subset \mathbb{R}$ and for $t \in I$, $\omega(t) : \overline{\Omega} \to \mathbb{R}$ and $y(t) : \overline{\Omega} \to \mathbb{R}^2$ are defined by requiring $\omega(t)(x) = \omega(t,x)$ and $y(t)(x) = y(t,x), \forall x \in \overline{\Omega}$. More precisely, the definition of a solution of the system Σ follows.

DEFINITION 13.3. *Let I be an interval. A function $\omega : I \to C^0(\overline{\Omega})$ is a solution of the system Σ if*

(i) $\omega \in C^0(I; C^0(\overline{\Omega}))(\cong C^0(I \times \overline{\Omega}))$,

(ii) *for $y \in C^0(I \times \overline{\Omega}; \mathbb{R}^2)$ defined by requiring (13.68) and (13.69) in the sense of distributions and (13.70), one has (13.67) in the sense of distributions,*

(iii) *in the sense of distributions on the open manifold $\{t \in \overset{\circ}{I};\, \omega(t) \neq 0\} \times \Gamma_0^-$, one has $\partial\omega/\partial t = -M\,|\omega(t)|_0\, \omega$.*

Our first theorem says that, for M large enough, the Cauchy problem for the system Σ has at least one solution defined on $[0, +\infty)$ for every initial data in $C^0(\overline{\Omega})$. More precisely one has:

THEOREM 13.4 ([**114**]). *There exists $M_0 > 0$ such that, for every $M \geqslant M_0$, the following two properties hold:*

(i) *For every $\omega_0 \in C^0(\overline{\Omega})$, there exists a solution of the system Σ defined on $[0, +\infty)$ such that $\omega(0) = \omega_0$.*

(ii) *Every maximal solution of system Σ defined at time 0 is defined on $[0,+\infty)$ (at least).*

REMARK 13.5. a. In this theorem, property (i) is in fact implied by property (ii) and Zorn's lemma. We state (i) in order to emphasize the existence of a solution to the Cauchy problem for the system Σ.

b. We do not know if the solution to the Cauchy problem is unique for positive time. (For negative time, one does not have uniqueness since there are solutions ω of system Σ defined on $[0,+\infty)$ such that $\omega(0) \neq 0$ and $\omega(T) = 0$ for some $T \in (0,+\infty)$ large enough.)

However, let us emphasize that, already for control systems in finite dimension, we have considered feedback laws which are merely continuous (see Chapter 11 and Chapter 12). With these feedback laws, the Cauchy problem for the closed-loop system may have many solutions. It turns out that this lack of uniqueness is not a real problem. Indeed, in finite dimension at least, if a point is asymptotically stable for a continuous vector field, then there exists, as in the case of regular vector fields, a (smooth) strict Lyapunov function. This result is due to Jaroslav Kurzweil [**289**].

It is tempting to conjecture that a similar result holds in infinite dimension under reasonable assumptions. The existence of this Lyapunov function ensures some robustness to perturbations. It is precisely this robustness which makes the interest of feedback laws compared to open-loop controls. We will see that, for our feedback laws, there also exists a strict Lyapunov function (see Proposition 13.9 on the next page) and therefore our feedback laws provide some kind of robustness.

Our next theorem shows that, at least for M large enough, our feedback laws globally and strongly asymptotically stabilize the origin in $C^0(\overline{\Omega})$ for the system Σ.

THEOREM 13.6. *There exists a positive constant $M_1 \geqslant M_0$ such that, for every $\varepsilon \in (0,1]$, every $M \geqslant M_1/\varepsilon$ and every maximal solution ω of system Σ defined at time 0,*

$$(13.72) \qquad |\omega(t)|_0 \leqslant \operatorname{Min}\left\{|\omega(0)|_0, \frac{\varepsilon}{t}\right\}, \, \forall t > 0.$$

REMARK 13.7. Due to the term $|\omega(t)|_0$ appearing in (13.70) and in (13.71), our feedback laws do not depend only on the value of ω on Γ_0. Let us point out that there is no asymptotically stabilizing feedback law depending only on the value of ω on Γ_0 such that the origin is asymptotically stable for the closed-loop system. In fact, given a nonempty open subset Ω_0 of Ω, there is no stabilizing feedback law which does not depend on the values of ω on Ω_0. This phenomenon is due to the existence of "phantom vortices": there are smooth stationary solutions $\bar{y} : \overline{\Omega} \to \mathbb{R}^2$ of the 2-D Euler equations such that Support $\bar{y} \subset \Omega_0$ and $\bar{\omega} := \operatorname{curl} \bar{y} \neq 0$; see, e.g., [**341**]. Then $\omega(t) = \bar{\omega}$ is a solution of the closed-loop system if the feedback law does not depend on the values of ω on Ω_0, and vanishes for $\omega = 0$.

REMARK 13.8. Let us emphasize that (13.72) implies that

$$(13.73) \qquad |\omega(t)|_0 \leqslant \varepsilon, \, \forall t \geqslant 1,$$

for every maximal solution ω of system Σ defined at time 0 (whatever $\omega(0)$ is). It would be interesting to know if one could have a similar result for the 2-D Navier-Stokes equations of viscous incompressible flows, that is, if, given $\varepsilon > 0$, there exists a feedback law such that (13.73) holds for every solution of the closed-loop Navier-Stokes system.

Note that $y = 0$ on Γ_0 is a feedback which leads to asymptotic stabilization of the null solution of the Navier-Stokes control system. However, this feedback does not have the required property. One may ask a similar question for the Burgers control system; for the null asymptotic stabilization of this control system, see the paper [**287**] by Miroslav Krstić and the references therein. For local stabilization of the Navier-Stokes equations, let us mention in particular [**29**] by Viorel Barbu, [**31**] by Viorel Barbu and Roberto Triggiani, [**30**] by Viorel Barbu, Irena Lasiecka and Roberto Triggiani, [**183**] by Andrei Fursikov, [**399**] by Jean-Pierre Raymond and [**492**]. These papers use in a strong way the viscous term $\nu \Delta y$ (see equation (6.23)); the Euler part $(y \cdot \nabla)y$ is essentially treated as a perturbation term which is more an annoying term than a useful term. Note that Theorem 13.4 and Theorem 13.6 seem to show that the Euler term might be useful in some situations, even for the Navier-Stokes control system.

Sketch of the proof of Theorem 13.6. (The detailed proofs of Theorem 13.4 and of Theorem 13.6 are given in [**114**].) Let us just mention that the proof relies on an explicit Lyapunov function $V : C^0(\overline{\Omega}) \to [0, +\infty)$. This function V is defined by

$$V(\omega) = |\omega \exp(-\theta)|_0, \tag{13.74}$$

where $\theta \in C^\infty(\overline{\Omega})$ satisfies

$$\Delta \theta = 0 \text{ in } \overline{\Omega}, \tag{13.75}$$

$$\frac{\partial \theta}{\partial n} = g \text{ on } \partial \Omega. \tag{13.76}$$

(Let us point out that the existence of θ follows from (13.66).) Theorem 13.6 is an easy consequence of the following proposition.

PROPOSITION 13.9. *There exist $M_2 \geqslant M_0$ and $\mu > 0$ such that, for every $M \geqslant M_2$ and every solution $\omega : [0, +\infty) \to C^0(\overline{\Omega})$ of system Σ, one has, for every $t \in [0, +\infty)$,*

$$[-\infty, 0] \ni \dot{V}(t) := \frac{d}{dt^+} V(\omega(t)) \leqslant -\mu M V^2(\omega(t)), \tag{13.77}$$

where $d/dt^+ V(\omega(t)) := \lim_{\varepsilon \to 0^+}(V(\omega(t+\varepsilon)) - V(\omega(t)))/\varepsilon$. ∎

Let us end this section by some comments on the case where Ω is not simply connected. In this case, in order to define the state, one adds to ω the real numbers $\lambda_1, \ldots, \lambda_g$ defined by

$$\lambda_i = \int_\Omega y \cdot \nabla^\perp \tau_i,$$

where, if one denotes by $\mathcal{C}_0, \mathcal{C}_1, \ldots, \mathcal{C}_g$ the connected components of Γ, the functions $\tau_i \in C^\infty(\overline{\Omega}), i \in \{1, \ldots, g\}$ are defined by

$$\begin{aligned}
\Delta \tau_i &= 0, \\
\tau_i &= 0 \text{ on } \partial \Omega \setminus \mathcal{C}_i, \\
\tau_i &= 1 \text{ on } \mathcal{C}_i,
\end{aligned}$$

and where $\nabla^\perp \tau_i$ denotes $\nabla \tau_i$ rotated by $\pi/2$. One has the following open problem.

OPEN PROBLEM 13.10. *Assume that $g \geqslant 1$ and that Γ_0 meets every connected component of Γ. Does there always exist a feedback law such that $0 \in C^0(\overline{\Omega}) \times \mathbb{R}^g$ is globally asymptotically stable for the closed-loop system?*

Recently Olivier Glass has got in [**199**] a positive answer to this open problem if one allows to add an integrator on the vorticity ω (in particular he has proved the global asymptotic stabilizability by means of *dynamic* feedback laws (see Definition 11.5 on page 292).

Brockett's necessary condition [**74**] for the existence of asymptotically stabilizing feedback laws cannot be directly applied to our situation since our control system is of infinite dimension. However, it naturally leads to the following question.

Question. *Assume that Γ_0 meets every connected component of Γ. Let $f \in C^\infty(\overline{\Omega})$. Do there exist $y \in C^\infty(\overline{\Omega}; \mathbb{R}^2)$ and $p \in C^\infty(\overline{\Omega})$ such that*

$$(y.\nabla)y + \nabla p = f \text{ in } \overline{\Omega}, \tag{13.78}$$

$$\operatorname{div} y = 0 \text{ in } \overline{\Omega}, \tag{13.79}$$

$$y \cdot n = 0 \text{ on } \Gamma \setminus \Gamma_0? \tag{13.80}$$

Let us point out that, by scaling arguments, one does not have to assume that f is "small" in this question. It turns out that the answer to this question is indeed positive. This has been proved in [**115**] if Ω is simply connected and by Olivier Glass in [**197**] for the general case. For many results on feedback in flow control, see the book [**1**].

13.4. A strict Lyapunov function for boundary control of hyperbolic systems of conservation laws

This section is borrowed from [**121**]. It concerns some systems of conservation laws that are described by partial differential equations, with an independent time variable $t \in [0, +\infty)$ and an independent space variable x on a finite interval $[0, L]$. For such systems, the boundary control problem that we consider is the problem of designing control actions at the boundaries (i.e., at $x = 0$ and $x = L$) in order to ensure that the smooth solution of the Cauchy problem converges to a desired steady-state. One can use an entropy of the system as a Lyapunov function (see [**120**] and [**125**]). Unfortunately, this Lyapunov function has only a semi-negative definite time derivative (i.e., one has only $\dot{V} \leqslant 0$). Since LaSalle's invariance principle seems to be difficult to apply (due to the problem of the precompactness of the trajectories), one cannot conclude easily the stability of the closed-loop control system.

In this section, assuming that the system can be diagonalized with the Riemann invariants, we exhibit a strict Lyapunov function which is an extension of the entropy but whose time derivative can be made strictly negative definite by an appropriate choice of the boundary controls. We give a theorem which shows that the boundary control allows us to have the local convergence of the system trajectories towards a desired steady-state. Furthermore, the control can be implemented as a feedback of the state only measured at the boundaries. The control design method is illustrated with an hydraulic application: the regulation of the level and the flow in an horizontal reach of an open channel.

For the sake of simplicity, our presentation is restricted to 2×2 systems of conservation laws. From our analysis, it is however very clear that the approach can be directly extended to any system of conservation laws which can be diagonalized with Riemann invariants. It is in particular the case for networks channels where the flux on each arc is modeled by a system of two conservation laws (see, e.g., [**210**] by Martin Gugat, Günter Leugering and Georg Schmidt, [**214**] by Jonathan de Halleux, Christophe Prieur and Georges Bastin, as well as [**215**]).

13.4.1. Boundary control of hyperbolic systems of conservation laws: statement of the problem. Let Ω be a nonempty connected open set in $\mathbb{R}^2$ and let $L > 0$. We consider a system of two conservation laws of the general form

$$(13.81) \qquad Y_t + f(Y)_x = 0,\, t \in [0, +\infty),\, x \in (0, L)$$

where

1. $Y = (y_1, y_2)^{\text{tr}} : [0, +\infty) \times [0, L] \to \Omega$,
2. $f \in C^3(\Omega; \mathbb{R}^2)$ is the flux density.

We are concerned with the smooth solutions of the Cauchy problem associated to the system (13.81) over $[0, +\infty) \times [0, L]$ under an initial condition

$$Y(0, x) = Y^0(x),\, x \in [0, L],$$

and two boundary conditions of the form

$$B_0(Y(t, 0), u_0(t)) = 0,\, t \in [0, +\infty),$$
$$B_L(Y(t, L), u_L(t)) = 0,\, t \in [0, +\infty),$$

with $B_0, B_L : \Omega \times \mathbb{R} \to \mathbb{R}$. This is a control system where, at time t, the state is $Y(t, \cdot)$ and the control is $(u_0(t), u_L(t))^{\text{tr}} \in \mathbb{R}^2$.

For constant control actions $u_0(t) = \bar{u}_0$ and $u_L(t) = \bar{u}_L$, a steady-state solution is a constant solution

$$Y(t, x) = \bar{Y},\, \forall t \in [0, +\infty),\, \forall x \in [0, L],$$

which satisfies (13.81) and the boundary conditions $B_0(\bar{Y}, \bar{u}_0) = 0$ and $B_L(\bar{Y}, \bar{u}_L) = 0$. In other words, $(\bar{Y}, \bar{u}_0, \bar{u}_L)$ is an equilibrium of our control system (see Definition 3.1 on page 125).

The boundary stabilization problem we are concerned with is the problem of finding localized feedback laws $Y(t, 0) \to u_0(Y(t, 0))$ and $Y(t, L) \to u_L(Y(t, L))$ such that, for any smooth enough initial condition $Y^0(x)$, the Cauchy problem for the system (13.81) has a unique smooth solution converging towards the desired given steady-state $\bar{Y}$.

In this section, we consider the special case where the system (13.81) is strictly hyperbolic, i.e., the Jacobian matrix $f'(Y)$ has two real distinct eigenvalues for every $Y \in \Omega$. We also assume that the two eigenvalues of $f'(Y)$ have opposite signs: $\lambda_2(Y) < 0 < \lambda_1(Y)$, $\forall Y \in \Omega$.

The system (13.81) can be "diagonalized", at least locally, using the Riemann invariants (see, e.g., [**305**, pages 34–35] by Peter Lax). This means that, for every $\bar{Y} \in \Omega$, there exist an open neighborhood $\tilde{\Omega}$ of $\bar{Y}$ and a change of coordinates $\xi(Y) = (a(Y), b(Y))^{\text{tr}}$ whose Jacobian matrix is denoted $D(Y)$,

$$D(Y) = \xi'(Y),$$

and which diagonalizes $f'(Y)$ in $\tilde{\Omega}$:

(13.82) $$D(Y)f'(Y) = \Lambda(Y)D(Y), \; Y \in \tilde{\Omega},$$

with
$$\Lambda(Y) = \mathrm{diag}(\lambda_1(Y), \lambda_2(Y)),$$

where $\lambda_1(Y), \lambda_2(Y)$ are the eigenvalues of $f'(Y)$. As it is shown by Peter Lax in [**305**, pages 34–35], the first order partial differential equation (13.82) can be reduced to the integration of ordinary differential equations. Moreover, in many cases, these ordinary differential equations can be solved explicitly by using separation of variables, homogeneity or symmetry properties (see, e.g., [**443**, pages 146–147, page 152] for examples of computations of Riemann invariants). In the coordinates $\xi = (a,b)^{\mathrm{tr}}$, the system (13.81) can then be rewritten in the following (diagonal) characteristic form:

$$\xi_t + \Lambda(\xi)\xi_x = 0,$$

or

(13.83) $$a_t + c(a,b)a_x = 0, \; b_t - d(a,b)b_x = 0,$$

where $c(a,b) = \lambda_1(\xi) > 0$ and $-d(a,b) = \lambda_2(\xi) < 0$ are the eigenvalues of $f'(Y)$ expressed in the $\xi = (a,b)^{\mathrm{tr}}$ coordinates. Throughout Section 13.4, we assume that c and d are two functions of class C^2 on a neighborhood of $(0,0) \in \mathbb{R}^2$ such that

$$c(0,0) > 0, \; d(0,0) > 0.$$

We observe that the quantity $a_t + c(a,b)a_x$ (resp. $b_t - d(a,b)b_x$) can be in fact viewed as the total time derivative $\mathrm{d}a/\mathrm{d}t$ (resp. $\mathrm{d}b/\mathrm{d}t$) of the function $a(t,x)$ (resp. $b(t,x)$) at a point (t,x) of the plane, along the curves having slopes

$$\frac{\mathrm{d}x}{\mathrm{d}t} = c(a,b) \; (\text{resp. } \frac{\mathrm{d}x}{\mathrm{d}t} = -d(a,b)).$$

These curves are called characteristic curves. Since $\mathrm{d}a/\mathrm{d}t = 0$ and $\mathrm{d}b/\mathrm{d}t = 0$ on the characteristic curves, it follows that $a(t,x)$ and $b(t,x)$ are constant along the characteristic curves. This explains why the characteristic solutions are called Riemann invariants.

The change of coordinates $\xi(Y)$ can be selected in such a way that $\xi(\bar{Y}) = 0$. We assume that such a selection is made. Then the control problem can be restated as the problem of determining the control actions in such a way that the characteristic solutions converge towards the origin.

Our goal in this section is to propose and analyze a control design method based on a strict Lyapunov function that is presented in the next section.

13.4.2. A strict Lyapunov function for boundary control design. Let us consider first the linear approximation of the characteristic form (13.83) around the origin:

(13.84) $$a_t + \bar{c}a_x = 0, \; b_t - \bar{d}b_x = 0,$$

with $\bar{c} = c(0,0) > 0$ and $\bar{d} = d(0,0) > 0$.

With a view to the boundary control design, let us propose the following candidate for a Lyapunov function:

$$U(t) := U_1(t) + U_2(t),$$

with

$$U_1(t) := \frac{A}{\bar{c}} \int_0^L a^2(t,x) e^{-(\mu/\bar{c})x} dx, \; U_2(t) := \frac{B}{\bar{d}} \int_0^L b^2(t,x) e^{+(\mu/\bar{d})x} dx,$$

where A, B and μ are positive constant coefficients.

This type of Lyapunov function is related to the Lyapunov function used in Section 13.3 to stabilize the Euler equation (see (13.74) and [**114**]). It has been introduced by Cheng-Zhong Xu and Gauthier Sallet in [**506**] for quite general linear hyperbolic systems.

The time derivative of this function along the trajectories of the linear approximation (13.84) is

$$\dot{U}(t) = -\mu U(t) - \Big[A e^{-(\mu/\bar{c})L} a^2(t,L) - A a^2(t,0)\Big] \\ - \Big[B b^2(t,0) - B e^{(\mu/\bar{d})L} b^2(t,L)\Big].$$

It can be seen that the two last terms depend only on the Riemann invariants at the two boundaries, i.e., at $x = 0$ and at $x = L$. The control laws $u_0(t)$ and $u_L(t)$ can then be defined in order to make these terms negative along the system trajectories.

A simple solution is to select $u_0(t)$ such that

(13.85) $$a(t,0) = k_0 b(t,0),$$

and $u_L(t)$ such that

(13.86) $$b(t,L) = k_L a(t,L),$$

with $|k_0 k_L| < 1$. The time derivative of the Lyapunov function is then written as

(13.87) $$\dot{U}(t) = -\mu U(t) + (B k_L^2 e^{(\mu/\bar{d})L} - A e^{-(\mu/\bar{c})L}) a^2(t,L) \\ + (A k_0^2 - B) b^2(t,0).$$

Since $|k_0 k_L| < 1$, we can select $\mu > 0$ such that

$$k_0^2 k_L^2 \leqslant |k_0 k_L| < e^{-\mu L\big((1/\bar{c}) + (1/\bar{d})\big)}.$$

Then we can select $A > 0$ and $B > 0$ such that

$$\frac{k_L^2}{e^{-\mu L\big((1/\bar{c}) + (1/\bar{d})\big)}} < \frac{A}{B} < \frac{1}{k_0^2}$$

(with the convention $1/k_0^2 = +\infty$ if $k_0 = 0$), which readily implies that

(13.88) $$A k_0^2 - B < 0 \text{ and } B k_L^2 e^{(\mu/\bar{d})L} - A e^{-(\mu/\bar{c})L} < 0.$$

Then it can be seen that $\dot{U}(t) \leqslant -\mu U(t)$ along the trajectories of the linear approximation (13.84) and that $\dot{U}(t) = 0$ if and only if $a(t,x) = b(t,x) = 0$ (i.e., at the system steady-state).

In the next section, we show that such boundary controls for the linearized system (13.84) can also be applied to the nonlinear system (13.83) with the guarantee that the trajectories locally converge to the origin.

13.4.3. Convergence analysis. In the previous section (Section 13.4.2), the inequality $\dot{U}(t) \leqslant -\mu U$ ensures the convergence in the $L^2(0,L)$-norm of the solutions of the linear system (13.84). As we shall see hereafter, in order to extend the analysis to the case of the nonlinear system (13.83), we will need to prove a convergence in $H^2(0,L)$-norm (see for instance [**476**, Chapter 16, Section 1]).

In the previous section, we considered linear boundary conditions (13.85)-(13.86). Here we assume more general nonlinear boundary conditions. More precisely, we assume that the boundary control functions $u_0(t)$ and $u_L(t)$ are chosen such that the boundary conditions have the form

(13.89) $$a(t,0) = \alpha(b(t,0)) \text{ and } b(t,L) = \beta(a(t,L)),$$

with C^1 functions $\alpha, \beta : \mathbb{R} \to \mathbb{R}$, and we denote

$$k_0 = \alpha'(0) \text{ and } k_L = \beta'(0).$$

The goal of this section is to prove the following theorem.

THEOREM 13.11. *Let us assume that $|k_0 k_L| < 1$. Then there exist positive real constants K, δ, λ such that, for every initial conditions (a^0, b^0) in $H^2(0,L)^2$ satisfying the compatibility conditions*

(13.90) $$a^0(0) = \alpha(b^0(0)), \, b^0(L) = \beta(a^0(L)),$$

(13.91) $$c(a^0(0), b^0(0))a_x^0(0) = -\alpha'(b^0(0))d(a^0(0), b^0(0))b_x^0(0),$$

(13.92) $$d(a^0(L), b^0(L))b_x^0(L) = -\beta'(a^0(L))c(a^0(L), b^0(L))a_x^0(L),$$

and the inequality

$$\|a^0\|_{H^2(0,L)} + \|b^0\|_{H^2(0,L)} < \delta,$$

the closed-loop system (13.83) with boundary conditions (13.89) has a unique solution in $C^0([0,+\infty); H^2(0,L)^2)$ and this solution satisfies, for every $t \in [0,+\infty)$,

$$\|a(t,\cdot)\|_{H^2(0,L)} + \|b(t,\cdot)\|_{H^2(0,L)} < K\left(\|a^0\|_{H^2(0,L)} + \|b^0\|_{H^2(0,L)}\right)e^{-\lambda t}.$$

One can also get Theorem 13.11 (in the C^1-norm instead of the H^2-norm) by using theorems due to James Greenberg and Ta-tsien Li [**203**] and to Ta-tsien Li [**316**, Theorem 1.3, Chapter V, page 173]. The proofs of [**203**] and of [**316**, Theorem 1.3, Chapter V, page 173] rely on direct estimates of the evolution of the Riemann invariants and their derivatives along the characteristic curves. Note that a strict Lyapunov function might be interesting for robustness issues which may be difficult to handle with these direct estimates. For example, using our strict Lyapunov function and proceeding as in the proof of Theorem 13.11, one can get the following theorem on the systems

(13.93) $$a_t + c(a,b)a_x + g(x,a,b) = 0, \, b_t - d(a,b)b_x + h(x,a,b) = 0,$$

THEOREM 13.12. *Let us assume that $|k_0 k_L| < 1$. Then there exists a positive real number ν, such that, for every g and g of class C^2 on a neighborhood of $[0,L] \times \{0\} \times \{0\}$ in $[0,L] \times \mathbb{R} \times \mathbb{R}$ satisfying*

$$\alpha(x,0,0) = 0, \, \beta(x,0,0) = 0, \, \forall x \in [0,L],$$

$$\left|\frac{\partial g}{\partial a}(x,0,0)\right| + \left|\frac{\partial g}{\partial b}(x,0,0)\right| + \left|\frac{\partial h}{\partial a}(x,0,0)\right| + \left|\frac{\partial h}{\partial a}(x,0,0)\right| \leqslant \nu, \, \forall x \in [0,L],$$

there exist $K > 0$, $\lambda > 0$ and $\delta > 0$ such that, for every initial condition (a^0, b^0) in $H^2(0, L)^2$ satisfying the compatibility conditions

$$a^0(0) = \alpha(b^0(0)),\ b^0(L) = \beta(a^0(L)),$$

$$c(a^0(0), b^0(0))a_x^0(0) + g(0, a^0(0), b^0(0))$$
$$= -\alpha'(b^0(0))\big(d(a^0(0), b^0(0))b_x^0(0) - h(0, a^0(0), b^0(0))\big),$$

$$d(a^0(L), b^0(L))b_x^0(L) - h(L, a^0(L), b^0(L))$$
$$= -\beta'(a^0(L))\big(c(a^0(L), b^0(L))a_x^0(L) + g(L, a^0(L), b^0(L))\big),$$

and the inequality

$$\|a^0\|_{H^2(0,L)} + \|b^0\|_{H^2(0,L)} < \delta,$$

the closed-loop system (13.93) with boundary conditions (13.89) has a unique solution in $C^0([0, +\infty); H^2(0, L)^2)$ and this solution satisfies, for every $t \in [0, +\infty)$,

$$\|a(t, \cdot)\|_{H^2(0,L)} + \|b(t, \cdot)\|_{H^2(0,L)} < K\left(\|a^0\|_{H^2(0,L)} + \|b^0\|_{H^2(0,L)}\right)e^{-\lambda t}.$$

We could also consider the cases where c and d depend "slightly" on $x \in [0, L]$ (in the sense that $c_x(x, 0, 0)$ and $d_x(x, 0, 0)$ are small). For robustness studies using direct estimates of the evolution of the Riemann invariants and their derivatives along the characteristic curves, see [**316**, Theorem 1.3, Chapter 5, page 173] by Ta-tsien Li, and [**396**] by Christophe Prieur, Joseph Winkin and Georges Bastin.

Proof of Theorem 13.11. By computing the time derivative of $U(t)$ along the solutions of the system (13.83), we get

$$\dot{U} = -\left[A\frac{a^2 c}{\bar{c}}e^{-(\mu x/\bar{c})}\right]_0^L + \left[B\frac{b^2 d}{\bar{d}}e^{+(\mu x/\bar{d})}\right]_0^L$$
$$- \int_0^L \left[A\frac{a^2 c\mu}{\bar{c}^2}e^{-(\mu x/\bar{c})} + B\frac{b^2 d\mu}{\bar{d}^2}e^{+(\mu x/\bar{d})}\right]dx$$
$$+ \int_0^L \left[A\frac{a^2 \phi}{\bar{c}}e^{-(\mu x/\bar{c})} - B\frac{b^2 \psi}{\bar{d}}e^{+(\mu x/\bar{d})}\right]dx,$$

with

$$\phi := c_x = a_x \frac{\partial c}{\partial a} + b_x \frac{\partial c}{\partial b},\ \psi := d_x = a_x \frac{\partial d}{\partial a} + b_x \frac{\partial d}{\partial b}.$$

Let us introduce the following notations:

$$v(t, x) := a_x(t, x)\ \text{and}\ w(t, x) := b_x(t, x).$$

Then straightforward computations using integrations by parts lead to the following lemma.

LEMMA 13.13. *If $|k_0 k_L| < 1$ and if the positive real constants μ, A, B satisfy inequalities (13.88), there exist positive real constants K_1, δ_1, λ_1 such that, if $|a(t, x)| + |b(t, x)| < \delta_1,\ \forall x \in [0, L]$, then*

$$\dot{U} \leqslant -\lambda_1 U + K_1 \int_0^L \left[a^2(t, x) + b^2(t, x)\right]\left[|v(t, x)| + |w(t, x)|\right]dx$$

along the solutions of the system (13.83) with the boundary conditions (13.89).

In contrast to the linear analysis of Section 13.4.2 on page 363, it appears readily from Lemma 13.13 on the preceding page that we cannot just complete the Lyapunov stability analysis with the function U but that we have to examine the dynamics of the variables $v(t,x)$ and $w(t,x)$ and consequently to extend the definition of the Lyapunov function.

By a time differentiation and using the model equations (13.83), it is readily shown that $v(t,x)$ and $w(t,x)$ satisfy the dynamics

$$(13.94) \qquad v_t + c(a,b)v_x + v\phi = 0, \; w_t - d(a,b)w_x - w\psi = 0,$$

and the associated boundary conditions

$$(13.95) \qquad \begin{cases} c(t,0)v(t,0) = -\alpha'(b(t,0))d(t,0)w(t,0), \\ d(t,L)w(t,L) = -\beta'(a(t,L))c(t,L)v(t,L). \end{cases}$$

Here $c(t,0)$, $d(t,0)$, $c(t,L)$ and $d(t,L)$ are compact notations for the values of the functions c and d evaluated at $(a(t,0),b(t,0))$ and $(a(t,L),b(t,L))$ respectively.

In fact, we also need to examine the dynamics of the spatial second order derivatives of $a(t,x)$ and $b(t,x)$, which are denoted as

$$q(t,x) := v_x(t,x) = a_{xx}(t,x), \; r(t,x) := w_x(t,x) = b_{xx}(t,x).$$

They satisfy the dynamics

$$(13.96) \qquad \begin{cases} q_t + c(a,b)q_x + 2q\phi + v\phi_x = 0, \\ r_t - d(a,b)r_x - 2r\psi - w\psi_x = 0, \end{cases}$$

with the associated boundary conditions
(13.97)
$$\begin{cases} c(t,0)q(t,0) + v(t,0)\phi(t,0) = \eta'(t)w(t,0) \\ \qquad\qquad\qquad\qquad\qquad + \eta(t)[d(t,0)r(t,0) + w(t,0)\psi(t,0)] \\ d(t,L)r(t,L) + w(t,L)\psi(t,L) = -\chi'(t)v(t,L) \\ \qquad\qquad\qquad\qquad\qquad + \chi(t)[c(t,L)q(t,L) + v(t,L)\phi(t,L)]. \end{cases}$$

Here $\phi(t,0)$, $\psi(t,0)$, are compact notations for the values of the functions ϕ and ψ evaluated at $(a(t,0),b(t,0),v(t,0),w(t,0))$, and $\phi(t,L)$ and $\psi(t,L)$ are similarly compact notations for the values of the functions ϕ and ψ evaluated at $(a(t,L),b(t,L),v(t,L),w(t,L))$. Finally the functions $\eta(t)$ and $\chi(t)$ are defined as

$$\eta(t) := \frac{\alpha'(b(t,0))d(t,0)}{c(t,0)}, \; \chi(t) := \frac{\beta'(a(t,L))c(t,L)}{d(t,L)}.$$

A key point for the proof of Theorem 13.11 is the fact that the linear approximations (around zero) of systems (13.94) and (13.96) have the following form:

$$v_t + \bar{c}v_x = 0, \; w_t - \bar{d}w_x = 0,$$
$$q_t + \bar{c}q_x = 0, \; r_t - \bar{d}r_x = 0.$$

Both systems have exactly the same form as the linear approximation (13.84) of the original system (13.83). Then, in order to prove that the solutions of the global system (13.83), (13.94), (13.96) converge to zero, it is quite natural to consider an extended Lyapunov function of the form

$$S(t) = U(t) + V(t) + W(t),$$

where $V(t)$ and $W(t)$ have the same form as $U(t)$:
$$V(t) := V_1(t) + V_2(t),$$
$$V_1(t) := \bar{c}A \int_0^L v^2(t,x) e^{-(\mu/\bar{c})x} dx, \ V_2(t) := \bar{d}B \int_0^L w^2(t,x) e^{(\mu/\bar{d})x} dx,$$
and
$$W(t) := W_1(t) + W_2(t),$$
$$W_1(t) := \bar{c}^3 A \int_0^L q^2(t,x) e^{-(\mu/\bar{c})x} dx, \ W_2(t) := \bar{d}^3 B \int_0^L r^2(t,x) e^{(\mu/\bar{d})x} dx.$$

Let us now examine the time derivatives of the functions $V(t)$ and $W(t)$ along the solutions of the closed-loop system (13.83), (13.89), (13.94)-(13.95)-(13.96)-(13.97). Straightforward (but lengthy) computations using integrations by parts lead to the following lemmas.

LEMMA 13.14. *If $|k_0 k_L| < 1$ and if the positive real constants μ, A, B satisfy inequalities (13.88), there exist positive real constants K_2, δ_2, λ_2 such that, if $|a(t,x)| + |b(t,x)| < \delta_2, \forall x \in [0, L]$, then*

$$\dot{V} \leqslant -\lambda_2 V + K_2 \int_0^L \left[v^2(t,x) + w^2(t,x) \right]^{3/2} dx$$

along the solutions of the systems (13.83), (13.94) with the boundary conditions (13.89) and (13.95).

LEMMA 13.15. *If $|k_0 k_L| < 1$ and if the positive real constants μ, A, B satisfy inequalities (13.88), there exist positive real constants K_3, δ_3, λ_3 such that, if $|a(t,x)| + |b(t,x)| < \delta_3, \forall x \in [0, L]$, then*

$$\dot{W} \leqslant -\lambda_3 W + K_3 \int_0^L \left(q^2(t,x) + r^2(t,x) \right) \left(|v(t,x)| + |w(t,x)| \right) dx$$
$$+ K_3 \int_0^L \left(v^2(t,x) + w^2(t,x) \right) \left(|q(t,x)| + |r(t,x)| \right) dx$$

along the solutions of the systems (13.83), (13.94), (13.96) with the boundary conditions (13.89), (13.95), (13.97).

We are now in a position to complete our Lyapunov convergence analysis. We start with the analysis of the global Lyapunov function $S = U + V + W$. From Lemma 13.13 on page 366, Lemma 13.14 and Lemma 13.15, we readily get the following lemma.

LEMMA 13.16. *If $|k_0 k_L| < 1$ and if the positive real constants μ, A, B satisfy inequalities (13.88), there exist positive real constants λ^0 and δ_0 such that, if $S(t) < \delta_0$, then*

$$\dot{S} \leqslant -\lambda_0 S$$

along the solutions of the closed-loop system (13.83), (13.89), (13.94), (13.95), (13.96), (13.97).

From [**476**, Chapter 16, Proposition 1.18, p. 364], we know that two solutions in
$$C^0([0,T]; H^2(0,L))^2 \cap C^1([0,T]; H^1(0,L))^2 \cap C^2([0,T]; L^2(0,L))^2$$

of the Cauchy problem (13.83), (13.89) with the same initial condition are equal on $[0, T]$. (Actually [**476**] deals with $\mathbb{R}$ instead of $[0, L]$ but the proof can be easily adapted).

Furthermore, concerning the existence of solutions to the Cauchy problem, we have the following result from [**476**, Chapter 16, Proposition 1.5, p. 365]. There exists $\delta_4 > 0$ such that, for every initial condition $(a^0(x), b^0(x)) \in H^2(0, L)^2$ satisfying the above compatibility conditions (13.90)-(13.91)-(13.92), if every solution

$$(a, b) \in C^0([0, T], H^2(0, L))^2 \cap C^1([0, T], H^1(0, L))^2 \cap C^2([0, T], L^2(0, L))^2$$

of the Cauchy problem (13.83), (13.89) with the initial condition $(a(0, \cdot), b(0, \cdot)) = (a^0, b^0)$ satisfies $S(t) < \delta_4$ for every $t \in [0, T)$, then this Cauchy problem has a solution defined on $[0, +\infty)$ and in

$$C^0([0, +\infty), H^2(0, L))^2 \cap C^1([0, +\infty), H^1(0, L))^2 \cap C^2([0, +\infty), L^2(0, L))^2.$$

(Again the proof in [**476**] deals with $\mathbb{R}$ instead of $[0, L]$ but it can be easily adapted).

Let $M > 0$ be such that

$$\frac{1}{M} S(t) \leqslant \|a(t, .)\|_{H^2(0,L)} + \|b(t, .)\|_{H^2(0,L)} \leqslant M S(t).$$

Then it follows from Lemma 13.16 that Theorem 13.11 holds with

$$\delta := \frac{1}{M} \min(\delta_0, \delta_4), \; K := M^2, \; \lambda := \lambda_0.$$

∎

REMARK 13.17. In the special case where $\mu = 0$, the Lyapunov function U is just an entropy function of the system in characteristic form linearized in the space of the Riemann coordinates. In references [**120**] and [**125**], the interested reader will find an alternative approach to the boundary control design where the entropy is used as a Lyapunov function. It must, however, be emphasized that the entropy is not a strict Lyapunov function because its time derivative is not negative definite but only semi-negative definite (as we can see by setting $\mu = 0$ in equality (13.87)). In fact, the Lyapunov analysis made in references [**120**] and [**125**], was only used as a guide to construct potentially stabilizing feedback laws. The proof of the asymptotic stability was obtained by using theorems due to James Greenberg and Ta-tsien Li [**203**] and to Ta-tsien Li [**316**, Theorem 1.3, Chapter V, page 173].

13.4.4. Application to level and flow control in a horizontal reach of an open channel. We consider a reach of an open channel delimited by two overflow spillways as represented in Figure 2 on the next page. We assume that

1. the channel is horizontal,
2. the channel is prismatic with a constant rectangular section and a unit width,
3. the friction effects are neglected.

The flow dynamics are described by a system of two conservation laws (shallow water equations, due to Saint-Venant [**430**]; see also Section 6.3), namely a mass conservation law

$$H_t + (HV)_x = 0,$$

and a momentum conservation law

$$V_t + \left(gH + \frac{V^2}{2}\right)_x = 0.$$

Here $H(t, x)$ represents the water level and $V(t, x)$ the water velocity in the reach, while g denotes the gravitation constant.

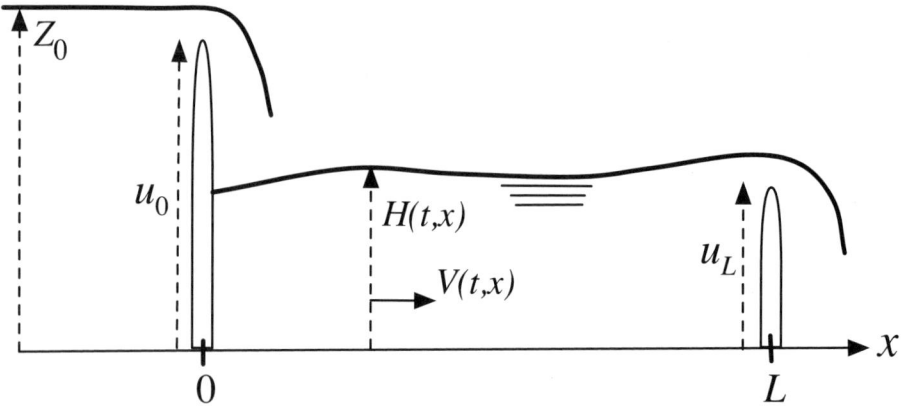

FIGURE 2. A reach of an open channel delimited by two adjustable overflow spillways

One has
$$f\begin{pmatrix}H\\V\end{pmatrix} := \begin{pmatrix}HV\\gH + \frac{V^2}{2}\end{pmatrix},$$
$$f'\begin{pmatrix}H\\V\end{pmatrix} = \begin{pmatrix}V & H\\g & V\end{pmatrix}.$$

The eigenvalues of the Jacobian matrix $f'(H, V)^{\text{tr}}$ are
$$\lambda_1(H, V) = V + \sqrt{gH}, \ \lambda_2(H, V) = V - \sqrt{gH}.$$

They are real and distinct if $H > 0$. They are generally called characteristic velocities. The flow is said to be fluvial (or subcritical) when the characteristic velocities have opposite signs:
$$\lambda_2(H, V) < 0 < \lambda_1(H, V).$$

The Riemann invariants can be defined as follows:
$$a = V - \bar{V} + 2(\sqrt{gH} - \sqrt{g\bar{H}}),$$
$$b = V - \bar{V} - 2(\sqrt{gH} - \sqrt{g\bar{H}}).$$

The control actions are the positions u_0 and u_L of the two spillways located at the extremities of the pool and related to the state variables H and V by the following expressions:

(13.98) $$H(t, 0)V(t, 0) - C_0(Z_0 - u_0(t))^{3/2} = 0,$$

(13.99) $$H(t, L)V(t, L) - C_0(H(t, L) - u_L(t))^{3/2} = 0,$$

where Z_0 denotes the water level above the pool (see Figure 2) and C_0 is a characteristic constant of the spillways.

For constant spillway positions $\bar{u}_0$ and $\bar{u}_L$, there is a unique steady-state solution which satisfies the following relations:
$$\bar{H} = Z_0 - \bar{u}_0 + \bar{u}_L, \quad \bar{V} = \frac{C_0(Z_0 - \bar{u}_0)^{3/2}}{Z_0 - \bar{u}_0 + \bar{u}_L}.$$
The control objective is to regulate the level H and the velocity V (or the flow rate $Q = HV$) at the steady-state $\bar{H}$ and $\bar{V}$ (or $\bar{Q} = \bar{H}\bar{V}$), by acting on the spillway positions u_0 and u_L.

By using the relations (13.85) and (13.86) for the control definition, combined with the spillway characteristics (13.98)-(13.99), the following boundary control laws are obtained:
$$u_0 = Z_0 - \sqrt[3]{\left(\frac{H_0}{C_0}\right)^2 \left(\bar{V} - 2\sqrt{g}\frac{1+k_0}{1-k_0}(\sqrt{H_0} - \sqrt{\bar{H}})\right)^2},$$
$$u_L = H_L - \sqrt[3]{\left(\frac{H_L}{C_0}\right)^2 \left(\bar{V} + 2\sqrt{g}\frac{1+k_L}{1-k_L}(\sqrt{H_L} - \sqrt{\bar{H}})\right)^2}$$
where $H_0 = H(t,0)$, $H_L = H(t,L)$.

Let us emphasize that u_0 (resp. u_L) depends only on the state at $x = 0$ (resp. $x = L$). In addition, let us also point out that the implementation of the control is particularly simple since only measurements of the levels $H(t,0)$ and $H(t,L)$ at the two spillways are required. This means that the feedback implementation does not require level measurements inside the pool or any velocity or flow rate measurements.

REMARK 13.18. In order to solve the control problem, we selected the particular simple boundary conditions (13.85) and (13.86). But obviously many other forms are admissible provided they make $\dot{U}$ negative. For instance, it can be interesting to use controls at a boundary which depend on the state at the other boundary, hence introducing some useful feedforward action in the control (see e.g. [37]).

REMARK 13.19. There have been recently many results on the controllability of channels and network of channels. Let us mention in particular [317, 318] by Ta-tsien Li, [320] by Ta-tsien Li and Bopeng Rao, [210] by Martin Gugat, Günter Leugering and Georg Schmidt.

APPENDIX A

Elementary results on semigroups of linear operators

In this appendix we recall some classical results on semigroups generated by linear operators with applications to evolution equations. We omit the proofs but we give precise references where they can be found. Let H be a complex Hilbert space. We denote by $\langle x, y \rangle$ the scalar product of two elements x and y in H. The norm of $x \in H$ is denoted $\|x\|$. Let

$$A : D(A) \subset H \to H$$
$$x \mapsto Ax$$

be a linear map from the linear subspace $D(A)$ of H into H. Throughout this section we assume that A is densely defined, i.e.,

(A.1) $\quad D(A)$ is dense in H.

Let us first recall some definitions.

DEFINITION A.1. The operator A is a *closed* operator if its graph (i.e., the set $\{(x, Ax); x \in H\}$) is a closed subset of $H \times H$.

DEFINITION A.2. The operator A is *dissipative* if

$$\Re \langle Ax, x \rangle \leqslant 0, \, \forall x \in D(A).$$

(Let us recall that, for $z \in \mathbb{C}$, $\Re z$ denotes the real part of z.)

DEFINITION A.3. The *adjoint* A^* of A is the linear operator

$$A^* : D(A^*) \subset H \to H$$
$$x \mapsto A^*x,$$

where

- $D(A^*)$ is the set of all y such that the linear map

$$D(A) \subset H \to \mathbb{C}$$
$$x \mapsto \langle Ax, y \rangle$$

is continuous, i.e., there exists $C > 0$ depending on y such that

(A.2) $\quad |\langle Ax, y \rangle| \leqslant C \|x\|, \, \forall x \in D(A).$

- For every $y \in D(A^*)$, A^*y is the unique element of H such that

(A.3) $\quad \langle Ax, y \rangle = \langle x, A^*y \rangle, \, \forall x \in D(A).$

(The uniqueness of A^*y such that (A.3) holds follows from (A.1), the existence follows from (A.2).)

We can now state the well-known Lumer-Phillips theorem (see for example [**381**, Corollary 4.4, Chapter 1, page 15]). One could alternatively state the Hille-Yosida theorem; see, for example, [**71**, Theorem VII.4, page 105] or [**381**, Theorem 3.1, Chapter 1, page 8 and Theorem 1.3 page 102].

THEOREM A.4. *Let us assume that A is densely defined and closed. If both A and A^* are dissipative, then, for every $x^0 \in D(A)$, there exists a unique*

(A.4) $$x \in C^1([0, +\infty); H) \cap C^0([0, +\infty); D(A))$$

such that

(A.5) $$\frac{dx}{dt} = Ax \text{ on } [0, +\infty),$$

(A.6) $$x(0) = x^0.$$

One has

(A.7) $$\|x(t)\| \leqslant \|x^0\|, \forall t \in [0, +\infty),$$

(A.8) $$\left\|\frac{dx}{dt}(t)\right\| = \|Ax(t)\| \leqslant \|Ax^0\|, \forall t \in [0, +\infty).$$

In (A.4) and throughout the whole book, the linear space $D(A)$ is equipped with the graph norm $\|y\|_{D(A)} := \|y\| + \|Ay\|$.

One also uses the following definition (see, for example, [**381**, Definition 1.1, page 1]).

DEFINITION A.5. A one-parameter family $S(t)$, $0 \leqslant t \leqslant \infty$, of continuous linear operators from H into H is a *semigroup of continuous linear operators on H* if

$$S(0) = \text{Id},$$
$$S(t_1 + t_2) = S(t_1) \circ S(t_2), \forall (t_1, t_2) \in [0, +\infty)^2.$$

In Definition A.5 and in the following, Id denotes the identity map on H. By (A.1) and (A.7), for every $t \geqslant 0$, the linear map $x^0 \in D(A) \mapsto x(t) \in H$ can be uniquely extended to an element of $\mathcal{L}(H; H)$, i.e., a linear continuous map from H into H. This extension is denoted $S(t)$. One has

$$S(0) = \text{Id},$$
$$S(t_1 + t_2) = S(t_1) \circ S(t_2), \forall (t_1, t_2) \in [0, +\infty)^2.$$

Because of these two properties, $S : [0, +\infty) \to \mathcal{L}(H, H)$ is called the semigroup *associated to the linear operator A*. Note that (A.7) implies that $\|S(t)\|_{\mathcal{L}(H;H)} \leqslant 1$, i.e.,

(A.9) $$S(t) \text{ is a contraction for every } t \in [0, +\infty).$$

Concerning the regularity of a semigroup, one has the following definition (see, for example, [**381**, Definition 2.1, page 4]).

DEFINITION A.6. Let H be a Hilbert space and let $S(t)$, $t \in [0, +\infty)$, be a semigroup of continuous linear operators from H into H. One says that S is a *strongly continuous* semigroup of continuous linear operators on H if

$$\lim_{t \to 0^+} S(t)x = x, \forall x \in H.$$

With this definition, it follows from (A.4) that the semigroup S associated to A is strongly continuous.

Concerning the inhomogeneous Cauchy problem, one has the following theorem (see, e.g., [**139**, Theorem 3.1.3, page 103]).

THEOREM A.7. *Let us assume that A is densely defined and closed. If both A and A^* are dissipative, then, for every $x^0 \in D(A)$, for every $T \in [0, +\infty)$ and for every $f \in C^1([0,T]; H)$, there exists a unique*

(A.10) $$x \in C^1([0, +\infty); H) \cap C^0([0, +\infty); D(A))$$

such that

(A.11) $$\frac{dx}{dt} = Ax + f(t) \text{ on } [0,T],$$

(A.12) $$x(0) = x^0.$$

One has (compare with Proposition 1.9 on page 6)

(A.13) $$x(t) = S(t)x^0 + \int_0^t S(t-\tau)f(\tau)d\tau, \forall t \in [0,T].$$

Concerning the behavior of the norm $\|S(t)\|_{\mathcal{L}(H;H)}$ of the continuous linear operators $S(t) : H \to H$, one has the following theorem (see, for example, [**381**, Theorem 2.2, Chapter 1, page 4]).

THEOREM A.8. *Let $S(t)$, $t \in [0, +\infty)$, be a strongly continuous semigroup of continuous linear operators on H. Then, there exist $C > 0$ and $\lambda \in \mathbb{R}$ such that*

$$\|S(t)\|_{\mathcal{L}(H;H)} \leqslant Ce^{\lambda t}, \forall t \in [0, +\infty).$$

Conversely, to every strongly continuous semigroup of continuous linear operators, one can associate a linear operator by the following definition (see, for example, [**381**, page 1]).

DEFINITION A.9. *Let $S(t)$, $t \in [0, +\infty)$, be a semigroup of continuous linear operators on H. Then the* infinitesimal generator *of S is the linear operator $A : D(A) \subset H \to H$ defined by*

$$D(A) = \left\{ x \in H; \lim_{t \to 0^+} \frac{S(t)x - x}{t} \text{ exists} \right\},$$

$$Ax = \lim_{t \to 0^+} \frac{S(t)x - x}{t}, \forall x \in D(A).$$

One has the following theorem (see, e.g., the proof of [**381**, Corollary 4.4, Chapter 1, page 15]).

THEOREM A.10. *Let us assume that A is densely defined and closed and that A and A^* are both dissipative. Then A is the infinitesimal generator of the strongly continuous semigroup associated to the operator A.*

The main properties of the infinitesimal generator of a strongly continuous semigroup of continuous linear operators on H are summarized in the following theorem (see [**381**, Theorem 2.4, Chapter 1, pages 4–5], [**381**, Corollary 10.6, Chapter 1, page 41], [**381**, Theorem 1.3, Chapter 4, page 102]).

THEOREM A.11. *Let $S(t)$, $t \in [0, +\infty)$, be a strongly continuous semigroup of continuous linear operators on H. Then*

(i) *The domain $D(A)$ is dense in H and A is a closed linear operator.*
(ii) *For every x^0 in $D(A)$, there exists one and only one*

$$x \in C^1([0, +\infty); H) \cap C^0([0, +\infty); D(A))$$

such that

$$x(t) \in D(A), \forall t \in [0, +\infty),$$
$$\frac{dx}{dt}(t) = Ax(t), \forall t \in [0, +\infty),$$
$$x(0) = x^0.$$

This solution satisfies

$$x(t) = S(t)x^0, \forall t \in [0, +\infty).$$

(iii) *$S(t)^*$, $t \in [0, +\infty)$, is a strongly continuous semigroup of continuous linear operators and the infinitesimal generator of this semigroup is the adjoint A^* of A.*

Some of the previous results and definitions can be extended to the case of groups of continuous linear operators. First, one has the following definitions (compare with Definition A.5, Definition A.6 and Definition A.9).

DEFINITION A.12. (See e.g. [**381**, Definition 6.1, page 22].) A one-parameter family $S(t)$, $t \in \mathbb{R}$, of continuous linear operators from H into H is a *group of continuous linear operators* on H if

$$S(0) = \text{Id},$$
$$S(t_1 + t_2) = S(t_1) \circ S(t_2), \forall (t_1, t_2) \in \mathbb{R}.$$

DEFINITION A.13. (See e.g. [**381**, Definition 6.1, Chapter 1, page 22].) Let $S(t)$, $t \in \mathbb{R}$, be a group of continuous linear operators from H into H. One says that S is a *strongly continuous* group of continuous linear operators on H if

$$\lim_{t \to 0} S(t)x = x, \forall x \in H.$$

DEFINITION A.14. (See e.g. [**381**, Definition 6.2, Chapter 1, page 22].) Let $S(t)$, $t \in \mathbb{R}$, be a group of continuous linear operators on H. Then the *infinitesimal generator of S* is the linear operator $A : D(A) \subset H \to H$ defined by

$$D(A) = \left\{ x \in H; \lim_{t \to 0} \frac{S(t)x - x}{t} \text{ exists} \right\},$$
$$Ax = \lim_{t \to 0} \frac{S(t)x - x}{t}, \forall x \in D(A).$$

The main properties of the infinitesimal generator of a strongly continuous group of continuous linear operators on H are summarized in the following theorem (see [**381**, Theorem 6.3, Chapter 1, page 23], and apply Theorem A.11 to $t \in [0, +\infty) \mapsto S(t)$ and to $t \in [0, +\infty) \mapsto S(-t)$).

THEOREM A.15. *Let $S(t)$, $t \in \mathbb{R}$, be a strongly continuous group of continuous linear operators on H. Then*

(i) *The domain $D(A)$ is dense in H and A is a closed linear operator.*

(ii) *For every x^0 in $D(A)$, there exists one and only one*
$$x \in C^1(\mathbb{R}; H) \cap C^0(\mathbb{R}; D(A))$$
such that
$$x(t) \in D(A), \forall t \in \mathbb{R},$$
$$\frac{dx}{dt}(t) = Ax(t), \forall t \in \mathbb{R},$$
$$x(0) = x^0.$$
This solution satisfies
$$x(t) = S(t)x^0, \forall t \in \mathbb{R}.$$

(iii) $S(t)^*$, $t \in \mathbb{R}$, *is a strongly continuous group of continuous linear operators and the infinitesimal generator of this semigroup is the adjoint A^* of A.*

One also has the following theorem (apply Theorem A.10 to A and $-A$).

THEOREM A.16. *Let us assume that A is densely defined and that $A = -A^*$. Then A is the infinitesimal generator of the strongly continuous group of isometries on H associated to the operator A.*

Of course, in Theorem A.16, the strongly continuous group of isometries associated to A is the group $S(t)$, $t \in \mathbb{R}$, of isometries of H such that, for every $x^0 \in D(A)$, $t \mapsto x(t) := S(t)x^0$ is in $C^1(\mathbb{R}; H) \cap C^0(\mathbb{R}; D(A))$ and is the solution of the Cauchy problem

$$\frac{\mathrm{d}x}{\mathrm{d}t} = Ax,$$
$$x(0) = x^0$$

(see page 374).

REMARK A.17. There are powerful generalizations to the nonlinear case of most of the results of this section. In this case, the operators A are no longer linear but maximal monotone. See, in particular, the book [**70**] by Haïm Brezis.

APPENDIX B

Degree theory

In this appendix we construct the degree of a map and prove the properties of the degree we use in this book. As an application of the degree, we also prove the Brouwer and Schauder fixed-point theorems that we also use in this book. For more properties and/or more applications of the degree, we refer to [333] by Noel Lloyd, [373] by Mitio Nagumo, [438, Chapter III] by Jacob Schwartz.

Throughout this appendix n is a given positive integer. Let us start with a theorem which defines the degree. Let D be the set of (ϕ, Ω, b) such that

1. Ω is a nonempty bounded open subset of $\mathbb{R}^n$,
2. ϕ is a continuous map from $\overline{\Omega}$ into $\mathbb{R}^n$,
3. $b \in \mathbb{R}^n \setminus \phi(\partial \Omega)$.

THEOREM B.1. *There exists a unique map*

$$\begin{aligned} \text{degree}: \quad D &\to \quad \mathbb{Z} \\ (\phi, \Omega, b) &\mapsto \text{degree}\,(\phi, \Omega, b) \end{aligned}$$

satisfying the following four properties:

(i) *For every nonempty bounded open subset Ω of $\mathbb{R}^n$ and for every $b \in \Omega$,*

$$\text{degree}\,(\text{Id}_n|_{\overline{\Omega}}, \Omega, b) = 1.$$

(We recall that Id_n denotes the identity map of $\mathbb{R}^n$.)

(ii) *Let $(\phi, \Omega, b) \in D$. Then, for every disjoint nonempty open subsets Ω_1 and Ω_2 of Ω such that*

$$b \notin \phi(\overline{\Omega} \setminus (\Omega_1 \cup \Omega_2)),$$

one has

(B.1) $$\text{degree}\,(\phi, \Omega, b) = \text{degree}\,(\phi|_{\overline{\Omega}_1}, \Omega_1, b) + \text{degree}\,(\phi|_{\overline{\Omega}_2}, \Omega_2, b).$$

(iii) *For every $(\phi, \Omega, b) \in D$, there exists $\varepsilon > 0$ such that, for every $\psi \in C^0(\overline{\Omega}; \mathbb{R}^n)$ satisfying*

$$\|\phi - \psi\|_{C^0(\overline{\Omega}; \mathbb{R}^n)} \leqslant \varepsilon,$$

one has

$$\text{degree}\,(\psi, \Omega, b) = \text{degree}\,(\phi, \Omega, b).$$

(iv) *For every $(\phi, \Omega, b) \in D$,*

$$\text{degree}\,(\phi, \Omega, b) = \text{degree}\,(\phi - b, \Omega, 0),$$

where $\phi - b$ is the map from $\overline{\Omega}$ into $\mathbb{R}^n$ defined by

$$(\phi - b)(x) = \phi(x) - b, \ \forall x \in \overline{\Omega}.$$

Moreover, this degree map is such that, for every $(\phi, \Omega, b) \in D$ satisfying

(B.2) $\qquad\qquad\qquad\qquad \phi$ *is of class C^1 in Ω,*

(B.3) $\qquad\qquad (\phi(x) = b,\ x \in \Omega) \Rightarrow (\phi'(x) \in \mathcal{L}(\mathbb{R}^n; \mathbb{R}^n)$ *is invertible*$)$,

one has

(B.4) $$\text{degree}\,(\phi, \Omega, b) = \sum_{x \in \phi^{-1}(b)} \text{Sign}\,(\det \phi'(x)),$$

with the convention that, if $\phi^{-1}(b)$ is empty, then the right hand side of (B.4) is 0.

Note that $(\phi, \Omega, b) \in D$, (B.2), (B.3) and the inverse function theorem imply that $\phi^{-1}(b) := \{x \in \overline{\Omega};\ \phi(x) = b\}$ is a discrete set and therefore a finite set since $\overline{\Omega}$ is compact. Hence the right hand side of (B.4) is well defined.

Proof of Theorem B.1. We only construct a map degree : $D \to \mathbb{Z}$ satisfying (i) to (iv), together with (B.4) if (B.2) and (B.3) hold. For the uniqueness statement, which is not used in this book, we refer to [**333**, Theorem 5.3.2 page 88].) Our proof is organized as follows:

- In Step 1, we define some integer $d_1(\phi, \Omega, b)$ when ϕ is of class C^1 in Ω and if b is a regular value, i.e., if (B.3) holds. This integer will turn out to be equal to degree (ϕ, Ω, b) under these assumptions.
- In Step 2, we prove a vanishing result for certain integrals.
- In Step 3, using this vanishing result, we define some integer $d_2(\phi, \Omega, b)$ by an integral formula when ϕ is of class C^1 in Ω. This integer will again turn out to be equal to degree (ϕ, Ω, b) under this smoothness assumption.
- Finally, in Step 4, we treat the case where ϕ is only continuous and define degree (ϕ, Ω, b).

Step 1: The regular case. In this step, $(\phi, \Omega, b) \in D$ is such that (B.2) and (B.3) hold. Then one defines $d_1(\phi, \Omega, b) \in \mathbb{Z}$ by

(B.5) $$d_1(\phi, \Omega, b) := \sum_{x \in \phi^{-1}(b)} \text{Sign}(\det \phi'(x)),$$

with the convention that,

(B.6) $\qquad$ if $\phi^{-1}(b)$ is empty, then the right hand side of (B.5) is 0.

(Note that, since we want to have (B.4), this is a reasonable choice!)

Step 2: An integral formula. In this step, $(\phi, \Omega, b) \in D$ is such that (B.2) holds. We do not assume that (B.3) holds. For $r > 0$, let $C(b, r)$ be the hypercube

$$C(b, r) := \{y := (y_1, \ldots, y_n)^{\text{tr}} \in \mathbb{R}^n;\ |y_i - b_i| < r,\ \forall i \in \{1, \ldots, n\}\},$$

where $(b_1, \ldots, b_n)^{\text{tr}} := b$. Let $\rho(\phi, \Omega, b)$ be the supremum of the set of $r > 0$ such that $C(b, r) \subset \mathbb{R}^n \setminus \phi(\partial\Omega)$. We define the hypercube $Q(\phi, \Omega, b)$ by

$$Q(\phi, \Omega, b) := C(b, \rho(\phi, \Omega, b)) \subset \mathbb{R}^n \setminus \phi(\partial\Omega).$$

Let us denote by J_ϕ the Jacobian determinant of ϕ, i.e., the determinant of ϕ'. One has the following proposition.

PROPOSITION B.2. *Let $F \in C^0(\mathbb{R}^n)$ be such that*

(B.7) $\qquad$ *the support of F is included in $Q(\phi, \Omega, b)$,*

(B.8) $$\int_{\mathbb{R}^n} F(y)dy = 0.$$

Then

(B.9) $$\int_\Omega F(\phi(x))J_\phi(x)dx = 0.$$

Proof of Proposition B.2. By density arguments, we may assume that

(B.10) $\qquad\qquad\qquad \phi$ is of class C^2 in Ω,

(B.11) $\qquad\qquad\qquad F \in C^1(\mathbb{R}^n)$.

Let us assume the following proposition, which has its own interest and will be proved later on.

PROPOSITION B.3. *Let $(a_i)_{i \in \{1,\ldots,n\}}$ and $(b_i)_{i \in \{1,\ldots,n\}}$ be two sequences of n real numbers such that*

$$a_i < b_i, \ \forall i \in \{1,\ldots,n\}.$$

Consider the hypercube $Q := \{y := (y_1, \ldots, y_n)^{tr}; y_i \in (a_i, b_i), \ \forall i \in \{1,\ldots,n\}\}$. Let $F \in C^1(\mathbb{R}^n)$ be such that (B.8) holds and

(B.12) $\qquad\qquad$ *the support of F is included in Q.*

Then there exists $Y = (Y_1, \ldots, Y_n)^{tr} \in C^1(\mathbb{R}^n; \mathbb{R}^n)$, $y := (y_1, \ldots, y_n)^{tr} \mapsto Y(y)$ such that

(B.13) $\qquad\qquad$ *the support of Y is included in Q,*

(B.14) $$\operatorname{div} Y := \sum_{i=1}^n \frac{\partial Y_i}{\partial y_i} = F.$$

We apply this proposition with $Q := Q(\phi, \Omega, b)$. Let

$$X := (X_1, \ldots, X_n)^{\operatorname{tr}} \in C^1(\Omega; \mathbb{R}^n)$$

be defined by

(B.15) $$\begin{aligned} X_1 &:= \det(Y(\phi), \phi_{x_2}, \phi_{x_3}, \ldots, \phi_{x_n}), \\ X_2 &:= \det(\phi_{x_1}, Y(\phi), \phi_{x_3}, \ldots, \phi_{x_n}), \\ &\vdots \\ X_n &:= \det(\phi_{x_1}, \phi_{x_2}, \ldots, \phi_{x_{n-1}}, Y(\phi)). \end{aligned}$$

Clearly,

(B.16) $\qquad$ the support of X is a compact subset of Ω.

Let us check that

(B.17) $\qquad\qquad \operatorname{div} X(x) = \operatorname{div} Y(\phi(x)) J_\phi(x), \ \forall x \in \Omega.$

Straightforward computations show that

$$\begin{aligned}
\text{div } X \; = \; & \det\left(Y'(\phi)\phi_{x_1}, \phi_{x_2}, \phi_{x_3}, \ldots, \phi_{x_n}\right) \\
& + \det\left(\phi_{x_1}, Y'(\phi)\phi_{x_2}, \phi_{x_3}, \ldots, \phi_{x_n}\right) \\
& \quad \vdots \\
& + \det\left(\phi_{x_1}, \phi_{x_2}, \ldots, \phi_{x_{n-1}}, Y'(\phi)\phi_{x_n}\right).
\end{aligned} \quad (B.18)$$

For $M \in \mathcal{M}_{n,n}(\mathbb{R})$ and $a := (a_1, \ldots, a_n) \in (\mathbb{R}^n)^n$, let

$$\begin{aligned}
\Delta(M, a) \; := \; & \det\left(Ma_1, a_2, a_3, \ldots, a_n\right) \\
& + \det\left(a_1, Ma_2, a_3, \ldots, a_n\right) \\
& \quad \vdots \\
& + \det\left(a_1, a_2, \ldots, a_{n-1}, Ma_n\right).
\end{aligned}$$

Clearly, $\Delta(M, \cdot)$ is multilinear in $a_1, a_2, \ldots, a_n$ and vanishes if two of the a_i's are equal. Hence there exists $K(M) \in \mathbb{R}$ such that

$$\Delta(M, a) = K(M) \det a, \; \forall a \in (\mathbb{R}^n)^n = \mathcal{M}_{n,n}(\mathbb{R}).$$

Taking for a the canonical basis of $\mathbb{R}^n$, one gets that $K(M)$ is the trace of M. Therefore,

(B.19) $\quad \Delta(M, a) = (\text{trace } M)(\det a), \; \forall a \in (\mathbb{R}^n)^n = \mathcal{M}_{n,n}(\mathbb{R}), \; \forall M \in \mathcal{M}_{n,n}(\mathbb{R}).$

Using (B.18) and applying (B.19) to $M := Y'(\phi)$ and $a_i := \phi_{x_i}$, $i \in \{1, \ldots, n\}$, one gets (B.17).

Let us point out that if O is a nonempty open subset of $\mathbb{R}$ and $\theta \in C^1(O)$ is such that its support is a compact subset of O, then

$$\int_O \theta'(x) dx = 0. \quad (B.20)$$

Indeed, it suffices to extend θ to $\mathbb{R}$ by requiring $\theta(x) = 0$ for every $x \in \mathbb{R} \setminus O$. This new θ is of class C^1 in $\mathbb{R}$ and has a compact support. In particular,

$$\int_\mathbb{R} \theta'(x) dx = 0,$$

which implies (B.20). Therefore, using (B.14), (B.16), (B.17) and the Fubini theorem, we get

$$\int_\Omega F(\phi(x)) J_\phi(x) dx = \int_\Omega \text{div } Y(\phi(x)) J_\phi(x) dx = \int_\Omega \text{div } X(x) dx = 0.$$

This concludes the proof of Proposition B.2 assuming Proposition B.3. ∎

Proof of Proposition B.3. Our proof will not use any geometrical tool. For a similar proof, but written in the language of differential forms, see, e.g. [51, Lemme 7.1.1, Chapitre 7, page 263] by Marcel Berger and Bernard Gostiaux (for an English translation, see [52]).

We proceed by induction on n. For $n = 1$, it suffices to take

$$Y(y) := \int_{-\infty}^{y} F(t) dt.$$

(By (B.8), the support of Y is included in $Q = (a_1, b_1)$.) We assume that Proposition B.3 holds for n and prove it for $n+1$. Let $(a_i)_{i \in \{1, \ldots, n\}}$ and $(b_i)_{i \in \{1, \ldots, n\}}$ be

two sequences of $n+1$ real numbers such that
$$a_i < b_i, \forall i \in \{1, \ldots, n, n+1\}.$$
Let $Q := \{y := (y_1, \ldots, y_n, y_{n+1})^{\text{tr}}; y_i \in (a_i, b_i), \forall i \in \{1, \ldots, n, n+1\}\}$. Let $F \in C^1(\mathbb{R}^{n+1})$ be such that (B.12) holds and
$$\int_{\mathbb{R}^{n+1}} F(y) dy = 0.$$
We write, for $y \in \mathbb{R}^{n+1}$,
$$y = \begin{pmatrix} \tilde{y} \\ y_{n+1} \end{pmatrix},$$
with $\tilde{y} \in \mathbb{R}^n$. Let $\tilde{F} \in C^1(\mathbb{R}^n)$ be defined by
$$\tilde{F}(\tilde{y}) := \int_{\mathbb{R}} F\begin{pmatrix} \tilde{y} \\ y_{n+1} \end{pmatrix} dy_{n+1}.$$
Then
$$\text{the support of } \tilde{F} \text{ is included in } \tilde{Q} \subset \mathbb{R}^n,$$
$$\int_{\mathbb{R}^n} \tilde{F}(\tilde{y}) d\tilde{y} = 0,$$
where $\tilde{Q} := \{y := (y_1, \ldots, y_n)^{\text{tr}}; y_i \in (a_i, b_i), \forall i \in \{1, \ldots, n\}\}$. Hence, applying the induction assumption, there exists $\tilde{Y} := (\tilde{Y}_1, \ldots, \tilde{Y}_n)^{\text{tr}} \in C^1(\mathbb{R}^n, \mathbb{R}^n)$, $\tilde{y} := (\tilde{y}_1, \ldots, \tilde{y}_n)^{\text{tr}} \mapsto \tilde{Y}(\tilde{y})$, such that
$$\text{the support of } \tilde{Y} \text{ is included in } \tilde{Q} \subset \mathbb{R}^n,$$
$$\text{div } \tilde{Y} := \sum_{i=1}^n \frac{\partial \tilde{Y}_i}{\partial \tilde{y}_i} = \tilde{F}.$$
Let $\rho \in C^1(\mathbb{R})$ be such that
$$\text{the support of } \rho \text{ is included in } (a_{n+1}, b_{n+1}),$$
$$\int_{\mathbb{R}} \rho(t) dt = 1.$$
Then, if we define $Y := (Y_1, \ldots, Y_n, Y_{n+1})^{\text{tr}} \in C^1(\mathbb{R}^{n+1}; \mathbb{R}^{n+1})$ by
$$Y_i \begin{pmatrix} \tilde{y} \\ y_{n+1} \end{pmatrix} := \rho(y_{n+1}) \tilde{Y}_i(\tilde{y}), \forall i \in \{1, \ldots, n\}, \forall \tilde{y} \in \mathbb{R}^n, \forall y_{n+1} \in \mathbb{R},$$
$$Y_{n+1}\begin{pmatrix} \tilde{y} \\ y_{n+1} \end{pmatrix} := \int_{-\infty}^{y_{n+1}} \left(F\begin{pmatrix} \tilde{y} \\ t \end{pmatrix} - \rho(t) \tilde{F}(\tilde{y}) \right) dt, \forall \tilde{y} \in \mathbb{R}^n, \forall y_{n+1} \in \mathbb{R},$$
one easily checks that
$$\text{the support of } Y \text{ is included in } Q,$$
$$\text{div } Y := \sum_{i=1}^{n+1} \frac{\partial Y_i}{\partial y_i} = F.$$
This concludes the proof of Proposition B.3. ∎

Step 3: The case of maps of class C^1. Throughout this step, $(\phi, \Omega, b) \in D$ and ϕ is assumed to be of class C^1 in Ω. Let $f \in C^0(\mathbb{R}^n)$ be such that

(B.21) \qquad the support of f is included in $Q(\phi, \Omega, b)$,

(B.22) $$\int_{\mathbb{R}^n} f(y)dy = 1.$$

One defines

(B.23) $$d_2(\phi, \Omega, b) := \int_\Omega f(\phi(x)) J_\phi(x) dx.$$

From Proposition B.2 we get that $d_2(\phi, \Omega, b)$ is independent of the choice of $f \in C^0(\mathbb{R}^n)$ satisfying (B.21) and (B.22).

One has the following proposition.

PROPOSITION B.4. *If (B.3) holds, then*

(B.24) $$d_1(\phi, \Omega, b) = d_2(\phi, \Omega, b).$$

Proof of Proposition B.4. As mentioned above on page 380, (B.3), which is assumed, implies that $\phi^{-1}(b)$ is a finite set. Let us first treat the case where $\phi^{-1}(b) = \emptyset$. Then, by (B.5) and (B.6),

(B.25) $$d_1(\phi, \Omega, b) = 0.$$

Moreover, there exists an open neighborhood V of b in $\mathbb{R}^n$ such that

(B.26) $$V \subset Q(\phi, \Omega, b), \ \phi(\overline{\Omega}) \cap V = \emptyset.$$

Let $f \in C^0(\mathbb{R}^n)$ be such that

(B.27) \qquad the support of f is included in V,

(B.28) $$\int_{\mathbb{R}^n} f(y)dy = 1.$$

From (B.23), (B.26), (B.27) and (B.28),

$$d_2(\phi, \Omega, b) = \int_\Omega f(\phi(x)) J_\phi(x) dx = 0,$$

which, together with (B.25), gives (B.24).

Let us now deal with the case where $\phi^{-1}(b)$ is not empty. One writes

$$\phi^{-1}(b) = \{x_1, x_2, \ldots, x_k\},$$

with $k \in \mathbb{N} \setminus \{0\}$ and $x_i \neq x_j$ if $i \neq j$. By the inverse function theorem, there exist k open subsets $\omega_1, \ldots, \omega_k$ of $\mathbb{R}^n$ and $\varepsilon > 0$ such that, defining $B(b, \varepsilon) := \{y \in \mathbb{R}^n; |y - b| < \varepsilon\}$,

$$x_i \in \omega_i, \ \forall i \in \{1, \ldots, k\},$$
$$\omega_i \cap \omega_j = \emptyset, \ \forall (i,j) \in \{1, \ldots, k\}^2 \text{ such that } i \neq j,$$
$$B(b, \varepsilon) \subset Q(\phi, \Omega, b),$$
$$\phi^{-1}(B(b, \varepsilon)) = \bigcup_{i=1}^k \omega_i,$$

$\phi|_{\omega_i}$ is a diffeomorphism between ω_i and $B(b, \varepsilon)$, $\forall i \in \{1, \ldots, k\}$.

Let $f \in C^0(\mathbb{R}^n)$ be such that
$$\text{support } f \subset B(b, \varepsilon),$$
$$\int_{\mathbb{R}^n} f(y) dy = 1.$$

One has, using (B.23) and the changes of variables $y = \phi|_{\omega_i}(x)$, where $x \in \omega_i$, $i \in \{1, \ldots, k\}$ and $y \in B(b, \varepsilon)$,

$$\begin{aligned}
d_2(\phi, \Omega, b) &= \int_\Omega f(\phi(x)) J_\phi(x) dx \\
&= \sum_{i=1}^k \int_{\omega_i} f(\phi(x)) J_\phi(x) dx \\
&= \sum_{i=1}^k \text{Sign}(J_\phi(x_i)) \int_{B(b,\varepsilon)} f(y) dy \\
&= \sum_{i=1}^k \text{Sign}(J_\phi(x_i)),
\end{aligned}$$

which, together with (B.5), concludes the proof of Proposition B.4. ∎

Our next proposition follows.

PROPOSITION B.5. *The map $b \in \mathbb{R}^n \setminus \phi(\partial\Omega) \to \mathbb{R}$, $b \mapsto d_2(\phi, \Omega, b)$, is continuous. Moreover, the real number $d_2(\phi, \Omega, b)$ is an integer.*

Proof of Proposition B.5. The continuity statement follows from the fact that if $f \in C^0(\mathbb{R}^n)$ has a support included in the bounded open set $Q(\phi, \Omega, b)$, then, for every $b' \in \mathbb{R}^n$ close enough to b, $f \in C^0(\mathbb{R}^n)$ has a support included in $Q(\phi, \Omega, b')$. In order to prove that the real number $d_2(\phi, \Omega, b)$ is an integer, it suffices to point out that, by the Sard theorem (see, e.g., [**164**, Theorem 3.4.3, page 316] or [**202**, Theorem 1.12 and Note (1) page 34]), there exists a sequence $(b_k)_{k \in \mathbb{N}}$ of elements of $\mathbb{R}^n \setminus \phi(\partial\Omega)$ such that

(B.29) $\quad (\phi(x) = b_k) \Rightarrow (\phi'(x) \in \mathcal{L}(R^n; \mathbb{R}^n) \text{ is invertible}), \forall k \in \mathbb{N},$

(B.30) $\quad b_k \to b$ as $k \to +\infty$.

By Proposition B.4 and the definition of d_1 (see (B.5) and (B.6)),
$$d_2(\phi, \Omega, b_k) = d_1(\phi, \Omega, b_k) \in \mathbb{Z},$$

which, together with (B.30) and the continuity of $d_2(\phi, \Omega, \cdot)$, shows that $d_2(\phi, \Omega, b) \in \mathbb{Z}$. This concludes the proof of Proposition B.5. ∎

The next proposition establishes the properties of d_2 analogous to Properties (i) to (iv) of the degree map $(\phi, \Omega, b) \in D \mapsto \text{degree}(\phi, \Omega, b)$ required in Theorem B.1 on page 379.

PROPOSITION B.6. *The map d_2 satisfies the following four properties.*

(i) For every nonempty bounded open subset Ω of $\mathbb{R}^n$ and for every $b \in \Omega$,
$$d_2(\text{Id}_n|_{\overline{\Omega}}, \Omega, b) = 1.$$

(ii) Let $(\phi, \Omega, b) \in D$ be such that ϕ is of class C^1 in Ω. Then, for every disjoint nonempty open subsets Ω_1 and Ω_2 of Ω such that

$$b \notin \phi(\overline{\Omega} \setminus (\Omega_1 \cup \Omega_2)),$$

one has

(B.31) $$d_2(\phi, \Omega, b) = d_2(\phi|_{\overline{\Omega}_1}, \Omega_1, b) + d_2(\phi|_{\overline{\Omega}_2}, \Omega_2, b).$$

(iii) Let Ω be a nonempty open subset of $\mathbb{R}^n$ and $b \in \mathbb{R}^n$. Let

$$H \in C^0([0,1] \times \overline{\Omega}; \mathbb{R}^n)$$

be such that

$$H(t,x) \neq b, \, \forall (t,x) \in [0,1] \times \partial\Omega,$$
$$H(t, \cdot) \text{ is of class } C^1 \text{ in } \Omega, \text{ for every } t \in [0,1],$$
$$\text{the map } (t,x) \in [0,1] \times \Omega \mapsto \frac{\partial H}{\partial x} \in \mathcal{L}(\mathbb{R}^n; \mathbb{R}^n) \text{ is continuous.}$$

Then

(B.32) $$d_2(H(0, \cdot), \Omega, b) = d_2(H(1, \cdot), \Omega, b).$$

(iv) For every $(\phi, \Omega, b) \in D$ such that ϕ is of class C^1 in Ω,

$$d_2(\phi, \Omega, b) = d_2(\phi - b, \Omega, 0).$$

Proof of Proposition B.6. Property (i) follows directly from (B.5) and Proposition B.4 on page 384. Property (ii) clearly holds if b is a regular value, i.e, if (B.3) holds. One deduces the general case from this particular case by using the continuity of d_2 with respect to b (Proposition B.5) and the density of the regular values in $\mathbb{R}^n$. (This density, as mentioned above, follows from the Sard theorem.) Concerning Property (iii), proceeding as in the proof of the continuity of d_2 with respect to b, one gets that

(B.33) $$t \in [0,1] \mapsto d_2(H(t, \cdot), \Omega, b) \text{ is continuous.}$$

Equality (B.32) then follows from (B.33) and the fact that, for every $t \in [0,1]$,

$$d_2(H(t, \cdot), \Omega, b) \in \mathbb{Z}$$

(see Proposition B.5). Finally, Property (iv) readily follows from the definition of d_2. This concludes the proof of Proposition B.6. ∎

Step 4: The case of maps of class C^0. Let $(\phi, \Omega, b) \in D$. Let

$$\delta(\phi, \Omega, b) := \text{dist } (b, \phi(\partial\Omega)) := \min\{|b - \phi(x)|; \, x \in \partial\Omega\}.$$

Since $(\phi, \Omega, b) \in D$, $\delta(\phi, \Omega, b) > 0$. Using standard smoothing processes, one gets the existence of $\psi \in C^0(\overline{\Omega}; \mathbb{R}^n)$ of class C^1 in Ω such that

(B.34) $$\|\phi - \psi\|_{C^0(\overline{\Omega})} < \delta(\phi, \Omega, b).$$

We define degree (ϕ, Ω, b) by

(B.35) $$\text{degree } (\phi, \Omega, b) := d_2(\psi, \Omega, b) \in \mathbb{Z}.$$

Our next lemma shows that this integer is independent of the choice of ψ.

LEMMA B.7. *Let ψ_i, $i \in \{1,2\}$, be two elements of $C^0(\overline{\Omega};\mathbb{R}^n)$ which are of class C^1 in Ω. Assume that*
$$\|\phi - \psi_i\|_{C^0(\overline{\Omega})} < \delta(\phi,\Omega,b), \ \forall i \in \{1,2\}.$$
Then
$$d_2(\psi_1,\Omega,b) = d_2(\psi_2,\Omega,b).$$

Proof of Lemma B.7. It suffices to apply Property (iii) of Proposition B.6 to the map $H \in C^0([0,1] \times \overline{\Omega};\mathbb{R}^n)$ defined by
$$H(t,x) := t\psi_1(x) + (1-t)\psi_2(x), \ \forall (t,x) \in [0,1] \times \overline{\Omega}.$$
∎

Note that, by taking $\psi = \phi$ if ϕ is of class C^1 in Ω, one has

(B.36) $\qquad \text{degree}\,(\phi,\Omega,b) = d_2(\phi,\Omega,b), \ \text{if } \phi \text{ is of class } C^1 \text{ in } \Omega.$

In particular, using (B.5) and Proposition B.4, one sees that (B.2) and (B.3) imply (B.4). It is easy to check that degree : $D \to \mathbb{Z}$ satisfies all the other properties mentioned in Theorem B.1 on page 379. This concludes the proof of this theorem.
∎

Let us now give some immediate applications of Theorem B.1, applications that we have previously used in this book. As a corollary of Property (iii) of Theorem B.1, one has the following proposition.

PROPOSITION B.8 (Homotopy invariance of the degree). *Let Ω be a nonempty open subset of $\mathbb{R}^n$, $H \in C^0([0,1] \times \overline{\Omega};\mathbb{R}^n)$, and $b \in \mathbb{R}^n$. We assume that*
$$H(t,x) \neq b, \ \forall (t,x) \in [0,1] \times \partial\Omega.$$
Then
$$\text{degree}\,(H(0,\cdot),\Omega,b) = \text{degree}\,(H(1,\cdot),\Omega,b).$$

Similarly, as a corollary of Properties (iii) and (iv) of Theorem B.1, one has the following proposition.

PROPOSITION B.9. *Let $(\phi,\Omega,b) \in D$. There exists $\varepsilon > 0$ such that, for every $b' \in \mathbb{R}^n$ such that $|b - b'| \leqslant \varepsilon$,*
$$(\phi,\Omega,b') \in D \ \text{and}\ \text{degree}\,(\phi,\Omega,b') = \text{degree}\,(\phi,\Omega,b).$$

As a consequence of Property (ii) of Theorem B.1 on page 379, one has the following two propositions (Propositions B.10 and B.11).

PROPOSITION B.10. *Let $(\phi,\Omega,b) \in D$ be such that $b \notin \phi(\Omega)$. Then*

(B.37) $\qquad \text{degree}\,(\phi,\Omega,b) = 0.$

Proof of Proposition B.10. We give a proof relying only on Property (ii) of Theorem B.1 (we could also give a simpler proof using our construction of the degree). Let ω_1, ω_2 and ω_3 be three nonempty disjoint open subsets of Ω. Writing (B.1) with $\Omega_1 := \omega_1$ and $\Omega_2 := \omega_2$, one has

(B.38) $\qquad \text{degree}\,(\phi,\Omega,b) = \text{degree}\,(\phi|_{\overline{\omega}_1},\omega_1,b) + \text{degree}\,(\phi|_{\overline{\omega}_2},\omega_2,b).$

Writing equality (B.1) with $\Omega_1 := \omega_1 \cup \omega_2$ and $\Omega_2 := \omega_3$, one has

(B.39) $\quad$ degree $(\phi, \Omega, b) = $ degree $(\phi|_{\overline{\omega_1 \cup \omega_2}}, \omega_1 \cup \omega_2, b) + $ degree $(\phi|_{\overline{\omega}_3}, \omega_2, b)$.

However, writing (B.1) for $(\phi, \Omega, b) := (\phi|_{\overline{\omega_1 \cup \omega_2}}, \omega_1 \cup \omega_2, b)$ and $(\Omega_1, \Omega_2) := (\omega_1, \omega_2)$, we get

$$\text{degree}\,(\phi|_{\overline{\omega_1 \cup \omega_2}}, \omega_1 \cup \omega_2, b) = \text{degree}\,(\phi|_{\overline{\omega}_1}, \omega_1, b) + \text{degree}\,(\phi|_{\overline{\omega}_2}, \omega_2, b),$$

which, together with (B.38) and (B.39), implies that

$$\text{degree}\,(\phi|_{\overline{\omega}_3}, \omega_3, b) = 0.$$

Similarly, one has

(B.40) $\quad$ degree $(\phi|_{\overline{\omega}_1}, \omega_1, b) = $ degree $(\phi|_{\overline{\omega}_2}, \omega_2, b) = 0$,

which, together with (B.38), concludes the proof of Proposition B.10. ∎

PROPOSITION B.11 (Excision property). *Let $(\phi, \Omega, b) \in D$. Let Ω_0 be a nonempty open subset of Ω such that $b \notin \phi(\Omega \setminus \Omega_0)$. Then*

(B.41) $\quad$ degree $(\phi, \Omega, b) = $ degree $(\phi|_{\overline{\Omega_0}}, \Omega_0, b)$.

Proof of Proposition B.11. Since $b \notin \phi(\partial \Omega)$, there exist disjoint nonempty open subsets Ω_1 and Ω_2 of Ω_0 such that

$$b \notin \phi(\Omega \setminus (\Omega_1 \cup \Omega_2)).$$

Using Property (ii) of Theorem B.1 for (ϕ, Ω, b) and the open sets Ω_1, Ω_2, one gets

(B.42) $\quad$ degree $(\phi, \Omega, b) = $ degree $(\phi|_{\overline{\Omega}_1}, \Omega_1, b) + $ degree $(\phi|_{\overline{\Omega}_2}, \Omega_2, b)$.

Using Property (ii) of Theorem B.1 for $(\phi|_{\overline{\omega_0}}, \Omega_0, b)$ and the open sets Ω_1, Ω_2, one gets

(B.43) $\quad$ degree $(\phi|_{\overline{\Omega_0}}, \Omega_0, b) = $ degree $(\phi|_{\overline{\Omega}_1}, \Omega_1, b) + $ degree $(\phi|_{\overline{\Omega}_2}, \Omega_2, b)$.

Equality (B.41) follows from (B.42) and (B.43). ∎

Let us now give two propositions that we have used in this book to compute the degree of a function. The first one is:

PROPOSITION B.12. *Let $(\phi, \Omega, b) \in D$. Let $A \in \mathcal{L}(\mathbb{R}^n; \mathbb{R}^n)$ be invertible. Then*

(B.44) $\quad$ degree $(A \circ \phi, \Omega, Ab) = $ Sign $(\det A)$ degree (ϕ, Ω, b).

REMARK B.13. From (B.4), one has, for every nonempty bounded open subset Ω of $\mathbb{R}^n$ and for every $b \notin \partial \Omega$,

$$\text{degree}\,(A|_{\overline{\Omega}}, \Omega, Ab) = \begin{cases} 0 & \text{if } b \notin \Omega, \\ \text{Sign}\,(\det A) & \text{if } b \in \Omega. \end{cases}$$

Hence (B.44) is a (very) special case of the multiplication theorem, due to Jean Leray [**313, 314**] (see, e.g., [**333**, Theorem 2.3.1, page 29] or [**438**, Theorem 3.20, page 95]), which gives the degree of the composition of two maps.

Proof of Proposition B.12. Using (B.4), one first proves (B.44) if ϕ is of class C^1 in Ω and if b is a regular value (i.e., if (B.3) holds). Then, using the continuity of degree $(\psi, \Omega, \cdot)$ (see Proposition B.9) and the Sard theorem as in the proof of Proposition B.5 on page 385, one deduces that (B.44) also holds if ϕ is of class C^1 in Ω even if b is not a regular value. Finally, the case where ϕ is only continuous on $\overline{\Omega}$ follows from this preceding case by a density argument and the continuity of the degree with respect to the function (Property (iii) of Theorem B.1 on page 379). ∎

The second proposition is:

PROPOSITION B.14. *Let n_1 and n_2 be two positive integers. Let Ω_1 (resp. Ω_2) be a nonempty bounded open subset of $\mathbb{R}^{n_1}$ (resp. $\mathbb{R}^{n_2}$). Let*

$$\Omega := \left\{ \begin{pmatrix} x_1 \\ x_2 \end{pmatrix} ; x_1 \in \Omega_1, x_2 \in \Omega_2 \right\} \subset \mathbb{R}^{n_1+n_2}.$$

Let $\phi \in C^0(\overline{\Omega_2}; \mathbb{R}^{n_1})$ and $\psi \in C^0(\overline{\Omega}; \mathbb{R}^{n_2})$. Let $b_1 \in \mathbb{R}^{n_1}$ and $b_2 \in \mathbb{R}^{n_2}$ be such that

$$\phi(\overline{\Omega_2}) + b_1 \subset \Omega_1,$$

$$\psi \begin{pmatrix} \phi(x_2) + b_1 \\ x_2 \end{pmatrix} \neq b_2, \forall x_2 \in \partial \Omega_2.$$

Let $b \in \mathbb{R}^{n_1+n_2}$ be defined by

$$b := \begin{pmatrix} b_1 \\ b_2 \end{pmatrix} \in \mathbb{R}^{n_1+n_2}.$$

Let $U \in C^0(\overline{\Omega}; \mathbb{R}^{n_1+n_2})$ be defined by

$$U \begin{pmatrix} x_1 \\ x_2 \end{pmatrix} := \begin{pmatrix} x_1 - \phi(x_2) \\ \psi \begin{pmatrix} x_1 \\ x_2 \end{pmatrix} \end{pmatrix}, \forall x_1 \in \overline{\Omega_1}, \forall x_2 \in \overline{\Omega_2}.$$

Finally, let $V \in C^0(\overline{\Omega_2}; \mathbb{R}^{n_2})$ be defined by

$$V(x_2) := \psi \begin{pmatrix} \phi(x_2) + b_1 \\ x_2 \end{pmatrix}, \forall x_2 \in \overline{\Omega_2}.$$

Then

(B.45) $$\text{degree}(U, \Omega, b) = \text{degree}(V, \Omega_2, b_2).$$

Proof of Proposition B.14. One proceeds as in the proof of Proposition B.12: One first treats the case where ϕ and ψ are of class C^1 in Ω_1 and Ω respectively and b is a regular value (for U). For this case, one points outs that if $(x_1, x_2) \in \overline{\Omega}_1 \times \overline{\Omega}_2$ and $x := (x_1, x_2)^{\text{tr}}$, one has

$$U'(x) = \begin{pmatrix} \text{Id}_{n_1} & -\phi'(x_2) \\ \dfrac{\partial \psi}{\partial x_1}(x) & \dfrac{\partial \psi}{\partial x_2}(x) \end{pmatrix}.$$

Hence
$$\det U'(x) = \det \begin{pmatrix} \mathrm{Id}_{n_1} & 0 \\ \dfrac{\partial \psi}{\partial x_1}(x) & \dfrac{\partial \psi}{\partial x_2}(x) + \dfrac{\partial \psi}{\partial x_1}(x)\phi'(x_2) \end{pmatrix}$$
$$= \det \left(\dfrac{\partial \psi}{\partial x_2}(x) + \dfrac{\partial \psi}{\partial x_1}(x)\phi'(x_2) \right).$$

In particular, if $x_1 = b_1 + \phi(x_2)$,

(B.46) $$\det U'(x) = \det V'(x_2).$$

Using (B.46), one readily gets (B.45) if ϕ and ψ are of class C^1 in Ω_1 and Ω respectively and b is a regular value (for U). Then one removes successively the assumption that b is a regular value and the assumption that ϕ and ψ are of class C^1 in Ω_1 and Ω respectively. ∎

Let us now deduce from Theorem B.1 on page 379, the homotopy invariance of the degree (Proposition B.8) and Proposition B.10, the Brouwer fixed-point theorem for a closed ball, which have been used in this book.

THEOREM B.15 (Brouwer fixed-point theorem for a closed ball). *Let $R \in (0,+\infty)$, $B_R := \{x \in \mathbb{R}^n; |x| < R\}$ and $\overline{B}_R = \{x \in \mathbb{R}^n; |x| \leqslant R\}$ be the closure of B_R in $\mathbb{R}^n$. Let $\mathcal{F}$ be a continuous map from $\overline{B}_R$ into itself. Then $\mathcal{F}$ has a fixed point.*

Proof of Theorem B.15. If $\mathcal{F}$ has a fixed point on ∂B_R, then $\mathcal{F}$ has a fixed point in $\overline{B}_R$. Therefore, we may assume that $\mathcal{F}$ has no fixed point on ∂B_R. Let $H : [0,1] \times \overline{B}_R \to \mathbb{R}^n$ be defined by
$$H(t,x) = x - t\mathcal{F}(x),\ \forall (t,x) \in [0,1] \times \overline{B}_R.$$
Note that $H(0,\cdot) = \mathrm{Id}_n|_{\overline{B}_R}$ and $H(1,\cdot) = \mathrm{Id}_n|_{\overline{B}_R} - \mathcal{F}$. Let us check that

(B.47) $$H(t,x) \neq 0,\ \forall (t,x) \in [0,1] \times \partial B_R.$$

Indeed, since, for every $x \in \overline{B}_R$, $|\mathcal{F}(x)| \leqslant R$,
$$(H(t,x) = 0 \text{ and } |x| = R) \Rightarrow (t = 1 \text{ and } \mathcal{F}(x) = x),$$
which, since $\mathcal{F}$ has no fixed point on ∂B_R, proves (B.47). Using (B.47) and the homotopy invariance of the degree (Proposition B.8), we have
$$\begin{aligned} \mathrm{degree}\,(\mathrm{Id}_n|_{\overline{B}_R}, B_R, 0) &= \mathrm{degree}\,(H(0,\cdot), B_R, 0) \\ &= \mathrm{degree}\,(H(1,\cdot), B_R, 0) \\ &= \mathrm{degree}\,(\mathrm{Id}_n|_{\overline{B}_R} - \mathcal{F}, B_R, 0), \end{aligned}$$
which, together with Property (i) of Theorem B.1 on page 379, implies that

(B.48) $$\mathrm{degree}\,(\mathrm{Id}_n|_{\overline{B}_R} - \mathcal{F}, B_R, 0) = 1.$$

However, from Proposition B.10 and (B.48), one gets
$$0 \in (\mathrm{Id}_n|_{\overline{B}_R} - \mathcal{F})(B_R),$$
i.e., $\mathcal{F}$ has a fixed point in B_R. This concludes the proof of Theorem B.15. ∎

More generally, one has the following theorem.

THEOREM B.16 (Brouwer fixed-point theorem). *Let K be a nonempty bounded closed convex subset of $\mathbb{R}^n$. Let $\mathcal{F}$ be a continuous map from K into itself. Then $\mathcal{F}$ has a fixed point.*

Proof of Theorem B.16. Since K is compact, there exists $R > 0$ such that
$$K \subset \overline{B}_R := \{x \in \mathbb{R}^n;\, x \in \mathbb{R}^n\}.$$
Let P_K be the projection on the nonempty closed convex subset K of $\mathbb{R}^n$. Let $\tilde{\mathcal{F}} : \overline{B}_R \to \mathbb{R}^n$ be defined by
$$\tilde{\mathcal{F}}(x) = \mathcal{F}(P_K(x)),\, \forall x \in \overline{B}_R.$$
One has $\tilde{\mathcal{F}}(\overline{B}_R) \subset K \subset \overline{B}_R$. Hence, using the Brouwer fixed point for a closed ball (Theorem B.15), $\tilde{\mathcal{F}}$ has a fixed point, i.e. there exists $x \in \overline{B}_R$ such that

(B.49) $$\tilde{\mathcal{F}}(x) = x.$$

However, (B.49) implies that $x \in \tilde{\mathcal{F}}(\overline{B}_R) \subset K$. Hence, since $\tilde{\mathcal{F}} = \mathcal{F}$ on K, $x \in K$ and is a fixed point of $\mathcal{F}$. This concludes the proof of Theorem B.16. ∎

Let us now prove two Schauder fixed-point theorems used in this book (see, e.g., [347] by Jean Mawhin or [512] by Eberhard Zeidler for many other applications of these theorems). The first one is the following.

THEOREM B.17. *Let E be a Banach space. Let B be either E or a nonempty (bounded) closed ball of E. Let $\mathcal{F}$ be a continuous map from B into itself. Let us assume that the image of $\mathcal{F}$ is included in a compact subset of E. Then $\mathcal{F}$ has a fixed point.*

Proof of Theorem B.17. Since a compact set is included in a (bounded) closed ball, we may consider only the case where $\mathcal{F}$ is a continuous map from a nonempty (bounded) closed ball B into itself. Let us denote by $\|\cdot\|_E$ the norm of the Banach space E. Let us assume, for the moment, that the following lemma holds.

LEMMA B.18. *For every positive integer n, there exist a finite dimensional vector subspace E_n and a continuous map $\mathcal{F}_n$ from B into $B \cap E_n$ such that*

(B.50) $$\|\mathcal{F}(x) - \mathcal{F}_n(x)\|_E \leqslant \frac{1}{n},\, \forall x \in B.$$

Applying the Brouwer fixed-point theorem (Theorem B.16) to $\mathcal{F}_n|_{B \cap E_n}$, there exists $x_n \in B \cap E_n$ such that

(B.51) $$\mathcal{F}_n(x_n) = x_n.$$

From (B.50) and (B.51), we get

(B.52) $$\|x_n - \mathcal{F}(x_n)\|_E \leqslant \frac{1}{n}.$$

Since $\mathcal{F}(B)$ is included in a compact subset of E, we may assume, without loss of generality, that $(\mathcal{F}(x_n))_{n \in \mathbb{N} \setminus \{0\}}$ is a convergent sequence. Then, by (B.52), $(x_n)_{n \in \mathbb{N} \setminus \{0\}}$ is also a convergent sequence, i.e., there exists $x \in B$ such that

(B.53) $$x_n \to x \text{ as } n \to +\infty.$$

Letting $n \to +\infty$ in (B.52) and using (B.53), we get that x is a fixed point of $\mathcal{F}$. This concludes the proof of Theorem B.17 assuming Lemma B.18. ■

It remains to prove Lemma B.18. Let n be given in $\mathbb{N} \setminus \{0\}$. Since $\mathcal{F}(B)$ is included in a compact set of B, there exist $k \in \mathbb{N} \setminus \{0\}$ and k elements $b_1, \ldots, b_k$ of B such that

$$\tag{B.54} \mathcal{F}(B) \subset \bigcup_{i=1}^{k} B(b_i, 1/n),$$

where, for $b \in E$ and $\rho > 0$, $B(b, \rho) := \{x \in E; \|x - b\|_E < \rho\}$. Let E_n be the finite dimensional vector subspace of E generated by the set $\{b_1, \ldots, b_k\}$. For $i \in \{1, \ldots, k\}$, let $\lambda_i \in C^0(B)$ be defined by

$$\lambda_i(x) := \max\left\{0, \frac{1}{n} - \|\mathcal{F}(x) - b_i\|_E\right\}, \forall x \in B.$$

By (B.54),

$$\sum_{j=1}^{k} \lambda_j(x) > 0, \forall x \in B.$$

Then, the map $\mathcal{F}_n : B \to E_n$ defined by

$$\mathcal{F}_n(x) := \sum_{i=1}^{k} \frac{\lambda_i(x)}{\sum_{j=1}^{k} \lambda_j(x)} b_i, \forall x \in B,$$

is well defined, continuous and takes its values in B. Moreover,

$$\tag{B.55} \begin{aligned} \|\mathcal{F}(x) - \mathcal{F}_n(x)\|_E &= \left\|\sum_{i=1}^{k} \frac{\lambda_i(x)}{\sum_{j=1}^{k} \lambda_j(x)} (\mathcal{F}(x) - b_i)\right\|_E \\ &\leqslant \sum_{i=1}^{k} \frac{\lambda_i(x)}{\sum_{j=1}^{k} \lambda_j(x)} \|\mathcal{F}(x) - b_i\|_E, \forall x \in B. \end{aligned}$$

However, since $\lambda_i(x) = 0$ if $\|\mathcal{F}(x) - b_i\|_E \geqslant 1/n$,

$$\lambda_i(x)\|\mathcal{F}(x) - b_i\|_E \leqslant \frac{1}{n}\lambda_i(x), \forall i \in \{1, \ldots, k\}, \forall x \in B,$$

which, together with (B.55), implies that

$$\|\mathcal{F}(x) - \mathcal{F}_n(x)\|_E \leqslant \frac{1}{n}, \forall x \in B.$$

This concludes the proof of Lemma B.18 and of Theorem B.17. ■

The second Schauder fixed-point theorem is the following one.

THEOREM B.19. *Let E be a Banach space, K a nonempty compact convex subset of E, and $\mathcal{F}$ a continuous map from K into K. Then $\mathcal{F}$ has a fixed point.*

Proof of Theorem B.19. Proceeding as in the proof of the Brouwer fixed-point theorem (Theorem B.16), we get Theorem B.19 as a corollary of the previous Schauder fixed-point theorem (Theorem B.17 on the preceding page) and of the following extension theorem due to James Dugundji [**151**].

THEOREM B.20. *Let E be a normed space and C a convex subset of E. Let X be a metric space, Y a nonempty closed subset of X, and f a continuous map from Y into E such that $f(Y) \subset C$. Then there exists a continuous map $\tilde{f}$ from X into E such that*

(B.56) $$\tilde{f}(y) = f(y), \, \forall y \in Y,$$

(B.57) $$\tilde{f}(X) \subset C.$$

REMARK B.21. Theorem B.20 generalizes the Tietze-Urysohn extension theorem, which deals with the case $E = \mathbb{R}$.

Indeed, let us apply Theorem B.20 to $E := E$, $C := K$, $X := E$, $Y := K$ and $f := \mathcal{F}$. Since K is compact, there exists $R > 0$ such that $K \subset \overline{B}_R := \{x \in E; \|x\|_E \leqslant R\}$, where $\|\cdot\|_E$ denotes the norm on the Banach space E. Then, if $\tilde{\mathcal{F}} := \tilde{f}|_{\overline{B}_R}$, $\tilde{\mathcal{F}}$ is a continuous map from $\overline{B}_R$ into itself and $\tilde{\mathcal{F}}(\overline{B}_R)$ is included in the compact set K. Hence, using the previous Schauder fixed-point theorem (Theorem B.17), $\tilde{\mathcal{F}}$ has a fixed point, i.e., there exists $x \in \overline{B}_R$ such that

(B.58) $$\tilde{\mathcal{F}}(x) = x.$$

However, (B.58) implies that $x \in \tilde{\mathcal{F}}(\overline{B}_R) \subset K$. Hence, since $\tilde{\mathcal{F}} = \mathcal{F}$ on K, $x \in K$ and is a fixed point of $\mathcal{F}$. This concludes the proof of Theorem B.16 assuming Theorem B.20.

Proof of Theorem B.20. Let us denote by d the distance of the metric space X and by $\|\cdot\|_E$ the norm of E. For a nonempty subset A of X, let us denote by $d(\cdot, A)$ the distance to A:

$$d(x, A) := \inf\{d(x, a); a \in A\}.$$

Let $z \in X \setminus Y$. Since Y is a closed subset of X, $d(z, Y) > 0$. Let ω_z be the open ball of radius of $d(z, Y)/2$ centered at z:

(B.59) $$\omega_z := \left\{ x \in X; d(z, x) < \frac{d(z, Y)}{2} \right\}.$$

Since $z \in \omega_z \subset X \setminus Y$,

$$X \setminus Y = \bigcup_{z \in X \setminus Y} \omega_z.$$

One can associate to this covering of $X \setminus Y$ by open subsets a partition of unity, i.e., there exist a set I of indices and a family $(\lambda_i)_{i \in I}$ of continuous functions from $X \setminus Y$ into $[0, 1]$ such that the following three properties hold.

(i) For every $z \in X \setminus Y$, there exist an open neighborhood V_z of z and a finite subset I_z of I such that

$$(z' \in V_z \text{ and } i \in I \setminus I_z) \Rightarrow \lambda_i(z') = 0.$$

(ii) For every $z \in X \setminus Y$,

(B.60) $$\sum_{i \in I} \lambda_i(z) = 1.$$

(Note that, by (i), the sum in (B.60) has only a finite number of terms which are not 0: For $i \in I \setminus I_z$, $\lambda_i(z) = 0$.)

(iii) For every $i \in I$, there exists $z_i \in X \setminus Y$ such that the support of λ_i is included in ω_{z_i}.

(See, for example, [**65**, Corollaire 1, Section 4 and Théorème 4, Section 5], or [**158**] together with [**418**].) For each $z \in X \setminus Y$, let us choose $y_z \in Y$ such that

(B.61) $$d(y_z, \omega_z) < 2 \inf\{d(y, \omega_z); y \in Y\}.$$

Then, one defines $\tilde{f}$ by

(B.62) $$\tilde{f}(x) = \begin{cases} f(x) & \text{if } x \in Y, \\ \sum_{i \in I} \lambda_i(x) f(y_{z_i}) & \text{if } x \in X \setminus Y. \end{cases}$$

Clearly, $\tilde{f}$ satisfies (B.56), (B.57), and is continuous in $X \setminus \partial Y$. Let $y \in \partial Y$. The only nonobvious property to check is that, if $(x_n)_{n \in \mathbb{N}}$ is a sequence of elements in $X \setminus Y$ converging to y as $n \to \infty$, then $\tilde{f}(x_n) \to f(y)$ as $n \to \infty$. Let $x \in X \setminus Y$ and let $i \in I$ be such that $\lambda_i(x) \neq 0$. Then, by (iii), $x \in \omega_{z_i}$ and $z_i \in X \setminus Y$. We have, using (B.59) and (B.61),

(B.63) $$\begin{aligned} d(y_{z_i}, x) & \leqslant d(y_{z_i}, \omega_{z_i}) + 2 \times \text{the radius of the ball } \omega_{z_i} \\ & \leqslant 2 \inf\{d(y', \omega_{z_i}); y' \in Y\} + d(z_i, Y) \\ & \leqslant 2d(y, x) + d(z_i, Y). \end{aligned}$$

Moreover, using again (B.59), we get

$$d(z_i, Y) \leqslant d(x, Y) + d(x, z_i) \leqslant d(x, Y) + \frac{d(z_i, Y)}{2},$$

which implies that

(B.64) $$d(z_i, Y) \leqslant 2d(x, Y) \leqslant 2d(x, y).$$

From (B.63) and (B.64), we get

(B.65) $$d(y_{z_i}, y) \leqslant d(y_{z_i}, x) + d(x, y) \leqslant 4d(x, y) + d(x, y) = 5d(x, y).$$

Coming back to the definition of $\tilde{f}$, we obtain, using (B.60) and (B.65),

$$\|\tilde{f}(x) - f(y)\|_E \leqslant \sum_{i \in I} \lambda_i(x) \|f(y_{z_i}) - f(y)\|_E \leqslant \delta(x),$$

with

$$\delta(x) := \sup\{\|f(y') - f(y)\|_E; y' \in Y, d(y', y) \leqslant 5d(x, y)\}.$$

Since $\delta(x) \to 0$ as $x \to y$, this concludes the proof of Theorem B.20. ∎

REMARK B.22. If X is a Hilbert space (or more generally a uniformly convex Banach space) and Y is a nonempty closed convex subset of X, one can simply define $\tilde{f}$ by

$$\tilde{f}(x) := f(P_Y(x)), \forall x \in X,$$

where P_Y is the projection on Y.

Let us end this appendix with the two following exercises.

EXERCISE B.23. *We take $n = 1$.*

1. *In this question Ω is a nonempty bounded open interval:*

$$\Omega := (\alpha, \beta), \text{ with } -\infty < \alpha < \beta < +\infty.$$

Let $\phi \in C^0(\overline{\Omega}) = C^0(\overline{\Omega};\mathbb{R})$ and let $b \in \mathbb{R} \setminus \{\phi(\alpha), \phi(\beta)\}$. Prove that

(B.66) $\quad \text{degree}\,(\phi, \Omega, b) = \begin{cases} 1 & \text{if } \phi(\alpha) < b < \phi(\beta), \\ -1 & \text{if } \phi(\beta) < b < \phi(\alpha), \\ 0 & \text{if } b \notin (\phi(\alpha), \phi(\beta)) \cup (\phi(\beta), \phi(\alpha)). \end{cases}$

2. In this question Ω is a nonempty bounded open subset of $\mathbb{R}$. It is the countable disjoint union of its connected components. In other words, there exist a set I which is either finite or equal to $\mathbb{N}$, two families of real numbers $(\alpha_i)_{i \in I}$, $(\beta_i)_{i \in I}$ such that

$$\alpha_i < \beta_i, \forall i \in I,$$
$$(\alpha_i, \beta_i) \cap (\alpha_j, \beta_j) = \emptyset, \forall (i, j) \in I^2 \text{ such that } i \neq j,$$
$$\Omega = \bigcup_{i \in I} (\alpha_i, \beta_i).$$

Prove that
$$\partial \Omega = \overline{\bigcup_{i \in I} \{\alpha_i, \beta_i\}}.$$

Let $\phi \in C^0(\overline{\Omega})$ and $b \in \mathbb{R} \setminus \phi(\partial \Omega)$. Let
$$I(b) := \{i \in I; b \in \phi((\alpha_i, \beta_i))\}.$$
Prove that $I(b)$ is a finite set and that
$$\text{degree}\,(\phi, \Omega, b) = \sum_{i \in I(b)} \text{degree}\,(\phi|_{[\alpha_i, \beta_i]}, (\alpha_i, \beta_i), b).$$

EXERCISE B.24 (Generalization of the integral formula (B.36)). Let $(\phi, \Omega, b) \in D$ be such that ϕ is of class C^1 in Ω. Let $f \in C^0(\mathbb{R}^n)$ satisfying (B.28) be such that the support of f is included in the connected component of $\mathbb{R}^n \setminus \phi(\partial \Omega)$ containing b. Prove that
$$\text{degree}\,(\phi, \Omega, b) = \int_\Omega f(\phi(x)) J_\phi(x) dx.$$

(**Hint.** Note that, by Proposition B.9 on page 387, $\text{degree}\,(\phi, \Omega, \cdot)$ is constant on the connected component of $\mathbb{R}^n \setminus \phi(\partial \Omega)$ containing b. Use a suitable partition of unity, (B.23) with f adapted to the partition of unity, and (B.36). For partitions of unity on open subsets of $\mathbb{R}^n$, see, e.g., [**342**, Théorème 1.4.1, page 61] or [**419**, Theorem 6.20, page 147].)

Bibliography

1. Ole Morten Aamo and Miroslav Krstić, *Flow control by feedback, stabilization and mixing*, Springer-Verlag, London, Berlin, Heidelberg, 2003.
2. Ralph Abraham and Jerrold E. Marsden, *Foundations of mechanics*, Benjamin/Cummings Publishing Co. Inc. Advanced Book Program, Reading, Mass., 1978, Second edition, revised and enlarged, With the assistance of Tudor Raţiu and Richard Cushman. MR 515141 (81e:58025)
3. Robert A. Adams, *Sobolev spaces*, Academic Press [A subsidiary of Harcourt Brace Jovanovich, Publishers], Elsevier, Oxford, 2005, Pure and Applied Mathematics, Vol. 140, Second edition. MR 0450957 (56 #9247)
4. Dirk Aeyels and Jacques L. Willems, *Pole assignment for linear time-invariant systems by periodic memoryless output feedback*, Automatica J. IFAC **28** (1992), no. 6, 1159–1168. MR 1196779 (93j:93058)
5. Andrei A. Agrachev, *Newton diagrams and tangent cones to attainable sets*, Analysis of controlled dynamical systems (Lyon, 1990), Progr. Systems Control Theory, vol. 8, Birkhäuser Boston, Boston, MA, 1991, pp. 11–20. MR 1131980 (92h:93012)
6. Andrei A. Agrachev and Revaz V. Gamkrelidze, *Local controllability and semigroups of diffeomorphisms*, Acta Appl. Math. **32** (1993), no. 1, 1–57. MR 1232941 (94i:93009)
7. _____, *Local controllability for families of diffeomorphisms*, Systems Control Lett. **20** (1993), no. 1, 67–76. MR 1198474 (94f:93008)
8. Andrei A. Agrachev and Yuri L. Sachkov, *Control theory from the geometric viewpoint*, Encyclopaedia of Mathematical Sciences, vol. 87, Springer-Verlag, Berlin, 2004, Control Theory and Optimization, II. MR 2062547 (2005b:93002)
9. Andrei A. Agrachev and Andrei V. Sarychev, *Navier-Stokes equations: controllability by means of low modes forcing*, J. Math. Fluid Mech. **7** (2005), no. 1, 108–152. MR 2127744
10. Serge Alinhac and Patrick Gérard, *Opérateurs pseudo-différentiels et théorème de Nash-Moser*, Savoirs Actuels. [Current Scholarship], InterEditions, Paris, 1991. MR 1172111 (93g:35001)
11. Fabio Ancona, *Decomposition of homogeneous vector fields of degree one and representation of the flow*, Ann. Inst. H. Poincaré Anal. Non Linéaire **13** (1996), no. 2, 135–169. MR 1378464 (97c:34065)
12. Fabio Ancona and Alberto Bressan, *Patchy vector fields and asymptotic stabilization*, ESAIM Control Optim. Calc. Var. **4** (1999), 445–471 (electronic). MR 1693900 (2000d:93053)
13. _____, *Flow stability of patchy vector fields and robust feedback stabilization*, SIAM J. Control Optim. **41** (2002), no. 5, 1455–1476 (electronic). MR 1971958 (2004d:93123)
14. _____, *Stability rates for patchy vector fields*, ESAIM Control Optim. Calc. Var. **10** (2004), no. 2, 168–200 (electronic). MR 2083482 (2005d:34016)
15. Fabio Ancona and Giuseppe Maria Coclite, *On the attainable set for Temple class systems with boundary controls*, SIAM J. Control Optim. **43** (2005), no. 6, 2166–2190 (electronic). MR 2179483 (2006f:93008)
16. Fabio Ancona and Andrea Marson, *On the attainable set for scalar nonlinear conservation laws with boundary control*, SIAM J. Control Optim. **36** (1998), no. 1, 290–312. MR 1616586 (99h:93008)
17. Brian D. O. Anderson, Robert R. Bitmead, C. Richard Johnson, Jr., Petar V. Kokotović, Robert L. Kosut, Iven M. Y. Mareels, Laurent Praly, and Bradley D. Riedle, *Stability of adaptive systems*, MIT Press Series in Signal Processing, Optimization, and Control, 8, MIT Press, Cambridge, MA, 1986, Passivity and averaging systems. MR 846209 (87i:93001)

18. Brigitte d'Andréa-Novel, *Commande non linéaire des robots*, Traité des nouvelles technologies, série Automatique, Hermès, Paris, 1988.
19. Brigitte d'Andréa-Novel and Michel Cohen de Lara, *Commande linéaire des systèmes dynamiques*, Modélisation. Analyse. Simulation. Commande. [Modeling. Analysis. Simulation. Control], Masson, Paris, 1994, With a preface by A. Bensoussan. MR 1272681 (95b:93001)
20. Zvi Artstein, *Stabilization with relaxed controls*, Nonlinear Anal. **7** (1983), no. 11, 1163–1173. MR 721403 (85h:93054)
21. Jean-Pierre Aubin, *Un théorème de compacité*, C. R. Acad. Sci. Paris **256** (1963), 5042–5044. MR 0152860 (27 #2832)
22. Sergei A. Avdonin and Sergei A. Ivanov, *Families of exponentials*, Cambridge University Press, Cambridge, 1995, The method of moments in controllability problems for distributed parameter systems, Translated from the Russian and revised by the authors. MR 1366650 (97b:93002)
23. Andrea Bacciotti, *Local stabilizability of nonlinear control systems*, Series on Advances in Mathematics for Applied Sciences, vol. 8, World Scientific Publishing Co. Inc., River Edge, NJ, 1992. MR 1148363 (92k:93005)
24. Andrea Bacciotti and Lionel Rosier, *Liapunov functions and stability in control theory*, second ed., Communications and Control Engineering Series, Springer-Verlag, Berlin, 2005. MR 2146587 (2005m:93001)
25. John Baillieul and Mark Levi, *Rotational elastic dynamics*, Phys. D **27** (1987), no. 1-2, 43–62. MR 912850 (89a:70009)
26. _____, *Constrained relative motions in rotational mechanics*, Arch. Rational Mech. Anal. **115** (1991), no. 2, 101–135. MR 1106071 (92g:73058)
27. Claudio Baiocchi, Vilmos Komornik, and Paola Loreti, *Ingham-Beurling type theorems with weakened gap conditions*, Acta Math. Hungar. **97** (2002), no. 1-2, 55–95. MR 1932795 (2003i:42011)
28. John M. Ball, Jerrold E. Marsden, and Marshall Slemrod, *Controllability for distributed bilinear systems*, SIAM J. Control Optim. **20** (1982), no. 4, 575–597. MR 661034 (84h:49079)
29. Viorel Barbu, *Feedback stabilization of Navier-Stokes equations*, ESAIM Control Optim. Calc. Var. **9** (2003), 197–206 (electronic). MR 1957098 (2003i:93063)
30. Viorel Barbu, Irena Lasiecka, and Roberto Triggiani, *Tangential boundary stabilization of Navier-Stokes equations*, Mem. Amer. Math. Soc. **181** (2006), no. 852, x+128. MR 2215059
31. Viorel Barbu and Roberto Triggiani, *Internal stabilization of Navier-Stokes equations with finite-dimensional controllers*, Indiana Univ. Math. J. **53** (2004), no. 5, 1443–1494. MR 2104285 (2005k:93169)
32. Claude Bardos, François Golse, and C. David Levermore, *Fluid dynamic limits of kinetic equations. I. Formal derivations*, J. Statist. Phys. **63** (1991), no. 1-2, 323–344. MR 1115587 (92d:82079)
33. _____, *Fluid dynamic limits of kinetic equations. II. Convergence proofs for the Boltzmann equation*, Comm. Pure Appl. Math. **46** (1993), no. 5, 667–753. MR 1213991 (94g:82039)
34. Claude Bardos, Gilles Lebeau, and Jeffrey Rauch, *Sharp sufficient conditions for the observation, control, and stabilization of waves from the boundary*, SIAM J. Control Optim. **30** (1992), no. 5, 1024–1065. MR 1178650 (94b:93067)
35. Guy Barles, *Solutions de viscosité des équations de Hamilton-Jacobi*, Mathématiques & Applications (Berlin) [Mathematics & Applications], vol. 17, Springer-Verlag, Paris, 1994. MR 1613876 (2000b:49054)
36. Tamer Başar and Pierre Bernhard, H^∞-*optimal control and related minimax design problems*, second ed., Systems & Control: Foundations & Applications, Birkhäuser Boston Inc., Boston, MA, 1995, A dynamic game approach. MR 1353236 (96f:93002)
37. Georges Bastin, Jean-Michel Coron, Brigitte d'Andréa-Novel, and Luc Moens, *Boundary control for exact cancellation of boundary disturbances in hyperbolic systems of conservation laws*, IEEE Conf. on Dec. and Cont. and Eur. Cont. Conf. (CDC-ECC'05), Sevilla, Spain, 2005.
38. George Keith Batchelor, *An introduction to fluid dynamics*, paperback ed., Cambridge Mathematical Library, Cambridge University Press, Cambridge, 1999. MR 1744638 (2000j:76001)
39. A. I. Baz', Yakov B. Zel'dovich, and Askold M. Perelomov, *Scattering, reactions and decay in non-relativistic quantum mechanics*, Nauka, Moscow (Russian), 1971.

40. Karine Beauchard, *Local controllability of a 1-D Schrödinger equation*, J. Math. Pures Appl. (9) **84** (2005), no. 7, 851–956. MR 2144647
41. _____, *Local controllability of a 1D beam equation*, Preprint, CMLA, Submitted for publication (2005).
42. _____, *Controllability of a quantum particle in a 1D infinite square potential with variable length*, ESAIM Control Optim. Calc. Var. (to appear).
43. Karine Beauchard and Jean-Michel Coron, *Controllability of a quantum particle in a moving potential well*, J. Funct. Anal. **232** (2006), no. 2, 328–389. MR 2200740
44. Karine Beauchard, Jean-Michel Coron, Mazyar Mirrahimi, and Pierre Rouchon, *Stabilization of a finite dimensional Schrödinger equation*, Systems Control Lett. (to appear in 2007).
45. Jerrold Bebernes and David Eberly, *Mathematical problems from combustion theory*, Applied Mathematical Sciences, vol. 83, Springer-Verlag, New York, 1989. MR 1012946 (91d:35165)
46. Lotfi Beji, Azgal Abichou, and Yasmina Bestaoui, *Position and attitude control of an underactuated autonomous airship*, Internat. J. Diff. Equations Appl. **8** (2003), no. 3, 231–255.
47. Alain Bensoussan, *Stochastic control of partially observable systems*, Cambridge University Press, Cambridge, 1992. MR 1191160 (93i:93001)
48. _____, *An introduction to the Hilbert uniqueness method*, Analysis and optimization of systems: state and frequency domain approaches for infinite-dimensional systems (Sophia-Antipolis, 1992), Lecture Notes in Control and Inform. Sci., vol. 185, Springer, Berlin, 1993, pp. 184–198. MR 1208270 (94h:93041)
49. Alain Bensoussan, Giuseppe Da Prato, Michel C. Delfour, and Sanjoy K. Mitter, *Representation and control of infinite-dimensional systems. Vol. I*, Systems & Control: Foundations & Applications, Birkhäuser Boston Inc., Boston, MA, 1992. MR 1182557 (94b:49003)
50. _____, *Representation and control of infinite-dimensional systems. Vol. II*, Systems & Control: Foundations & Applications, Birkhäuser Boston Inc., Boston, MA, 1993. MR 1246331 (94m:49001)
51. Marcel Berger and Bernard Gostiaux, *Géométrie différentielle*, Librairie Armand Colin, Paris, 1972, Maîtrise de mathématiques, Collection U/Série "Mathématiques". MR 0494180 (58 #13102)
52. _____, *Differential geometry: manifolds, curves, and surfaces*, Graduate Texts in Mathematics, vol. 115, Springer-Verlag, New York, 1988, Translated from the French by Silvio Levy. MR 917479 (88h:53001)
53. Arne Beurling, *The collected works of Arne Beurling. Vol. 2*, Contemporary Mathematicians, Birkhäuser Boston Inc., Boston, MA, 1989, Harmonic analysis, Edited by L. Carleson, P. Malliavin, J. Neuberger and J. Wermer. MR 1057614 (92k:01046b)
54. Sanjay P. Bhat and Dennis S. Bernstein, *Geometric homogeneity with applications to finite-time stability*, Math. Control Signals Systems **17** (2005), no. 2, 101–127. MR 2150956 (2006c:34093)
55. Rosa Maria Bianchini, *High order necessary optimality conditions*, Rend. Sem. Mat. Univ. Politec. Torino **56** (1998), no. 4, 41–51 (2001), Control theory and its applications (Grado, 1998). MR 1845743 (2002f:49041)
56. Rosa Maria Bianchini and Gianna Stefani, *Sufficient conditions for local controllability*, Proc. 25th IEEE Conf. Decision and Control, Athens (1986), 967–970.
57. _____, *Controllability along a trajectory: a variational approach*, SIAM J. Control Optim. **31** (1993), no. 4, 900–927. MR 1227538 (94d:93009)
58. Anthony Bloch and Sergey Drakunov, *Stabilization and tracking in the nonholonomic integrator via sliding modes*, Systems Control Lett. **29** (1996), no. 2, 91–99. MR 1420406 (97i:93082)
59. Anthony M. Bloch and Edriss S. Titi, *On the dynamics of rotating elastic beams*, New trends in systems theory (Genoa, 1990), Progr. Systems Control Theory, vol. 7, Birkhäuser Boston, Boston, MA, 1991, pp. 128–135. MR 1125101
60. Jerry Bona and Ragnar Winther, *The Korteweg-de Vries equation, posed in a quarter-plane*, SIAM J. Math. Anal. **14** (1983), no. 6, 1056–1106. MR 718811 (85c:35076)
61. Bernard Bonnard, *Contrôle de l'attitude d'un satellite rigide*, R.A.I.R.O. Automatique/ Systems Analysis and Control **16** (1982), 85–93.
62. Bernard Bonnard and Monique Chyba, *Singular trajectories and their role in control theory*, Mathématiques & Applications (Berlin) [Mathematics & Applications], vol. 40, Springer-Verlag, Berlin, 2003. MR 1996448 (2004f:93001)

63. Bernard Bonnard, Ludovic Faubourg, and Trélat Emmanuel, *Mécanique céleste et contrôle des véhicules spatiaux*, Mathématiques & Applications (Berlin) [Mathematics & Applications], vol. 51, Springer-Verlag, Berlin, 2006.
64. Ugo Boscain and Benedetto Piccoli, *Optimal syntheses for control systems on 2-D manifolds*, Mathématiques & Applications (Berlin) [Mathematics & Applications], vol. 43, Springer-Verlag, Berlin, 2004. MR 2031058
65. Nicolas Bourbaki, *Éléments de mathématique. I: Les structures fondamentales de l'analyse. Fascicule VIII. Livre III: Topologie générale. Chapitre 9: Utilisation des nombres réels en topologie générale*, Deuxième édition revue et augmentée. Actualités Scientifiques et Industrielles, No. 1045, Hermann, Paris, 1958. MR 0173226 (30 #3439)
66. Frédéric Bourquin, *Approximation for the fast stabilization of the wave equation from the boundary*, Proceedings of MMAR2000, Poland, 2000.
67. Frédéric Bourquin, Michel Joly, Manuel Collet, and Louis Ratier, *An efficient feedback control algorithm for beams: experimental investigations*, Journal of Sound and Vibration **278** (2004), 181–206.
68. Alberto Bressan, *Hyperbolic systems of conservation laws*, Oxford Lecture Series in Mathematics and its Applications, vol. 20, Oxford University Press, Oxford, 2000, The one-dimensional Cauchy problem. MR 1816648 (2002d:35002)
69. Alberto Bressan and Giuseppe Maria Coclite, *On the boundary control of systems of conservation laws*, SIAM J. Control Optim. **41** (2002), no. 2, 607–622 (electronic). MR 1920513 (2003f:93007)
70. Haïm Brezis, *Opérateurs maximaux monotones et semi-groupes de contractions dans les espaces de Hilbert*, North-Holland Publishing Co., Amsterdam, 1973, North-Holland Mathematics Studies, No. 5. Notas de Matemática (50). MR 0348562 (50 #1060)
71. _____, *Analyse fonctionnelle*, Collection Mathématiques Appliquées pour la Maîtrise. [Collection of Applied Mathematics for the Master's Degree], Masson, Paris, 1983, Théorie et applications. [Theory and applications]. MR 697382 (85a:46001)
72. Jean-Séverin Briffaut, *Méthodes numériques pour le contrôle et la stabilisation rapide des grandes structures flexibles*, PhD Thesis, École Nationale des Ponts et Chaussées (advisors: Frédéric Bourquin and Roland Glowinski) (1999).
73. Roger W. Brockett, *Volterra series and geometric control theory*, Automatica—J. IFAC **12** (1976), no. 2, 167–176. MR 0398607 (53 #2458)
74. _____, *Asymptotic stability and feedback stabilization*, Differential geometric control theory (Houghton, Mich., 1982) (R.W. Brockett, R.S. Millman, and H.J. Sussmann, eds.), Progr. Math., vol. 27, Birkhäuser Boston, Boston, MA, 1983, pp. 181–191. MR 708502 (85e:93034)
75. Roger W. Brockett and Elmer G. Gilbert, *An addendum to: "Volterra series and geometric control theory" (Automatica–J. IFAC **12** (1976), no. 2, 167–176) by Brockett*, Automatica–J. IFAC **12** (1976), no. 6, 635. MR 0432276 (55 #5265)
76. Nicolas Burq, *Contrôlabilité exacte des ondes dans des ouverts peu réguliers*, Asymptot. Anal. **14** (1997), no. 2, 157–191. MR 1451210 (98e:93005)
77. Nicolas Burq and Patrick Gérard, *Condition nécessaire et suffisante pour la contrôlabilité exacte des ondes*, C. R. Acad. Sci. Paris Sér. I Math. **325** (1997), no. 7, 749–752. MR 1483711 (98j:93052)
78. Anatoliy G. Butkovskiy and Yu. I. Samoĭlenko, *Control of quantum-mechanical processes and systems*, Mathematics and its Applications (Soviet Series), vol. 56, Kluwer Academic Publishers Group, Dordrecht, 1990. MR 1070712 (91f:93001)
79. Christopher I. Byrnes and Alberto Isidori, *New results and examples in nonlinear feedback stabilization*, Systems Control Lett. **12** (1989), no. 5, 437–442. MR 1005310 (90f:93038)
80. _____, *On the attitude stabilization of rigid spacecraft*, Automatica J. IFAC **27** (1991), no. 1, 87–95. MR 1087144
81. Piermarco Cannarsa and Carlo Sinestrari, *Semiconcave functions, Hamilton-Jacobi equations, and optimal control*, Progress in Nonlinear Differential Equations and their Applications, 58, Birkhäuser Boston Inc., Boston, MA, 2004. MR 2041617 (2005e:49001)
82. Carlos Canudas-de Wit (ed.), *Advanced robot control*, Lecture Notes in Control and Information Sciences, vol. 162, Berlin, Springer-Verlag, 1991. MR 1180966 (93d:70002)
83. Torsten Carleman, *Sur un problème d'unicité pur les systèmes d'équations aux dérivées partielles à deux variables indépendantes*, Ark. Mat., Astr. Fys. **26** (1939), no. 17, 9. MR 0000334 (1,55f)

84. Thierry Cazenave and Alain Haraux, *Introduction aux problèmes d'évolution semi-linéaires*, Mathématiques & Applications (Paris) [Mathematics and Applications], vol. 1, Ellipses, Paris, 1990. MR 1299976 (95f:35002)
85. Eduardo Cerpa, *Exact controllability of a nonlinear Korteweg-de Vries equation on a critical spatial domain*, submitted to SIAM J. Control Optim. (2006).
86. A. Chang, *An algebraic characterization of controllability*, IEEE Trans. Automat. Control **68** (1965), 112–113.
87. Marianne Chapouly, *Global controllability of nonviscous Burgers type equations*, Preprint, Université de Paris-Sud 11 (2006).
88. Jacques Chazarain and Alain Piriou, *Introduction à la théorie des équations aux dérivées partielles linéaires*, Gauthier-Villars, Paris, 1981. MR 598467 (82i:35001)
89. Hua Chen and Luigi Rodino, *General theory of PDE and Gevrey classes*, General theory of partial differential equations and microlocal analysis (Trieste, 1995), Pitman Res. Notes Math. Ser., vol. 349, Longman, Harlow, 1996, pp. 6–81. MR 1429633 (98e:35004)
90. Kuo-Tsai Chen, *Integration of paths, geometric invariants and a generalized Baker-Hausdorff formula*, Ann. of Math. (2) **65** (1957), 163–178. MR 0085251 (19,12a)
91. Yacine Chitour, Jean-Michel Coron, and Mauro Garavello, *On conditions that prevent steady-state controllability of certain linear partial differential equations*, Discrete Contin. Dyn. Syst. **14** (2006), no. 4, 643–672. MR 2177090 (2006e:93007)
92. Yacine Chitour, Jean-Michel Coron, and Laurent Praly, *Une nouvelle approche pour le transfert orbital à l'aide de moteurs ioniques*, CNES (1997).
93. Yacine Chitour and Emmanuel Trélat, *Controllability of partial differential equations*, Advanced topics in control systems theory (F. Lamnabhi-Lagarrigue, A. Loría, and E. Panteley, eds.), Lecture Notes in Control and Information Sciences, vol. 328, Springer-Verlag London Ltd., London, 2006, Lecture notes from the Graduate School on Automatic Control (FAP) held in Paris, February–March 2005, pp. 171–198.
94. Wei-Liang Chow, *Über Systeme von linearen partiellen Differentialgleichungen erster Ordnung*, Math. Ann. **117** (1939), 98–105. MR 0001880 (1,313d)
95. Marco Cirinà, *Boundary controllability of nonlinear hyperbolic systems*, SIAM J. Control **7** (1969), 198–212. MR 0254408 (40 #7617)
96. Francis H. Clarke, *Necessary conditions in dynamic optimization*, Mem. Amer. Math. Soc. **173** (2005), no. 816, x+113. MR 2117692
97. Francis H. Clarke, Yuri S. Ledyaev, Ludovic Rifford, and Ronald J. Stern, *Feedback stabilization and Lyapunov functions*, SIAM J. Control Optim. **39** (2000), no. 1, 25–48 (electronic). MR 1780907 (2001f:93056)
98. Francis H. Clarke, Yuri S. Ledyaev, Eduardo D. Sontag, and Andrei I. Subbotin, *Asymptotic controllability implies feedback stabilization*, IEEE Trans. Automat. Control **42** (1997), no. 10, 1394–1407. MR 1472857 (98g:93003)
99. Francis H. Clarke, Yuri S. Ledyaev, and Ronald J. Stern, *Asymptotic stability and smooth Lyapunov functions*, J. Differential Equations **149** (1998), no. 1, 69–114. MR 1643670 (99k:34109)
100. Francis H. Clarke, Yuri S. Ledyaev, Ronald J. Stern, and Peter R. Wolenski, *Nonsmooth analysis and control theory*, Graduate Texts in Mathematics, vol. 178, Springer-Verlag, New York, 1998. MR 1488695 (99a:49001)
101. François Coron, *Derivation of slip boundary conditions for the Navier-Stokes system from the Boltzmann equation*, J. Statist. Phys. **54** (1989), no. 3-4, 829–857. MR 988561 (90c:76093)
102. Jean-Michel Coron, *A necessary condition for feedback stabilization*, Systems Control Lett. **14** (1990), no. 3, 227–232. MR 1049357
103. _____, *Global asymptotic stabilization for controllable systems without drift*, Math. Control Signals Systems **5** (1992), no. 3, 295–312. MR 1164379 (93m:93084)
104. _____, *Contrôlabilité exacte frontière de l'équation d'Euler des fluides parfaits incompressibles bidimensionnels*, C. R. Acad. Sci. Paris Sér. I Math. **317** (1993), no. 3, 271–276. MR 1233425 (94g:93067)
105. _____, *Links between local controllability and local continuous stabilization*, IFAC Nonlinear Control Systems Design (Bordeaux, France, 24-26 June 1992) (Michel Fliess, ed.), 1993, pp. 165–171.
106. _____, *Linearized control systems and applications to smooth stabilization*, SIAM J. Control Optim. **32** (1994), no. 2, 358–386. MR 1261144 (95a:93105)

107. _____, *On the stabilization of controllable and observable systems by an output feedback law*, Math. Control Signals Systems **7** (1994), no. 3, 187–216. MR 1359027 (96j:93052)
108. _____, *Relations entre commandabilité et stabilisations non linéaires*, Nonlinear partial differential equations and their applications. Collège de France Seminar, Vol. XI (Paris, 1989–1991), Pitman Res. Notes Math. Ser., vol. 299, Longman Sci. Tech., Harlow, 1994, pp. 68–86. MR 1268900 (95c:93064)
109. _____, *On the stabilization in finite time of locally controllable systems by means of continuous time-varying feedback law*, SIAM J. Control Optim. **33** (1995), no. 3, 804–833. MR 1327239 (96f:93064)
110. _____, *Stabilizing time-varying feedback*, IFAC Nonlinear Control Systems Design (Tahoe, USA, 26-28 June 1995), 1995.
111. _____, *On the controllability of the 2-D incompressible Navier-Stokes equations with the Navier slip boundary conditions*, ESAIM Control Optim. Calc. Var. **1** (1995/96), 35–75 (electronic). MR 1393067 (97e:93005)
112. _____, *On the controllability of 2-D incompressible perfect fluids*, J. Math. Pures Appl. (9) **75** (1996), no. 2, 155–188. MR 1380673 (97b:93010)
113. _____, *Stabilization of controllable systems*, Sub-Riemannian geometry (André Bellaïche and Jean-Jacques Risler, eds.), Progr. Math., vol. 144, Birkhäuser, Basel, 1996, pp. 365–388. MR 1421826 (97h:93068)
114. _____, *On the null asymptotic stabilization of the two-dimensional incompressible Euler equations in a simply connected domain*, SIAM J. Control Optim. **37** (1999), no. 6, 1874–1896. MR 1720143 (2000j:93084)
115. _____, *Sur la stabilisation des fluides parfaits incompressibles bidimensionnels*, Seminaire: Équations aux Dérivées Partielles, 1998–1999, Sémin. Équ. Dériv. Partielles, École Polytech., Palaiseau, 1999, pp. Exp. No. VII, 17. MR 1721325 (2000j:93083)
116. _____, *Local controllability of a 1-D tank containing a fluid modeled by the shallow water equations*, ESAIM Control Optim. Calc. Var. **8** (2002), 513–554, A tribute to J. L. Lions. MR 1932962 (2004a:93009)
117. _____, *On the small-time local controllability of a quantum particle in a moving one-dimensional infinite square potential well*, C. R. Math. Acad. Sci. Paris **342** (2006), no. 2, 103–108. MR 2193655 (2006g:93006)
118. _____, *Some open problems on the control of nonlinear partial differential equations*, Perspectives in Nonlinear Partial Differential Equations: In honor of Haim Brezis (Henri Berestycki, Michiel Bertsch, Bert Peletier, and Laurent Véron, eds.), Contemporary Mathematics, American Mathematical Society, Providence, RI, to appear in 2007.
119. Jean-Michel Coron and Brigitte d'Andréa-Novel, *Stabilization of a rotating body beam without damping*, IEEE Trans. Automat. Control **43** (1998), no. 5, 608–618. MR 1618052 (2000e:93048)
120. Jean-Michel Coron, Brigitte d'Andréa-Novel, and Georges Bastin, *A Lyapunov approach to control irrigation canals modeled by the Saint Venant equations*, European Control Conference (Karlruhe, Germany, September 1999), no. F1008-5, 1999.
121. _____, *A strict Lyapunov function for boundary control of hyperbolic systems of conservation laws*, IEEE Trans. Automat. Control (to appear in 2007).
122. Jean-Michel Coron and Emmanuelle Crépeau, *Exact boundary controllability of a nonlinear KdV equation with critical lengths*, J. Eur. Math. Soc. (JEMS) **6** (2004), no. 3, 367–398. MR 2060480 (2005b:93016)
123. Jean-Michel Coron and Andrei V. Fursikov, *Global exact controllability of the 2D Navier-Stokes equations on a manifold without boundary*, Russian J. Math. Phys. **4** (1996), no. 4, 429–448. MR 1470445 (98k:93057)
124. Jean-Michel Coron and Sergio Guerrero, *Singular optimal control: a linear 1-D parabolic-hyperbolic example*, Asymptot. Anal. **44** (2005), no. 3-4, 237–257. MR 2176274
125. Jean-Michel Coron, Jonathan de Halleux, Georges Bastin, and Brigitte d'Andréa-Novel, *On boundary control design for quasi-linear hyperbolic systems with entropies as Lyapunov functions*, Proceedings 41-th IEEE Conference on Decision and Control, Las Vegas, USA, December 2002, 2002, pp. 3010–3014.
126. Jean-Michel Coron and El-Yazid Keraï, *Explicit feedbacks stabilizing the attitude of a rigid spacecraft with two control torques*, Automatica J. IFAC **32** (1996), no. 5, 669–677. MR 1392057

127. Jean-Michel Coron and Jean-Baptiste Pomet, *A remark on the design of time-varying stabilizing feedback laws for controllable systems without drift*, IFAC Nonlinear Control Systems Design (Bordeaux, France, 24-26 June 1992) (Michel Fliess, ed.), 1993, pp. 397–401.
128. Jean-Michel Coron and Laurent Praly, *Adding an integrator for the stabilization problem*, Systems Control Lett. **17** (1991), no. 2, 89–104. MR 1120754 (92f:93099)
129. _____, *Transfert orbital à l'aide de moteurs ioniques*, Science & Tec/PF/R 1442 (1996).
130. Jean-Michel Coron, Laurent Praly, and Andrew Teel, *Feedback stabilization of nonlinear systems: sufficient conditions and Lyapunov and input-output techniques*, Trends in control (Rome, 1995), Springer, Berlin, 1995, pp. 293–348. MR 1448452 (98e:93083)
131. Jean-Michel Coron and Lionel Rosier, *A relation between continuous time-varying and discontinuous feedback stabilization*, J. Math. Systems Estim. Control **4** (1994), no. 1, 67–84. MR 1298548 (95h:93068)
132. Jean-Michel Coron and Emmanuel Trélat, *Global steady-state controllability of one-dimensional semilinear heat equations*, SIAM J. Control Optim. **43** (2004), no. 2, 549–569 (electronic). MR 2086173 (2005f:93009)
133. _____, *Global steady-state stabilization and controllability of 1D semilinear wave equations*, Commun. Contemp. Math. **8** (2006), no. 4, 535–567.
134. Richard Courant and David Hilbert, *Methods of mathematical physics. Vol. II*, Wiley Classics Library, John Wiley & Sons Inc., New York, 1989, Partial differential equations, Reprint of the 1962 original, A Wiley-Interscience Publication. MR 1013360 (90k:35001)
135. Michael G. Crandall and Pierre-Louis Lions, *Viscosity solutions of Hamilton-Jacobi equations*, Trans. Amer. Math. Soc. **277** (1983), no. 1, 1–42. MR 690039 (85g:35029)
136. Emmanuelle Crépeau, *Exact boundary controllability of the Korteweg-de Vries equation around a non-trivial stationary solution*, Internat. J. Control **74** (2001), no. 11, 1096–1106. MR 1848887 (2002d:93007)
137. Peter E. Crouch, *Spacecraft attitude control and stabilization: applications of geometric control theory to rigid body models*, IEEE Trans. Automatic Control **AC-29** (1984), no. 4, 321–331.
138. Ruth F. Curtain and George Weiss, *Well posedness of triples of operators (in the sense of linear systems theory)*, Control and estimation of distributed parameter systems (Vorau, 1988), Internat. Ser. Numer. Math., vol. 91, Birkhäuser, Basel, 1989, pp. 41–59. MR 1033051 (91d:93027)
139. Ruth F. Curtain and Hans Zwart, *An introduction to infinite-dimensional linear systems theory*, Texts in Applied Mathematics, vol. 21, Springer-Verlag, New York, 1995. MR 1351248 (96i:93001)
140. René Dáger and Enrique Zuazua, *Wave propagation, observation and control in 1-d flexible multi-structures*, Mathématiques & Applications (Berlin) [Mathematics & Applications], vol. 50, Springer-Verlag, Berlin, 2006. MR 2169126
141. Robert Dautray and Jacques-Louis Lions, *Analyse mathématique et calcul numérique pour les sciences et les techniques. Vol. 5*, INSTN: Collection Enseignement. [INSTN: Teaching Collection], Masson, Paris, 1988, Spectre des opérateurs. [The operator spectrum], With the collaboration of Michel Artola, Michel Cessenat, Jean Michel Combes and Bruno Scheurer, Reprinted from the 1984 edition. MR 944303 (89m:00003a)
142. Wijesuriya P. Dayawansa and Clyde F. Martin, *Two examples of stabilizable second order systems*, Computation and Control (K. Bowers and J. Lund, eds.), Birkhäuser, Boston, 1989, pp. 53–63.
143. Wijesuriya P. Dayawansa, Clyde F. Martin, and Sandra L. Samelson, Samelson, *Asymptotic stabilization of a generic class of three-dimensional homogeneous quadratic systems*, Systems Control Lett. **24** (1995), no. 2, 115–123. MR 1312348 (95k:93080)
144. Lokenath Debnath, *Nonlinear water waves*, Academic Press Inc., Boston, MA, 1994. MR 1266390 (95c:76011)
145. Jesús Ildefonso Díaz, *Sobre la controlabilidad approximada de problemas no lineales disipativos*, Proceedings of *Jornadas Hispano-Francesas sobre Control de Sistemas Distribuidos* (A. Valle, ed.), 1990, pp. 41–48.
146. Szymon Dolecki and David L. Russell, *A general theory of observation and control*, SIAM J. Control Optimization **15** (1977), no. 2, 185–220. MR 0451141 (56 #9428)

147. Anna Doubova and Enrique Fernández-Cara, *Some control results for simplified one-dimensional models of fluid-solid interaction*, Math. Models Methods Appl. Sci. **15** (2005), no. 5, 783–824. MR 2139944 (2006d:93018)
148. Ronald G. Douglas, *On majorization, factorization, and range inclusion of operators on Hilbert space*, Proc. Amer. Math. Soc. **17** (1966), 413–415. MR 0203464 (34 #3315)
149. John C. Doyle, Bruce A. Francis, and Allen R. Tannenbaum, *Feedback control theory*, Macmillan Publishing Company, New York, 1992. MR 1200235 (93k:93002)
150. François Dubois, Nicolas Petit, and Pierre Rouchon, *Motion planning and nonlinear simulations for a tank containing a fluid*, European Control Conference (Karlruhe, Germany, September 1999), 1999.
151. James Dugundji, *An extension of Tietze's theorem*, Pacific J. Math. **1** (1951), 353–367. MR 0044116 (13,373c)
152. William B. Dunbar, Nicolas Petit, Pierre Rouchon, and Philippe Martin, *Motion planning for a nonlinear Stefan problem*, ESAIM Control Optim. Calc. Var. **9** (2003), 275–296 (electronic). MR 1966534 (2004c:80007)
153. Rachida El Assoudi, Jean-Paul Gauthier, and Ivan Kupka, *On subsemigroups of semisimple Lie groups*, Ann. Inst. H. Poincaré Anal. Non Linéaire **13** (1996), no. 1, 117–133. MR 1373474 (96m:22030)
154. Abdelhaq El Jaï and Anthony J. Pritchard, *Sensors and controls in the analysis of distributed systems*, Ellis Horwood Series: Mathematics and its Applications, Ellis Horwood Ltd., Chichester, 1988, Translated from the French by Catrin Pritchard and Rhian Pritchard. MR 933327 (89b:93002)
155. Caroline Fabre, *Uniqueness results for Stokes equations and their consequences in linear and nonlinear control problems*, ESAIM Contrôle Optim. Calc. Var. **1** (1995/96), 267–302 (electronic). MR 1418484 (97k:35193)
156. Caroline Fabre, Jean-Pierre Puel, and Enrique Zuazua, *Approximate controllability of the semilinear heat equation*, Proc. Roy. Soc. Edinburgh Sect. A **125** (1995), no. 1, 31–61. MR 1318622 (96a:93007)
157. Nicolas Chung Siong Fah, *Input-to-state stability with respect to measurement disturbances for one-dimensional systems*, ESAIM Control Optim. Calc. Var. **4** (1999), 99–121 (electronic). MR 1680756 (2000a:93109)
158. Albert Fathi, *Partitions of unity for countable covers*, Amer. Math. Monthly **104** (1997), no. 8, 720–723. MR 1476756
159. Hector O. Fattorini, *Control in finite time of differential equations in Banach space*, Comm. Pure Appl. Math. **19** (1966), 17–34. MR 0194223 (33 #2436)
160. _____, *Infinite-dimensional optimization and control theory*, Encyclopedia of Mathematics and its Applications, vol. 62, Cambridge University Press, Cambridge, 1999. MR 1669395 (2000d:49001)
161. _____, *Infinite dimensional linear control systems*, North-Holland Mathematics Studies, vol. 201, Elsevier Science B.V., Amsterdam, 2005, The time optimal and norm optimal problems. MR 2158806
162. Hector O. Fattorini and David L. Russell, *Exact controllability theorems for linear parabolic equations in one space dimension*, Arch. Rational Mech. Anal. **43** (1971), 272–292. MR 0335014 (48 #13332)
163. Ludovic Faubourg and Jean-Baptiste Pomet, *Control Lyapunov functions for homogeneous "Jurdjevic-Quinn" systems*, ESAIM Control Optim. Calc. Var. **5** (2000), 293–311 (electronic). MR 1765428 (2001g:93060)
164. Herbert Federer, *Geometric measure theory*, Die Grundlehren der mathematischen Wissenschaften, Band 153, Springer-Verlag New York Inc., New York, 1969. MR 0257325 (41 #1976)
165. Luis A. Fernández and Enrique Zuazua, *Approximate controllability for the semilinear heat equation involving gradient terms*, J. Optim. Theory Appl. **101** (1999), no. 2, 307–328. MR 1684673 (2000j:93015)
166. Enrique Fernández-Cara and Sergio Guerrero, *Remarks on the null controllability of the Burgers equation*, C. R. Math. Acad. Sci. Paris **341** (2005), no. 4, 229–232. MR 2164677 (2006b:93025)

167. Enrique Fernández-Cara, Sergio Guerrero, Oleg Yu. Imanuvilov, and Jean-Pierre Puel, *Local exact controllability of the Navier-Stokes system*, J. Math. Pures Appl. (9) **83** (2004), no. 12, 1501–1542. MR 2103189 (2005g:93013)
168. Enrique Fernández-Cara and Enrique Zuazua, *The cost of approximate controllability for heat equations: the linear case*, Adv. Differential Equations **5** (2000), no. 4-6, 465–514. MR 1750109 (2001e:93007)
169. _____, *Null and approximate controllability for weakly blowing up semilinear heat equations*, Ann. Inst. H. Poincaré Anal. Non Linéaire **17** (2000), no. 5, 583–616. MR 1791879 (2001j:93009)
170. Alexey F. Filippov, *Differential equations with discontinuous right-hand side*, Mat. Sb. (N.S.) **51 (93)** (1960), 99–128. MR 0114016 (22 #4846)
171. Franco Flandoli, *Hyperbolic dynamics with boundary control*, Distributed parameter systems (Vorau, Styria, 1986) (F. Kappel, K. Kunisch, and W. Schappacher, eds.), Lecture notes in control and information sciences, vol. 102, Springer Verlag, Berlin, Heidelberg, 1987, pp. 89–111.
172. Franco Flandoli, Irena Lasiecka, and Roberto Triggiani, *Algebraic Riccati equations with non-smoothing observation arising in hyperbolic and Euler-Bernoulli boundary control problems*, Ann. Mat. Pura Appl. (4) **153** (1988), 307–382 (1989). MR 1008349 (90f:49003)
173. Michel Fliess, *Séries de Volterra et séries formelles non commutatives*, C. R. Acad. Sci. Paris Sér. A-B **280** (1975), Aii, A965–A967. MR 0381784 (52 #2673)
174. _____, *Fonctionnelles causales non linéaires et indéterminées non commutatives*, Bull. Soc. Math. France **109** (1981), no. 1, 3–40. MR 613847 (82h:93037)
175. Michel Fliess, Jean Lévine, Philippe Martin, and Pierre Rouchon, *Flatness and defect of non-linear systems: introductory theory and examples*, Internat. J. Control **61** (1995), no. 6, 1327–1361. MR 1613557
176. Halina Frankowska, *An open mapping principle for set-valued maps*, J. Math. Anal. Appl. **127** (1987), no. 1, 172–180. MR 904219 (88i:49011)
177. _____, *Local controllability of control systems with feedback*, J. Optim. Theory Appl. **60** (1989), no. 2, 277–296. MR 984985 (90b:93009)
178. Randy A. Freeman and Petar V. Kokotović, *Robust nonlinear control design*, Systems & Control: Foundations & Applications, Birkhäuser Boston Inc., Boston, MA, 1996, State-space and Lyapunov techniques. MR 1396307 (98b:93001)
179. Bernard Friedland, *Control system design: An introduction to state-space methods*, Dover Publications, NY, Mineola, 2005.
180. Hongchen Fu, Sonia G. Schirmer, and Allan I. Solomon, *Complete controllability of finite-level quantum systems*, J. Phys. A **34** (2001), no. 8, 1679–1690. MR 1818760 (2001j:93011)
181. Andrei V. Fursikov, *Exact boundary zero controllability of three-dimensional Navier-Stokes equations*, J. Dynam. Control Systems **1** (1995), no. 3, 325–350. MR 1354539 (96m:93015)
182. _____, *Optimal control of distributed systems. Theory and applications*, Translations of Mathematical Monographs, vol. 187, American Mathematical Society, Providence, RI, 2000, Translated from the 1999 Russian original by Tamara Rozhkovskaya. MR 1726442 (2000m:49001)
183. _____, *Stabilization for the 3D Navier-Stokes system by feedback boundary control*, Discrete Contin. Dyn. Syst. **10** (2004), no. 1-2, 289–314, Partial differential equations and applications. MR 2026196 (2004k:93107)
184. Andrei V. Fursikov and Oleg Yu. Imanuvilov, *On exact boundary zero-controllability of two-dimensional Navier-Stokes equations*, Acta Appl. Math. **37** (1994), no. 1-2, 67–76, Mathematical problems for Navier-Stokes equations (Centro, 1993). MR 1308746 (95m:93008)
185. _____, *On controllability of certain systems simulating a fluid flow*, Flow control (Minneapolis, MN, 1992), IMA Vol. Math. Appl., vol. 68, Springer, New York, 1995, pp. 149–184. MR 1348646 (96f:93013)
186. _____, *Controllability of evolution equations*, Lecture Notes Series, vol. 34, Seoul National University Research Institute of Mathematics Global Analysis Research Center, Seoul, 1996. MR 1406566 (97g:93002)
187. _____, *Local exact controllability of the Navier-Stokes equations*, C. R. Acad. Sci. Paris Sér. I Math. **323** (1996), no. 3, 275–280. MR 1404773 (97k:93005)
188. _____, *Exact controllability of the Navier-Stokes and Boussinesq equations*, Russian Math. Surveys **54** (1999), 565–618. MR 1728643 (2000j:93016)

189. Jean-Paul Gauthier, *Structure des systèmes non linéaires*, Éditions du Centre National de la Recherche Scientifique (CNRS), Paris, 1984. MR 767635 (87g:93039)
190. Sophie Geffroy, *Les techniques de moyennisation en contrôle optimal – Applications aux transferts orbitaux à poussée faible continue*, Rapport de stage ENSEEIHT, Directeur de stage : Richard Épenoy (1994).
191. Maurice Gevrey, *Sur la nature analytique des solutions des équations aux dérivées partielles. Premier mémoire*, Ann. Sci. École Norm. Sup. (3) **35** (1918), 129–190. MR 1509208
192. Giuseppe Geymonat and Enrique Sánchez-Palencia, *On the vanishing viscosity limit for acoustic phenomena in a bounded region*, Arch. Rational Mech. Anal. **75** (1980/81), no. 3, 257–268. MR 605891 (82c:76089)
193. Elmer G. Gilbert, *Functional expansions for the response of nonlinear differential systems*, IEEE Trans. Automatic Control **AC-22** (1977), no. 6, 909–921. MR 0490201 (58 #9553)
194. Olivier Glass, *Contrôlabilité exacte frontière de l'équation d'Euler des fluides parfaits incompressibles en dimension 3*, C. R. Acad. Sci. Paris Sér. I Math. **325** (1997), no. 9, 987–992. MR 1485616 (99a:35200)
195. _____, *Exact boundary controllability of 3-D Euler equation*, ESAIM Control Optim. Calc. Var. **5** (2000), 1–44 (electronic). MR 1745685 (2001a:93011)
196. _____, *An addendum to a J. M. Coron theorem concerning the controllability of the Euler system for 2D incompressible inviscid fluids. "On the controllability of 2-D incompressible perfect fluids"* [J. Math. Pures Appl. (9) **75** (1996), no. 2, 155–188; MR1380673 (97b:93010)], J. Math. Pures Appl. (9) **80** (2001), no. 8, 845–877. MR 1860818 (2002i:93012)
197. _____, *Existence of solutions for the two-dimensional stationary Euler system for ideal fluids with arbitrary force*, Ann. Inst. H. Poincaré Anal. Non Linéaire **20** (2003), no. 6, 921–946. MR 2008684 (2004k:76016)
198. _____, *On the controllability of the Vlasov-Poisson system*, J. Differential Equations **195** (2003), no. 2, 332–379. MR 2016816 (2005a:93016)
199. _____, *Asymptotic stabilizability by stationary feedback of the two-dimensional Euler equation: the multiconnected case*, SIAM J. Control Optim. **44** (2005), no. 3, 1105–1147 (electronic). MR 2178059 (2006g:93109)
200. _____, *On the controllabilty of the 1-D isentropic Euler equation*, J. Eur. Math. Soc. (JEMS) (to appear).
201. Roland Glowinski, *Ensuring well-posedness by analogy: Stokes problem and boundary control for the wave equation*, J. Comput. Phys. **103** (1992), no. 2, 189–221. MR 1196839 (93j:65179)
202. Martin Golubitsky and Victor Guillemin, *Stable mappings and their singularities*, Springer-Verlag, New York, 1973, Graduate Texts in Mathematics, Vol. 14. MR 0341518 (49 #6269)
203. James M. Greenberg and Ta-tsien Li, *The effect of boundary damping for the quasilinear wave equation*, J. Differential Equations **52** (1984), no. 1, 66–75. MR 737964 (85k:35138)
204. Pierre Grisvard, *Singularities in boundary value problems*, Recherches en Mathématiques Appliquées [Research in Applied Mathematics], vol. 22, Masson, Paris, 1992. MR 1173209 (93h:35004)
205. Mikhael Gromov, *Smoothing and inversion of differential operators*, Mat. Sb. (N.S.) **88(130)** (1972), 382–441. MR 0310924 (46 #10022)
206. _____, *Partial differential relations*, Ergebnisse der Mathematik und ihrer Grenzgebiete (3) [Results in Mathematics and Related Areas (3)], vol. 9, Springer-Verlag, Berlin, 1986. MR 864505 (90a:58201)
207. Sergio Guerrero, *Local exact controllability to the trajectories of the Navier-Stokes system with nonlinear Navier-slip boundary conditions*, ESAIM Control Optim. Calc. Var. **12** (2006), no. 3, 484–544 (electronic). MR 2224824
208. Sergio Guerrero and Oleg Imanuvilov, *Remarks on global controllability for the Burgers equation with two control forces*, Ann. Inst. H. Poincaré Anal. Non Linéaire (to appear).
209. Sergio Guerrero and Gilles Lebeau, *Singular optimal control for a transport-diffusion equation*, Preprint, Université de Nice (2005).
210. Martin Gugat, Günter Leugering, and E. J. P. Georg Schmidt, *Global controllability between steady supercritical flows in channel networks*, Math. Methods Appl. Sci. **27** (2004), no. 7, 781–802. MR 2055319 (2005a:93017)
211. Robert Gulliver and Walter Littman, *Chord uniqueness and controllability: the view from the boundary. I*, Differential geometric methods in the control of partial differential equations

(Boulder, CO, 1999), Contemp. Math., vol. 268, Amer. Math. Soc., Providence, RI, 2000, pp. 145–175. MR 1804794 (2002h:93013)
212. Matthias Günther, *On the perturbation problem associated to isometric embeddings of Riemannian manifolds*, Ann. Global Anal. Geom. **7** (1989), no. 1, 69–77. MR 1029846 (91a:58023)
213. _____, *Isometric embeddings of Riemannian manifolds*, Proceedings of the International Congress of Mathematicians, Vol. I, II (Kyoto, 1990) (Tokyo), Math. Soc. Japan, 1991, pp. 1137–1143. MR 1159298 (93b:53049)
214. Jonathan de Halleux, Christophe Prieur, and Georges Bastin, *Boundary control design for cascades of hyperbolic 2×2 pde systems via graph theory*, Proceedings 43rd IEEE Conference on Decision and Control, Nassau, The Bahamas, December 2004, 2004, pp. 3313 – 3318.
215. Jonathan de Halleux, Christophe Prieur, Jean-Michel Coron, Brigitte d'Andréa-Novel, and Georges Bastin, *Boundary feedback control in networks of open channels*, Automatica J. IFAC **39** (2003), no. 8, 1365–1376. MR 2141681
216. Richard S. Hamilton, *The inverse function theorem of Nash and Moser*, Bull. Amer. Math. Soc. (N.S.) **7** (1982), no. 1, 65–222. MR 656198 (83j:58014)
217. Robert M. Hardt, *Semi-algebraic local-triviality in semi-algebraic mappings*, Amer. J. Math. **102** (1980), no. 2, 291–302. MR 564475 (81d:32012)
218. Matheus L. J. Hautus, *A simple proof of Heymann's lemma*, IEEE Trans. Automatic Control **AC-22** (1977), no. 5, 885–886. MR 0456654 (56 #14878)
219. J. William Helton and Orlando Merino, *Classical control using H^∞ methods*, Society for Industrial and Applied Mathematics (SIAM), Philadelphia, PA, 1998, Theory, optimization, and design. MR 1642633 (99i:93002)
220. Jacques Henry, *Étude de la contrôlabilité de certaines équations paraboliques non linéaires*, Thèse, Paris (1977).
221. Robert Hermann, *On the accessibility problem in control theory*, Internat. Sympos. Nonlinear Differential Equations and Nonlinear Mechanics, Academic Press, New York, 1963, pp. 325–332. MR 0149402 (26 #6891)
222. Robert Hermann and Arthur J. Krener, *Nonlinear controllability and observability*, IEEE Trans. Automatic Control **AC-22** (1977), no. 5, 728–740. MR 0476017 (57 #15597)
223. Henry Hermes, *Controllability and the singular problem*, J. Soc. Indust. Appl. Math. Ser. A Control **2** (1965), 241–260. MR 0173572 (30 #3785)
224. _____, *Discontinuous vector fields and feedback control*, Differential Equations and Dynamical Systems (Proc. Internat. Sympos., Mayaguez, P. R., 1965) (J.K. Hale and J.P. La Salle, eds.), Academic Press, New York, 1967, pp. 155–165. MR 0222424 (36 #5476)
225. _____, *Local controllability and sufficient conditions in singular problems*, J. Differential Equations **20** (1976), no. 1, 213–232. MR 0405214 (53 #9008)
226. _____, *Controlled stability*, Ann. Mat. Pura Appl. (4) **114** (1977), 103–119. MR 0638354 (58 #30677)
227. _____, *On the synthesis of a stabilizing feedback control via Lie algebraic methods*, SIAM J. Control Optim. **18** (1980), no. 4, 352–361. MR 579546 (81h:93070)
228. _____, *Control systems which generate decomposable Lie algebras*, J. Differential Equations **44** (1982), no. 2, 166–187, Special issue dedicated to J. P. LaSalle. MR 657777 (84h:49082)
229. _____, *Homogeneous coordinates and continuous asymptotically stabilizing feedback controls*, Differential equations (Colorado Springs, CO, 1989), Lecture Notes in Pure and Appl. Math., vol. 127, Dekker, New York, 1991, pp. 249–260. MR 1096761 (92c:93065)
230. _____, *Smooth homogeneous asymptotically stabilizing feedback controls*, ESAIM Control Optim. Calc. Var. **2** (1997), 13–32 (electronic). MR 1440077 (98b:93057)
231. Henry Hermes and Matthias Kawski, *Local controllability of a single input, affine system*, Proc. 7th Int. Conf. Nonlinear Analysis, Dallas (1986), 235–248.
232. Michael Heymann, *Comments "On pole assignment in multi-input controllable linear systems"*, IEEE Trans. Automatic Control **AC-13** (1968), 748–749. MR 0274084 (42 #8960)
233. Heisuke Hironaka, *Subanalytic sets*, Number theory, algebraic geometry and commutative algebra, in honor of Yasuo Akizuki, Kinokuniya, Tokyo, 1973, pp. 453–493. MR 0377101 (51 #13275)
234. Lop Fat Ho, *Observabilité frontière de l'équation des ondes*, C. R. Acad. Sci. Paris Sér. I Math. **302** (1986), no. 12, 443–446. MR 838598 (87d:93017)

235. Bertina Ho-Mock-Qai and Wijesuriya P. Dayawansa, *Simultaneous stabilization of linear and nonlinear systems by means of nonlinear state feedback*, SIAM J. Control Optim. **37** (1999), no. 6, 1701–1725 (electronic). MR 1720133 (2000i:93065)
236. Lars Hörmander, *On the Nash-Moser implicit function theorem*, Ann. Acad. Sci. Fenn. Ser. A I Math. **10** (1985), 255–259. MR 802486 (87a:58025)
237. _____, *The Nash-Moser theorem and paradifferential operators*, Analysis, et cetera, Academic Press, Boston, MA, 1990, pp. 429–449. MR 1039355 (91k:58009)
238. _____, *Lectures on nonlinear hyperbolic differential equations*, Mathématiques & Applications (Berlin) [Mathematics & Applications], vol. 26, Springer-Verlag, Berlin, 1997. MR 1466700 (98e:35103)
239. Thierry Horsin, *On the controllability of the Burgers equation*, ESAIM Control Optim. Calc. Var. **3** (1998), 83–95 (electronic). MR 1612027 (99c:93007)
240. _____, *Applications of the exact null controllability of the heat equation to moving sets*, C. R. Acad. Sci. Paris Sér. I Math. (to appear).
241. _____, *Local exact Lagrangian controllability of the Burgers viscous equation*, Ann. Inst. H. Poincaré Anal. Non Linéaire (to appear).
242. Oleg Yu. Imanuvilov, *Boundary controllability of parabolic equations*, Uspekhi Mat. Nauk **48** (1993), no. 3(291), 211–212. MR 1243631 (95e:35027)
243. _____, *Controllability of parabolic equations*, Mat. Sb. **186** (1995), no. 6, 109–132. MR 1349016 (96g:93005)
244. _____, *On exact controllability for the Navier-Stokes equations*, ESAIM Control Optim. Calc. Var. **3** (1998), 97–131 (electronic). MR 1617825 (99e:93037)
245. _____, *Remarks on exact controllability for the Navier-Stokes equations*, ESAIM Control Optim. Calc. Var. **6** (2001), 39–72 (electronic). MR 1804497 (2001k:93013)
246. Albert Edward Ingham, *Some trigonometrical inequalities with applications to the theory of series*, Math. Z. **41** (1936), 367–369.
247. Alberto Isidori, *Nonlinear control systems*, third ed., Communications and Control Engineering Series, Springer-Verlag, Berlin, 1995. MR 1410988 (97g:93003)
248. Victor Ja. Ivriĭ, *The second term of the spectral asymptotics for a Laplace-Beltrami operator on manifolds with boundary*, Funktsional. Anal. i Prilozhen. **14** (1980), no. 2, 25–34. MR 575202 (82m:58057)
249. Howard Jacobowitz, *Implicit function theorems and isometric embeddings*, Ann. of Math. (2) **95** (1972), 191–225. MR 0307127 (46 #6248)
250. David H. Jacobson, *Extensions of linear-quadratic control, optimization and matrix theory*, Academic Press [Harcourt Brace Jovanovich Publishers], London, 1977, Mathematics in Science and Engineering, Vol. 133. MR 0459795 (56 #17985)
251. Stéphane Jaffard and Sorin Micu, *Estimates of the constants in generalized Ingham's inequality and applications to the control of the wave equation*, Asymptot. Anal. **28** (2001), no. 3-4, 181–214. MR 1878794 (2002j:93047)
252. Stéphane Jaffard, Marius Tucsnak, and Enrique Zuazua, *On a theorem of Ingham*, J. Fourier Anal. Appl. **3** (1997), no. 5, 577–582, Dedicated to the memory of Richard J. Duffin. MR 1491935 (2000e:42004)
253. _____, *Singular internal stabilization of the wave equation*, J. Differential Equations **145** (1998), no. 1, 184–215. MR 1620290 (99g:93073)
254. Mrdjan Janković, Rodolphe Sepulchre, and Petar V. Kokotović, *Constructive Lyapunov stabilization of nonlinear cascade systems*, IEEE Trans. Automat. Control **41** (1996), no. 12, 1723–1735. MR 1421408 (97k:93054)
255. Hamadi Jerbi, *On the stabilizability of homogeneous systems of odd degree*, ESAIM Control Optim. Calc. Var. **9** (2003), 343–352 (electronic). MR 1966537 (2004a:93073)
256. Zhong-Ping Jiang and Henk Nijmeijer, *A recursive technique for tracking control of nonholonomic systems in chained form*, IEEE Trans. Automat. Control **44** (1999), no. 2, 265–279. MR 1669982 (99k:93068)
257. Victor Iosifovich Judovič, *A two-dimensional non-stationary problem on the flow of an ideal incompressible fluid through a given region*, Mat. Sb. (N.S.) **64 (106)** (1964), 562–588. MR 0177577 (31 #1840)
258. Velimir Jurdjevic, *Geometric control theory*, Cambridge Studies in Advanced Mathematics, vol. 52, Cambridge University Press, Cambridge, 1997. MR 1425878 (98a:93002)

259. Velimir Jurdjevic and Ivan Kupka, *Control systems on semisimple Lie groups and their homogeneous spaces*, Ann. Inst. Fourier (Grenoble) **31** (1981), no. 4, vi, 151–179. MR 644347 (84a:93014)
260. _____, *Control systems subordinated to a group action: accessibility*, J. Differential Equations **39** (1981), no. 2, 186–211. MR 607781 (82f:93009)
261. Velimir Jurdjevic and John P. Quinn, *Controllability and stability*, J. Differential Equations **28** (1978), no. 3, 381–389. MR 0494275 (58 #13181)
262. Jean-Pierre Kahane, *Pseudo-périodicité et séries de Fourier lacunaires*, Ann. Sci. École Norm. Sup. (3) **79** (1962), 93–150. MR 0154060 (27 #4019)
263. Rudolph E. Kalman, *Contributions to the theory of optimal control*, Bol. Soc. Mat. Mexicana (2) **5** (1960), 102–119. MR 0127472 (23 #B518)
264. _____, *Mathematical description of linear dynamical systems*, J. SIAM Control Ser. A **1** (1963), 152–192. MR 0152167 (27 #2147)
265. Rudolph E. Kalman, Yu-Chi Ho, and Kumpati S. Narendra, *Controllability of linear dynamical systems*, Contributions to Differential Equations **1** (1963), 189–213. MR 0155070 (27 #5012)
266. Tosio Kato, *On the Cauchy problem for the (generalized) Korteweg-de Vries equation*, Studies in applied mathematics, Adv. Math. Suppl. Stud., vol. 8, Academic Press, New York, 1983, pp. 93–128. MR 759907 (86f:35160)
267. _____, *Perturbation theory for linear operators*, Classics in Mathematics, Springer-Verlag, Berlin, 1995, Reprint of the 1980 edition. MR 1335452 (96a:47025)
268. Matthias Kawski, *Stabilization and nilpotent approximations*, Proc. 27th IEEE Conference Decision and Control (Austin 1988), IEEE, New York, 1988, pp. 1244–1248.
269. _____, *Stabilization of nonlinear systems in the plane*, Systems Control Lett. **12** (1989), no. 2, 169–175. MR 985567 (90i:93096)
270. _____, *High-order small-time local controllability*, Nonlinear controllability and optimal control (H.J. Sussmann, ed.), Monogr. Textbooks Pure Appl. Math., vol. 133, Dekker, New York, 1990, pp. 431–467. MR 1061394 (91j:93010)
271. Alexandre Kazhikov, *Note on the formulation of the problem of flow through a bounded region using equations of perfect fluid*, PMM USSR **44** (1981), 672–674.
272. El-Yazid Keraï, *Analysis of small-time local controllability of the rigid body model*, Proc. IFAC, System Structure and Control, Nantes, July 5-7 (1995).
273. Hassan K. Khalil, *Nonlinear systems*, Macmillan Publishing Company, New York, 1992. MR 1201326 (93k:34001)
274. Pramod P. Khargonekar, Antonio M. Pascoal, and R. Ravi, *Strong, simultaneous, and reliable stabilization of finite-dimensional linear time-varying plants*, IEEE Trans. Automat. Control **33** (1988), no. 12, 1158–1161. MR 967400 (90a:93078)
275. David L. Kleinman, *An easy way to stabilize a linear control system*, IEEE Trans. Automat. Control **AC-15** (1970), no. 1, 692.
276. Daniel E. Koditschek, *Adaptive techniques for mechanical systems*, Proc. 5th. Yale University Conference, New Haven (1987), Yale University, New Haven, 1987, pp. 259–265.
277. Vilmos Komornik, *Exact controllability and stabilization*, RAM: Research in Applied Mathematics, Masson, Paris, 1994, The multiplier method. MR 1359765 (96m:93003)
278. _____, *Rapid boundary stabilization of linear distributed systems*, SIAM J. Control Optim. **35** (1997), no. 5, 1591–1613. MR 1466918 (98h:93022)
279. Vilmos Komornik and Paola Loreti, *A further note on a theorem of Ingham and simultaneous observability in critical time*, Inverse Problems **20** (2004), no. 5, 1649–1661. MR 2109141 (2005i:93051)
280. _____, *Fourier series in control theory*, Springer Monographs in Mathematics, Springer-Verlag, New York, 2005. MR 2114325 (2006a:93001)
281. Paul Koosis, *The logarithmic integral. I*, Cambridge Studies in Advanced Mathematics, vol. 12, Cambridge University Press, Cambridge, 1998, Corrected reprint of the 1988 original. MR 1670244 (99j:30001)
282. Werner Krabs, *On moment theory and controllability of one-dimensional vibrating systems and heating processes*, Lecture Notes in Control and Information Sciences, vol. 173, Springer-Verlag, Berlin, 1992. MR 1162111 (93f:93002)
283. Mark Aleksandrovich Krasnosel′skiĭ, *The operator of translation along the trajectories of differential equations*, Translations of Mathematical Monographs, Vol. 19. Translated from

the Russian by Scripta Technica, American Mathematical Society, Providence, R.I., 1968. MR 0223640 (36 #6688)

284. Mark Aleksandrovich Krasnosel′skiĭ and Petr P. Zabreĭko, *Geometrical methods of nonlinear analysis*, Grundlehren der Mathematischen Wissenschaften [Fundamental Principles of Mathematical Sciences], vol. 263, Springer-Verlag, Berlin, 1984, Translated from the Russian by Christian C. Fenske. MR 736839 (85b:47057)

285. Heinz-Otto Kreiss, *Initial boundary value problems for hyperbolic systems*, Comm. Pure Appl. Math. **23** (1970), 277–298. MR 0437941 (55 #10862)

286. Arthur J. Krener, *Local approximation of control systems*, J. Differential Equations **19** (1975), no. 1, 125–133. MR 0429224 (55 #2243)

287. Miroslav Krstić, *On global stabilization of Burgers' equation by boundary control*, Systems Control Lett. **37** (1999), no. 3, 123–141. MR 1751258 (2001m:93101)

288. Miroslav Krstić and Hua Deng, *Stabilization of nonlinear uncertain systems*, Communications and Control Engineering Series, Springer-Verlag London Ltd., London, 1998. MR 1639235 (99h:93001)

289. Jaroslav Kurzweil, *On the inversion of Lyapunov's second theorem on stability of motion*, Ann. Math. Soc. Trans. Ser.2, **24** (1956), 19–77.

290. Huibert Kwakernaak and Raphael Sivan, *Linear optimal control systems*, Wiley-Interscience [John Wiley & Sons], New York, 1972. MR 0406607 (53 #10394)

291. Olga Alexandrovna Ladyzhenskaya, *The boundary value problems of mathematical physics*, Applied Mathematical Sciences, vol. 49, Springer-Verlag, New York, 1985, Translated from the Russian by Jack Lohwater [Arthur J. Lohwater]. MR 793735 (87f:35001)

292. Olga Alexandrovna Ladyzhenskaya, Vsevolod Alekseevich Solonnikov, and Nina Nikolaevna Ural′ceva, *Linear and quasilinear equations of parabolic type*, Translated from the Russian by S. Smith. Translations of Mathematical Monographs, Vol. 23, American Mathematical Society, Providence, R.I., 1967. MR 0241822 (39 #3159b)

293. John E. Lagnese, *The Hilbert uniqueness method: a retrospective*, Optimal control of partial differential equations (Irsee, 1990), Lecture Notes in Control and Inform. Sci., vol. 149, Springer, Berlin, 1991, pp. 158–181. MR 1178298 (93j:93020)

294. John E. Lagnese and Günter Leugering, *Domain decomposition methods in optimal control of partial differential equations*, International Series of Numerical Mathematics, vol. 148, Birkhäuser Verlag, Basel, 2004. MR 2093789 (2005g:49002)

295. Béatrice Laroche, Philippe Martin, and Pierre Rouchon, *Motion planning for the heat equation*, Internat. J. Robust Nonlinear Control **10** (2000), no. 8, 629–643, Nonlinear adaptive and linear systems (Mexico City, 1998). MR 1776232 (2002g:93011)

296. Joseph Pierre LaSalle, *The time optimal control problem*, Contributions to the theory of nonlinear oscillations, Vol. V, Princeton Univ. Press, Princeton, N.J., 1960, pp. 1–24. MR 0145169 (26 #2704)

297. Irena Lasiecka and Roberto Triggiani, *Regularity of hyperbolic equations under $L_2(0, T; L_2(\Gamma))$-Dirichlet boundary terms*, Appl. Math. Optim. **10** (1983), no. 3, 275–286. MR 722491 (85j:35111)

298. _____, *Sharp regularity theory for second order hyperbolic equations of Neumann type. I. L_2 nonhomogeneous data*, Ann. Mat. Pura Appl. (4) **157** (1990), 285–367. MR 1108480 (92e:35102)

299. _____, *Differential and algebraic Riccati equations with application to boundary/point control problems: continuous theory and approximation theory*, Lecture Notes in Control and Information Sciences, vol. 164, Springer-Verlag, Berlin, 1991. MR 1132440 (92k:93009)

300. _____, *Exact controllability of semilinear abstract systems with application to waves and plates boundary control problems*, Appl. Math. Optim. **23** (1991), no. 2, 109–154. MR 1086465 (92a:93021)

301. _____, *Control theory for partial differential equations: continuous and approximation theories. I*, Encyclopedia of Mathematics and its Applications, vol. 74, Cambridge University Press, Cambridge, 2000, Abstract parabolic systems. MR 1745475 (2001m:93002)

302. _____, *Control theory for partial differential equations: continuous and approximation theories. II*, Encyclopedia of Mathematics and its Applications, vol. 75, Cambridge University Press, Cambridge, 2000, Abstract hyperbolic-like systems over a finite time horizon. MR 1745476 (2001m:93003)

303. Jean-Paul Laumond (ed.), *Robot motion planning and control*, Lecture Notes in Control and Information Sciences, vol. 229, Springer-Verlag London Ltd., London, 1998. MR 1603373 (98k:70012)
304. Brian E. Launder and Brian D. Spalding, *Mathematical models of turbulence*, Academic Press, New York, 1972.
305. Peter D. Lax, *Hyperbolic systems of conservation laws and the mathematical theory of shock waves*, Society for Industrial and Applied Mathematics, Philadelphia, Pa., 1973, Conference Board of the Mathematical Sciences Regional Conference Series in Applied Mathematics, No. 11. MR 0350216 (50 #2709)
306. Claude Le Bris, *Control theory applied to quantum chemistry: some tracks*, Contrôle des systèmes gouvernés par des équations aux dérivées partielles (Nancy, 1999), ESAIM Proc., vol. 8, Soc. Math. Appl. Indust., Paris, 2000, pp. 77–94 (electronic). MR 1807561 (2001k:81372)
307. Gilles Lebeau and Luc Robbiano, *Contrôle exact de l'équation de la chaleur*, Comm. Partial Differential Equations **20** (1995), no. 1-2, 335–356. MR 1312710 (95m:93045)
308. Yuri S. Ledyaev and Eduardo D. Sontag, *A Lyapunov characterization of robust stabilization*, Nonlinear Anal. **37** (1999), no. 7, Ser. A: Theory Methods, 813–840. MR 1695080 (2000c:93092)
309. Ernest Bruce Lee and Lawrence Markus, *Foundations of optimal control theory*, second ed., Robert E. Krieger Publishing Co. Inc., Melbourne, FL, 1986. MR 889459 (88b:49001)
310. Kyun K. Lee and Aristotle Arapostathis, *Remarks on smooth feedback stabilization of nonlinear systems*, Systems Control Lett. **10** (1988), no. 1, 41–44. MR 920803
311. Naomi Ehrich Leonard, *Periodic forcing, dynamics and control of underactuated spacecraft and underwater vehicles*, Proceedings 34th IEEE Conference on Decision and Control (New Orleans, USA, December 1995), vol. 4, 2002, pp. 3980–3985.
312. Naomi Ehrich Leonard and P. S. Krishnaprasad, *Motion control of drift-free, left-invariant systems on Lie groups*, IEEE Trans. Automat. Control **40** (1995), no. 9, 1539–1554. MR 1347834 (96i:93038)
313. Jean Leray, *Topologie des espaces de M. Banach*, C. R. Math. Acad. Sci. Paris **200** (1935), 1082–1084.
314. _____, *La théorie des points fixes et ses applications en analyse*, Proceedings of the International Congress of Mathematicians, Cambridge, Mass., 1950, vol. 2 (Providence, R. I.), Amer. Math. Soc., 1952, pp. 202–208. MR 0047318 (13,859a)
315. Boris Yakovlevich Levin, *Lectures on entire functions*, Translations of Mathematical Monographs, vol. 150, American Mathematical Society, Providence, RI, 1996, In collaboration with and with a preface by Yu. Lyubarskii, M. Sodin and V. Tkachenko, Translated from the Russian manuscript by Tkachenko. MR 1400006 (97j:30001)
316. Ta-tsien Li, *Global classical solutions for quasilinear hyperbolic systems*, RAM: Research in Applied Mathematics, vol. 32, Masson, Paris, 1994. MR 1291392 (95m:35115)
317. _____, *Exact boundary controllability of unsteady flows in a network of open canals*, Differential equations & asymptotic theory in mathematical physics, Ser. Anal., vol. 2, World Sci. Publ., Hackensack, NJ, 2004, pp. 310–329. MR 2161977
318. _____, *Exact controllability for quasilinear hyperbolic systems and its application to unsteady flows in a network of open canals*, Math. Methods Appl. Sci. **27** (2004), no. 9, 1089–1114. MR 2063097 (2005b:76030)
319. Ta-tsien Li and Bo-Peng Rao, *Exact boundary controllability for quasi-linear hyperbolic systems*, SIAM J. Control Optim. **41** (2003), no. 6, 1748–1755 (electronic). MR 1972532 (2004b:93016)
320. Ta-tsien Li and Bopeng Rao, *Exact boundary controllability of unsteady flows in a tree-like network of open canals*, Methods Appl. Anal. **11** (2004), no. 3, 353–365. MR 2214680
321. Ta-tsien Li and Wen Ci Yu, *Boundary value problems for quasilinear hyperbolic systems*, Duke University Mathematics Series, V, Duke University Mathematics Department, Durham, NC, 1985. MR 823237 (88g:35115)
322. Ta-tsien Li and Bing-Yu Zhang, *Global exact controllability of a class of quasilinear hyperbolic systems*, J. Math. Anal. Appl. **225** (1998), no. 1, 289–311. MR 1639252 (99h:93018)
323. Jacques-Louis Lions, *Contrôle optimal de systèmes gouvernés par des équations aux dérivées partielles*, Avant propos de P. Lelong, Dunod, Paris, 1968. MR 0244606 (39 #5920)

324. _____, *Contrôle des systèmes distribués singuliers*, Méthodes Mathématiques de l'Informatique [Mathematical Methods of Information Science], vol. 13, Gauthier-Villars, Montrouge, 1983. MR 712486 (85c:93002)

325. _____, *Contrôlabilité exacte, perturbations et stabilisation de systèmes distribués. Tome 1*, Recherches en Mathématiques Appliquées [Research in Applied Mathematics], vol. 8, Masson, Paris, 1988, Contrôlabilité exacte. [Exact controllability], With appendices by E. Zuazua, C. Bardos, G. Lebeau and J. Rauch. MR 953547 (90a:49040)

326. _____, *Exact controllability, stabilization and perturbations for distributed systems*, SIAM Rev. **30** (1988), no. 1, 1–68. MR 931277 (89e:93019)

327. _____, *Exact controllability for distributed systems. Some trends and some problems*, Applied and industrial mathematics (Venice, 1989), Math. Appl., vol. 56, Kluwer Acad. Publ., Dordrecht, 1991, pp. 59–84. MR 1147191 (92m:93007)

328. _____, *Are there connections between turbulence and controllability?*, 9th INRIA International Conference, Antibes (June 12-15, 1990).

329. Jacques-Louis Lions and Enrico Magenes, *Problèmes aux limites non homogènes et applications. Vol. 1*, Travaux et Recherches Mathématiques, No. 17, Dunod, Paris, 1968. MR 0247243 (40 #512)

330. Jacques-Louis Lions and Enrique Zuazua, *Approximate controllability of a hydro-elastic coupled system*, ESAIM Contrôle Optim. Calc. Var. **1** (1995/96), 1–15 (electronic). MR 1382513 (97a:93012)

331. Pierre-Louis Lions, *Generalized solutions of Hamilton-Jacobi equations*, Research Notes in Mathematics, vol. 69, Pitman (Advanced Publishing Program), Boston, Mass., 1982. MR 667669 (84a:49038)

332. Walter Littman, *Boundary control theory for hyperbolic and parabolic partial differential equations with constant coefficients*, Ann. Scuola Norm. Sup. Pisa Cl. Sci. (4) **5** (1978), no. 3, 567–580. MR 507002 (80a:35023)

333. Noel Glynne Lloyd, *Degree theory*, Cambridge University Press, Cambridge, 1978, Cambridge Tracts in Mathematics, No. 73. MR 0493564 (58 #12558)

334. Claude Lobry, *Controllability of nonlinear systems on compact manifolds*, SIAM J. Control **12** (1974), 1–4. MR 0338881 (49 #3645)

335. Rogelio Lozano, *Robust adaptive regulation without persistent excitation*, IEEE Trans. Automat. Control **34** (1989), no. 12, 1260–1267.

336. Dahlard L. Lukes, *Stabilizability and optimal control*, Funkcial. Ekvac. **11** (1968), 39–50. MR 0238589 (38 #6865)

337. _____, *Global controllability of nonlinear systems*, SIAM J. Control **10** (1972), 112–126. MR 0304004 (46 #3140)

338. _____, *Erratum: "Global controllability of nonlinear systems" (SIAM J. Control 10 (1972), 112–126)*, SIAM J. Control **11** (1973), 186. MR 0685287 (58 #33253)

339. Uwe Mackenroth, *Robust control systems*, Springer-Verlag, Berlin, 2004, Theory and case studies. MR 2080385 (2005c:93002)

340. Andrew Majda, *Compressible fluid flow and systems of conservation laws in several space variables*, Applied Mathematical Sciences, vol. 53, Springer-Verlag, New York, 1984. MR 748308 (85e:35077)

341. _____, *Vorticity and the mathematical theory of incompressible fluid flow*, Comm. Pure Appl. Math. **39** (1986), no. S, suppl., S187–S220, Frontiers of the mathematical sciences: 1985 (New York, 1985). MR 861488 (87j:76041)

342. Paul Malliavin, *Intégration et probabilités. Analyse de Fourier et analyse spectrale*, Masson, Paris, 1982, Collection: Maîtrise de Mathématiques Pures. [Collection: Masters in Pure Mathematics]. MR 662563 (83j:28001)

343. Lawrence Markus, *Controllability of nonlinear processes*, J. Soc. Indust. Appl. Math. Ser. A Control **3** (1965), 78–90. MR 0186487 (32 #3947)

344. Lawrence Markus and George Sell, *Capture and control in conservative dynamical systems*, Arch. Rational Mech. Anal. **31** (1968/1969), 271–287. MR 0234087 (38 #2406)

345. Sonia Martínez, Jorge Cortés, and Francesco Bullo, *Analysis and design of oscillatory control systems*, IEEE Trans. Automat. Control **48** (2003), no. 7, 1164–1177. MR 1988087 (2004c:93101)

346. José L. Massera, *Contributions to stability theory*, Ann. of Math. (2) **64** (1956), 182–206. MR 0079179 (18,42d)

347. J. Mawhin, *Topological degree methods in nonlinear boundary value problems*, CBMS Regional Conference Series in Mathematics, vol. 40, American Mathematical Society, Providence, R.I., 1979, Expository lectures from the CBMS Regional Conference held at Harvey Mudd College, Claremont, Calif., June 9–15, 1977. MR 525202 (80c:47055)
348. Frédéric Mazenc, Kristin Pettersen, and Henk Nijmeijer, *Global uniform asymptotic stabilization of an underactuated surface vessel*, IEEE Trans. Automat. Control **47** (2002), no. 10, 1759–1762. MR 1929956 (2003h:93069)
349. Frédéric Mazenc and Laurent Praly, *Adding integrations, saturated controls, and stabilization for feedforward systems*, IEEE Trans. Automat. Control **41** (1996), no. 11, 1559–1578. MR 1419682 (97i:93087)
350. Fréréric Mazenc and Laurent Praly, *Global stabilization for nonlinear systems*, Preprint, CAS, ENSMP (1993).
351. Robert T. M'Closkey and Richard M. Murray, *Nonholonomic systems and exponential convergence: some analysis tools*, Proc. 32nd IEEE Conf. Decision and Control (San Antonio, Texas, 1993), IEEE, New York, 1993, pp. 943–948.
352. Frank Merle and Hatem Zaag, *Stability of the blow-up profile for equations of the type $u_t = \Delta u + |u|^{p-1}u$*, Duke Math. J. **86** (1997), no. 1, 143–195. MR 1427848 (98d:35098)
353. Albert Messiah, *Mécanique quantique*, 2 vols, Dunod, Paris, 1965, Nouveau tirage. MR 0129304 (23 #B2340)
354. Sorin Micu and Enrique Zuazua, *On the lack of null-controllability of the heat equation on the half-line*, Trans. Amer. Math. Soc. **353** (2001), no. 4, 1635–1659 (electronic). MR 1806726 (2001k:93019)
355. _____, *On the lack of null-controllability of the heat equation on the half space*, Port. Math. (N.S.) **58** (2001), no. 1, 1–24. MR 1820835 (2002a:93011)
356. _____, *On the controllability of a fractional order parabolic equation*, SIAM J. Control Optim. **44** (2006), no. 6, 1950–1972, (electronic).
357. Luc Miller, *Geometric bounds on the growth rate of null-controllability cost for the heat equation in small time*, J. Differential Equations **204** (2004), no. 1, 202–226. MR 2076164 (2005f:93018)
358. John Milnor, *Singular points of complex hypersurfaces*, Annals of Mathematics Studies, No. 61, Princeton University Press, Princeton, N.J., 1968. MR 0239612 (39 #969)
359. Mazyar Mirrahimi and Pierre Rouchon, *Controllability of quantum harmonic oscillators*, IEEE Trans. Automat. Control **49** (2004), no. 5, 745–747. MR 2057808 (2005a:93023)
360. Mazyar Mirrahimi, Pierre Rouchon, and Gabriel Turinici, *Lyapunov control of bilinear Schrödinger equations*, Automatica J. IFAC **41** (2005), no. 11, 1987–1994. MR 2168664
361. Richard Montgomery, *A tour of subriemannian geometries, their geodesics and applications*, Mathematical Surveys and Monographs, vol. 91, American Mathematical Society, Providence, RI, 2002. MR 1867362 (2002m:53045)
362. Cathleen S. Morawetz, *The limiting amplitude principle*, Comm. Pure Appl. Math. **15** (1962), 349–361. MR 0151712 (27 #1696)
363. Pascal Morin, Jean-Baptiste Pomet, and Claude Samson, *Design of homogeneous time-varying stabilizing control laws for driftless controllable systems via oscillatory approximation of Lie brackets in closed loop*, SIAM J. Control Optim. **38** (1999), no. 1, 22–49 (electronic). MR 1740609 (2000j:93091)
364. Pascal Morin and Claude Samson, *Time-varying exponential stabilization of a rigid spacecraft with two control torques*, IEEE Trans. Automat. Control **42** (1997), no. 4, 528–534. MR 1442588
365. _____, *A characterization of the Lie algebra rank condition by transverse periodic functions*, SIAM J. Control Optim. **40** (2001/02), no. 4, 1227–1249 (electronic). MR 1882731 (2002m:93022)
366. _____, *Practical stabilization of driftless systems on Lie groups: The transverse function approach*, IEEE Trans. Automat. Control **48** (2003), no. 9, 1496–1508. MR 2000107 (2004f:93124)
367. Pascal Morin, Claude Samson, Jean-Baptiste Pomet, and Zhong-Ping Jiang, *Time-varying feedback stabilization of the attitude of a rigid spacecraft with two controls*, Systems Control Lett. **25** (1995), no. 5, 375–385. MR 1343223
368. Jürgen Moser, *A new technique for the construction of solutions of nonlinear differential equations*, Proc. Nat. Acad. Sci. U.S.A. **47** (1961), 1824–1831. MR 0132859 (24 #A2695)

369. _____, *A rapidly convergent iteration method and non-linear differential equations. II*, Ann. Scuola Norm. Sup. Pisa (3) **20** (1966), 499–535. MR 0206461 (34 #6280)
370. _____, *A rapidly convergent iteration method and non-linear partial differential equations. I*, Ann. Scuola Norm. Sup. Pisa (3) **20** (1966), 265–315. MR 0199523 (33 #7667)
371. Hugues Mounier, Joachim Rudolph, Michel Fliess, and Pierre Rouchon, *Tracking control of a vibrating string with an interior mass viewed as delay system*, ESAIM Control Optim. Calc. Var. **3** (1998), 315–321 (electronic). MR 1644431 (99h:73057)
372. Tadashi Nagano, *Linear differential systems with singularities and an application to transitive Lie algebras*, J. Math. Soc. Japan **18** (1966), 398–404. MR 0199865 (33 #8005)
373. Mitio Nagumo, *A theory of degree of mapping based on infinitesimal analysis*, Amer. J. Math. **73** (1951), 485–496. MR 0042696 (13,150a)
374. John Nash, *The imbedding problem for Riemannian manifolds*, Ann. of Math. (2) **63** (1956), 20–63. MR 0075639 (17,782b)
375. Claude Louis Marie Henri Navier, *Sur les lois du mouvement des fluides*, Mem. Acad. R. Sci. Inst. France **6** (1823), 389–440.
376. Henk Nijmeijer and Arjan van der Schaft, *Nonlinear dynamical control systems*, Springer-Verlag, New York, 1990. MR 1047663 (91d:93024)
377. Axel Osses and Jean-Pierre Puel, *Approximate controllability for a hydro-elastic model in a rectangular domain*, Optimal control of partial differential equations (Chemnitz, 1998), Internat. Ser. Numer. Math., vol. 133, Birkhäuser, Basel, 1999, pp. 231–243. MR 1723989 (2000j:93019)
378. _____, *Approximate controllability for a linear model of fluid structure interaction*, ESAIM Control Optim. Calc. Var. **4** (1999), 497–513 (electronic). MR 1713527 (2000j:93020)
379. Alain Oustaloup, *La dérivation non entière: théorie, synthèse et applications*, Hermes, Paris, 1995.
380. Ademir Fernando Pazoto, *Unique continuation and decay for the Korteweg-de Vries equation with localized damping*, ESAIM Control Optim. Calc. Var. **11** (2005), no. 3, 473–486 (electronic). MR 2148854 (2006b:35292)
381. Amnon Pazy, *Semigroups of linear operators and applications to partial differential equations*, Applied Mathematical Sciences, vol. 44, Springer-Verlag, New York, 1983. MR 710486 (85g:47061)
382. Gustavo Perla Menzala, Carlos Frederico Vasconcellos, and Enrique Zuazua, *Stabilization of the Korteweg-de Vries equation with localized damping*, Quart. Appl. Math. **60** (2002), no. 1, 111–129. MR 1878262 (2002j:35273)
383. Ian R. Petersen, Valery A. Ugrinovskii, and Andrey V. Savkin, *Robust control design using H^∞-methods*, Communications and Control Engineering Series, Springer-Verlag London Ltd., London, 2000. MR 1834840 (2003f:93002)
384. Nicolas Petit and Pierre Rouchon, *Flatness of heavy chain systems*, SIAM J. Control Optim. **40** (2001), no. 2, 475–495 (electronic). MR 1857359 (2002f:93031)
385. _____, *Dynamics and solutions to some control problems for water-tank systems*, IEEE Trans. Automat. Control **47** (2002), no. 4, 594–609. MR 1893517 (2003b:93061)
386. V. N. Polotskiĭ, *Estimation of the state of single-output linear systems by means of observers*, Automat. Remote Control **12** (1980), 18–28. MR 649138 (83c:93012)
387. Jean-Baptiste Pomet, *Explicit design of time-varying stabilizing control laws for a class of controllable systems without drift*, Systems Control Lett. **18** (1992), no. 2, 147–158. MR 1149359 (92m:93043)
388. Jean-Baptiste Pomet and Claude Samson, *Exponential stabilization of nonholonomic systems in power form*, IFAC Symposium on Robust Control Design (Rio de Janeiro 1994), Pergamon, Oxford, 1994, pp. 447–452.
389. Lev S. Pontryagin, *Optimal regulation processes*, Uspehi Mat. Nauk **14** (1959), no. 1 (85), 3–20. MR 0120435 (22 #11189)
390. Lev S. Pontryagin, Vladimir G. Boltyanskii, Revaz V. Gamkrelidze, and Evgenii Frolovich Mishchenko, *The mathematical theory of optimal processes*, Translated from the Russian by K. N. Trirogoff; edited by L. W. Neustadt, Interscience Publishers John Wiley & Sons, Inc. New York-London, 1962. MR 0166037 (29 #3316b)
391. Laurent Praly, *An introduction to forwarding*, Chapter 4 in: Control of complex systems (Karl Åström, Pedro Albertos, Mogens Blanke, Alberto Isidori, Walter Schaufelberger, and Ricardo Sanz, eds.), Springer-Verlag, London, 2001.

392. Laurent Praly, Brigitte d'Andréa-Novel, and Jean-Michel Coron, *Lyapunov design of stabilizing controllers for cascaded systems*, IEEE Trans. Automat. Control **36** (1991), no. 10, 1177–1181. MR 1125898 (92i:93084)
393. Christophe Prieur, *Asymptotic controllability and robust asymptotic stabilizability*, SIAM J. Control Optim. **43** (2005), no. 5, 1888–1912 (electronic). MR 2137506
394. Christophe Prieur and Emmanuel Trélat, *Robust optimal stabilization of the Brockett integrator via a hybrid feedback*, Math. Control Signals Systems **17** (2005), no. 3, 201–216. MR 2160777
395. _____, *Quasi-optimal robust stabilization of control systems*, SIAM J. Control Optim. (to appear).
396. Christophe Prieur, Joseph Winkin, and Georges Bastin, *Boundary control of non-homogeneous systems of conservation laws*, Preprint, LAAS (2006).
397. Jean-Pierre Ramis, *Séries divergentes et théories asymptotiques*, Bull. Soc. Math. France **121** (1993), no. Panoramas et Syntheses, suppl., 74. MR 1272100 (95h:34074)
398. Petr K. Rashevski, *About connecting two points of complete nonholonomic space by admissible curve*, Uch Zapiski Ped. Inst. Libknexta **2** (1938), 83–94.
399. Jean-Pierre Raymond, *Feedback boundary stabilization of the two-dimensional Navier-Stokes equations*, SIAM J. Control Optim. **45** (2006), no. 3, 790–828 (electronic). MR 2247716
400. Ray M. Redheffer, *Remarks on incompleteness of $\{e^{i\lambda_n x}\}$, nonaveraging sets, and entire functions*, Proc. Amer. Math. Soc. **2** (1951), 365–369. MR 0041270 (12,823a)
401. Michael Reed and Barry Simon, *Methods of modern mathematical physics. IV. Analysis of operators*, Academic Press [Harcourt Brace Jovanovich Publishers], New York, 1978. MR 0493421 (58 #12429c)
402. Mahmut Reyhanoglu, Arjan van der Schaft, N. Harris McClamroch, and Ilya Kolmanovsky, *Dynamics and control of a class of underactuated mechanical systems*, IEEE Trans. Automat. Control **44** (1999), no. 9, 1663–1671. MR 1709867 (2000f:93064)
403. Ludovic Rifford, *Semiconcave control-Lyapunov functions and stabilizing feedbacks*, SIAM J. Control Optim. **41** (2002), no. 3, 659–681 (electronic). MR 1939865 (2004b:93128)
404. _____, *Stratified semiconcave control-Lyapunov functions and the stabilization problem*, Ann. Inst. H. Poincaré Anal. Non Linéaire **22** (2005), no. 3, 343–384. MR 2136728 (2006a:93086)
405. Lionel Rosier, *Homogeneous Lyapunov function for homogeneous continuous vector fields*, Systems Control Lett. **19** (1992), no. 6, 467–473. MR 1195304 (94h:34058)
406. _____, *Étude de quelques problèmes de stabilisation*, PhD Thesis, ENS de Cachan (advisors: Jean-Michel Coron and Laurent Praly) (1993), 1–99.
407. _____, *Exact boundary controllability for the Korteweg-de Vries equation on a bounded domain*, ESAIM Control Optim. Calc. Var. **2** (1997), 33–55 (electronic). MR 1440078 (98d:93016)
408. _____, *Exact boundary controllability for the linear Korteweg-de Vries equation—a numerical study*, Control and partial differential equations (Marseille-Luminy, 1997), ESAIM Proc., vol. 4, Soc. Math. Appl. Indust., Paris, 1998, pp. 255–267 (electronic). MR 1663665 (99i:93008)
409. _____, *Control of the Korteweg-de Vries equation*, First Nonlinear Control Network Pedagogical School (K. Kyriakopoulos, F. Lamnabhi-Lagarrigue, and J. Tsinias, eds.), 1999, pp. 285–305.
410. _____, *Exact boundary controllability for the linear Korteweg-de Vries equation on the half-line*, SIAM J. Control Optim. **39** (2000), no. 2, 331–351 (electronic). MR 1788062 (2001j:93012)
411. _____, *Control of the surface of a fluid by a wavemaker*, ESAIM Control Optim. Calc. Var. **10** (2004), no. 3, 346–380 (electronic). MR 2084328 (2005h:93091)
412. _____, *Control of partial differential equations: an introduction*, ICTP Lect. Notes, Abdus Salam Int. Cent. Theoret. Phys., Trieste, to appear.
413. Lionel Rosier and Bing-Yu Zhang, *Global stabilization of the generalized Korteweg-de Vries equation posed on a finite domain*, SIAM J. Control Optim. **45** (2006), no. 3, 927–956 (electronic). MR 2247720
414. Nicolas Rouche, Patrick Habets, and M. Laloy, *Stability theory by Liapunov's direct method*, Springer-Verlag, New York, 1977, Applied Mathematical Sciences, Vol. 22. MR 0450715 (56 #9008)

415. Pierre Rouchon, *On the control of quantum oscillators*, Technical report, Centre Automatique et Systèmes, École des Mines de Paris **A/320** (2002).
416. _____, *Control of a quantum particle in a moving potential well*, Lagrangian and Hamiltonian methods for nonlinear control 2003, IFAC, Laxenburg, 2003, pp. 287–290. MR 2082989
417. _____, *Control of a quantum particule in a moving potential well*, 2nd IFAC Workshop on Lagrangian and Hamiltonian Methods for Nonlinear Control, Seville (2003).
418. Mary Ellen Rudin, *A new proof that metric spaces are paracompact*, Proc. Amer. Math. Soc. **20** (1969), 603. MR 0236876 (38 #5170)
419. Walter Rudin, *Functional analysis*, McGraw-Hill Book Co., New York, 1973, McGraw-Hill Series in Higher Mathematics. MR 0365062 (51 #1315)
420. _____, *Real and complex analysis*, second ed., McGraw-Hill Book Co., New York, 1974, McGraw-Hill Series in Higher Mathematics. MR 0344043 (49 #8783)
421. _____, *Principles of mathematical analysis*, third ed., McGraw-Hill Book Co., New York, 1976, International Series in Pure and Applied Mathematics. MR 0385023 (52 #5893)
422. David L. Russell, *Nonharmonic Fourier series in the control theory of distributed parameter systems*, J. Math. Anal. Appl. **18** (1967), 542–560. MR 0211044 (35 #1926)
423. _____, *On boundary-value controllability of linear symmetric hyperbolic systems*, Mathematical Theory of Control (Proc. Conf., Los Angeles, Calif., 1967), Academic Press, New York, 1967, pp. 312–321. MR 0258500 (41 #3147)
424. _____, *Control theory of hyperbolic equations related to certain questions in harmonic analysis and spectral theory*, J. Math. Anal. Appl. **40** (1972), 336–368. MR 0324228 (48 #2580)
425. _____, *Exact boundary value controllability theorems for wave and heat processes in star-complemented regions*, Differential games and control theory (Proc. NSF—CBMS Regional Res. Conf., Univ. Rhode Island, Kingston, R.I., 1973), Dekker, New York, 1974, pp. 291–319. Lecture Notes in Pure Appl. Math., Vol. 10. MR 0467472 (57 #7329)
426. _____, *Controllability and stabilizability theory for linear partial differential equations: recent progress and open questions*, SIAM Rev. **20** (1978), no. 4, 639–739. MR 508380 (80c:93032)
427. _____, *Mathematics of finite-dimensional control systems*, Lecture Notes in Pure and Applied Mathematics, vol. 43, Marcel Dekker Inc., New York, 1979, Theory and design. MR 531035 (80d:93007)
428. David L. Russell and Bing Yu Zhang, *Exact controllability and stabilizability of the Korteweg-de Vries equation*, Trans. Amer. Math. Soc. **348** (1996), no. 9, 3643–3672. MR 1360229 (96m:93025)
429. Eugene P. Ryan, *On Brockett's condition for smooth stabilizability and its necessity in a context of nonsmooth feedback*, SIAM J. Control Optim. **32** (1994), no. 6, 1597–1604. MR 1297100 (95h:93075)
430. Adhémar Jean Claude Barré de Saint-Venant, *Théorie du mouvement non permanent des eaux, avec applications aux crues des rivières et à l'introduction des marées dans leur lit*, C. R. Acad. Sci. Paris Sér. I Math. **53** (1871), 147–154.
431. Reiko Sakamoto, *Mixed problems for hyperbolic equations. I. Energy inequalities*, J. Math. Kyoto Univ. **10** (1970), 349–373. MR 0283400 (44 #632a)
432. _____, *Mixed problems for hyperbolic equations. II. Existence theorems with zero initial datas and energy inequalities with initial datas*, J. Math. Kyoto Univ. **10** (1970), 403–417. MR 0283401 (44 #632b)
433. Yoshiyuki Sakawa, *Observability and related problems for partial differential equations of parabolic type*, SIAM J. Control **13** (1975), 14–27. MR 0368843 (51 #5081)
434. Dietmar Salamon, *Infinite-dimensional linear systems with unbounded control and observation: a functional analytic approach*, Trans. Amer. Math. Soc. **300** (1987), no. 2, 383–431. MR 876460 (88d:93024)
435. Claude Samson, *Velocity and torque feedback control of a nonholonomic cart*, Advanced robot control (Grenoble, 1990) (C. Canudas de Wit, ed.), Lecture Notes in Control and Inform. Sci., vol. 162, Springer, Berlin, 1991, pp. 125–151. MR 1180972 (93g:70032)
436. Claude Samson, Michel Le Borgne, and Bernard Espiau, *Robot control: the task function approach*, Oxford engineering series, vol. 22, Clarendon Press, Oxford, UK, 1991.
437. Michael Schmidt and Emmanuel Trélat, *Controllability of Couette flows*, Commun. Pure Appl. Anal. **5** (2006), no. 1, 201–211. MR 2190785

438. Jacob T. Schwartz, *Nonlinear functional analysis*, Gordon and Breach Science Publishers, New York, 1969, Notes by H. Fattorini, R. Nirenberg and H. Porta, with an additional chapter by Hermann Karcher, Notes on Mathematics and its Applications. MR 0433481 (55 #6457)

439. Laurent Schwartz, *Étude des sommes d'exponentielles. 2ième éd*, Publications de l'Institut de Mathématique de l'Université de Strasbourg, V. Actualités Sci. Ind., Hermann, Paris, 1959. MR 0106383 (21 #5116)

440. Atle Seierstad and Knut Sydsæter, *Optimal control theory with economic applications*, Advanced Textbooks in Economics, vol. 24, North-Holland Publishing Co., Amsterdam, 1987. MR 887536 (88h:49002)

441. Rodolphe Sepulchre and Dirk Aeyels, *Stabilizability does not imply homogeneous stabilizability for controllable homogeneous systems*, SIAM J. Control Optim. **34** (1996), no. 5, 1798–1813. MR 1404857 (98b:93059)

442. Rodolphe Sepulchre, Mrdjan Janković, and Petar V. Kokotović, *Constructive nonlinear control*, Communications and Control Engineering Series, Springer-Verlag, Berlin, 1997. MR 1481435 (98k:93002)

443. Denis Serre, *Systèmes de lois de conservation. I, Fondations.* [Foundations], Diderot Editeur, Paris, 1996, Hyperbolicité, entropies, ondes de choc. [Hyperbolicity, entropies, shock waves]. MR 1459988 (99b:35139)

444. Jean-Pierre Serre, *A course in arithmetic*, Springer-Verlag, New York, 1973, Translated from the French, Graduate Texts in Mathematics, No. 7. MR 0344216 (49 #8956)

445. _____, *Cours d'arithmétique*, Presses Universitaires de France, Paris, 1977, Deuxième édition revue et corrigée, Le Mathématicien, No. 2. MR 0498338 (58 #16473)

446. Armen Shirikyan, *Approximate controllability of three-dimensional Navier-Stokes equations*, Comm. Math. Phys. **266** (2006), no. 1, 123–151. MR 2231968

447. Leonard M. Silverman and Henry E. Meadows, *Controllability and time-variable unilateral networks*, IEEE Trans. Circuit Theory **CT-12** (1965), 308–314. MR 0246709 (39 #8013)

448. _____, *Controllability and observability in time-variable linear systems*, SIAM J. Control **5** (1967), 64–73. MR 0209043 (34 #8851)

449. Jacques Simon, *Compact sets in the space $L^p(0,T;B)$*, Ann. Mat. Pura Appl. (4) **146** (1987), 65–96. MR 916688 (89c:46055)

450. Marshall Slemrod, *A note on complete controllability and stabilizability for linear control systems in Hilbert space*, SIAM J. Control **12** (1974), 500–508. MR 0353107 (50 #5593)

451. Eduardo D. Sontag, *Conditions for abstract nonlinear regulation*, Inform. and Control **51** (1981), no. 2, 105–127. MR 686833 (84d:93040)

452. _____, *Finite-dimensional open-loop control generators for nonlinear systems*, Internat. J. Control **47** (1988), no. 2, 537–556. MR 929174 (89h:93029)

453. _____, *Remarks on stabilization and input-to-state stability*, Proc. 28th IEEE Conf. Decision and Control (Tampa 1989), IEEE, New York, 1989, pp. 1376–1378.

454. _____, *Smooth stabilization implies coprime factorization*, IEEE Trans. Automat. Control **34** (1989), no. 4, 435–443. MR 987806 (90b:93092)

455. _____, *A "universal" construction of Artstein's theorem on nonlinear stabilization*, Systems Control Lett. **13** (1989), no. 2, 117–123. MR 1014237 (90g:93069)

456. _____, *Universal nonsingular controls*, Systems Control Lett. **19** (1992), no. 3, 221–224. MR 1180510 (93f:93024)

457. _____, *Control of systems without drift via generic loops*, IEEE Trans. Automat. Control **40** (1995), no. 7, 1210–1219. MR 1344033 (96i:93008)

458. _____, *Mathematical control theory*, second ed., Texts in Applied Mathematics, vol. 6, Springer-Verlag, New York, 1998, Deterministic finite-dimensional systems. MR 1640001 (99k:93001)

459. _____, *Clocks and insensitivity to small measurement errors*, ESAIM Control Optim. Calc. Var. **4** (1999), 537–557 (electronic). MR 1746166 (2001c:93091)

460. Eduardo D. Sontag and Héctor J. Sussmann, *Remarks on continuous feedback*, Proc. IEEE Conf. Decision and Control, Albuquerque (1980), IEEE, New York, 1980, pp. 916–921.

461. Edwin H. Spanier, *Algebraic topology*, McGraw-Hill Book Co., New York, 1966. MR 0210112 (35 #1007)

462. Olof Staffans, *Well-posed linear systems*, Encyclopedia of Mathematics and its Applications, vol. 103, Cambridge University Press, Cambridge, 2005. MR 2154892

463. Gianna Stefani, *On the local controllability of a scalar-input control system*, Theory and applications of nonlinear control systems (Stockholm, 1985) (C.I. Byrns and A. Lindquist, eds.), North-Holland, Amsterdam, 1986, pp. 167–179. MR 935375 (89c:49030)
464. Elias M. Stein, *Singular integrals and differentiability properties of functions*, Princeton Mathematical Series, No. 30, Princeton University Press, Princeton, N.J., 1970. MR 0290095 (44 #7280)
465. Héctor J. Sussmann, *Orbits of families of vector fields and integrability of distributions*, Trans. Amer. Math. Soc. **180** (1973), 171–188. MR 0321133 (47 #9666)
466. _____, *Single-input observability of continuous-time systems*, Math. Systems Theory **12** (1979), no. 4, 371–393. MR 541865 (81f:93025)
467. _____, *Subanalytic sets and feedback control*, J. Differential Equations **31** (1979), no. 1, 31–52. MR 524816 (80b:34085)
468. _____, *Lie brackets and local controllability: a sufficient condition for scalar-input systems*, SIAM J. Control Optim. **21** (1983), no. 5, 686–713. MR 710995 (85j:49029)
469. _____, *A general theorem on local controllability*, SIAM J. Control Optim. **25** (1987), no. 1, 158–194. MR 872457 (88f:93025)
470. _____, *New theories of set-valued differentials and new versions of the maximum principle of optimal control theory*, Nonlinear control in the year 2000, Vol. 2 (Paris), Lecture Notes in Control and Inform. Sci., vol. 259, Springer, London, 2001, pp. 487–526. MR 1806192 (2002e:49040)
471. _____, *High-order open mapping theorems*, Directions in mathematical systems theory and optimization, Lecture Notes in Control and Inform. Sci., vol. 286, Springer, Berlin, 2003, pp. 293–316. MR 2014799 (2004k:58012)
472. Héctor J. Sussmann and Wensheng Liu, *Motion planning and approximate tracking for controllable systems without drift*, Proceedings of the 25th Annual Conference on Information Sciences and Systems (Johns Hopkins University, USA, 1991), 1991, pp. 547–551.
473. Daniel Tataru, *On the regularity of boundary traces for the wave equation*, Ann. Scuola Norm. Sup. Pisa Cl. Sci. (4) **26** (1998), no. 1, 185–206. MR 1633000 (99e:35129)
474. Abdelhamid Tayebi and Ahmed Rachid, *A time-varying based robust control for the parking problem of a wheeled mobile robot*, Proceedings of IEEE International Conference on Robotics and Automation, Minneapolis, USA, 1996, pp. 3099–3104.
475. Michael E. Taylor, *Partial differential equations. I*, Applied Mathematical Sciences, vol. 115, Springer-Verlag, New York, 1996, Basic theory. MR 1395148 (98b:35002b)
476. _____, *Partial differential equations. III*, Applied Mathematical Sciences, vol. 117, Springer-Verlag, New York, 1997, Nonlinear equations, Corrected reprint of the 1996 original. MR 1477408 (98k:35001)
477. Andrew R. Teel, *Using saturation to stabilize a class of single-input partially linear composite systems*, IFAC Nonlinear Control Systems Design (Bordeaux, France, 24-26 June 1992) (Michel Fliess, ed.), 1993, pp. 165–171.
478. _____, *A nonlinear small gain theorem for the analysis of control systems with saturation*, IEEE Trans. Automat. Control **41** (1996), no. 9, 1256–1270. MR 1409471 (98b:93051)
479. Roger Temam, *Behaviour at time $t = 0$ of the solutions of semilinear evolution equations*, J. Differential Equations **43** (1982), no. 1, 73–92. MR 645638 (83c:35058)
480. Bernard Thuilot, Brigitte d'Andréa-Novel, and Alain Micaelli, *Modeling and feedback control of mobile robots equipped with several steering wheels*, IEEE Robotics and Automation **12** (1996), no. 3, 375–390.
481. Emmanuel Trélat, *Contrôle optimal, Théorie et applications*, Mathématiques concrètes, Vuibert, Paris, 2005.
482. Alexander I. Tret′yak, *Necessary conditions for optimality of odd order in a time-optimality problem for systems that are linear with respect to control*, Mat. Sb. **181** (1990), no. 5, 625–641. MR 1055978 (92e:49034)
483. Roberto Triggiani, *On the stabilizability problem in Banach space*, J. Math. Anal. Appl. **52** (1975), no. 3, 383–403. MR 0445388 (56 #3730)
484. _____, *The dual algebraic Riccati equations: additional results under isomorphism of the Riccati operator*, Appl. Math. Lett. **18** (2005), no. 9, 1001–1008. MR 2156994 (2006e:49075)
485. John Tsinias, *Sufficient Lyapunov-like conditions for stabilization*, Math. Control Signals Systems **2** (1989), no. 4, 343–357. MR 1015672 (90h:93082)

486. Marius Tucsnak, *Wellposedness, controllability and stabilizability of systems governed by partial differential equations*, Cours de DEA, Université de Nancy (2004).
487. Marius Tucsnak and George Weiss, *Passive and conservative linear systems*, Preliminary version, Université de Nancy, 2006.
488. Gabriel Turinici, *On the controllability of bilinear quantum systems*, Mathematical models and methods for ab initio quantum chemistry (M. Defranceschi and C. Le Bris, eds.), Lecture Notes in Chem., vol. 74, Springer, Berlin, 2000, pp. 75–92. MR 1855575
489. José Manuel Urquiza, *Contrôle d'équations des ondes linéaires et quasilinéaires*, PhD Thesis, Université de Paris VI (advisor: Frédéric Bourquin) (2000).
490. _____, *Rapid exponential feedback stabilization with unbounded control operators*, SIAM J. Control Optim. **43** (2005), no. 6, 2233–2244 (electronic). MR 2179485
491. Vadim I. Utkin, *Sliding modes in control and optimization*, Communications and Control Engineering Series, Springer-Verlag, Berlin, 1992, Translated and revised from the 1981 Russian original. MR 1295845 (95f:93005)
492. Rafael Vázquez, Jean-Michel Coron, and Emmanuel Trélat, *Control for fast and stable laminar-to-high-Reynolds-number transfer in a 2D Navier-Stokes system*, Preprint (2006).
493. Marilena Vendittelli, Giuseppe Oriolo, Frédéric Jean, and Jean-Paul Laumond, *Nonhomogeneous nilpotent approximations for nonholonomic systems with singularities*, IEEE Trans. Automat. Control **49** (2004), no. 2, 261–266. MR 2034349 (2004j:93003)
494. Mathukumalli Vidyasagar, *Decomposition techniques for large-scale systems with nonadditive interactions: stability and stabilizability*, IEEE Trans. Automat. Control **25** (1980), no. 4, 773–779. MR 583455 (81h:93075)
495. Richard Vinter, *Optimal control*, Systems & Control: Foundations & Applications, Birkhäuser Boston Inc., Boston, MA, 2000. MR 1756410 (2001c:49001)
496. Claire Voisin, *Hodge theory and complex algebraic geometry. I*, Cambridge Studies in Advanced Mathematics, vol. 76, Cambridge University Press, Cambridge, 2002, Translated from the French original by Leila Schneps. MR 1967689 (2004d:32020)
497. Gregory C. Walsh, Richard Montgomery, and Shankar S. Sastry, *Orientation control of the dynamic satellite*, Proc. Amer. Contr. Conf., Baltimore, MD (1994), 138–142.
498. Jan C. Wang, *Stabilization of decentralized control systems via time-varying controllers*, IEEE Trans. Automat. Control **36** (1982), 741–744.
499. George Weiss, *Admissibility of unbounded control operators*, SIAM J. Control Optim. **27** (1989), no. 3, 527–545. MR 993285 (90c:93060)
500. _____, *The representation of regular linear systems on Hilbert spaces*, Control and estimation of distributed parameter systems (Vorau, 1988), Internat. Ser. Numer. Math., vol. 91, Birkhäuser, Basel, 1989, pp. 401–416. MR 1033074 (91d:93026)
501. George W. Whitehead, *Elements of homotopy theory*, Graduate Texts in Mathematics, vol. 61, Springer-Verlag, New York, 1978. MR 516508 (80b:55001)
502. Gerald Beresford Whitham, *Linear and nonlinear waves*, Wiley-Interscience [John Wiley & Sons], New York, 1974, Pure and Applied Mathematics. MR 0483954 (58 #3905)
503. Jan C. Willems, *Paradigms and puzzles in the theory of dynamical systems*, IEEE Trans. Automat. Control **36** (1991), no. 3, 259–294. MR 1092818 (92h:93006)
504. W. Murray Wonham, *On pole assignment in multi-input controllable linear systems*, IEEE Trans. Automat. Control **12** (1967), no. 6, 660–665.
505. Cheng-Zhong Xu and John Baillieul, *Stabilizability and stabilization of a rotating body-beam system with torque control*, IEEE Trans. Automat. Control **38** (1993), no. 12, 1754–1765. MR 1254313 (94m:93059)
506. Cheng-Zhong Xu and Gauthier Sallet, *Exponential stability and transfer functions of processes governed by symmetric hyperbolic systems*, ESAIM Control Optim. Calc. Var. **7** (2002), 421–442 (electronic). MR 1925036 (2003f:93052)
507. Hatem Zaag, *A remark on the energy blow-up behavior for nonlinear heat equations*, Duke Math. J. **103** (2000), no. 3, 545–556. MR 1763658 (2001k:35160)
508. Jerzy Zabczyk, *Some comments on stabilizability*, Appl. Math. Optim. **19** (1989), no. 1, 1–9. MR 955087 (89f:93035)
509. _____, *Mathematical control theory: an introduction*, Systems & Control: Foundations & Applications, Birkhäuser Boston Inc., Boston, MA, 1992. MR 1193920 (93h:49001)

510. _____, *Topics in stochastic processes*, Scuola Normale Superiore di Pisa. Quaderni. [Publications of the Scuola Normale Superiore of Pisa], Scuola Normale Superiore, Pisa, 2004. MR 2101945 (2006a:60004)
511. Olivier Zarrouati, *Trajectoires spatiales*, Cépaduès-Éditions, Toulouse, 1987.
512. Eberhard Zeidler, *Nonlinear functional analysis and its applications. I*, Springer-Verlag, New York, 1986, Fixed-point theorems, Translated from the German by Peter R. Wadsack. MR 816732 (87f:47083)
513. Bing-Yu Zhang, *Exact boundary controllability of the Korteweg-de Vries equation*, SIAM J. Control Optim. **37** (1999), no. 2, 543–565 (electronic). MR 1670653 (2000b:93010)
514. Xu Zhang and Enrique Zuazua, *Polynomial decay and control of a 1-d model for fluid-structure interaction*, C. R. Math. Acad. Sci. Paris **336** (2003), no. 9, 745–750. MR 1988314 (2004e:93050)
515. Enrique Zuazua, *Exact boundary controllability for the semilinear wave equation*, Nonlinear partial differential equations and their applications. Collège de France Seminar, Vol. X (Paris, 1987–1988) (Haïm Brezis and Jacques-Louis Lions, eds.), Pitman Res. Notes Math. Ser., vol. 220, Longman Sci. Tech., Harlow, 1991, pp. 357–391. MR 1131832 (92k:93029)
516. _____, *Exact controllability for semilinear wave equations in one space dimension*, Ann. Inst. H. Poincaré Anal. Non Linéaire **10** (1993), no. 1, 109–129. MR 1212631 (94d:93022)
517. _____, *Finite-dimensional null controllability for the semilinear heat equation*, J. Math. Pures Appl. (9) **76** (1997), no. 3, 237–264. MR 1441986 (98c:93018)
518. _____, *Approximate controllability for semilinear heat equations with globally Lipschitz nonlinearities*, Control Cybernet. **28** (1999), no. 3, 665–683, Recent advances in control of PDEs. MR 1782020 (2001h:93016)
519. _____, *Propagation, observation, and control of waves approximated by finite difference methods*, SIAM Rev. **47** (2005), no. 2, 197–243.
520. _____, *Control and numerical approximation of the wave and heat equations*, Proceedings of the International Congress of Mathematicians, Vol. I, II, III (Madrid, 2006) (Marta Sanz-Solé, Javier Soria, Varona Juan Luis, and Joan Verdera, eds.), vol. III, European Mathematical Society, 2006, pp. 1389–1417.
521. _____, *Controllability and observability of partial differential equations: Some results and open problems*, Handbook of differential equations: evolutionary differential equations (C. M. Dafermos and E. Feireisl, eds.), vol. 3, Elsevier/North-Holland, Amsterdam, 2006, pp. 527–621.

List of symbols

$*$	adjoint of a linear operator, page 28
$\mathcal{A}(x_e, u_e)$	strong jet accessibility subspace, page 133
ad_X	ad operator (Lie bracket with X), page 130
$\mathbb{C}$	set of complexes
$\mathfrak{C}$	controllability Gramian, page 6
$\cdot$	usual scalar product in $\mathbb{R}^l$, page 20
$[\cdot, \cdot]$	Lie bracket, page 129
$\overline{z}$	complex conjugate of the complex number z, page 48
$\mathcal{D}'(\Omega)$	set of distributions on the open subset Ω of $\mathbb{R}^n$
$H^{-1}(0, L)$	set of derivative of functions in $L^2(0, L)$, page 99
$H_0^1(0, L)$	set of β in $H^1(0, L)$ such that $\beta(0) = \beta(L) = 0$, page 100
$H_{(0)}^1(0, L)$	set of functions $f \in H^1(0, L)$ such that $f(0) = 0$, page 67
$H_0^1(I; \mathbb{C})$	set of φ in $H^1(I; \mathbb{C})$ such that $\varphi(0) = \varphi(L) = 0$, page 95
$H_0^1(\Omega)$	the closure in $H^1(\Omega)$ of the set of functions $\varphi \in C^\infty(\Omega)$ with compact support, page 77
$H_{(0)}^3(I; \mathbb{C})$	set of ψ in $H^3(I; \mathbb{C})$ such that $\psi(-1) = \psi^{(2)}(-1) = \psi(1) = \psi^{(2)}(1) = 0$, page 96
$H_{(0)}^5(I; \mathbb{C})$	set of ψ in $H^5(I; \mathbb{C})$ such that $\psi(-1) = \psi^{(2)}(-1) = \psi^{(4)}(-1) = \psi(1) = \psi^{(2)}(1) = \psi^{(4)}(1) = 0$, page 175
$H_{(0)}^7(I; \mathbb{C})$	set of ψ in $H^7(I; \mathbb{C})$ such that $\psi(-1) = \psi^{(2)}(-1) = \psi^{(4)}(-1) = \psi^{(6)}(-1) = \psi(1) = \psi^{(2)}(1) = \psi^{(4)}(1) = \psi^{(6)}(1) = 0$, page 248
$\langle \cdot, \cdot \rangle$	Hermitian scalar product in a complex Hilbert space, page 95
HUM	Hilbert Uniqueness Method, page 19
I	open interval (-1,1), page 95
Id_n	identity map of $\mathbb{R}^n$, page 5
$\Im z$	imaginary part of the complex number z, page 48
$\mathrm{Lie}(\mathcal{F})$	Lie algebra generated by the vector fields in $\mathcal{F}$, page 133
$\mathcal{L}(\mathbb{R}^k; \mathbb{R}^l)$	set of linear maps from $\mathbb{R}^k$ into $\mathbb{R}^l$, page 3
$L_X V$	(Lie) derivative of V in the direction of X, page 146
$L_X^k V$	iterated (Lie) derivatives of V in the direction of X, page 307
$\mathcal{M}_{k,l}(\mathbb{C})$	set of $k \times l$ matrices with complex coefficients, page 3
$\mathcal{M}_{k,l}(\mathbb{R})$	set of $k \times l$ matrices with real coefficients, page 3
$\mathbb{N}$	set of nonnegative integers
$\mathbb{N} \setminus \{0\}$	set of strictly positive integers
$\mathcal{O}$	nonempty open subset of $\mathbb{R}^n \times \mathbb{R}^m$, page 125
P_A	characteristic polynomial of the matrix A, page 10
$\mathbb{R}$	set of reals
$R(\cdot, \cdot)$	resolvent, page 4

$\Re z$	real part of the complex number z, page 95
$\mathbb{R}^k$	vector space of k-dimensional real column vector, page 3
$\mathbb{S}$	unit sphere of $L^2(I;\mathbb{C})$, page 95
$\sigma(L)$	set of eigenvalues of L, page 285
tr	transpose, page 6
$\rightharpoonup$	weak convergence, page 54
(x_e, u_e)	equilibrium of the control system considered, page 125
$\left(\frac{x}{k}\right)$	Legendre symbol, page 239
Y_T	set of $y : (0,T) \times (0,L) \to \mathbb{R}$ such that $y \in L^2((0,T); H^2(0,L))$ and $y_t \in L^2((0,T) \times (0,L))$, page 225

Index

abstract linear control system, 51–67
   Cauchy problem, 52–54
   controllability, 55–62
   Korteweg-de Vries equation, 65–67
   transport equation, 62–65
adjoint operator, 373
$ad_X^k Y$ (definition), 130
admissibility condition, 52, *see also* regularity property
approximate controllability, 55, 57, 61
Ascoli's theorem, 152, 170
attractor, 280
averaging, 332–333

backstepping, 334–337, 353
   desingularization, 335–336, 353
Banach fixed-point theorem, 163
Brockett condition (obstruction to stabilizability), 289, 292, 297, 341, 357, 361
Brouwer fixed-point theorem, 133, 236, 244, 390, 391
Burgers equation, 202, 360

Carleman inequality, 80, 113
Cayley-Hamilton theorem, 10, 133
characteristic curves, 363
characteristic polynomial, 10, 275
characteristic velocities, 370
closed operator, 373
closed-loop system (definition), 281
compact resolvent, 95, 228, 354
companion matrix, 276
compatibility conditions, 165, 200, 365, 366
control affine (definition), 131
control Lyapunov function, 313–315, 334, 336, 353, 355
   application to stabilization, 313–314
controllability
   abstract linear control system, 55–62
   Euler equations, 194–197
   finite-dimensional linear control system, 3–22

finite-dimensional nonlinear control system, 125–157
heat equation, 103–118, 225–233
hyperbolic system, 165–174
Korteweg-de Vries equation, 42–49
Navier-Stokes equations, 197–203
Saint-Venant equations, 204–221
Schrödinger equation, 96–99, 248–270
spacecraft, 128–129, 144–145
theorem
   abstract linear control system, 56, 57
   Euler equations, 195–197
   heat equation, 79, 87, 104, 105, 225
   hyperbolic system, 168
   Kalman rank condition, 9
   Korteweg-de Vries equation, 42, 161, 237
   linear finite-dimensional control system, 6, 9, 11
   Navier-Stokes equations, 199, 201
   necessary condition, 134, 145
   Saint-Venant equations, 204
   Schrödinger equation, 96, 248, 251
   sufficient condition, 135, 143
   transport equation, 29
transport equation, 29–37
controllability Gramian, 6, 151, 282
controller form, 275
Couette flows, 232
CRONE, 281

damping, 314–328
definition
   $ad_X^k Y$, 130
   adjoint operator, 373
   approximate controllability, 55
   closed operator, 373
   closed-loop system, 281
   control affine, 131
   control Lyapunov function, 313
   controllability Gramian, 6
   controllability of $\dot{x} = A(t)x + B(t)u$, 4
   degree of a function, 379
   dissipative operator, 373

423

drift, 130
driftless control system, 130
equilibrium, 125
exact controllability, 55
Gevrey function, 86
globally asymptotically stabilizable, 292, 296
group of continuous linear operators, 376
Hermes condition, 143
homogeneous, 328
infinitesimal generator, 375, 376
Lie algebra, 133
Lie algebra rank condition, 134
Lie bracket, 129
Lie null-observable, 306
linearized control system along a trajectory, 127
linearized control system at an equilibrium, 128
local controllability along a trajectory, 126
locally asymptotically stabilizable, 281, 292, 296, 306
locally asymptotically stable, 279, 296
locally continuously reachable, 299
locally stabilizable in small time, 302, 306
null controllable, 55
resolvent, 4
semigroup of continuous linear operators, 374
small-time local controllability, 125
steady-state, 225, 362
strong jet accessibility subspace, 133
strongly continuous group of continuous linear operators, 376
strongly continuous semigroup of continuous linear operators, 374
supple trajectory, 303
Sussmann condition, 143
trajectory, 126
degree of a function, 154, 155, 236, 290, 291, 302, 379–395
desingularization (for backstepping), 335
Dirichlet density theorem, 240
discontinuous feedback, 311–312
dissipative operator, 49, 373
drift (definition), 130
driftless control system (definition), 130
duality controllability/observavility, 19, 34–37
Duhamel's principle, 6
dynamic extension, 292
dynamic feedback laws, 292, 306
dynamic output feedback laws, 307

equilibrium, 125
Euler equations
 controllability, 194–197
 stabilization, 356–361
exact controllability, 55

flat control system, 85, 215
flat output, 85
fluvial, 370
forwarding, 320, 337–340
fractional power of the Laplacian, 94
Frobenius theorem, 141

Gevrey function, 86
global controllability
 driftless system, 134–141
 Euler equations, 194–197
 heat equation, 225–233
 homogeneity, 153–156
 Navier-Stokes equations, 199–203
 parabolic equation, 180
 perturbations of linear systems, 150–153
 spacecraft, 144–145
 theorem
  driftless system, 134
  Euler equations, 195
  heat equation, 225
  homogeneity, 154
  Navier-Stokes equations, 199, 201
  perturbations of linear systems, 151, 152
  wave equation, 178
 wave equation, 177–180
globally asymptotically stabilizable, 281, 292, 296
Gramian (controllability), 6, 151, 282
group of continuous linear operators, 376, 377

heat equation, 76–94, 225–233
 linear, 76–94
  Cauchy problem, 76–78
  controllability, 78–94, 99–118
  finite number of controls, 88–94
  singular optimal control, 99–118
 nonlinear, 225–233
Hermes condition (local controllability), 143, 144
hidden regularity, 31, 32, 43
 Korteweg-de Vries equation, 43
 transport equation, 31, 32
H^∞ control, 281
homogeneity, 328–332
homogeneous (definition), 328
homotopy invariance of the degree, 155, 291, 387
HUM, 19–22, 201, 243
 finite dimension, 19–22
 Korteweg-de Vries equation, 243
hyperbolic system, 165–174, 361–371
 controllability, 165–174

stabilization, 361–371

index condition, 289, 292, 293, 295
infinitesimal generator, 51, 347, 375, 376
  definition, 375, 376
isentropic Euler equations, 197

Jacobi identity, 129

Kalman rank condition, 9–19, 284
  for stationary systems, 9
  for time-varying systems, 11
Korteweg-de Vries equation, 38–50, 159–165, 237–246
  linear, 38–50, 65–67
    Cauchy problem, 38–42
    controllability, 42–49
  nonlinear, 159–165, 237–246
    Cauchy problem, 160–161
    controllability, 161–165, 237–246

Laplacian (fractional power of the), 94
LaSalle invariance principle, 284, 315, 316, 323, 326, 361
Lie algebra, 133
Lie algebra rank condition, 134, 135, 141, 143–145, 149, 183, 190, 296, 302, 303, 331, 341, 345
Lie bracket, 129, 181–185
Lie derivative, 146, 306, 307, 317
Lie null-observable, 306, 307
linear test, 127, 128
linearized control system along a trajectory, 127
linearized control system at an equilibrium, 128
local controllability, *see also* small-time local controllability
  heat equation, 103–118
  Hermes condition, 143, 144
  hyperbolic system, 165–174
  Korteweg-de Vries equation, 159–165, 237–246
  Saint-Venant equations, 204–221
  Schrödinger equation, 96–99, 248–270
  spacecraft, 128–129, 144–145
  Sussmann condition, 143
  theorem
    heat equation, 79, 87, 104, 105
    hyperbolic system, 168
    Korteweg-de Vries equation, 42, 161, 237
    Saint-Venant equations, 204
    Schrödinger equation, 96, 248, 251
    transport equation, 29
  transport equation, 29–37
local controllability along a trajectory
  definition, 126
  sufficient condition, 127

locally asymptotically stabilizable, 281, 292, 296, 302, 306, 307
locally asymptotically stable, 279, 296
locally continuously reachable, 299, 307
locally stabilizable in small time, 302, 306, 307
low-thrust, 319–324
Lumer-Phillips theorem, 374, 375, 377
Lyapunov function, 227
  converse theorem, 287

moments, 98
μ analysis, 281
multiplier method, 36–37, 75

Nash-Moser method, 159, 168, 172, 177, 259, 263
Navier-Stokes equations, 197–203, 360
  controllability, 197–203
  stabilization, 360
Newton method, 176
nonholonomic integrator, 135, 190, 289, 297, 312
  controllability, 135, 190
  stabilization, 289, 297, 312
null controllability, 30, 55–57, 61, 85, 88, 99
null observability, 306, 307

observability inequality, 35, 45, 50, 57, 65, 74, 79, 110
  abstract linear control system, 57
  heat equation, 79, 110
  Korteweg-de Vries equation, 45
  transport equation, 35, 65
  wave equation, 74
obstruction to small-time local controllability, 145
  Stefani condition, 145
  Sussmann condition, 145
output feedback laws, 305–311

parabolic equation, 76–94, 180, 225–233
  linear, 76–94
    Cauchy problem, 76–78
    controllability, 78–94, 99–118
    finite number of controls, 88–94
    singular optimal control, 99–118
  nonlinear, 180, 225–233
partition of unity, 314, 393, 395
phase variable canonical form, 275
pole-shifting theorem, 228, 231, 275–279, 281, 282, 356
power series expansion, 235–246, 257–259, 263
  description of the method, 235–237
  Korteweg-de Vries equation, 237–244
  Schrödinger equation, 257–259, 263

quadratic reciprocity law, 239

quasi-static deformations, 208–210, 215–217, 223–232, 260
   description of the method, 223–224
   heat equation, 225–232
   Saint-Venant equations, 215–217
   Schrödinger equation, 260
   toy model, 208–210

Rashevski-Chow theorem, 134
regularity property, 52, 66
   definition, 52
   Korteweg-de Vries equation, 66
   transport equation, 63
resolvent, 4
return method
   Burgers equation, 202
   description, 122, 187–192
   Euler equations, 195–197
   isentropic Euler equations, 197
   Korteweg-de Vries equation, 241
   Navier-Stokes equations, 200–201
   Saint-Venant equations, 211–221
   Schrödinger equation, 260–263
   toy model, 207–210
   Vlasov-Poisson system, 197
Riemann invariants, 362, 363, 370
rotating body beam, 351–356

Saint-Venant equations, 203–221, 369–371
   controllability, 203–221
   stabilization, 369–371
Schauder fixed-point theorem, 152, 170, 179, 391, 392
Schrödinger equation, 95–99, 174–177, 182–185, 247–270
   linear, 95–99
      Cauchy problem, 96
      controllability, 96–99
   nonlinear, 174–177, 182–185, 247–270
      controllability, 174–177, 248, 252–263
      noncontrollability, 182–185, 251, 263–270
self-adjoint, 95, 351, 354
semigroup associated to an operator, 374
semigroup of continuous linear operators, 374
shallow water equations, 203, see also Saint-Venant equations
singular optimal control, 99–118
sliding mode control, 312
small control property, 313
small-time local controllability, 125, 301
   definition, 125
   driftless systems, 135
   example, 129, 144, 145, 149, 190
   linear test, 128
   necessary condition, 134, 145
   sufficient condition, 143

small-time local reachability, 299–305, 307, 308
spacecraft
   global controllability, 144–145
   local controllability, 128–129, 144–145
   stabilization, 282, 289, 304–305, 330–333, 337
stabilization
   driftless systems, 296–298
   Euler equations, 356–361
   finite-dimensional linear control system, 275–279, 282–286
   general finite-dimensional systems, 299–305
   hyperbolic system, 361–371
   Navier-Stokes equations, 360
   output feedback laws, 305–311
   rotating body beam, 351–356
   Saint-Venant equations, 369–371
   spacecraft, 282, 289, 304–305, 330–333, 337
stable point, 280
stationary feedback laws, 281
steady-state, 225, 362
Stefani condition (obstruction to small-time local controllability), 145
strong jet accessibility subspace, 133
strongly continuous group of continuous linear operators, 376, 377
strongly continuous semigroup of continuous linear operators, 374
supple trajectory, 303
Sussmann condition (local controllability), 143
Sussmann condition (obstruction to small-time local controllability), 145

time-varying feedback laws, 296, 302, 306
   driftless systems, 296–298
   general finite-dimensional systems, 299–305
trajectory(definition), 126
transport equation, 24–37, 165–174
   linear, 24–37, 62–65
      Cauchy problem, 25–29
      controllability, 29–37, 99–118
   nonlinear, 165–174
      controllability, 165–174
transverse functions, 340–345

Vlasov-Poisson system, 197

wave equation, 67–76, 177–180
   linear, 67–76
      Cauchy problem, 67–71
      controllability, 71–76
   nonlinear, 177–180